D1228588
R0017273923

The Climate of the British Isles

The Climate of the British Isles

Edited by

T. J. Chandler

and S. Gregory

Longman
London and New York

Longman Group Limited

London and New York

Associated companies, branches and representatives throughout the world

Published in the United States of America by
Longman Inc., New York

First published 1976

Library of Congress Cataloging in Publication Data
Main entry under title:
The Climate of the British Isles.
Bibliography: p.
Includes index.
1. Great Britain — Climate — Addresses, essays, lectures.
I. Chandler, Tony John. II. Gregory, Stanley.
QC989.G69C56 551.6'9'41 76-18113
ISBN 0-582-48558-4

Printed by Lowe & Brydone Printers Limited, England

Contents

List of Figures

List of Tables

Preface

The history of this book can, we suppose, be traced back to the time when the two editors were undergraduates together in the Department of Geography, King's College London. After graduation, our careers moved their separate ways though we had both chosen to become research climatologists, one in the Universities of Liverpool and Sheffield and the other in those of London and Manchester. But we kept in contact with each other's work and with that of our colleagues in other universities and at the Meteorological Office, and in 1973 we realised that our collective interests covered most aspects of British Climatology. We therefore invited many of these colleagues to join us in writing a series of up-to-date studies of the Climate of the British Isles and this volume is the fruit of these endeavours.

Our plans would have come to nought, however, without the magnificent collaboration not only of the authors of the chapters, whom we thank for responding so well to our various requests and comments, but also to the many others who helped in a variety of ways – the supply of data, the typing of the text, the drawing of the diagrams, the compilation of the index and so on. There are so many of these people, without whose help the book could never have been written. We hope they will forgive us if we do not mention them all by name but trust that they will derive pleasure and recompense from seeing the book completed and knowing that they contributed in their various ways to whatever success it might have.

There are, however, a few people who have made such a major contribution to the compilation of the book that we would like to thank them individually. Mr J. S. Frampton and Miss S. G. Ottewell of the Drawing Office in the Department of Geography, University of Sheffield, drew practically all the maps, and Mrs Ann Barham, Research Assistant in that Department, compiled the list of references. We are very grateful indeed for the magnificent way in which they have done their work. We also wish to record our gratitude to the secretaries of both our Departments for coping with several rush typing assignments in their usual calm and efficient way, and to thank the editorial and processing staff at Messrs Longman Group Ltd for seeing the book through the press. Finally, however, we wish to thank those former teachers at London University, including Professors P. R. Crowe and G. Manley, who first introduced us to our hobby and profession, climatology.

Tony J. Chandler
Stanley Gregory
Pennine Way, July 1975.

Acknowledgements

We are grateful to the following for permission to reproduce copyright material:

Academic Press Inc. for data from *Radiation in the Atmosphere* by K. Y. Kondratyev, 1969; American Geographical Society for data from 'The Climates of the Earth' by C. W. Thornthwaite in *Geographical Review* Vol. 23, 1933. Reprinted by permission of the Geographical Review and the American Geographical Society; American Geophysical Union for data by V. Conrad in *Transactions of the American Geophysical Union* 27; Blackwell Scientific Publications Ltd. for data from 'A Map of Annual Average Potential Water Deficit in the British Isles' in *Journal of Applied Ecology* 1; British Association for the Advancement of Science for data based on the volumes prepared by the British Association for the Advancement of Science for the Annual Conference for the 25 year period 1949–74; Centre for Agricultural Publications and Documentation for data from 'Agro-Elimatic Atlas of Europe' by P. Thran & S. Broekhuizen, 1965 in *Agro-Ecological Atlas of Cereal Growing in Europe*; David & Charles (Holdings) Ltd. for data from 'Problems in the Measurement and Evaluation of the Climatic Resources of Upland Britain' by S. J. Harrison in *Climatic Resources and Economic Activity: A Symposium* ed. by J. A. Taylor; The Author for data from an unpublished dissertation *The Haar of North-East England* 1968 by K. A. Pratt; The Authors and Elsevier Scientific Publishing Company for data from 'Reservoir Storage and the Thermal Regime of Rivers with special Reference to the River Lune' by H. E. Lavis & K. Smith in *Science of the Total Environment* 1; Elsevier Scientific Publishing Company and respective authors for various data from *Climates of Northern and Western Europe* Vol. 5 by Prof. G. Manley and 'Meteorology and the Pattern of British Grassland Farming' by Dr. L. P. Smith in *Agricultural Meteorology* 4; Fischer Publishers for Data from *Klimadiagramm Weltatlas* by H. Walter and H. Leith 1967; Geographical Publications Ltd. for data from *The Land of Britain: The Report of the Land Utilization Survey of Britain* by L. D. Stamp (London Geographical Publications for Land Utilization Survey) 1937–46; Harvard University Press for data from *The Climate Near The Ground* by R. Geiger 1965; Ministry of Agriculture, Fisheries and Food for figures based on *Crown Copyright Data*; The Controller of Her Majesty's Stationery Office for various maps, diagrams and statistical data based on *Crown Copyright Data*; Hutchinson Publishing Group Ltd. for a table from *The Climate of London* by T. J. Chandler; The Author for data from a personal communication from M. D. Newson, Institute of Hydrology showing 'Rainfall and Altitude Relationships in the Upper Wye and Severn Catchments 1970–4'; World Meteorological Organization and Unesco for data by J. Grindley from 'Estimation and Mapping of Evaporation' in *World Water Balance: Proceedings of the Reading Symposium* (c) Unesco/IAHS/WMO 1972; Institute for Meteorology for data from *Der Trend der Meerestemperatur im Nordatlantik* by M. Rodewald 1973. Beilage zur Berliner Wetterkarte No. 108/73 and 119/73; Journal of the Institute of Fuel for data from 'Monitoring of the Environment' by J. S. S. Reay in *Fuel and the Environment: Conference Proceedings Vol. 1 Papers* (Eastbourne Conference, 1973); The Institute of British Geographers for data from 'Rainfall Excesses in the United Kingdom' by J. C. Rodda 1970 in *Transactions* Institute of British Geographers 49; Institute of Water Engineers and Scientists for data reproduced from a paper by A. Bleasdale entitled 'The Distribution of Exceptionally Heavy Falls of Rain in the United Kingdom 1963 to 1970' in *Journal* I.W.E.Vol.17 No. 1 1963; Longman Group Ltd. for data from *Concepts in Climatology* by P. R. Crowe; Macmillan Publishers Ltd. for data from *Techniques in Physical Geography* by D. J. Hanwell & M. D. Newson. Reprinted by permission of Macmillan London and Basingstoke; Irish Meteorological Service for data showing 'Mean Annual Rainfall 1931–1960'. Reprinted by permission of the Meteorological Service, Dublin; The Author for data by Dr. H. H. Lamb et al. in *Northern Hemisphere Monthly and Annual Mean-Sea-Level Pressure Distribution For 1951–66* and *Changes of Pressure and Temperature Compared with those of 1900–39* from Meteorological Office, Geophys. Mem. No. 118 and for data taken from *British Isles Weather Types* and a Register of the *Daily Sequence of Circulation Patterns* 1861–1971 from Meteorological Ghys. Mem. No. 116, 1972 and for data from *The English Climate* by London Universities Press, H. H. Lamb (2nd Edition) 1964. The Author and Nature Magazine for data by Dr. H. H. Lamb in 'Whither Climate Now?' from *Nature* 244; Societe Meteorologique de France for data by E. Ekhart in *Meteorologie* 1948; Thomas Nelson & Sons Ltd. for data from 'Climate' in *The British Isles: A Systematic Geography* 1964, ed. by Watson & Sissons; New Science Publications for data from 'Climate in Britain over 10,000 years' by G. Manley in Geographical Magazine. This article first appeared in The Geographical Magazine, London, the monthly review of geographical studies; Oliver & Boyd Publishers for data from *Climates of the U.S.S.R.* by A. A. Borisov translated by R. A. Medward; Pergamon Press Ltd. for data from 'Grass-Growing Days' by G. W. Hurst and L. P. Smith from *Weather and Agriculture* edited by J. A. Taylor; Royal Meteorological Society for data from the following issues of *Quarterly Journal of Royal Meteorological Society* 66, 81, 83, 92, 100 and for data from *Phenological Report* 1935; Weather Magazine for data from 'Climate and Water Supply in Great Britain' by S. Gregory 1959 in *Weather* 14; The Royal Society and the Author for data from 'Eddy Diffusion of Momentum Water Vapour and Heat Near the Ground' by N. E. Rider in *Phil Trans. Royal Society* 246, 1954; Springer-Verlag for data from an article by F. Defant in *Arch. Met. Geophys. Bioklim.* 1; The University College of Wales for data from an unpublished Ph.D. dissertation 'An Eco-climatic Gradient in North Cardiganshire, West Central Wales' by S. J. Harrison, Dept. of Geography 1973.

Whilst every effort has been made to trace the owners of copyrights, in a few cases this has proved impossible and we would appreciate any information that would enable us to do so.

The nature and vagaries of the weather are almost a British obsession. But why they should have become the subject of so much popular ridicule is a mystery to all those with some knowledge of conditions elsewhere. Though the variability of Britain's weather is great, it is no more than in many other countries, and though presenting many problems it has several advantages. The weather is, for instance, the basis for unfailing agreement and social contact between friends and strangers alike, generally through some banal comment upon the prevailing or possible future weather conditions. And because of the variability, there is the advantage that even in situations of repeated, perhaps daily encounter, one is almost certain never to be in danger of repeating oneself. But more seriously, the concern of the British for the day-to-day, month-to-month and year-to-year character of the weather has, in part at least, been borne out of an early recognition of its relevance to the nature and success of the socio-economic life form of individuals, communities and whole nations. The weather everywhere exercises a most important and often very close control upon our lives and on time scales ranging from minutes to centuries. Mason (1970) and Taylor (1972) have both emphasized the very considerable benefit to be obtained from the application of meteorological knowledge to forecasting for agriculture, transport and industry.

But there are also many very substantial benefits to be obtained from the adjustment of our socio-economic organizations to the opportunities and limitations of the longer term features of the British climate (Taylor, 1974a). For the farmer, of course, both the vagaries of day-to-day change and the specific details of the unfolding seasons in any particular year provide the essential framework within which his farming activities, and his very profitability, are constrained. An unusually hard or mild winter, a late spring, a dry early summer, a wet harvesting period, unseasonal frost or torrential rain, all form hazards with which he must cope. Moreover, the farming response to particular conditions will vary with the farm economy, the detailed features of the farm itself, the capital invested or available, and the extent to which later conditions can be expected to repair the immediate problems. Industry, too, is still partly at the mercy of the elements. This may be most evident in terms of the construction industry, where frost, snow, high winds or excessive rainfall may cause unforeseen and expensive delays, or in the transport field too. Few winters pass without winter snowfalls blocking roads and railways, or fogs leading to motorway multiple crashes. Climatic influences on transport are far more widespread and universal than these extremes suggest, however. Delays, slower journeys, accidents and increased costs are the experience of all, from the pedestrian on a weekly shopping expedition to the long-distance lorry driver – not to mention the delays that may afflict the air traveller due to fog. Climatic conditions have also proved important in the routing and design of major roads such as the M.62 across the Pennines.

In retailing also, the success or otherwise of particular marketing decisions may be closely affected by weather and climate. The sale of certain foods such as ice-cream is obviously weather sensitive, but there are many other less obvious relationships. The same is also true of the impact of climatic conditions upon the use of fuel for heating and lighting. A mild winter can reduce the domestic fuel bill considerably, thus affecting the demands being placed upon the power generating industries and upon coal, oil and gas supplies.

Additionally, there are also many personal activities in which the climate plays a not inconsiderable role. Certainly, the whole recreation field is critically influenced. Any British summer would seem a pale imitation of the real thing without such newspaper headings as 'Test abandoned: rain stopped play', whilst even the more robust winter sports face the hazards of dense fog, hard frozen grounds, waterlogging of pitches and deep snow. The impact of climatic hazards on hill-walking and mountaineering leads to periodic accidents and fatalities, while sudden storms or squalls have similar effects upon the sailing enthusiast. At a more leisurely pace, the day in the country or at the sea-side, even when the car is never left more than 20 m away from its passengers, is also weather dependent – the overnight weather forecast, the bright sunshine at breakfast, or the first fine Sunday in spring all stimulate the exodus to the beaches and the hills, though often unwisely! On a broader scale, the summer holiday itself may be conditioned largely by climatic expectation – the high sunshine figures returned by some south-coast resorts, the bracing climate advertised for the windier locations, or the decision to seek the more reliable Mediterranean sun rather than gamble that August will not be the wettest month in the year in Britain – all these influence the tourist industry and the actions of individuals.

There is also, of course, the whole field of general scientific interest in climatic conditions. The present-day concern with matters of environmental characteristics and quality, the role of climate in ecological considerations, and its essential impact upon the management of natural resources such as our water supplies and forest reserves, all mean that a detailed and scientifically sound knowledge and understanding of our climate is essential. It is to the climatologist, with his preoccupation with the study of climate in its own right as well as his concern with its impact upon so many fields and activities, that one must turn to obtain the detailed information and balanced assessment that is required.

The fundamental seminal text to which reference has been made in the past was that by Bilham (1938). This volume drew together the work of the previous half-century, and stamped upon it the interpretive skill and climatological understanding of its distinguished author. For at least a quarter of a century, this provided the authoritative source for successive studies in the field of the climatology of Britain, and the editors of this present volume must acknowledge their own personal debt to this work, both during their undergraduate days together at King's College, London, and during their early years as university lecturers and researchers. In the last 20 years or more, it has been complemented, but not replaced, by a number of more recent texts. In terms of factual information, especially in mapped form, *The Climatological Atlas of the British Isles* (Meteorological Office, 1952) presented more up-to-date material, analysed in a consistent manner over a wide range of elements. Most of the maps were related to the 30-year period 1901–30 which for a time had been accepted internationally as the standard period to replace the earlier 'norm' of 1881–1915. Additionally there have been several books that have described and discussed the British climate from a particular viewpoint. Perhaps the most influential has been *Climate and the British Scene* by Manley (1952), especially as it formed part of a popular series; whilst a later contribution (Manley, 1970b) provides a briefer statement. Other well-known authors in this area include Brooks (1954) and Lamb (1964), while in many more general climatological or regional volumes there have been useful summaries of the British climate.

Since Bilham, or for almost 40 years, there has not been a text that both drew together the existing literature and presented up-to-date information in a major way. Yet during this long period the growth in climatological literature on Britain has been considerable, building on previous analyses and also enlarging and modifying our factual knowledge, changing our ideas and concepts, and expanding and intensifying our understanding and appreciation. The purpose of this present book is therefore two-fold. Firstly, the intention is to draw together the maximum body of information, in a consolidated and compact form and for as recent a period as possible. Secondly, the aim is to present a critical evaluation of the literature, reviewing the ideas and approaches in the field, as they have developed and evolved over the past few decades. A number of publications and concepts that were basic in Bilham's work remain as essential today as they were then, but there is a vast body of new research the results of which have had to be integrated to ensure that this present volume adequately reflects the field of study as it is seen today. Achieving both these objectives is a major operation, and it has been possible to effect it in a relatively short period of time only by inviting a number of researchers to contribute on a pre-scheduled theme. The role of the editors, as well as acting as contributors, has been to structure the overall work and to attempt to integrate the individual contributions into a unified account.

If the book had been concerned simply with a literary review of the field it is likely that it would have taken a very different form. With the express need to incorporate an additional role as an information source, the structure has tended to retain traditional characteristics. This is emphasized by the fact that it reflects much of the structure of Bilham's book. This was deliberate, both because this permits basic factual information to be communicated in an organized way, and also because it allows the growth and development of ideas over the past 40 years to be appreciated more readily.

In outline, the book falls into three broad sections. Firstly, there is a major discussion of the basic circulation systems as they affect Britain, to provide a climatologically orientated evaluation of synoptic situations. Thus, Chapter 2 is fundamental to the whole of the rest of the book, providing the framework into which the other contributions may be fitted. The second broad component consists of Chapters 3 to 10, which focus upon a series of climatic elements. These begin with the primary energy source, radiation including sunshine, followed by the advecting medium of wind motion. These, along with the synoptic conditions, provide much of the background explanation for the succeeding chapters on atmospheric temperatures, precipitation, evaporation together with atmospheric humidity and the water balance, clouds and thunder, and atmospheric visibility. All these are considered in the light of the most recent data available, as is discussed below, and questions of temporal changes are discussed only in terms of variability within such a recent period. The broader theme of changes with time, beyond the immediate period, is considered in a separate presentation in Chapter 10, which thus draws together the relationships between changes in the different elements. The third structural section of the book (Chapters 11 to 15) is concerned with summarizing certain regional unities within the climate of the British Isles. Four general sets of conditions are evaluated – coastal climates, together with similar results around inland water bodies; upland climates, that affect such a large part of the area but for which an adequate station network does not exist; climatic differences resulting from local site characteristics, especially slopes, soils and vegetation; and the climatic features peculiar to urban areas, within which the majority of the population lives. The final chapter attempts the regional differentiation of climatic conditions.

The information and data on which these contributions are based has been derived from a variety of sources. Much has been drawn necessarily from previous published studies, especially very many of the maps and diagrams and parts of the tabulated material. In this way the provision of information overlaps the review of the literature, and it is obvious that for the factual content of this book we are heavily indebted to the many researchers who have published over the recent decades, and to the editors and publishers of the journals and books in which their findings first appeared. There has also been considerable abstraction of data from a number of official publications, especially the *Monthly Weather Report*, the *Daily Weather Report*, the volumes of *British Rainfall* (all published by the Meteorological Office), and the *Surface Water Yearbook of Great Britain*. However, in such cases, the compilations and analyses represent new contributions. The latter is also true in relation to the third major category of data, namely that which has been provided direct from manuscript sources by the UK Meteorological Office and the Irish Meteorological Services. The willingness of these official bodies to make such information available to our contributors, especially when such data may have been scheduled for analysis through their own channels at some time in the future, has been most gratefully appreciated. This action and the encouragement we have received from these and other official bodies and private persons has ensured that the objective of providing up-to-date information has been achieved to an extent that would have proved impossible otherwise.

The data thus obtained from the official services, and from such reports as are listed above, although of considerable detail and complexity, still cannot ensure a country-wide uniformity of data periods. For any one climatic element the spread of information over the British Isles also varies markedly in intensity from one area to another, while there are further differences between the data nets for different elements. This variation in the detail of the analyses is most commonly a reflection of the spread of recording stations over the United Kingdom and the Republic of Ireland, the pattern of which for the United Kingdom can be seen in Fig. 1.1 for a variety of station categories. What is most apparent is the relative sparsity of stations in upland areas, and even when they exist here they are mostly located at relatively low altitudes. This is true even in terms of rainfall for which upland stations are more numerous than for other climatic elements. This theme is discussed further in Chapter 12, where upland climates are reviewed, but it should be remembered that in all the other chapters the bulk of the data is for lowland sites.

Unfortunately, but unavoidably, it has not proved possible to ensure a common data period for all climatic elements, and often different considerations within the same chapter have had to move to different data periods. This is partly the result of the very different data sets used in the original publications from which much of the information is drawn. Also, however, even with officially provided data, the length of record available varies between the elements. Additionally, the policy decision by the editors that the contributions should, whenever possible, be concerned with the British Isles as a whole including the Republic of Ireland, and not simply with the United Kingdom, has created further problems. Data periods on which official or semi-official publications and tabulations are based are not always common between the two countries, while the format in which data are provided sometimes differs. Also, in some cases data were not available from Ireland for particular climatic aspects, so that a limitation to the United Kingdom is unavoidable. Despite these difficulties, however, the attempt to write about the British Isles as a whole is the academically sensible one, and – thanks to the official cooperation already mentioned – it has been successful in the main.

Throughout the book, a very large number of climatic stations have been mentioned. Some of them, with long-term records on which many studies have been based, occur very frequently. For example, Heathrow (London Airport), Kew, Lerwick, Oxford, Plymouth and Valentia are cases in point. Others may occur only in some specific context and may in fact be stations that no longer exist. All stations mentioned are located on the map in the Station Index, where they are also listed alphabetically together with the county in which they occur, with chapter references to where they appear in the text, and with reference to figures and tables in which they are specifically designated. It should be stressed, however, that for many of the maps, stations other than those listed were also used, these often, though not invariably, being stations included in the *Monthly*

Fig. 1.1 *Climatological stations in the United Kingdom, January 1969 (after Taylor, 1972).* (a) *Major and auxiliary synoptic stations;* (b) *Climatological stations;* (c) *Agro-meteorological and other stations;* (d) *Anemograph stations*

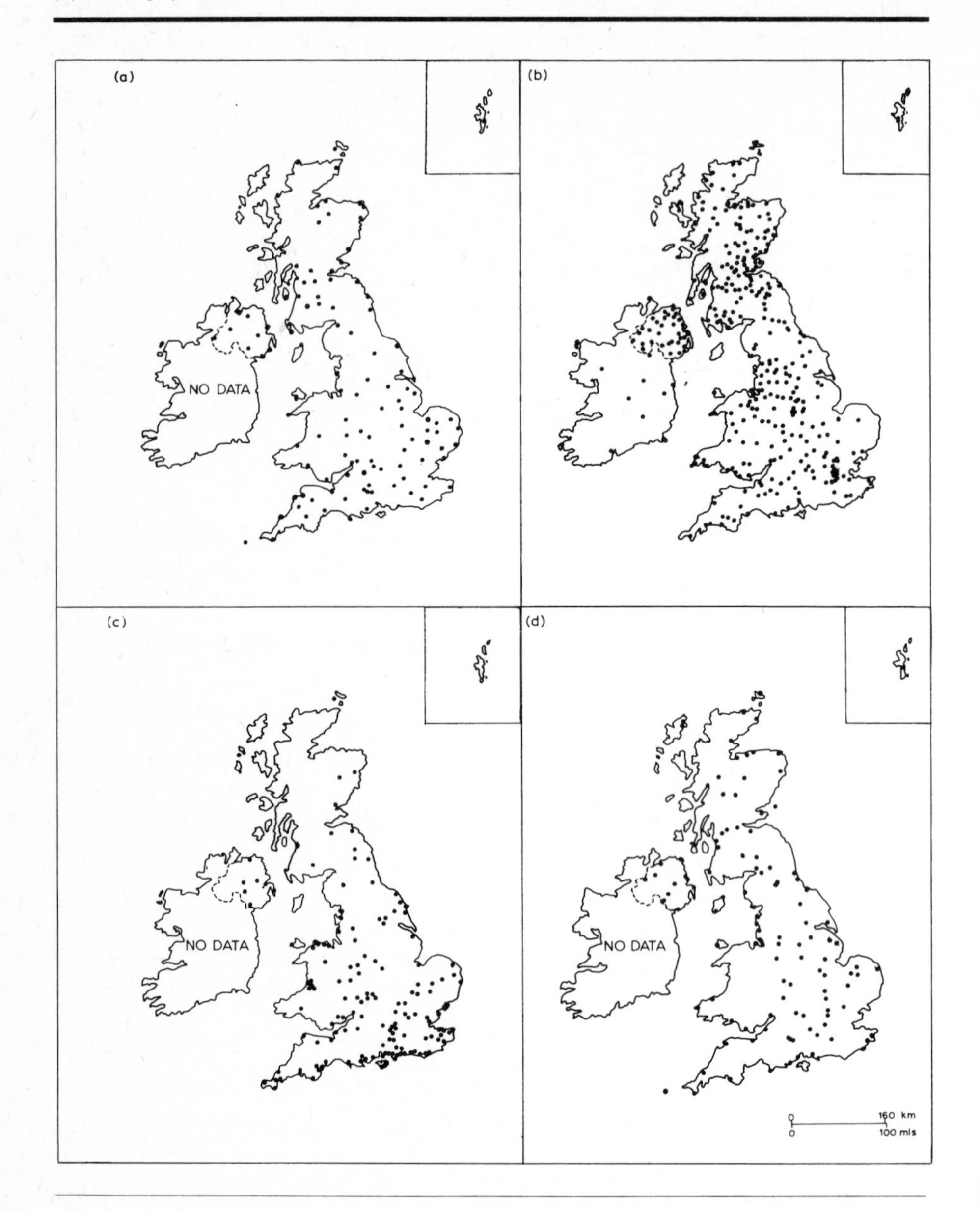

Weather Report, for which a location map is available from the Meteorological Office. In the interests of uniformity of presentation, the editors have also scheduled other items of standardization. Units of measurement are primarily in IS units, although if results and conclusions are being presented from an earlier paper, the original form of the data is sometimes given. All time references have also been converted to the 24-hour system, and all times should be assumed to be GMT unless otherwise specified – as a result, the term GMT has been omitted.

From this background has been drawn together a detailed and up-to-date picture of the climate of the British Isles as a whole. No such book can claim to be absolutely comprehensive, especially in terms of the themes covered, and in this context the editors are responsible for the sins of both omission and commission. Within each chapter, some of the major contributions of the last 30 to 40 years have been incorporated, and the most recent scene is set. It is hoped that the resulting volume will provide useful information and interesting ideas for a very wide range of readers, all of whom will find the climatological presentation relevant to their own particular interests and needs. If this volume succeeds in providing for climatologists in the 1980s the service that Bilham's book provided in the 1940s and 1950s, then the editors and the contributors will be more than satisfied.

Introduction

The basic aim of synoptic climatology has been defined (Barry and Perry, 1973) as relating local or regional climates to the atmospheric circulation, instead of using an arbitrary time base for assessing average or modal values. It is interesting to note that although the term was first proposed in 1942, there is in Bilham's (1938) classic text a whole chapter devoted to defining and describing characteristic pressure distributions affecting the British Isles and the associated different types of weather. In fact, such an approach, which would now be called synoptic, had been pioneered in the work of Abercromby (1883) and Gold (1920).

Since Bilham's day much work has been undertaken by Levick (1949), and particularly by Lamb (1950, 1972) in deriving long-period catalogues of surface airflow types. The impetus to undertake this work has come partly from the demands of long-range weather forecasters, who require extensive data banks (Craddock, 1970), if the analogue method of forecasting is to be used successfully. The recognition that climate, far from being stable and unchanging, has shown significant changes even during this century has prompted investigation into changes in the frequency of weather types over the British Isles (Lamb, 1965) and their characteristic durations (Perry, 1970).

The atmospheric circulation

The importance of the atmospheric circulation in shaping the climate of the British Isles has been ably outlined by Manley (1970b) when he stated: 'the climate is fundamentally an expression of increasing proximity to the principal route of travel of active frontal depressions, which is represented throughout the year on charts of mean surface pressure by the Icelandic low'. Of course the mean surface pressure distribution shown in Fig. 2.1 is merely an integration of a wide variety of real-time conditions and we need to note particularly that while the Azores high pressure area is a semi-permanent, quasi-stationary feature of the circulation, the Icelandic low is a statistical feature representing the area in which mobile depressions most frequently reach their greatest development. The two mid-season months represented in Fig. 2.1 suggest a weakening of the circulation from winter to summer, but it should be noted that the mean pressure gradient from south to north over the British Isles is in fact least in May.

In summer, the pressure in the Azores high reaches its maximum value. Over the North Atlantic between 40°W and the British Isles the centre of the high is further to the north than in winter, while the polar front, along which many depressions form and move, lies on average 1°–4° further south

Fig. 2.1 *Mean sea-level pressure distribution in mbs (in full lines) for 1950–9 (after Lamb, 1964), and mean map of 500 mb topography in geopotential decametres (in broken lines) for 1949–68 (after* Die Grosswetter Europas, *1970) for:* (a) *January;* (b) *July*

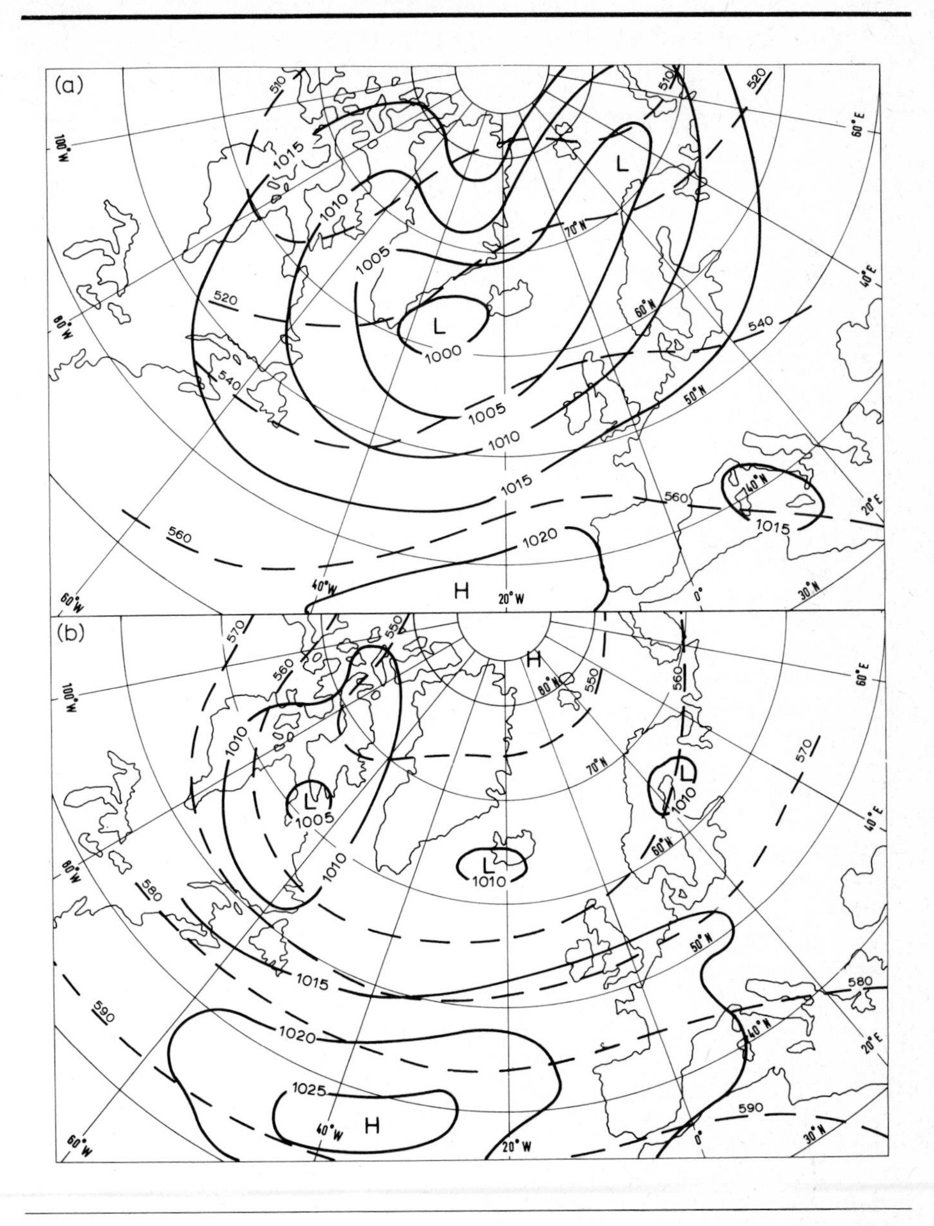

in July than in January (Lamb, 1972). Significantly for the day to day changeability of weather in the British Isles a zone where the alternation between high and low pressure (known also as the 'perturbation duct') is most rapid lies across the Central Atlantic and extends into our islands (Petterssen, 1950).

Although the main depression track lies between Scotland and

Iceland individual lows frequently cross the British Isles and Thomas (1960a, 1960b) has shown that in a normal year between 160 and 180 depressions contribute to the annual rainfall over the country. Table 2.1 gives details of types and amounts of precipitation typical from depressions passing along the major tracks. The frequency of frontal passages is also high and Eriksen (1971) has suggested that an annual average of 110 occlusions, cold and warm fronts cross the country. In terms of synoptic diversity and rapidity of weather type change Britain is thus characterized by an extremely mobile circulation.

In the upper air, variations in the strength of the westerly flow and its latitude, together with variations in the amplitude and eccentricity of the long or Rossby waves, can lead to periods of anomalous circulation and abnormal seasons. In winter one of the three major waves in the hemisphere is situated over the eastern Atlantic and Europe, with an associated ridge extending from the Azores to Norway and a trough at about 30–40°E, but occasionally as far west as 25°W. By summer the wave pattern weakens and the flow becomes practically zonal. The flow patterns vary between two extreme states – one with long wavelengths associated with strong zonal westerly flow in middle latitudes, and the other with large amplitude waves and a replacement of the mid-latitude westerlies by a blocking pattern. Thus, when a large high pressure area is established in the zone of the westerlies, the upper level jet stream becomes bifurcated and associated frontal depressions are forced to travel around the north or south side of the blocking high.

It is the relative dominance of these two forms of circulation – high and low index respectively – which frequently establishes the character of a particular month or season. Sumner (1959) has shown that blocking occurs most frequently in the longitudes just to the west of the British Isles, and especially in spring, but there remains a good deal of disagreement as to whether any notable periodicities in the occurrence of blocking takes place, with Brezowsky *et al.* (1951) claiming to find a 22–23 year oscillation in activity, and Conte *et al.* (1973) finding a 5-year 6-month recurrence interval. Blocking patterns, once established, tend to persist, having a median duration of about 12 days. During spring there is a tendency for retrogression from Europe to the Atlantic to take place, that is a westward movement, but in late summer and autumn most blocks shift eastwards with time. The weather pattern over the British Isles during a blocking spell will depend on the position of the high, but in general weather will be drier than normal.

A characteristic of the 500 mb trough and ridge pattern is the tendency for certain wavelengths to occur in spells. De la Mothe (1968) has shown that a spell of maximum wavelengths is common in early December and January with strong zonal flow, but a progressive shortening takes place during the spring so that the short wavelength pattern associated with blocking is prevalent by May. This is then followed by a certain renewed increase and abrupt though limited eastward shift of the circulation features in mid-late June at the time of the 'European Monsoon'. The position of the jet stream

Table 2.1 *Precipitation effects of the more frequent depression types in the British Isles (after Thomas, 1960b)*

Depression Track	Frontal Precipitation ⩾5 hours rainfall		Orographic Precipitation Light rain/drizzle for 1–3 days		Instability Precipitation Showers over 1–3 days	
	⩾5 hours rainfall	About 2 hours rainfall	Light rain/drizzle for 1–3 days	Intermitt. light rain 6–24 hours	Showers over 1–3 days	Showers over 12–24 hours
S.E. across Shetlands	W., N.W. Scotland.	Remainder of area except Cen., E., S.E. England.			N. England; N. Wales; N. Ireland and Scotland except S.E.	Remainder of area.
S.E. across Iceland and Cen. England	S.W. Scotland and rest of area except E., S.E. England.	S.E. Scotland; E., S.E. England.		S.W. Ireland; S.W. England; S. Wales.		N.W. Ireland; Ulster; N.W., W., S.W. Scotland; N.W., S.W. England; S. Wales.
N.E. to the North of Iceland	W., N.W. Scotland.	N.W. Ireland; Ulster; S.W. Scotland; N.W. England.	Wales; S.W. England; N.W., S.W. Ireland.	N.W., W., S.W. Scotland; Ulster; N.W. England.		N.W. Scotland; N.W. Ireland; N.W. England
N.E. between Shetland and Iceland	N.W., W., S.W. Scotland; N.E. England; N. Wales; N.W. Ireland and Ulster.	Remainder of Scotland and Ireland; and England except E., S.E.		Ireland, except Cen.; E., S.W. Scotland; N.W., S.W. England; Wales.		N.W., W., S.W. Scotland; Ireland except Cen.; E., N.E. England; Wales
East across N. Scotland	All areas except S.W., S., Cen., E., S.E. England.	Remainder of England.		S.W. Ireland; S.W. England.	N.W., W., S.W. Scotland; N.W. England; N.W. Ireland and Wales.	E., S.E. Scotland; Wales; S.W. Ireland; S. Cen., S.W. England.
N.E. across Cen. England	Ireland except N.W.; N.W., W., S.W., Cen., England; S. Wales	Scotland; N.W. Ireland; E., N.E. England; N. Wales				S.W., S.E. Scotland; Ireland; England; except S.W.
N.E. across Ireland	S.W. Scotland; Ireland; Wales; N.W., S.W., W., S-Cen., England.	Remainder of area				S.W., S.E. Scotland; Ireland; England; S. Wales.

across the North Atlantic and Europe also varies from month to month and its location can be used to predict rainfall over the British Isles.

Air masses

'The weather of the British Archipelago depends heavily upon imported air though its outstanding characteristics, moisture and moderation, may be regarded as at least partially home-brewed since they arise from the nature of the surrounding seas' (Crowe, 1971).

Fig. 2.2 *Trajectories of principal air masses affecting the British Isles*

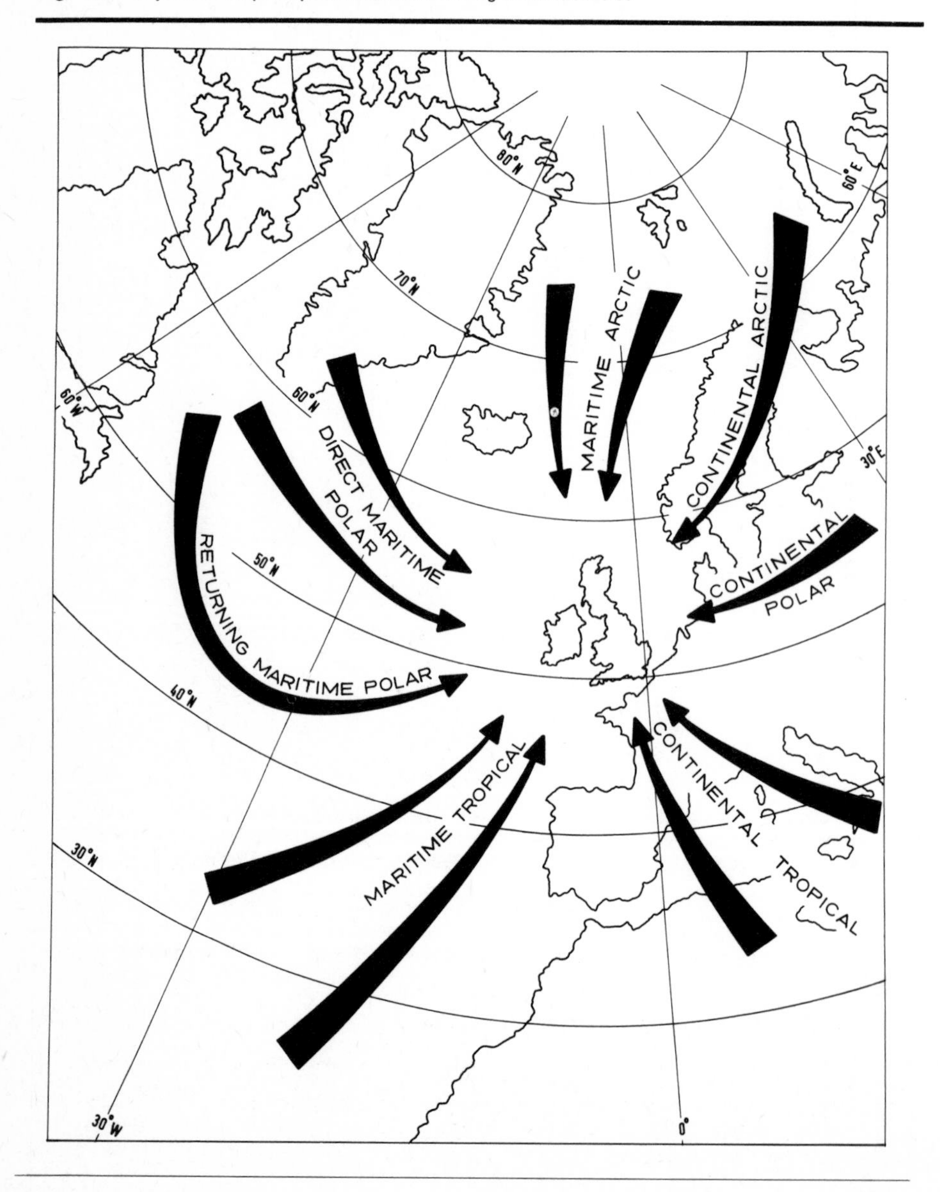

Although on a world scale the British Isles have been referred to as being occupied by mid-latitude air, in practice the airmasses are very variable in their origins (Fig. 2.2), and because of the great variety of origin and trajectory they are extremely difficult to classify. Belasco (1952) found it desirable to distinguish twenty-three types of air over Britain, but many of these are sub-types of the major categories:

1. Maritime Polar (mP) – cold or rather cold at all seasons; unstable with bright intervals and showers especially on windward coasts and over high ground; cloud and showers die out inland at night and in winter; origin in the northern part of the North Atlantic.
2. Maritime Tropical (mT) – mild in winter, warm and close in summer; stable air with dull skies, hill and coastal fog and drizzle in windward areas but often bright and sunny in sheltered eastern areas; origin in the subtropical North Atlantic.
3. Continental Tropical (cT) – very mild in winter, very warm in summer, but rare at all seasons; typically cloudy in winter and thundery in summer; area of origin in North Africa and over the Mediterranean.
4. Continental Polar (cP) and Continental Arctic (cA) – very cold in winter but warm in summer; cloud and showers, often of snow along the east coast in winter and spring; usually mainly dry with clear skies in western areas; origin over Western Russia (cP) and the Arctic ocean (cA).

Table 2.2 *Summary of air mass frequencies (percentage of days) over the British Isles, 1938–49 (Crowe 1971, after Belasco, 1952)*

(58°11′N) *STORNOWAY*	**Winter**		**Summer**	
I. Northern air	25		27	
II. Central North Atlantic air	18	=63	18·5	=60
III. Air with northern affinities	20		14·5	
IV. South-east European air	6	= 6	3	= 3
V. Southern air	6	=11	3·5	= 6
VI. Air with southern affinities	5		2·5	
VII. Air in high pressure areas	10	=20	15·5	=31
VIII. Air near fronts and lows	10		15·5	

SCILLY (49°56′N)	**Winter**		**Summer**	
I. Northern air	15		20	
II. Central North Atlantic air	19	=44	19	=47
III. Air with northern affinities	10		8	
IV. South-east European air	8	= 8	5	= 5
V. Southern air	15	=21·5	9	=12
VI. Air with southern affinities	6·5		3	
VII. Air in high pressure areas	15·5	=26·5	22	=36
VIII. Air near fronts and lows	11		14	

5. Maritime Arctic (mA) – cool or cold at all seasons; unstable and showery but with good visibilities; clear skies in leeward areas; origin in the polar seas.

Table 2.2 shows that at Scilly in both winter and summer approximately 50 per cent of the days had air in occupance from a part of the North Atlantic and this was true of over 60 per cent of the days at Stornoway. Air from the European continent is comparatively rare at these two westerly points, occurring on between one-sixth and one-eight of all winter days. Maritime tropical air is much more frequent at Scilly than at Stornoway. In some instances it may be possible to deduce the nature of the air mass from wind direction alone. Maritime tropical air, for example, frequently arrives with a south-west wind, but wind direction can be misleading. Maritime polar air circulating around a deep depression near southern Ireland can arrive as a southerly wind (returning mP air) or as a west or north-west wind (direct mP air) if the depression centre is further north over western Scotland.

Weather types over the British Isles

Each day's weather from 1861 to the present has been allocated to one of twenty-seven categories using the criteria established by Lamb (1950, 1972). This catalogue for the latitude and longitude 'square' 50°–60°N and 10°W–2°E recognizes the existence of eight directional types each subdivided into anticyclonic, cyclonic and undifferentiated categories according to the nature of the isobaric curvature over the British Isles. In addition, three non-directional types are recognized – anticyclonic, cyclonic and unclassifiable cases. Each of Lamb's weather types may result in several air masses affecting an area as it is crossed by depressions travelling within the general airflow, e.g. the westerly type may be associated with maritime polar or maritime tropical air.

The characteristics of the major synoptic patterns and the weather types that result from them will now be examined.

Changeable westerly type weather

When pressure is high to the south of the British Isles and low to the north, sequences of fronts and ridges of high pressure travel eastward producing unsettled and changeable conditions with spells of frontal rainfall, particularly in northern and western districts. The main depression centres pass just north of the British Isles, but secondary depressions formed on trailing cold fronts may pass across the country bringing severe gales in autumn and winter. In some winters blocking anticyclones develop and persist over eastern Europe and deep depressions commonly stagnate over the sea area between Iceland and Scotland with westerly winds on their southern flanks. Such conditions characterized much of the winter of 1966–7. At Sheffield, the 87-day period between 14 December 1966 and 10 March 1967 had measurable rainfall on 49 days and daily minimum temperatures fell below 0°C on only thirteen occasions.

In eastern and south-eastern England the maritime polar air behind an eastward moving cold front often gives sunny dry days in winter but in summer, associated convectional activity frequently leads to heavy, thundery showers, particularly if there are troughs embedded in the westerly flow, or if the general curvature of the isobars is cyclonic. Such days start bright and sunny but heavy cumulus and cumulonimbus clouds quickly develop during the morning with the showers falling mainly in the afternoon and evening. During cool, wet summers, weather of this type alternating with periods of rainy frontal weather often means there are few completely dry days, as in August 1963 when measurable rain fell on two-thirds of all days in south-eastern England. When the isobaric pattern over the British Isles shows anticyclonic curvature, often with high pressure over France and depressions well to the north of Scotland, the westerly weather type is frequently fine with small fair-weather cumulus in summer and temperatures near or above the seasonal normal; disturbed weather is then confined to the north-western coasts. In winter, the humid air frequently gives dull, grey skies and hill and coastal fog in the south-west, sunshine being confined to the lee of high ground – for example, the eastern coastal strip of Scotland where temperatures in such weather can occasionally rise above 15 °C in

Fig. 2.3 *Specific sea-level synoptic charts (with 500 mb ridges and troughs in broken lines):* (a) *12.00 hrs, 24 January 1974;* (b) *06.00 hrs, 28 June 1962;* (c) *06.00 hrs, 21 January 1958;* (d) *Key to Figs 2.3 and 2.4*

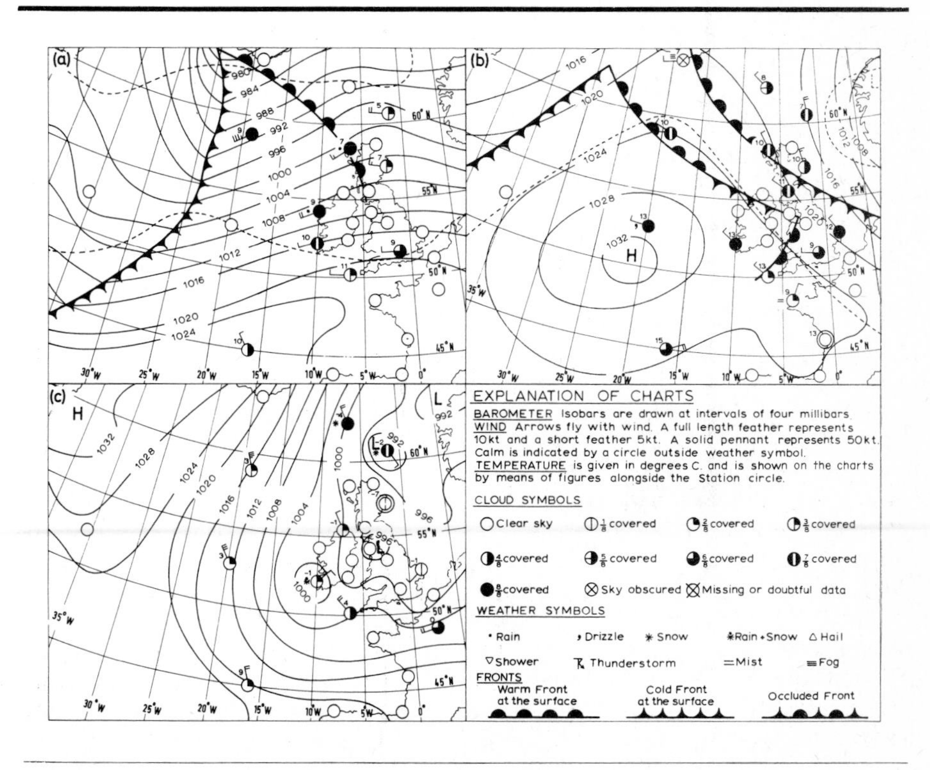

mid-winter, e.g. 16·5 °C recorded at Aberdeen on 21 February 1974. Relief can play a major part in determining the local susceptibility to advection fog (see Chapter 9) ; for example, Poole harbour is almost completely sheltered by the Purbeck hills and usually remains fog-free. During such weather, night-time minimum temperatures above 10 °C are occasionally recorded in the south, even in midwinter. Figure 2.3*a* represents a typically mild winter day with westerly winds and a weak frontal system affecting north-western areas.

Cool north-westerly type weather

When the Azores anticyclone is displaced north-eastward toward western Europe and depressions are steered south-eastward into the North Sea and Scandinavia, a north-westerly flow covers the British Isles. Temperatures at all seasons are likely to be below normal and the weather is frequently unsettled with a mixture of sunny intervals and showers. Along western coasts and over the hills the showers can be heavy and with relatively steep lapse rates, hail, sleet and snow can occur from October to May, from Snowdonia northwards. In sheltered areas like the Welsh Borderlands or the south coast of England, showers are much less frequent and consequently sunshine amounts are greater.

Sometimes small wave depressions moving around the northern flank of the high pass across the British Isles, giving cloudy weather with outbreaks of rain. The chart of 28 June 1962 (Fig. 2.3*b*) shows such a situation with a small low over Lancashire and cloudy skies nearly everywhere except along the south coast. A similar situation on the 12 December 1974 led to a maximum temperature of 1 °C at Wick with some snow, while in Torquay the temperature reached 13 °C in the mild air circulating around an anticyclone lying west of Britain.

This weather type normally lasts only a day or two, and very rarely for more than 5 days, giving way either to a westerly type if the anticyclone slips southwards or to a northerly type if the high builds northwards.

Cold northerly type weather

When a blocking anticyclone forms west of the British Isles, and especially when the high pressure belt assumes a north–south orientation sometimes extending from Greenland to the Azores, northerly winds invade the British Isles. At the same time pressure is low over the Continent. At all times of the year temperatures are below normal and in mid-winter may be continuously below freezing, especially in northern districts. The most severe winter weather in Scotland is associated with this type, giving instability showers of sleet and snow especially over high ground and on windward coasts.

The onset of this type is often accompanied by high winds, particularly on eastern coasts and associated with tight pressure gradients over the British Isles. In January 1953 widespread flooding on North Sea coasts occurred when severe gales were associated with abnormally high tides. Instability is frequently intensified by the formation of small polar-air depressions and troughs which move south in the airstream. Intense heating over the warm seas around Britain is often the cause of these depressions which may be less than 150 km in diameter. On 8 December 1967, 30 cm of snow fell in Brighton as a polar depression, forming near Ocean Weather Ship I, moved south-eastwards from Northern Ireland to Sussex. Late snowfalls can result from such disturbances as in mid-May 1955 in southern England and in April 1970 in northern England. Figure 2.3c shows an occasion when three polar lows were situated over different parts of the British Isles at the same time.

Northerly types in summer give cool, showery weather and, if as happened in the first week of August 1956, upper cold pools develop, heavy thunderstorms with hail occur, giving Tunbridge Wells on this occasion a fall of hail which blocked streets and dislocated traffic.

The spread back of rain and low cloud into eastern coastal districts is common when depressions, after crossing the British Isles, slow down over the North Sea giving persistent gradient winds between north and east in eastern districts. The rain associated with such a situation is generally slight but is usually prolonged and slow to clear and may aggregate to a substantial total amount. Periods of rain and extensive low cloud can persist even with sustained rises of pressure and several notable floods have occured under such conditions.

Easterly weather from the Continent

If pressure is high over Scandinavia, perhaps with an extension towards Iceland, and low in the Azores–Spain–Biscay region, easterly airflow from the Continent will affect most of the British Isles. The nature of the weather depends largely on which of the pressure systems is dominant and on the curvature of the isobars over the country. In winter, the easterly flow is likely to be very cold, especially when the sea-track is short as in south-east England. Figure 2.4a shows the weather pattern on one of the coldest days on record in southern England when a sharp frost was recorded even in the Scilly Isles. Further north, the greater width of the North Sea allows more warming of the air and slightly higher temperatures. Warming over the sea also leads to instability, and showers of snow and sleet are common along the east coast, especially in areas like Thanet and East Anglia, often accompanied by strong winds in the Straits of Dover and below freezing temperatures. Further west there are normally few if any showers and the longest hours of sunshine and highest temperatures are commonly recorded in the Western Isles. The most widespread and prolonged snowfalls in southern England usually occur when depressions over France and Biscay push frontal systems northwards which then become quasi-stationary before dying out or retreating away southwards again. Frequently the forward and subsequent retrogressive movements are caused by a depression moving along the front. With a weather pattern of this type, mild air from the south may spread into Cornwall where temperatures may be 8 °C higher than in the cold easterly flow north of the front.

Fig. 2.4 *Specific sea level synoptic charts (with 500 mb ridges and troughs in broken lines) the key is in Fig. 2.3.* (a) *06.00 hrs, 1 February 1956;* (b) *12.00 hrs, 15 September 1968;* (c) *12.00 hrs, 30 June 1968;* (d) *12.00 hrs, 13 January 1969*

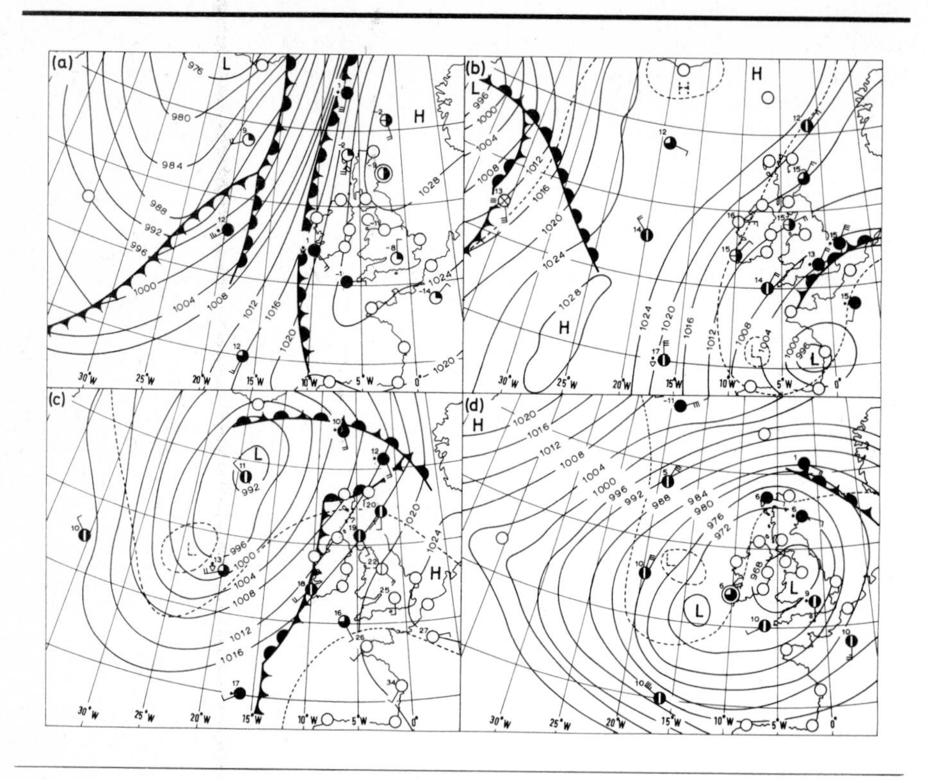

During spring and early summer, easterly flow often leads to cool, cloudy conditions near the east coast; these spread inland at night, but during the day time inland and western districts are frequently sunny with near normal temperatures. If the source of the air lies over the Continent, it is frequently warmer than the cold surface waters of the North Sea. It is chilled from below on its journey, leading to the formation of fog which affects the coast, especially from the Humber northwards, and which is known locally as 'haar' (see Chapters 9 and 11). The cold and dismal nature of easterly weather in spring in the north-east was well illustrated in 1974 when the mean maximum temperature at Tynemouth averaged 7·1 °C in April, compared with 7·7 °C in January, 3 months earlier.

Anticyclonic easterly types in summer lead to generally warm and fine conditions, although spells of this type often end with thundery outbreaks spreading north from France. In early summer the continental air is often brilliantly clear, visibility is excellent and the long hours of sunshine lead to ideal conditions except perhaps on the eastern littoral where it is usually more cloudy with sea breezes. If low pressure is persistent over the Low Countries and Germany, the generally cyclonic easterly flow can give most unpleasant conditions in the south and east, as happened in late July and early August 1968.

Cool, wet weather in London then contrasted with sunny, warm conditions in northern and western Scotland, near to the blocking high lying to the north-west, thus giving a reversal of the normal summer weather pattern. Later the same year, in mid-September, a depression moving up the English Channel brought heavy, continuous rain to the south-east again with falls of 200 mm in 3 days in parts of Kent and Surrey, leading to widespread and disastrous flooding. The synoptic chart on this occasion is shown in Fig. 2.4b.

Mild southerly weather type

When high pressure covers central and northern Europe and depressions are travelling north or north-east off our western coasts, southerly winds cover the British Isles. Weather is warm and thundery in spring and summer, but often dull and mild in winter. Even in early March, maximum temperatures can exceed 21 °C and on 12 March 1957 Cape Wrath in Sutherland recorded 23 °C, equalling the average annual maximum temperature there. Occasionally, dust from the Sahara can be transported northwards and deposited over Britain, as happened on 30 June–1 July 1968. The synoptic chart on this occasion (Fig. 2.4c) shows clear skies and high temperatures over central and eastern districts in contrast with the cooler and cloudier weather in the north and west associated with an invading frontal system. Occasionally in winter a rather persistent southerly type leads to several days of mild, dull and rather quiet weather as in late February 1964 when a warm anticyclone was slow to leave the North Sea area. Normally, southerly types are liable to undergo modification either through cyclonic activity near the west coast, or through anticyclonic curvature of the isobars developing because of a continental anticyclone thrusting west over England or forming a new centre over the country.

Cyclonic weather type

The weather pattern is designated cyclonic when either a mobile cyclonic sequence with successive centres is passing across the British Isles or a slow-moving, usually filling depression is situated over the country. In the latter case the weather is often showery, but if the low is occluded, the occlusion often moves forward faster than the depression centre and may display considerable curvature or even circumvent the centre giving a belt of rain along its course. Figure 2.4d shows an occasion when a deep and complex depression was slow moving and all districts lay within its circulation giving cloudy unsettled conditions with outbreaks of rain, and in the extreme north of Scotland cold weather with easterly and south-easterly winds.

Anticyclonic weather type

When anticyclones are centred over or near the British Isles, weather is normally mainly dry with light winds. Lamb (1972) notes that sometimes a frontal col or trough is present in the central anticyclonic region and these situations can give some rainfall, especially in summer if thundery outbreaks develop in the col region. Large detached portions of the Azores anticyclone drift slowly across the British Isles giving long hours of sunshine and cloudless skies in some summer months, and occasionally in winter too. A persistent high gives marked departures from average conditions over several weeks. In February 1965 anticyclones centred north-west of the British Isles were extremely persistent and many places from eastern Scotland to the south coast had less than 10 mm of rain during the month. Winter anticyclones can easily lead to the development of widespread and persistent fog and as such highs move eastward into the Continent, their influence can linger over south-eastern England, even when freshening southerly winds have lifted the fog elsewhere. If pressure builds over the country following a snowfall occasioned by a previous continental or arctic airstream, very cold nights can occur even in southern England, as happened around 23–24 January 1963 when temperatures in the northern Home Counties fell to −18°C. Small, transient cold highs normally make their influence felt for a relatively short time, often with the renewed onset of a westerly type after a few days.

Table 2.3 *Average frequency of Lamb's weather types, 1868–1967 (after Lamb, 1972)*

Year		J	F	M	A	M	J	J	A	S	O	N	D
25·7	W	34	27	23	19	15	22	26	28	27	27	27	33
4·7	NW	4	4	4	5	4	5	7	6	4	4	5	5
7·4	N	6	5	8	9	11	9	7	6	7	7	8	6
7·6	E	6	9	11	11	12	7	4	4	5	8	8	6
8·6	S	12	11	10	8	8	6	5	6	8	10	9	10
24·9	A	22	25	25	27	27	29	24	23	29	24	23	21
17·5	C	14	15	16	17	19	18	22	24	15	17	17	16

From Table 2.3 it can be seen that there are marked seasonal differences in the average frequency of each weather type, and this is particularly so in the case of the westerly type which is more than twice as frequent in the mid-winter months as in May. Over the period since 1861 there have also been changes in the frequency of the various types and again this has been especially marked in the case of the westerly type. After reaching a maximum frequency in the decade 1920–9, the westerly weather type has become in recent years less frequent than at any other time since 1861, with only half as many days in the year with this type as was the case in the years of peak occurrence in the 1920s. Belasco (1948) has shown that anticyclones are most frequently positioned to the south-east of the British Isles in January, to the north-east in May, to the south-west in June and July and to the east in October and December. The average duration of anticyclonic spells in the period 1889–1940 was greatest (8·2 days) in February and least (5·5 days) in August.

Master seasonal trends and pulses

From a study of long spells lasting over 25 days of one or other of the weather types, Lamb (1950) has been able to show that spells are by no means equally common at all times of the year, occurring in less than 20 per cent of years in early June but in nearly 50 per cent of years in both late July and early October. The same author (Lamb, 1964, 1973a) finds, from a study of the weather type catalogue, four kinds of events:

1. Broad imprecisely-dated climaxes of various seasonal trends, i.e., the year's greatest and least development of various weather types. Physical causes can often be suggested for these master seasonal trends associated with the strength and position of the main energy sources and sinks around the northern hemisphere.
2. Short periods of heightened tendency for particular types to occur. Their onset and ending may be sharply affected by geographical barriers which come into play when the seasonal trend has reached a certain point.
3. Abrupt, almost step-like changes in the frequency level of particular weather types about particular dates. Some of these pulse phenomena are linked to the oscillation of the pressure field; for example, the main centres of action of the atmosphere fluctuate in their position and intensity.
4. Random variations from day to day superimposed on 1–3.

Natural seasons

The occurrence of long spells of weather has been used by Lamb (1950) to define a series of five natural seasons:

1. High summer – 18 June to 9 September.
2. Autumn – 10 September to 19 November.
3. Early winter – 20 November to 19 January.
4. Late winter – 20 January to 29 March.
5. Spring – 30 March to 17 June.

The synoptic characteristics of each of the five seasons are as follows:

1. High summer The frequency of long spells of weather is marked at this season, with years affected by wet, cyclonic sequences being about twice as numerous as persistently anticyclonic ones. Depressions are often quite shallow and move less rapidly than in winter but sometimes remain quasi-stationary for several days. A strong correlation exists between coolness and wetness in high summer, although occasionally months which are dominated by blocking between Scotland and Iceland are both cool and dry, e.g. August 1920, June 1929.

In fine summers, offshoots of the Azores anticyclone move eastward across southern England or northern France interrupted by occasional mainly weak fronts. Sunny, warm days are much in evidence, except in the north-west where cloud and rain occur from passing fronts associated with depressions far to the north. Sometimes in the latter stages of an anticyclonic spell the high centre will move eastwards to Germany or Denmark allowing a very warm Continental air supply with a light south or south-east wind into the Midlands and south-eastern England (Perry, 1967). Temperatures may then exceed 34 °C and on average (1921–50) 32 °C is reached in the London area in about one July in three. Such spells are usually terminated when a cold front approaches from the west, often with pre-frontal troughs and heat lows moving north ahead of the trough and giving thundery rain and thunderstorms. These frequently develop over France and the Low Countries and move into southern England during the evening and night.

2. *Autumn* The first week is on average one of the driest of the year especially in central and eastern districts, but a mid-September cyclonic spell with persistent low pressure over Western and Central Europe can lead to heavy rainfall, e.g. in Kent in 1968. October, although often changeable, has given settled spells and high temperatures in several recent years with an excess of southerly anticyclonic weather types, especially in early and mid-month. However, between 23 October and 11 November, wet, stormy weather is very prevalent, and anticyclonic types decline to less than half their frequency in the first part of the month.

3. *Early winter* Long spells are distinctly less common than in high summer or autumn and only affect about half the years, mostly being of westerly, mild types. Except in 1962–3 none of the long spells established in this part of the winter during the last 150 years seems to have carried on without a break into the late winter. Most prevalent is the pattern of strong cyclonic types early and late in each month with quiet anticyclonic weather around mid-month. The post-Christmas stormy period is often associated with intense depression activity and if these depressions take an unusually southerly track they can bring heavy snowfalls to southern England, as in the closing days of 1961 and 1962. Lamb (1974) has suggested that the atmospheric circulation prevailing around the time of the New Year is a sensitive indicator of the weather types likely later in the winter. A significant association exists between cold weather at this time and cold winters and vice versa.

4. *Late winter* Long spells of widely differing types have affected about half the years. The coldest winters occur when long spells of easterly and anticyclonic types accompany persistent blocking in the Scandinavia–Iceland area, as in 1962–3 and in 1947. Cold, northerly spells are much more transitory, seldom lasting for more than 5–7 days. By contrast the mildest winters occur when the British Isles are repeatedly covered by maritime tropical and returning maritime polar air, the zonal index is high and the weather changeable. During most of January and February 1974 this type was in occupance with deep secondary depressions moving counter-clockwise around a large and deep primary Icelandic low. There is a strong correlation between wetness and mildness at this season, although occasional months dominated by high pressure over France give dry, mild conditions, e.g.,

March 1938, February 1949. The circulation types that dominate most years from about mid-February to mid-March onwards account for a greater frequency of droughts in spring than at other seasons and February has marked the beginning of long dry spells, e.g. in 1887 and 1911. Dry anticyclonic weather in mid-March very frequently extends over a fortnight or more particularly in the south, and in London the period 17–25 March is on average the driest part of the year.

5. Spring This is the time of year least given to long continued runs of similar weather and consequently changeability from day to day is very marked. Northerly and easterly types tend to be rather frequent especially during the second half of April, proverbially the month of showers, and late spring snowfalls often associated with small polar lows and troughs in northerly airstreams have occurred in some years. May is generally a rather quiet and often dry month with a good chance of spells of anticyclonic types particularly late in the month.

The seasonal calendar and the lateness or early arrival of particular seasons needs to be viewed in the context of the variation in pressure over the continental interior of Euro–Asia between summer and winter. During autumn, for example, pressure is rising over the Continent and falling over the ocean and the tendency for slow, eastward-moving, anticyclones to be situated over Europe with the British Isles lying within their circulation increases. In a year like 1969 warm, dry weather in October can give a prolongation of summer conditions, and in that year autumn can be regarded as having been very short since snow and frost were widespread by mid-November. The cold, wet weather of March and April 1970 was associated with a continuation of winter synoptic types and the suppression of the characteristic spring types, but 1970 had the warmest May since 1964.

An alternative method for the differentiation of seasons has been adopted by Davis (1972b) and Perry (1973), who have investigated the date on which certain temperature threshold values are crossed and then examined the synoptic conditions responsible for early or late triggering of these temperature values. Thus, if two or more consecutive days with a maximum temperature of 21 °C are taken as evidence of the onset of the first warm spell of early summer at Kew, this occurs on average (over the period 1881–1972) on 15 May. In years when the warm spell arrived early it occurred with highest pressure over Eastern Europe and low pressure near southern Greenland, while in years when the first warm spell was delayed it was initiated by a ridge from the Azores high across Western Europe and a light westerly anticyclonic type.

Indices of monthly and seasonal character

Considerable interest has been shown in recent years in developing indices to summarize the weather of a season over an area or at a location. Attention has particularly focused on devising a summer index which would reflect the 'goodness' or 'badness' of that season in particular years and enable comparisons between individual years or groups of years. An excellent review of this work by Cutler (1973) includes reference to Davis's (1968) optimum index which enables comparisons between different localities to be carried out accurately. When the index is mapped (Fig. 2.5*a*) it can be seen that values range from over 800 in south-eastern England to values below 500 in north and west Scotland. The index is derived from the formula:

$$I = 10Tm + 20Sm - Ri$$

where: Tm is the mean daily maximum temperature in the 3 months June, July and August (°F); Sm the daily mean sunshine during these months (hrs); and Ri the total rainfall (mm).

Fig. 2.5 *Indices of the characteristics of the summer season.* (a) *isopleths of an optimum summer weather index (after Davis, 1968);* (b) *the number of 'summer days' (maximum temperature ⩾25°C) and the associated basic weather types, 1947–66 (after Perry, 1968)*

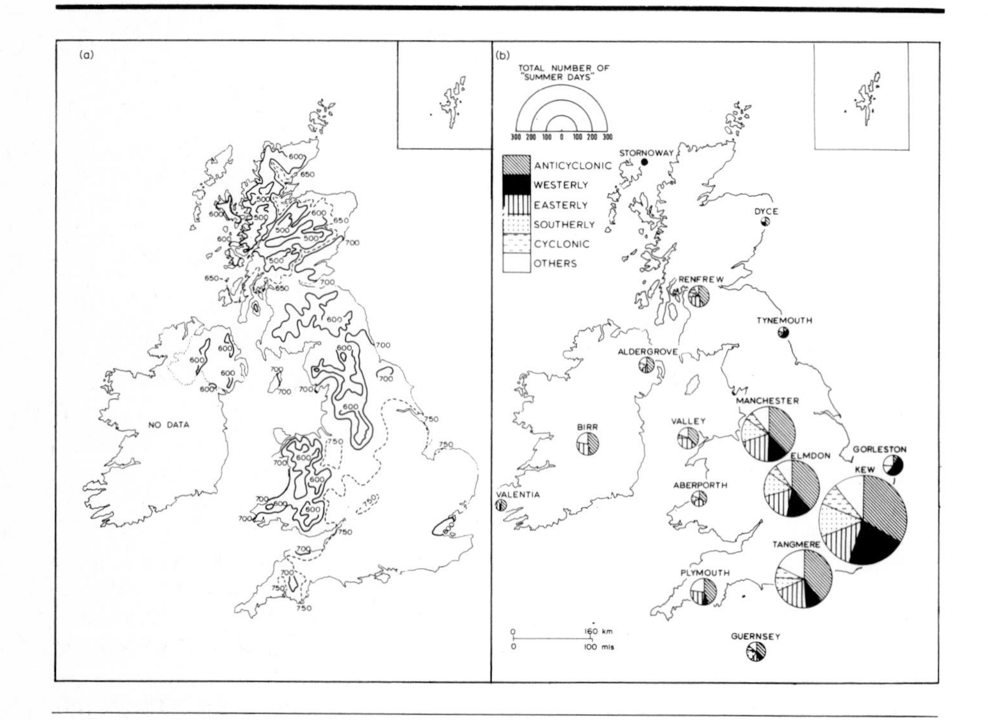

This same general regional trend in summer climate (Fig. 2.5*b*) is also in evidence in the number of 'summer days' (maximum temperature ⩾25 °C) in the British Isles (Perry, 1968).

The P.S.C.M. indices, as outlined under (a) to (d) below, were introduced by Murray and Lewis (1966) to measure in a succinct and meaningful way the main characteristics of the synoptic situation over periods of from about 10 days to a year. They are calculated objectively from the register of daily circulation types by giving each day a score in the range −2 to +2.

(*a*) **P**rogressive index–a measure of the frequency of days of westerly type, with negative scores implying a preponderance of easterly types;

(*b*) **S**outherly index – aims to measure the difference in frequency of southerly and of northerly days over or near the British Isles;

(*c*) **C**yclonicity index – a measure of the net cyclonic or anticyclonic character based on the algebraic sum of the daily scores;

(*d*) **M**eridional index – gives the frequency of all days with a meridional pressure gradient over the British Isles.

Murray and Benwell (1970) have indicated that P.S.C.M. indices are very useful as parameters for specifying large-scale anomalous circulation in long range forecasting work. The long period monthly means of the four indices (Table 2.4) draw attention to the anti-progressiveness of spring, the rapid increase in progressiveness between June and July, and the tendency for the southerly index to be negative in summer and positive in winter.

Table 2.4 *Monthly mean values of P.S.C.M. indices, 1873–1964 (after Murray and Lewis, 1966)*

	J	F	M	A	M	J	J	A	S	O	N	D
P	13	4	1	0	—9	7	16	14	10	5	8	15
S	6	4	3	—1	0	—2	—2	0	2	4	3	3
C	—3	—4	—4	—4	—6	—5	0	3	—6	—2	0	—1
M	14	14	14	13	13	12	10	10	11	15	13	13

For England and Wales the cyclonicity index has a highly significant correlation with precipitation in every month, but in Scotland rainfall is never closely related to the progressiveness of the synoptic types. There is general positive correlation between the southerly index and the temperature in all months. From November to April progressiveness and temperature are highly correlated, but between May and October this correlation is weakly negative in most places.

The P.S.C.M. indices are only one way of looking at the broad-scale circulation. Hay (1969) has employed monthly mean pressure maps for classifying synoptic types near the British Isles. A series of flow indices have been derived by him to yield a numerical measure of the cyclonicity or anticyclonicity of each month since 1874. The indices are dependent upon flow pattern, pressure and pressure gradient and a method of scoring was devised which allowed for the effect of these elements. The three indices are known as flow, progression and meridionality. The association between monthly mean flow in winter and five groups of temperature conditions (quintile 1 coldest group, quintile 5 mildest group) are shown in Table 2.5.

It is clear that a high proportion of mild winter months are associated with south-westerly and westerly pressure patterns, while flow patterns from north-west through north-east to south-east give predominantly cold winters (quintile 1 and 2). These results can be

Table 2.5 *Frequencies of direction of monthly mean flow near the British Isles in winter, subdivided according to temperature quintiles (after Hay, 1969)*

Winter temperature Quintile	Direction of flow							
	S.W.	W.	N.W.	N.	N.E.	E.	S.E.	S.
5—mildest	28	23	1	0	0	0	1	4
4	24	25	1	1	1	0	0	2
3	24	19	5	0	0	1	3	4
2	10	25	6	1	3	1	1	3
1—coldest	11	12	4	3	0	6	9	5

verified by examining the charts of mean pressure anomalies with different quintile values prepared by Murray and Lankester (1974).

Synoptic case studies

The second stage in synoptic climatological studies involves the assessment of weather elements in relation to the synoptic classification. Barry (1967) has pointed out that this work has been much neglected in Britain. Thus considerable interest surrounds his attempts at using the Lamb catalogue to study the climate of central south England (Barry, 1963, 1967). Averages of temperature, precipitation and sunshine were calculated for the major airflow types for 1921–50 and Fig. 2.6 illustrates two ways of displaying seasonal variability at a given locality. Maps of the spatial variation of precipitation for individual circulation types have also been prepared. In an extension of this approach to a large area in South Wales, Faulkner and Perry (1974) have calculated median precipitation amounts for each type and tested the statistical validity of differences between the types at each of the sixteen stations analysed. Lowndes (1968, 1969), studying heavy rainfalls in the Dee and Clwyd river basins of North Wales, has shown the importance of Lamb's westerly type in producing amounts exceeding 50 mm.

Table 2.6 *Percentage contribution of eight synoptic categories to total precipitation along a pluvial transect, 1956–60 (after Matthews, 1971)*

Station	Warm Front	Warm Sector	Cold Front	Occluded Front	Polar Low	mP air	cP air	Arctic air	Thunder-storms
Valley (Anglesey)	26	17	17	16	7	14	0·4	0·7	1·4
Cwm Dyli	18	30	13	10	5	22	0·1	0·8	0·8
Shawbury	25·5	13	12·5	14	10·5	19	0·4	1·1	4
Keele	22	13	14·5	14	10	19	0·5	1·6	5
Elmdon	26·5	12	12·5	17·5	11	16	0·8	0·7	2·4
Watnall	30	12·5	12	18	12	10·5	1·0	1·0	2·7
Cranwell (Lincs)	27	10·5	14	19	9	11	2·0	1·9	5

Using a rather elaborate synoptic classification consisting of nine categories, some of which are frontal and some air mass based, autographic rain records have been examined at individual stations (e.g. Beaver and Shaw, 1970), and along transect lines (Matthews, 1971) and average precipitation totals assessed. Orographic effects and especially the influence of shelter by large mountain masses can be seen in Table 2.6 to reduce considerably the importance particularly of warm sector precipitation at Shawbury in Shropshire, while continental polar air decreases in importance as a precipitation provider from east to west, along the transect line from Lincolnshire to Anglesey.

Fig. 2.6 *The contribution of Lamb's weather types to precipitation at Southampton, 1921–50 (after Barry and Perry, 1973.* (a) *percentage contribution to the monthly mean;* (b) *ratio of the mean daily rainfall for a type to the mean rainfall for all days*

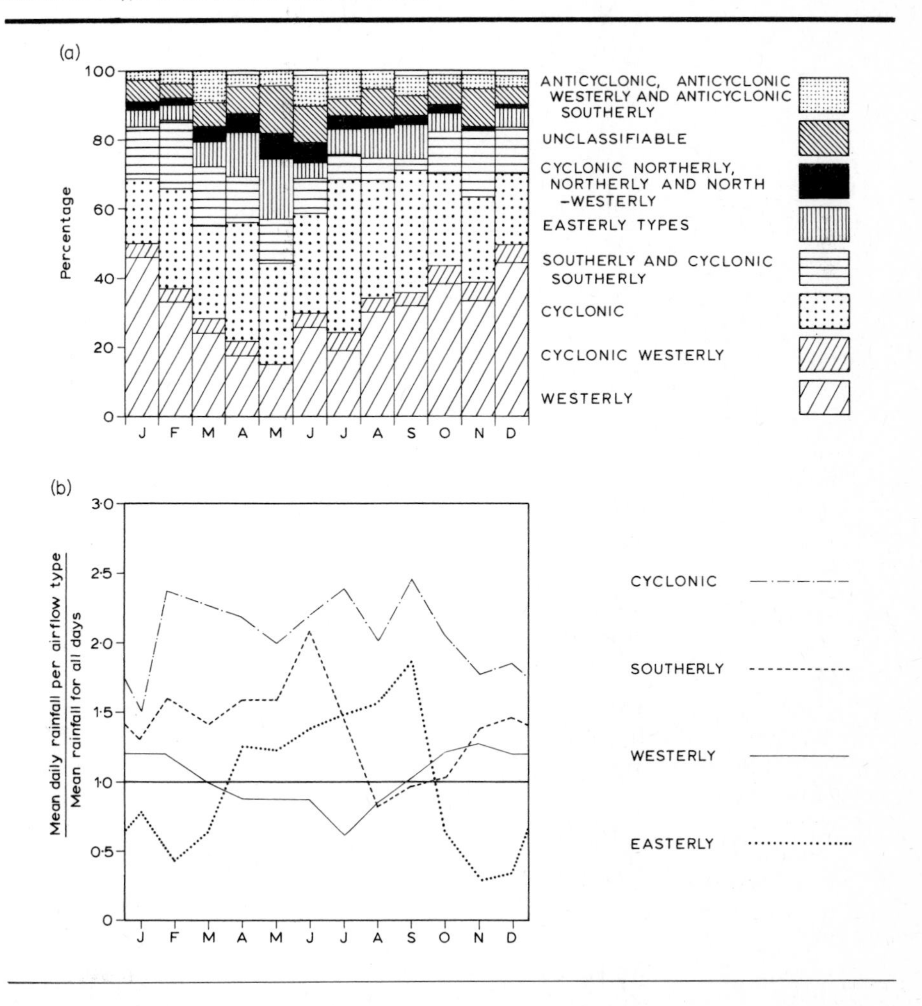

In a series of recent papers Lawrence (1971b, 1972, 1973a, b) has calculated rainfall averages over England and Wales for

Fig. 2.7 *Mean sea-level pressure, 1919–38, for 5-day periods (after Lamb, 1964).* (a) *21–25 May,* (b) *20–24 June;* (c) *13–17 October;* (d) *2–6 November*

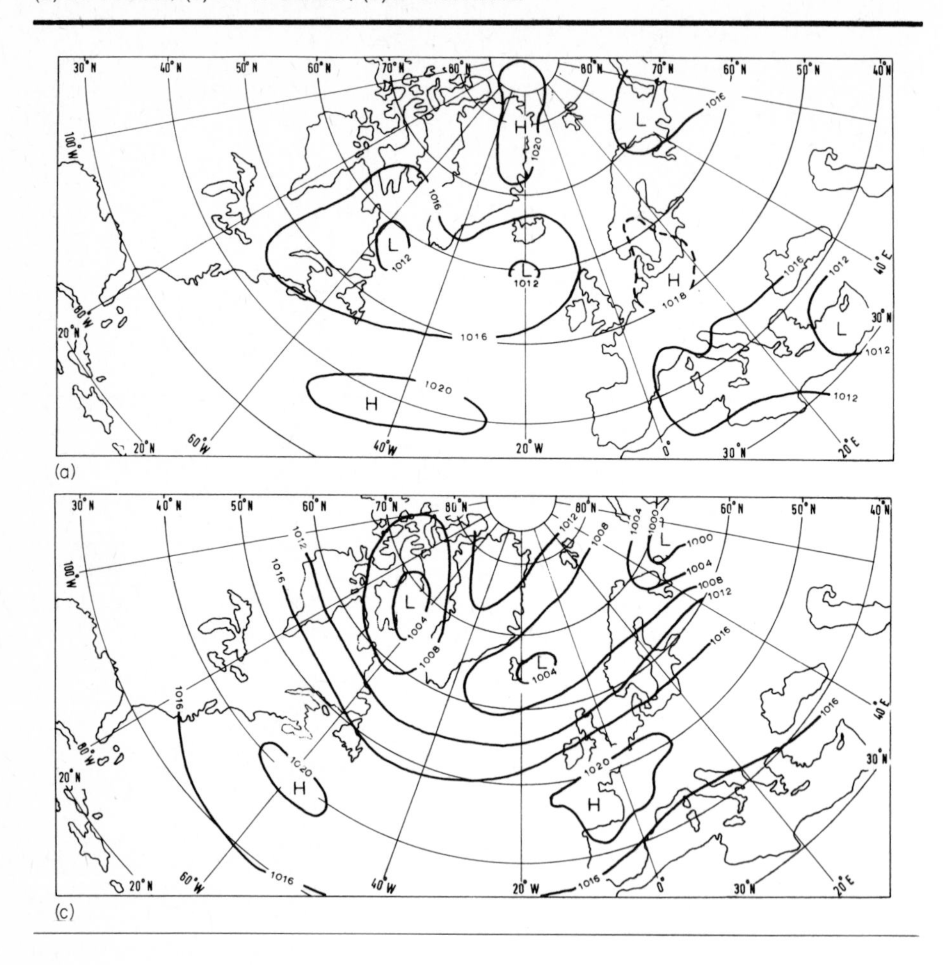

the main Lamb types. The work is based on over thirty stations and uses data for the period 1950–69. The results in Table 2.7 show that the cyclonic type exhibits an annual wave with the driest part on average in spring and the wettest in October and November, the westerly types are driest around June and wettest in winter, while the easterlies and south-easterlies tend to be generally wetter in summer. Lawrence has considered the cyclonic westerly types in detail and calculated long-term monthly and annual averages of the daily rainfall for 5 mb ranges of mean sea level pressure. Since 1950 there has been an increase in the frequency of cyclonic types (Lawrence, 1973a, b) and a corresponding increase in the number of days with high values of areal rainfall (> 20 mm) over England and Wales.

Singularities

A singularity may be defined as the recurrence tendency of some weather characteristic about a specified date in the year. Brooks

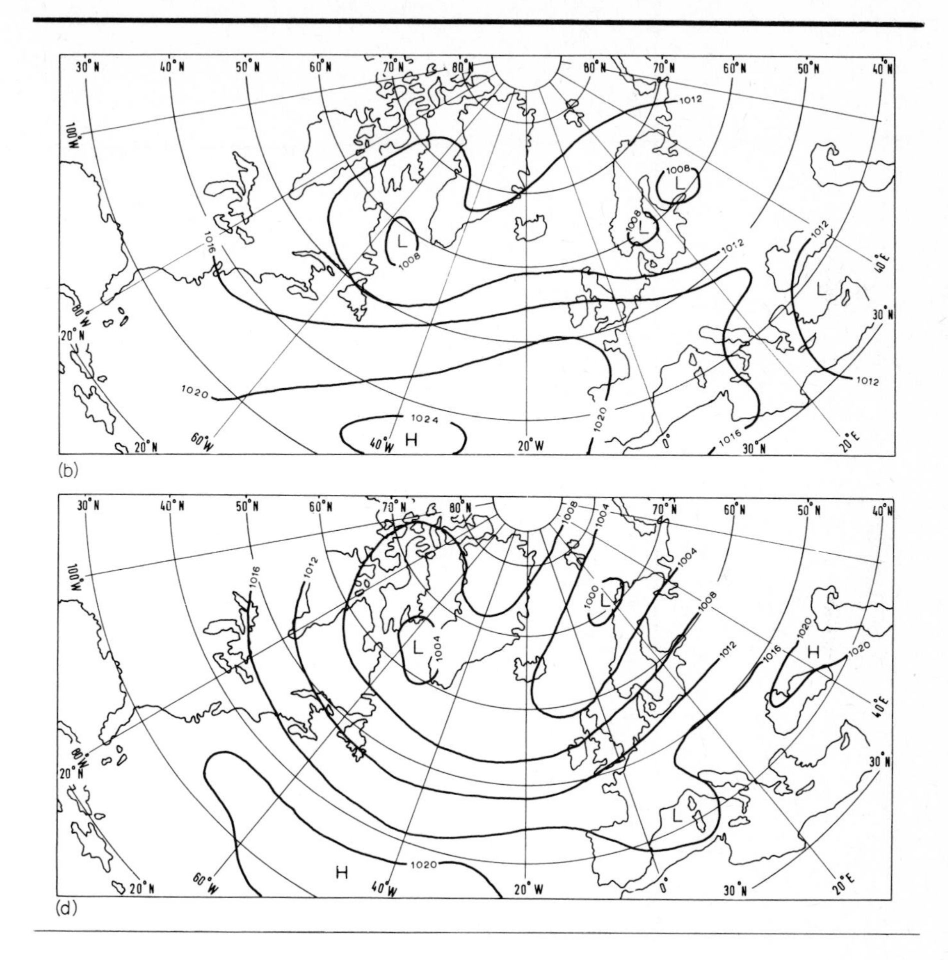

(b)
(d)

(1946) classed the pressure patterns over the British Isles for 1889–1940 into stormy and anticyclonic types and investigated the frequency of these types. The stormy spell from 24 November to 14 December was found to occur with a 98 per cent frequency in these years and the anticyclonic mid-January period with an 87 per cent frequency. A more complete singularity catalogue has been prepared by Lamb (1964) and the primary features are summarized in Table 2.8. Of course in any individual year certain singularities may simply not occur or be greatly accentuated. Lamb likens the singularity calendar to a steeplechase: if a particular long spell of set weather character succeeds in getting over a hurdle, i.e. in surviving dates which the calendar indicates as unfavourable for it, then the spell may be firmly enough established to last for some time longer. This undoubtedly happened during the exceptionally severe winter of 1962–3 when the stormy, westerly singularities just did not develop.

The significance of individual singularities seems to vary over

Table 2.7 *Synoptic type* averages over England and Wales, 1950–69, of (i) monthly and annual average daily rainfall (R mm/day); (ii) frequency (F days/month or year); (iii) product (RF mm) – (after Lawrence 1971b)*

		ANE	NE	CNE	AE	E	CE	ASE	SE	CSE	AS	S	CS	ASW	SW	CSW	AW	W	CW	ANW	NW	CNW	AN	N	CN	A	C	U
January	R	0·4	0·8	3·9	0·5	1·8	3·9	0·3	1·4	2·7	0·1	3·2	5·5	1·1	4·2	4·2	1·0	3·8	4·6	0·8	1·6	2·3	0·7	1·4	2·5	0·5	4·9	3·7
	F	0·36	0·33	0·13	0·73	1·52	0·29	0·26	0·54	0·14	0·34	1·62	0·31	0·26	0·82	0·20	1·09	5·92	1·39	0·37	1·51	0·28	0·75	1·69	0·54	4·89	3·30	1·42
	RF	0·14	0·26	0·51	0·37	2·74	1·13	0·08	0·76	0·38	0·03	5·18	1·71	0·29	3·44	0·84	1·09	22·50	6·39	0·30	2·42	0·64	0·53	2·37	1·35	2·45	16·17	5·25
February	R	0·4	1·2	1·8	0·4	1·6	3·2	0·5	1·5	3·5	0·2	3·7	6·1	0·7	3·8	4·3	0·7	3·0	4·1	0·9	1·3	2·3	0·7	1·2	2·2	0·4	4·7	4·3
	F	0·48	0·31	0·09	0·88	1·67	0·40	0·31	0·83	0·17	0·36	1·61	0·35	0·15	0·80	0·21	1·11	4·46	0·82	0·35	0·85	0·31	0·71	1·55	0·37	4·96	3·04	1·10
	RF	0·19	0·37	0·16	0·35	2·67	1·28	0·15	1·25	0·59	0·07	5·96	2·13	0·11	3·04	0·90	0·78	13·38	3·36	0·31	1·11	0·71	0·50	1·86	0·81	1·98	14·29	4·73
March	R	0·2	1·3	2·1	0·3	1·9	3·1	0·6	1·6	5·2	0·3	4·0	5·3	0.1	3·6	3·5	0·5	2·7	3·1	0·6	1·3	2·0	0·5	1·1	2·3	0·3	4·5	4·0
	F	0·55	0·36	0·09	1·11	1·84	0·52	0·41	1·17	0·21	0·52	1·82	0·43	0·14	0·71	0·20	1·34	4·59	0·55	0·45	0·81	0·39	0·84	1·70	0·39	5·64	3·07	1·15
	RF	0·11	0·47	0·19	0·33	3·50	1·61	0·25	1·87	1·09	0·16	7·28	2·28	0·01	2·56	0·70	0·67	12·39	1·71	0·27	1·05	0·78	0·42	1·87	0·90	1·69	13·81	4·60
April	R	0·2	1·4	2·4	0·4	2·8	3·7	1·0	2·3	6·7	0·5	3·9	4·1	0·3	3·4	3·0	0·4	2·5	3·4	0·5	1·2	1·8	0·3	1·2	2·3	0·3	4·5	3·1
	F	0·58	0·33	0·10	1·25	1·72	0·42	0·28	0·76	0·20	0·39	1·48	0·43	0·16	0·59	0·13	1·13	4·24	0·58	0·42	1·01	0·39	1·00	1·80	0·49	5·58	3·30	1·24
	RF	0·12	0·46	0·24	0·50	4·82	1·55	0·28	1·75	1·34	0·19	5·77	1·76	0·05	2·01	0·39	0·45	10·60	1·97	0·21	1·21	0·70	0·30	0·65	1·13	1·67	14·85	3·84
May	R	0·4	1·1	2·7	0·5	3·4	4·8	1·6	4·4	6·5	0·5	3·7	3·9	0·7	2·5	2·7	0·4	2·2	2·8	0·4	1·0	1·7	0·4	1·5	2·8	0·5	4·4	3·0
	F	0·58	0·34	0·08	1·37	1·61	0·37	0·23	0·55	0·15	0·24	1·18	0·37	0·24	0·51	0·06	1·11	4·38	0·87	0·46	1·07	0·34	1·04	1·88	0·48	5·59	4·46	1·44
	RF	0·23	0·37	0·22	0·69	5·47	1·78	0·37	2·42	0·97	0·12	4·37	1·44	0·17	1·27	0·16	0·44	9·64	2·44	0·18	1.07	0·58	0·42	2·82	1·34	2·79	19·62	4·32
June	R	0·6	1·5	3·3	0·5	3·4	7·6	1·1	5·4	5·5	0·5	3·5	5·3	1·3	2·6	3·8	0·6	2·0	2·6	0·3	0·9	2·0	0·4	1·7	4·4	0·7	4·6	3·6
	F	0·31	0·31	0·05	1·05	1·02	0·39	0·21	0·42	0·06	0·16	0·87	0·20	0·24	0·43	0·05	1·55	4·99	0·97	0·61	1·14	0·38	0·89	1·67	0·39	5·52	4·82	1·30
	RF	0·19	0·47	0·17	0·53	3·47	2·96	0·23	2·27	0·33	0·08	3·05	1·06	0·31	1·12	0·19	0·93	9·98	2·52	0·18	1·03	0·76	0·36	2·84	1·72	3·86	22·17	4·68

July	R	0·7	3·0	3·0	0·9	3·5	9·1	1·2	5·0	10·5	0·5	3·4	6·7	1·9	4·0	4·1	0·7	2·3	3·3	0·4	0·9	2·2	0·4	1·7	4·8	0·8	5·1	3·6
	F	0·18	0·38	0·04	0·69	0·52	0·43	0·19	0·18	0·06	0·20	0·69	0·14	0·18	0·45	0·11	1·98	5·57	1·05	0·67	1·46	0·53	0·87	1·65	0·52	5·40	5·58	1·28
	RF	0·13	1·14	0·12	0·62	1·82	3·91	0·23	0·90	0·63	0·10	2·35	0·94	0·34	1·80	0·45	1·39	12·81	3·47	0·27	1·31	1·17	0·35	2·81	2·50	4·32	28·46	4·61
August	R	1·1	3·3	4·6	0·9	4·3	7·5	1·5	2·0	10·7	1·1	3·6	4·8	1·0	4·7	4·7	0·8	2·9	3·8	0·5	1·1	2·0	0·6	1·9	4·3	0·7	5·3	3·0
	F	0·33	0·43	0·05	0·48	0·45	0·34	0·23	0·25	0·11	0·32	0·91	0·23	0·15	0·54	0·15	1·83	5·61	1·23	0·53	1·30	0·42	0·62	1·52	0·58	4·89	6·03	1·47
	RF	0·36	1·42	0·23	0·43	1·93	2·55	0·35	0·50	1·18	0·35	3·28	1·10	0·15	2·54	0·71	1·46	16·27	4·67	0·27	1·43	0·84	0·37	2·89	2·49	3·42	31·96	4·41
September	R	1·3	2·7	6·5	0·3	4·6	5·7	0·6	2·3	7·1	1·5	3·6	4·5	0·1	4·4	5·8	0·9	3·3	4·2	0·5	1·6	2·4	0·8	2·2	3·6	0·4	5·6	2·9
	F	0·34	0·28	0·09	0·39	0·57	0·30	0·31	0·58	0·12	0·43	1·24	0·30	0·17	0·82	0·17	1·62	5·98	1·32	0·41	1·00	0·22	0·34	1·25	0·49	5·22	4·59	1·45
	RF	0·44	0·76	0·59	0·12	2·62	1·71	0·19	1·33	0·85	0·65	4·46	1·35	0·02	3·61	0·99	1·46	19·73	5·54	0·21	1·60	0·53	0·27	2·75	1·76	2·09	25·70	4·21
October	R	1·6	1·9	6·9	0·4	3·5	5·8	0·1	2·5	4·8	1·5	3·4	5·1	0·1	4·1	5·8	1·0	3·5	4·5	0·5	1·8	3·7	0·6	1·9	3·0	0·4	6·6	3·3
	F	0·33	0·16	0·09	0·47	0·78	0·38	0·38	0·71	0·17	0·44	1·49	0·32	0·18	1·08	0·20	1·65	6·25	1·25	0·36	1·09	0·22	0·37	1·41	0·36	5·81	3·77	1·28
	RF	0·53	0·30	0·62	0·19	2·73	2·20	0·04	1·77	0·82	0·66	5·07	1·63	0·02	4·43	1·16	1·65	21·87	5·63	0·18	1·96	0·81	0·22	2·68	1·08	2·32	21·88	4·22
November	R	1·7	1·0	6·8	0·8	2·5	5·3	0·1	2·3	4·6	1·2	3·4	5·5	0·5	4·1	5·2	1·0	4·1	4·4	0·7	1·8	3·3	0·5	1·6	3·0	0·4	6·7	3·3
	F	0·38	0·27	0·09	0·49	0·98	0·42	0·32	0·48	0·20	0·41	1·31	0·36	0·20	0·86	0·17	1·29	6·14	1·20	0·34	1·47	0·28	0·41	1·63	0·43	4·88	3·75	1·24
	RF	0·65	0·27	0·61	0·39	2·45	2·23	0·03	1·10	0·92	0·49	4·45	1·98	0·10	3·53	0·88	1·29	25·17	5·28	0·24	2·65	0·92	0·21	2·61	1·29	1·95	25·13	4·09
December	R	1·1	0·7	6·5	0·7	2·1	4·6	0·1	1·4	3·6	0·5	3·2	5·2	1·0	4·4	4·0	1·0	4·3	4·5	0·8	1·9	2·6	0·6	1·6	3·1	0·5	5·7	3·3
	F	0·33	0·39	0·11	0·53	1·21	0·34	0·28	0·43	0·16	0·44	1·37	0·34	0·29	0·68	0·16	1·06	6·68	1·48	0·39	1·85	0·29	0·57	1·69	0·57	4·38	3·50	1·48
	RF	0·36	0·27	0·71	0·37	2·54	1·56	0·03	0·60	0·58	0·22	4·38	1·77	0·29	2·99	0·64	1·06	28·72	6·66	0·31	3·51	0·75	0·34	2·70	1·77	2·19	19·95	4·88
Year	R	0·73	1·69	4·33	0·52	2·65	5·32	0·65	2·39	5·53	0·73	3·57	5·13	0·79	3·90	4·43	0·76	3·13	3·91	0·55	1·40	2·27	0·51	1·48	3·23	0·49	5·16	3·40
	F	4·75	3·89	1·01	9·44	13·89	4·60	3·41	6·90	1·75	4·25	15·59	3·73	2·36	8·29	1·81	16·76	64·81	12·71	5·36	14·56	4·05	8·41	19·44	5·61	62·81	49·21	15·85
	RF	3·5	6·6	4·4	4·9	36·8	24·5	2·2	16·5	9·7	3·1	55·7	19·1	1·9	32·3	8·0	12·7	202·9	49·7	2·9	20·4	9·2	4·3	28·8	18·1	30·8	253·9	53.9

*A = anticyclonic, C = cyclonic, U = unclassifiable

time and this theme will be returned to in Chapter 10. To interpret and investigate the development of singularities Lamb (1973a) has prepared mean pressure maps for 5-day periods (pentads) throughout the year. Four of these maps are reproduced in Fig. 2.7. The first two maps show the 20-year average (1919–38) pressure fields for 21–25 May and 20–24 June. The end of May is typically a period of early summer anticyclonic weather, often with northerly or easterly winds flowing around a high centre over Scandinavia, and this can be contrasted with the situation near the end of June when the frequency of westerlies becomes much greater again. The 'return of the westerlies', after the period of blocking during spring, is sometimes known as the onset of the European monsoon and often sets in with the abrupt incursion of a cool north-westerly maritime polar airstream which can bring to an end fine, warm weather in mid-June. The chart for 13–17 October suggests that at this time anticyclones are common to the south of the British Isles. Southern England especially often lies under their influence and a period of quiet weather known as 'Indian Summer' or 'Old Wives Summer' is common with mist and fog at night. By the beginning of November, cyclonic, stormy weather is common with depressions passing close to the British Isles as they pass from south-west to north-east.

Persistence

Day to day

Some large-scale circulation patterns have a built-in tendency to persist or to evolve in one way in preference to another at certain times of the year. Lawrence (1957) has shown that at least in south-eastern England in summer the persistence factor varies with time in such a way that runs of dry days show positive persistence up to 8–10 days and anti-persistence over 30 days. Lowndes (1963) showed that in winter the probability of a cold spell continuing for a further day in London increased from 0·77 after 4 days to 0·83 after 6 days. The wider question of the difference in persistence between different areas or for weather sequences has received little attention. We do know that the general circulation appears to have moods in which persistence or repeated reappearance of similar broad-scale types characterizes a month or season and probably this is caused by anomalous development of the major heat sources and sinks – persistent sea temperature anomalies, or unusual excesses or deficits of ice and snow cover.

Month to Month

Probably because of the many different synoptic situations which contribute to the total precipitation of a month, there is little tendency for persistence of sequences of months with above or below normal precipitation, as has been demonstrated by Bilham (1934) and recently by Murray (1967b) and Stephenson (1967). Temperature persistence is much more in evidence and Craddock and Ward (1962) show that on several occasions during the year the correlation between

Table 2.8 *Singularities in the British Isles (Barry and Perry, 1973, after Lamb, 1964)*

Period	Circulation type	Characteristics	Type frequency % (and significance level)	Period
20–23 January	AC,S, and E together	Generally dry and sunny in central and southern England.	50	D
		Year's lowest frequency of C type (10–12%) 24–26 January.	(5% level)	D
12–23 March	AC,N and E together	Notable rainfall minimum in central and southern England.	70	D
		12–14 March peak of AC.	35 (1% level)	D
12–18 May	N type	Annual maximum about these dates; 14–20 May is sunniest week of the year in Ireland.	30	A,B,C
21 May–10 June	AC type	Annual maximum frequency, 40% or more on some days during most of this period; driest weeks of year in Scotland, Ireland; more year-to-year variations in southern half of England.	(5% level)	A,B,C
18–22 June	W,NW and AC together	Generally dry and sunny in southern England; cloudy and wet in Scotland and Ireland.	70	D
		W type frequency 52% on 20 June	(1% level)	D
31 July–4 August	C type	Sharp peak (replaced by twin maxima around 20 July and mid-August).	35% + (5% level)	B, A,C
17 August–2 September	W and NW together	Wet in most areas.	70	D
	C type	Peaks 19 and 28 August.	30 (5% level)	D
6–19 September	AC, N and NW together	Dry, especially east and central England.	80	A,B,C
		C type frequency, >20% between 6–12 September.	(5% level)	A,B,C
5–7 October	AC type	Slight check to seasonal cooling.	40 (5% level)	D
24–31 October	C, E and N types	Great decline to year's minimum frequency of AC type (<10%) about 28–31 October.	(1% level) (5% level)	A,B,C
		Stormy, wet weather.		A,B,C
17–20 November	AC type	Dry, foggy period in central and southern England.	30 (1% level)	A,B,C
3–11 December	W and NW together	Wet and stormy in most areas with 3–9 December generally wettest week of year on average.	70	A,B,C
17–21 December	AC type	Generally dry, foggy weather.	25	A,B,C

Period A = 1873–97; B = 1898–1937; C = 1938–61; D = 1890–1950 about 10 years.

successive months is significant with Student's 't' test at the 0·1 per cent level. In particular, persistence is notably strong in winter (December–March) and summer (June–September). Murray (1967a) was able to confirm that the non-persistent periods tended to be in late spring and late autumn and pointed out that the maxima and minima of temperature persistence are broadly in line with the maxima and minima of long synoptic spells found by Lamb (1950).

Sequential tendencies

The tendency for lag relationships between the weather types in one season and those in a following season have been analysed in the search for useful predictors on long-range forecasting work. Only occasionally are the physical explanations for such sequential tendencies as yet fully understood. The principal results of this work can now be summarized:

(*a*) Progressive, cyclonic Junes tend to be followed by Septembers which are more cyclonic than usual, whereas non-progressive cyclonic Junes tend to be associated with anticyclonic Septembers (Murray, 1967a).

(*b*) Wet autumns in Britain show a tendency to be followed by cold winters and dry autumns by mild winters (Hay, 1967). Anticyclonic weather over Iceland in mid-October can be a precursor of cold winters in the United Kingdom (Hay, 1970a).

(*c*) Warm summers appear to be followed by warm Septembers whenever a ridge from the Azores anticyclone extends across the British Isles to Scandinavia during the period June–August. Cold Septembers follow summers when the monthly mean low centre is found near to the west or north-west of the British Isles or over or near Scandinavia (Hay, 1968).

(*d*) Ratcliffe and Collison (1969) have demonstrated a correlation between the position of the European 500 mb trough in April and the following summer's rainfall in England and Wales. The wettest summers follow when troughs lie between 10° and 30°E, the driest when troughs are east of 30°E or west of 10°E.

Some success has been achieved by Murray (1972) in identifying, in the three months preceding the season in question, key areas in the northern hemisphere where anomalous circulation shows some association with the subsequent seasonal rainfall and temperature. Thus cold springs in England have been preceded by Februaries with pressure above normal over Iceland and Greenland and below normal from the Azores to the central Mediterranean. These pressure anomalies over N. Greenland are statistically different from zero at the 5 per cent level of significance with Student's 't' test (Murray, 1972). Warm springs were preceded by Februaries with pressure above average and statistically significant anomalies over the Mediterranean. Ratcliffe (1974), continuing this work with monthly mean 500 mb charts, has showed that prediction can be made of the category of 15-day and monthly rainfall and mean temperatures in terms of terciles and quintiles over England and Wales. The method selects those areas in the northern hemisphere where anomalies on the 500 mb monthly mean charts preceding warm dry months have been significantly different from the anomalies in the same area preceding cold wet months at the same time of the year.

Climatic cycles

Quasi-biennial

An atmospheric oscillation with a period of just over two years has been recognized as affecting the weather of particular seasons (e.g. Davis, 1967), and work is well advanced in analysing the synoptic effects of the cycle. Murray and Moffitt (1969) have scrutinized pressure in odd and even years and found that it has been consistently higher in odd-numbered years than in even years from June to October, although the pattern in the winter months has been more variable. Hay (1970b) showed, based on 100 years of records, that blocked pressure patterns near the British Isles are significantly more prevalent in odd year summers, and suggested that the good summers of odd years were associated with this excess of blocked anticyclonic types. In winter the tendency to blocking over and north of the British Isles is greater in odd years than even years especially in January and pressure averages are about 3 mbs higher over the United Kingdom in odd years.

Climate and the solar cycle

Considerable attention has focused on the effects of the eleven-year solar cycle on atmospheric circulation. Lamb (1972) quotes recent Meteorological Office work on monthly mean pressure anomalies associated with different phases of the cycle and which shows a tendency for a more zonal high index flow over the British Isles in the descending phase of the cycle in winter and a tendency for more blocking to affect Europe about the time of sunspot minima. Considering the cycles within the period 1874–1971 the mean surface pressure anomaly over the British Isles in winter ranges from +1·43 mb in sunspot minima years to −1·07 mbs in maxima years.

The literature includes a large number of papers which have claimed to have discerned periodicity based on a study of the cycles. In terms of summers, Poulter (1962) detected three 11- to 12-year cycles of good summers:

(1) 1887, 1899, 1911, 1921, 1933, 1943, 1955, 1967.

(2) 1893, 1904, 1914, 1925, 1937, 1949, 1959, 1970.

(3) 1897, 1908, 1919, 1929, 1940, 1952, 1964.

At least one well-marked 11- to 12-year cycle of poor summers has been noted: 1912, 1924, 1936, 1948, 1960, 1972. Even years have showed an oscillation of about 34 years (Hughes, 1972). Lyall (1971c) has claimed to have found a cycle of more severe winters at 10 or 12 years in odd numbered years and with an average 11-year separation, i.e. 1917, 1929, 1940, 1951, 1963.

Interaction between the effects of the quasi-biennial and solar cycles has been noted by Schove (1971), who suggests that the biennial oscillation is stronger at times of strong sunspot maxima, and when the solar cycles are weaker and longer-spaced a triennial oscillation is more in evidence.

Sea temperature and its anomalies

Sea temperature anomalies

Since so much of our weather arrives from the west, perhaps it is not surprising that changes in the sea temperature, and particularly the surface temperature, of the North Atlantic are likely to have repercussions on our climate. Anomalies of sea surface temperature can be expected to result in anomalies in the exchange of sensible and latent heat between ocean and atmosphere, and to shifts in the preferred depression tracks, to anomalous circulation patterns and hence to anomalous weather. The sea surface temperature in the area immediately south of Newfoundland has been found to be particularly important in affecting the atmospheric circulation a month later over the area near the British Isles (Ratcliffe and Murray, 1970). Warmer than normal water in this area seems to favour low pressure and progressive synoptic types over the British Isles, while colder than normal ocean temperatures are associated with blocked, anticyclonic atmospheric patterns the following month (see Fig. 2.8). Because, once established, sea temperature anomaly patterns tend to persist for several months, the weather of whole seasons can sometimes be affected. Namias (1964) pointed out that a colder than normal ocean in the vicinity of Ocean Weather Ship C in the central North Atlantic during the period from mid-1958 to early-1960 was probably the cause of persistent blocking in the north-east Atlantic over that period.

A good example of the effects of distant sea temperature anomalies on our climate occurred in the summer of 1968. Negative anomalies of sea surface temperature of more than 4 °C developed over a huge area covering thousands of square miles of ocean and centred to the south-east of Newfoundland near Ocean Weather Ship D in mid-late May and persisted throughout the summer (Murray and Ratcliffe, 1969). The degree of blocking near the British Isles and the unusually frequent occurrence of easterly and northerly winds was the main feature of the summer, and this led to long fine spells of weather in north-western districts, but to persistently dull, cool and wet weather in the south-east.

Sometimes the origins of sea temperature anomalies can be speculated upon, as in the spring of 1972, when abnormal north-westerly flow in the Davis Strait–western Atlantic area caused large numbers of icebergs to travel out into the ocean. As this ice melted a cold water area was formed which extended eastward. It was helped by above normal depression activity in May which would have caused up welling of more cold water and may have deflected the warm North Atlantic Drift to the south-east of its normal position. By June 1972 a belt of water 1 ° to 2 °C colder than normal extended across the Atlantic to the British Isles and this had the effect of pro-

Fig. 2.8 *Pressure, temperature and rainfall anomalies for nine Februaries following Januaries with colder than normal sea surface temperature south of Newfoundland (after Ratcliffe, 1973).* (a) *mean pressure anomalies in mbs;* (b) *mean monthly temperature in °C;* (c) *mean monthly rainfall percentages*

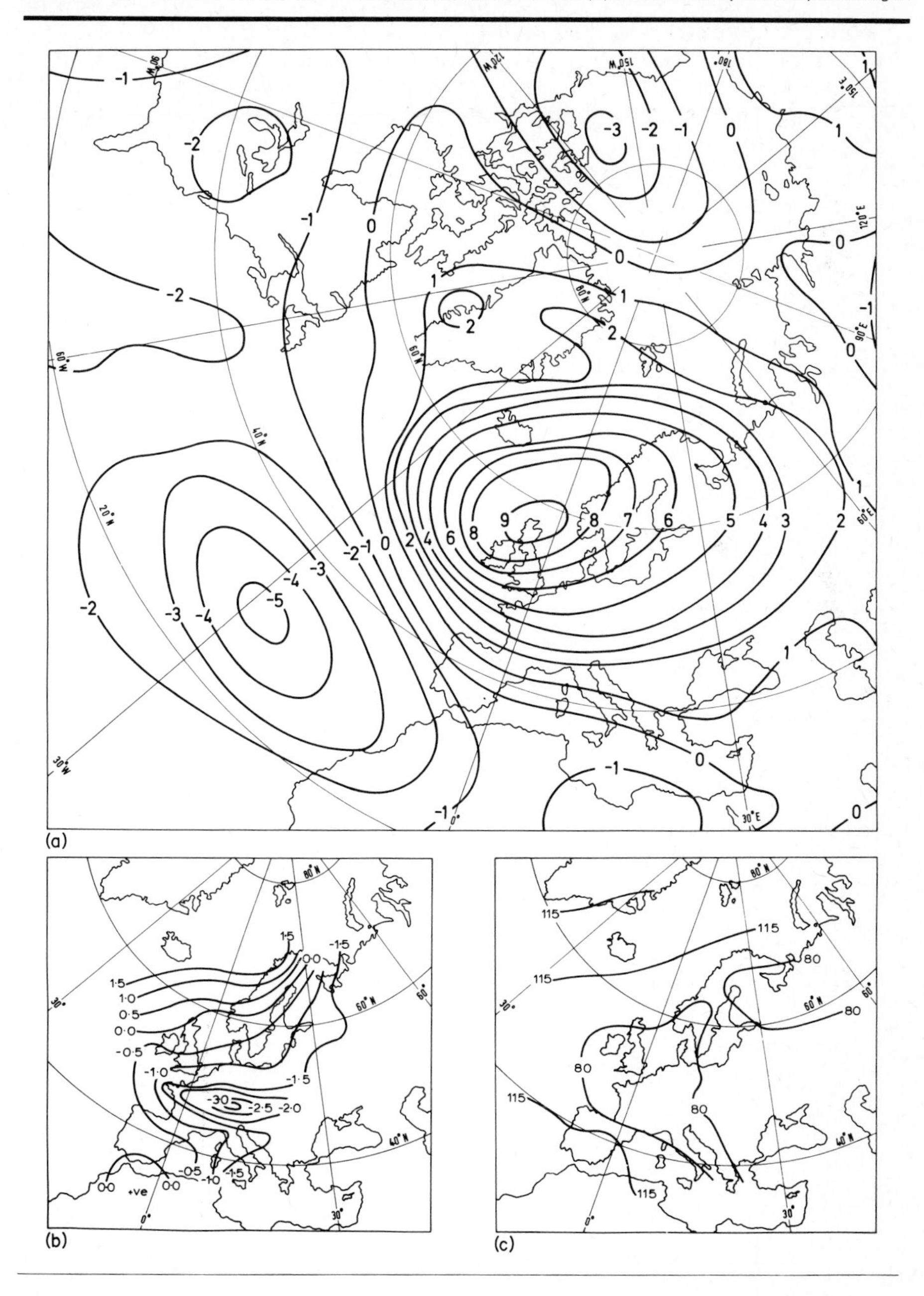

ducing a zone of strong thermal gradient in the atmosphere much further south than normal along which depressions formed and moved eastward over the British Isles (Perry, 1972a). The result was the coldest June this century over England and Wales with temperatures in most places failing to exceed 21 °C on any day, and mid-summer day that year was, in many places, colder than the previous Christmas Day.

Ratcliffe (1973) has suggested that sea temperature anomalies in other areas may offer some guidance as to the development of future weather patterns near the United Kingdom. A warm sea in the Bay of Biscay in July favours warmth in Britain the following month while a cold sea in this area during the autumn favours cool, dry months in England and Wales. Farmer (1973) showed that anomalies as far away as the North Pacific could also affect subsequently the European atmospheric circulation.

Anomalies of ice cover

Unusual excesses or deficits of snow and ice in the polar seas and the Baltic can also have an effect on British weather. Northerly winds in a year in which ice in the East Greenland Sea spreads south as far as the north coast of Iceland, as happened during some years in the 1960s (Marshall, 1968), reach the United Kingdom with a lower temperature than in those years in which the ice edge is further north. The frequency of late springs in the United Kingdom is closely linked with ice amounts (Davis, 1972b). There is a tendency in heavy ice years for there to be a positive monthly surface pressure anomaly to the north of Britain and a gradient for cold north or north-east winds over the British Isles which keep temperatures low and inhibit the commencement of spring growth. Excesses or deficits of ice are themselves caused by anomalous circulation patterns and wind components in previous months, so that complex feedback processes between the atmosphere, the ocean and the polar ice can occur.

Conclusion

In this chapter the commonest weather types affecting the British Isles have been recognized and their contribution to the changing kaleidoscope of seasonal weather assessed. Since there is a continuum of synoptic types and no two sequences of weather are entirely similar, any classificatory system which attempts to group together weather types on the basis of their similarity must generalize and be open to disagreement. Nevertheless climatological summaries made with reference to the circulation pattern are more useful and meaningful than those made with reference to calendar time.

Wind as a vector

Wind velocity is a vector quantity involving both direction and magnitude. The direction is that from which the wind blows relative to true north, nowadays usually expressed in degrees. The magnitude may be expressed either as wind speed, usually in knots or metres per second but sometimes in miles per hour, or as a force on a numerical scale ranging from 0 (calm) to 12 (hurricane) known as the Beaufort scale. The wind-speed equivalents of the Beaufort scale were originally based on an empirical relationship between estimated Beaufort force and measured speed $V = 1{\cdot}625\sqrt{B^3}$, where V is wind speed in knots and B is the corresponding Beaufort number. Because wind is a vector quantity it is desirable, whenever possible, to deal simultaneously with its two components, direction and speed. In summaries this can be done in two ways, either by means of a two-way frequency table of direction against speed or, diagrammatically, by constructing a wind rose. The tabular statistics can be made more detailed and are much more easily prepared nowadays with the aid of a computer, than was the case when they had to be prepared laboriously by hand. Unfortunately, however, such tables take up a good deal of space and for this reason many published wind statistics, e.g. those given in the *Monthly Weather Report*, still refer separately to speed and direction.

As examples of two-way or combined frequency tables of wind speed and direction, Tables 3.1 and 3.2 present average annual percentage frequencies of winds within stated speed ranges which came from each of twelve 30-degree direction ranges at Lerwick (Shetland Islands), 1957–70, and at Scilly (Cornwall), 1958–71. They represent conditions at well-exposed places in the extreme north and south of the British Isles respectively. The tables are based on observations made eight times a day at three-hourly intervals (at 00.00, 03.00, 06.00, 09.00, 12.00, 15.00, 18.00 and 21.00 hrs) the observations being made over ten-minute periods reduced to the standard height above the ground of 10 metres (33 ft). The speed ranges chosen, expressed in knots, correspond to the numbers 0 to 12 respectively of the Beaufort scale of wind force. A frequency of less than 0·05 per cent is shown as 0+. The highest frequency in each column appears in italic to indicate the speed range occurring most often from that direction. Taking all directions together it can be seen that force 4 (11–16 kt) is the most frequently observed wind force at both stations. This is the case, too, for all the individual direction ranges, except that force 3 is most frequent for winds from the east (080°–100°) at Lerwick. As the force increases further the frequencies tend to fall off rather rapidly but fairly regularly. The overall frequency of gale force winds (force 8 and above) is about 2 per cent at Lerwick and a little under 1 per cent at Scilly. At Lerwick force 9 winds occur from every direction but

Table 3.1 *Annual percentage frequency of winds with stated direction and force at Lerwick, 1957–70*

Beaufort force	Mean speed knots	Percentage number of hours with wind from												All directions Total
		350°–010°	**020°–040°**	**050°–070°**	**080°–100°**	**110°–130°**	**140°–160°**	**170°–190°**	**200°–220°**	**230°–250°**	**260°–280°**	**290°–310°**	**320°–340°**	
0	Calm													4·5
1	1–3	0·2	0·3	0·2	0·2	0·2	0·3	0·3	0·4	0·2	0·2	0·3	0·4	3·2
2	4–6	0·9	0·7	0·5	0·5	0·5	0·9	1·2	0·9	0·7	0·9	1·0	1·2	9·9
3	7–10	2·2	1·6	1·1	*1·1*	1·3	2·3	2·7	1·6	1·7	1·8	1·9	2·2	21·5
4	11–16	*2·6*	*2·2*	*1·2*	1·0	*1·7*	*2·8*	*3·8*	*2·4*	*3·3*	*2·5*	*2·1*	*2·5*	*28·0*
5	17–21	1·3	1·2	0·4	0·4	1·0	1·7	1·9	1·5	2·5	1·5	0·9	1·3	15·7
6	22–27	0·7	0·6	0·3	0·3	0·6	1·1	1·3	1·3	2·0	1·2	0·5	0·7	10·6
7	28–33	0·3	0·2	0·1	0·1	0·4	0·6	0·6	0·6	0·8	0·6	0·2	0·2	4·6
8	34–40	0·1	0·1	0+	0+	0+	0·2	0·2	0·3	0·3	0·2	0·1	0·1	1·7
9	41–47	0+	0+	0+	0+	0+	0+	0+	0+	0·1	0·1	0+	0+	0·2
10	48–55	0+	0+				0+	0+	0+	0+	0+	0+		0·1
11	56–63								0+		0+			0+
12	>63								0+					0+
Total		8·3	6·9	3·8	3·6	5·7	9·9	12·0	9·1	11·6	9·0	7·0	8·6	100·0

Table 3.2 *Annual percentage frequency of winds with stated direction and force at Scilly, 1958–71*

Beaufort force	Mean speed knots	Percentage number of hours with wind from												
		350°–010°	**020°–040°**	**050°–070°**	**080°–100°**	**110°–130°**	**140°–160°**	**170°–190°**	**200°–220°**	**230°–250°**	**260°–280°**	**290°–310°**	**320°–340°**	**All directions Total**
0	Calm													5·9
1	1–3	0·7	0·6	0·5	0·4	0·3	0·2	0·3	0·5	0·5	0·4	0·5	0·4	5·4
2	4–6	1·1	1·2	0·9	0·9	0·7	0·6	0·9	1·0	1·1	1·0	1·0	0·9	11·3
3	7–10	1·5	1·6	1·5	1·5	1·4	1·2	1·6	2·1	2·0	2·1	2·2	1·5	20·1
4	11–16	*1·9*	*1·8*	*2·0*	*2·5*	*2·5*	*1·7*	*2·6*	*3·7*	*3·5*	*4·1*	*3·9*	*2·4*	*32·5*
5	17–21	0·6	0·6	0·7	0·8	0·8	0·4	0·8	1·3	1·4	2·0	1·6	0·8	11·9
6	22–27	0·3	0·3	0·5	0·7	0·4	0·3	0·7	1·0	1·4	1·8	1·4	0·7	9·6
7	28–33	0·1	0·1	0·1	0·1	0·2	0+	0·1	0·2	0·4	0·6	0·4	0·2	2·4
8	34–40	0+		0+	0·1	0+	0+	0+	0·1	0·1	0·3	0·2	0·1	0·8
9	41–47	0+			0+	0+	0+	0+	0+	0+	0+	0+	0+	0·1
10	48–55							0+		0+	0+	0+		0+
11	56–63													
12	>63													
Total		6·2	6·2	6·2	7·0	6·3	4·4	7·0	9·9	10·4	12·3	11·2	7·0	100·0

Fig. 3.1 *Annual percentage frequency of the force (Beaufort scale) and direction of the wind at selected stations (for periods, see Table 3.3)*

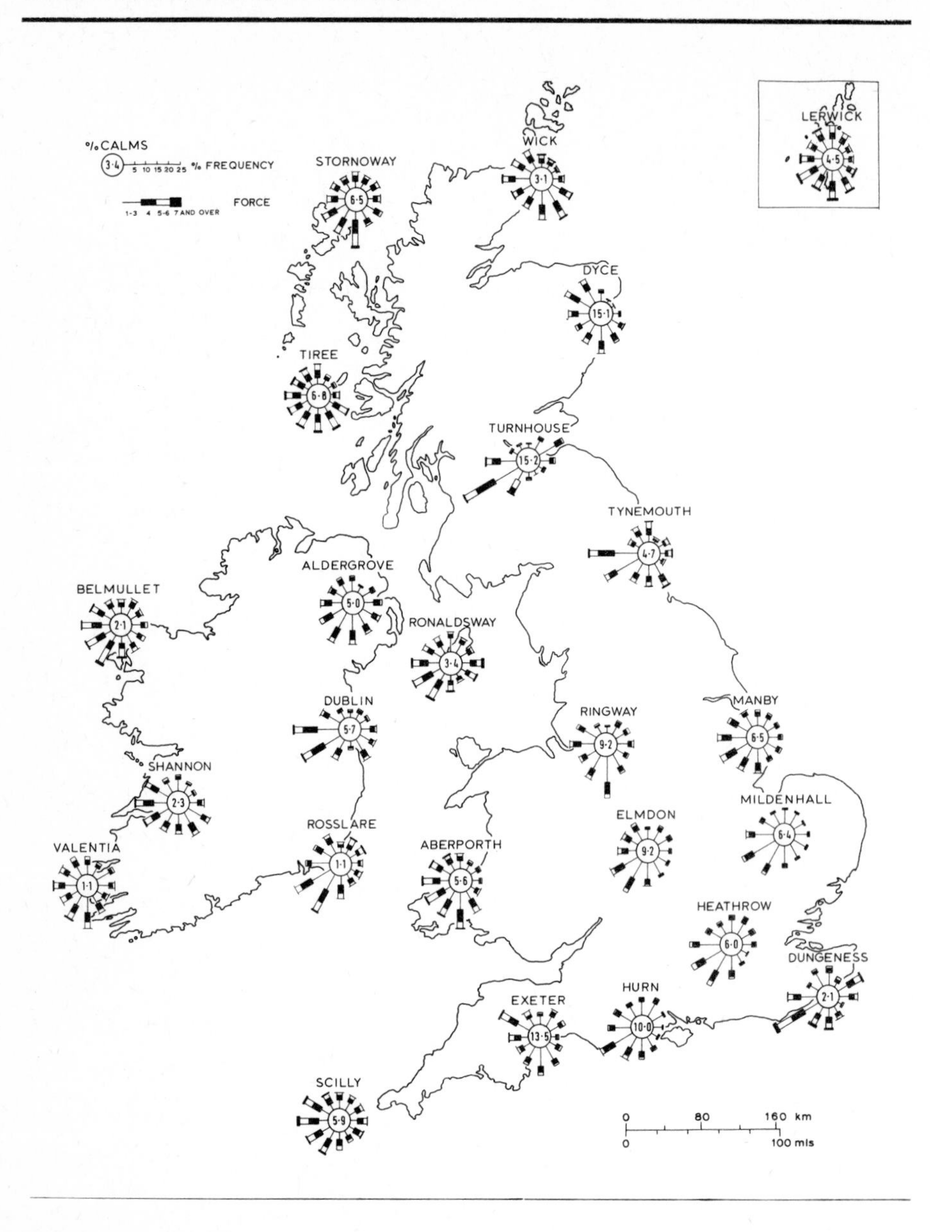

those of force 10 and above are confined to the ranges between 350° and 040° and between 140° and 310°. In fact there were twenty-nine observations of force 10 and above during the 14-year period, of which seven each came from 230°–250° and 260°–280°, five from 200°–220°, four from 170°–190°, three from 350°–010° and one each from 020°–040°, 140°–160° and 290°–310°. Of the six observations of force 11 and over, four came from 260°–280° and two from 200°–220°, one of the latter being force 12. At Scilly, force 8 winds occur from every direction except 020°–040°, but well over half of them are from between 260° and 300°. There were fifty-five observations of force 9 and above during the 14-year period, of which twenty were from 290°–310°, eighteen from 260°–280°, eight from 230°–250°, two from 170°–190° and one each from the other directions for which the entry in Table 3.2 is 0+. Of the seven observations of force 10 there were three from 260°–280°, two from 230°–250° and one each from 170°–190° and 290°–310°. At both Lerwick and Scilly the three forces 3 to 5 (7 to 22 kt) account for about 65 per cent of all the winds but whereas at Lerwick more than 17 per cent of winds are of force 6 and above, the corresponding figure for Scilly is a little under 13 per cent.

The data of Tables 3.1 and 3.2 are also presented diagrammatically in Fig. 3.1, in the form of 12-point wind roses, together with similar diagrams for a further twenty-two stations in the British Isles. For this purpose the wind speed frequencies for each station, originally in the same form as in Table 3.1 were simplified by grouping together forces 1 to 3 (light winds), 4 (moderate winds), 5 and 6 (fresh to strong winds), 7 and above (near gales and gales). The figure within the centre circle of a rose indicates the percentage frequency of calms, and the length of each part of an arrow, which flies with the wind, is proportional to the frequency of speeds in the range which it represents. For the nineteen stations in the United Kingdom the data are mostly for the 15-year period 1957–71 and are based on 10-minute means measured every 3 hours. For the five stations in Eire, they relate to the 10-year period 1961–70, are based on 10-minute means measured every hour and were taken from Butler and Farley (1973). In all cases speeds were reduced, where necessary, to the standard height of 10 metres above the ground. The stations are listed in Table 3.3 which gives their positions and altitudes and the exact period of the data used for each.

Figure 3.1 provides in compact form a broad picture of the wind distribution over our islands in an average year. The most common, or prevailing, wind directions are seen to be generally from between south and west, while the least frequent winds are those having easterly components. There are some stations, however, where directions other than south to west are predominant and these exceptions to the general rule can often be attributed to the effects of topography, as will be discussed further below. The frequencies of winds of force 5 and over are much greater in the north and west than they are in the south and east, while winds of force 7 and over are rather infrequent in England. Calms and light winds are much more common in the east than they are in the west. Space does not permit a detailed description of the annual wind regime at each station as was done for Lerwick and Scilly but the reader will be able to draw his own conclusions from a study of the wind roses of Fig. 3.1.

Maps similar to Fig. 3.1 could have been prepared for each of the twelve months of the year but lack of space prevents this. To indicate the main differences between winter and summer conditions, however, it is sufficient to confine our attention to the wind roses for

January and July which are shown in Fig. 3.2 and 3.3 respectively. Roses for the stations in Eire have had to be omitted because the necessary data were not readily available.

The January wind roses (Fig. 3.2) suggest that the prevailing winds in mid-winter, like those for the year as a whole, are generally from directions between south and west but here again with certain exceptions, which are usually explicable in terms of the regional and local topography. Winds of force 5 and above are considerably more frequent than they are over the year as a whole but the general patterns at individual stations are quite markedly similar. This suggests that

Table 3.3 *Particulars of stations for which data are presented in Figs 3.1, 3.2, 3.3 and 3.5*

Station	Lat. N. °	′	Long. °	′	Height above msl (m)	Period of data
Scotland						
Lerwick (Shetlands)	60	08	1	11W.	82	1957–70
Wick (Caithness)	58	27	3	05W.	36	1957–71
Stornoway Apt (Ross & Cromarty)	58	13	6	20W.	3	1957–71
Dyce (Aberdeen)	57	12	2	12W.	58	1957–70
Turnhouse (Midlothian)	55	57	3	21W.	35	1957–71
Tiree (Argyll)	56	30	6	53W.	9	1957–71
England and Wales						
Ronaldsway (Isle of Man)	54	05	4	38W.	17	1957–71
Tynemouth (Tyne & Wear)	55	01	1	25W.	33	1957–71
Manby (Lincoln)	53	21	0	05E.	17	1957–70
Mildenhall (Suffolk)	52	22	0	28E.	5	1957–68
Elmdon (West Midlands)	52	27	1	45W.	99	1957–71
Heathrow (Greater London)	51	29	0	27W.	25	1957–71
Dungeness (Kent)	50	55	0	57E.	3	1957–71
Hurn (Dorset)	50	47	1	50W.	10	1957–71
Ringway (Greater Manchester)	53	21	2	16W.	75	1957–71
Aberporth (Dyfed)	52	08	4	34W.	133	1957–71
Exeter (Devon)	50	38	3	24W.	32	1957–71
Scilly (Cornwall)	49	56	6	18W.	48	1958–71
Ireland						
Aldergrove (Antrim)	54	39	6	13W.	68	1957–71
Belmullet (Mayo)	54	14	10	00W.	9	1961–70
Shannon (Clare)	52	41	8	55W.	8	1961–70
Dublin Apt (Dublin)	53	26	6	14W.	65	1961–70
Rosslare (Wexford)	52	18	6	24W.	24	1961–70
Valentia (Kerry)	51	56	10	15W.	17	1961–70

Fig. 3.2 *January percentage frequency of the force (Beaufort scale) and direction of the wind at selected stations (for periods, see Table 3.3)*

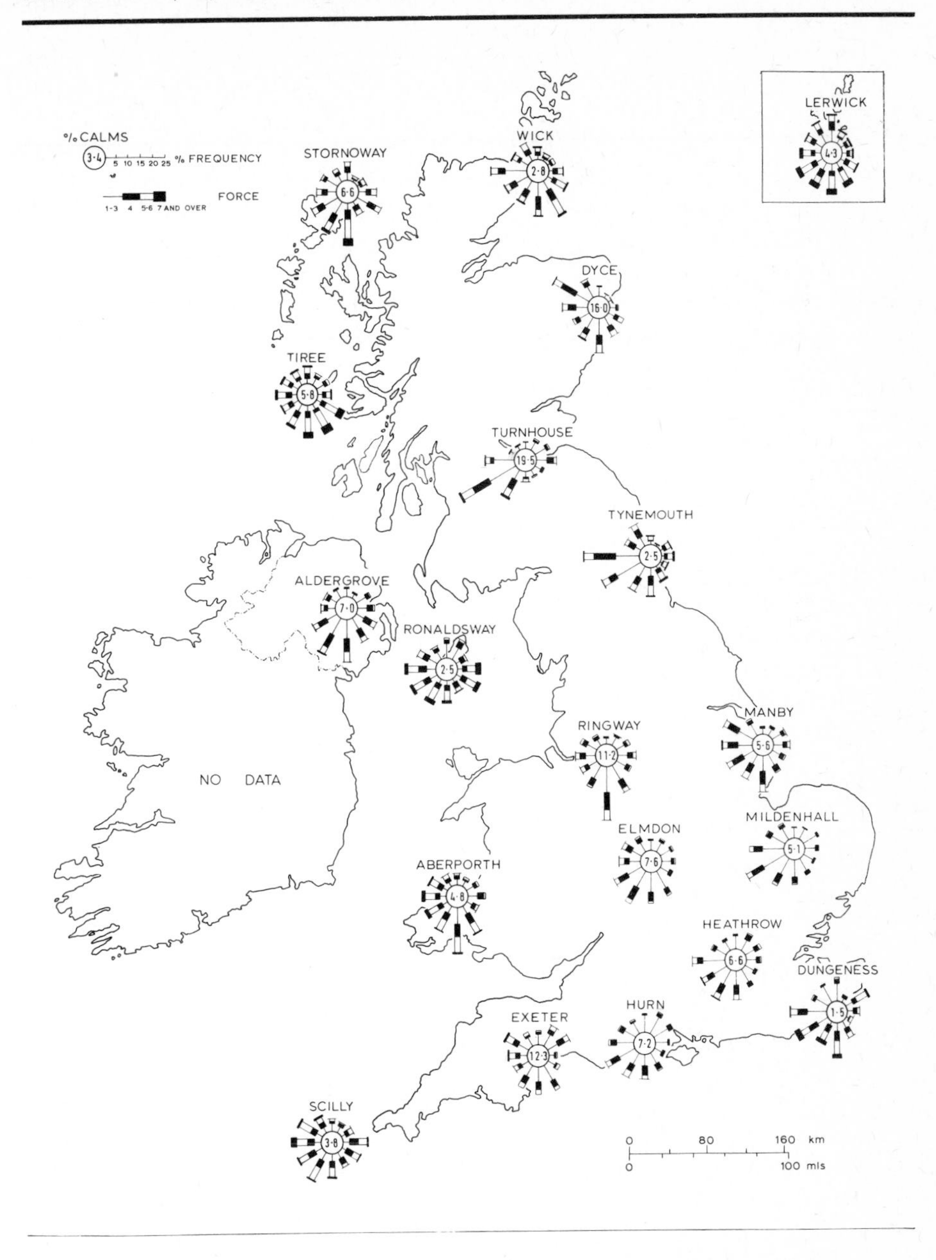

the distributions of wind direction at a given place are governed more by its position, particularly in relation to topographical features, than by the time of year. A detailed comparison of Figs 3.1 and 3.2 indicates, however, that in January there are fewer winds from between north and east and more winds from between south and west than there are over the entire year and this seems to apply fairly generally.

Fig. 3.3 *July percentage frequency of the force (Beaufort scale) and direction of the wind at selected stations (for periods, see Table 3.3)*

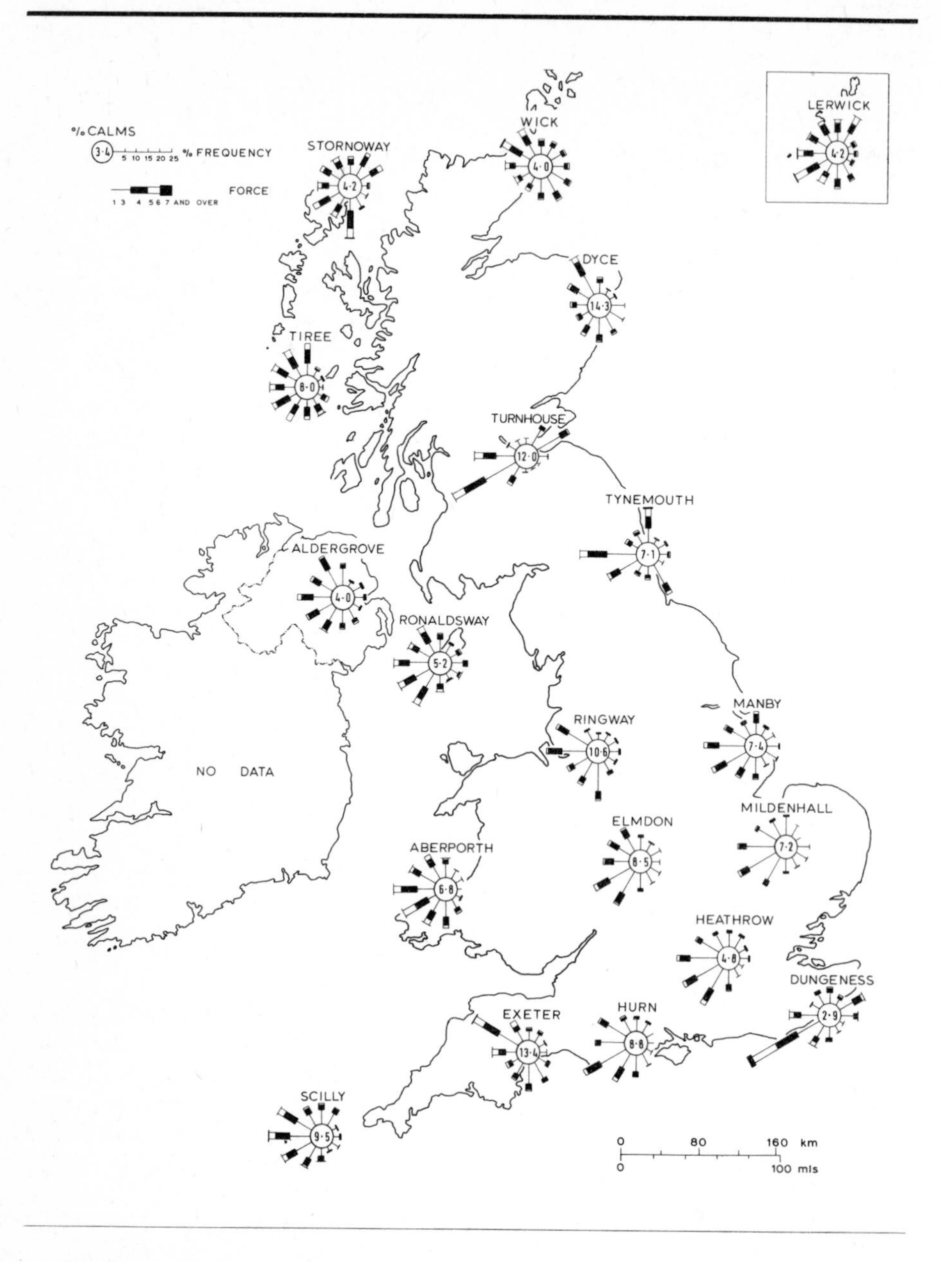

Figure 3.3 shows the wind roses for July for the same stations. Once more the patterns for individual stations show a broad similarity to those already seen in Figs 3.1 and 3.2, confirming what has been said about the importance of topography. Of course the frequencies of light winds are much greater and those of strong winds are much less in mid-summer than they are in mid-winter or over the year as a whole.

Comparison of the July wind roses with those shown in Fig. 3.1 suggests that there are fewer winds from between east and south and more from between north and west than there are over the entire year and once again this applies fairly generally, indicating that this is a feature of the broad airflow over the British Isles.

The frequency of winds from different directions

Surface wind directions were formerly observed according to a 32- or 16-point scale and subsequently reduced to an 8-point scale, N., N.E., E., S.E., etc., in official summaries. For many years now, however, directions have been measured and reported to the nearest ten degrees, i.e. on a 36-point scale, and it has therefore become the official practice to summarize them on a 12-point scale, i.e. to use twelve 30-degree ranges commencing with the one centred on true north, viz. 350°–010°. This is why the data shown in Tables 3.1 and 3.2 and in Figs 3.1 to 3.3 were presented in this way.

Annual values

Table 3.4 shows the annual percentage frequencies of winds from each of these 12 direction ranges, and of calms, at twenty-five representative stations, i.e., those for which details are given in Table 3.4 plus Prestwick in Scotland. It is clear from this table that the predominant wind directions are from between south (170°–190°) and west (260°–280°) although there are exceptions, e.g. at Wick and Tiree where 140°–160° is the most frequent direction and at Exeter where winds from 290°–310° are the most common. Some of these variations are almost certainly caused by topographical constraints, e.g. the westerly maximum at Tynemouth, the southerly one at Ringway and the aforementioned 290°–310° maximum at Exeter. At Turnhouse there is a marked maximum for winds from 230°–250° and a secondary one for winds from 050°–070°. At this station only about 14 per cent of winds blow from directions other than those between 200° and 280° and between 020° and 100°, suggesting that there is a marked tendency for channelling or canalization of the winds by the topography of the Firth of Forth. The figures for Dungeness too show a distinct tendency for its winds to flow along the English Channel, with highest frequencies from 230°–250° and 050°–070°. The directions which occur least often according to Table 3.4 are mostly between north (350°–010°) and east (080°–100°), the exceptions being Ronaldsway, where the least frequent direction is 140°–160°, and a number of stations – Heathrow, Dungeness, Hurn and Valentia – where it is 110°–130°.

Monthly values

In Table 3.5 are presented the percentage frequencies of winds from each of the twelve 30-degree direction ranges, and of calms, for six selected stations, Lerwick, Tynemouth, Heathrow, Ringway, Scilly and Aldergrove, for each month of the year, the periods of the

Table 3.4 *Annual percentage frequency of winds from stated directions, plus calms, at twenty-five stations*

Station	350°–010°	020°–040°	050°–070°	080°–100°	110°–130°	140°–160°	170°–190°	200°–220°	230°–250°	260°–280°	290°–310°	320°–340°	Calm
Lerwick	8·3	6·9	3·8	3·6	5·7	9·9	*12·0*	9·0	11·6	9·0	7·0	8·6	4·5
Wick	5·2	3·9	3·6	4·9	9·2	*13·2*	11·5	8·4	8·7	11·6	8·7	8·1	3·1
Stornoway Apt.	6·2	5·3	6·3	5·1	6·8	6·4	*14·9*	11·3	10·9	7·6	7·0	5·7	6·5
Dyce	4·8	1·8	2·0	4·1	6·4	8·7	*11·8*	8·7	6·7	8·2	11·4	10·3	15·1
Turnhouse	2·0	5·5	11·6	6·5	2·6	1·5	2·7	10·1	*25·1*	12·2	2·8	2·2	15·2
Tiree	8·7	4·3	3·5	4·1	9·5	*10·3*	10·1	9·7	8·5	9·1	7·5	7·8	6·8
Prestwick*	1·9	4·1	8·1	8·7	9·4	6·9	9·1	8·5	*12·7*	10·5	8·4	4·1	7·6
Ronaldsway	7·4	6·4	5·9	9·0	7·5	3·7	6·6	11·9	*12·6*	11·2	7·7	6·8	3·4
Tynemouth	8·4	2·3	4·6	4·9	2·5	10·0	8·1	7·8	15·4	*20·0*	4·5	6·9	4·7
Manby	5·7	5·3	4·7	5·3	6·1	5·8	10·0	10·3	*12·7*	12·0	8·9	6·7	6·5
Mildenhall	5·9	5·9	5·0	5·6	5·9	6·0	8·0	12·6	*14·8*	11·1	6·9	6·0	6·4
Elmdon	5·0	6·0	6·0	4·8	5·0	6·7	8·9	*13·7*	12·3	7·7	7·1	7·5	9·2
Heathrow	7·0	8·0	6·4	5·8	3·3	4·7	8·6	*13·6*	12·7	11·9	6·5	5·3	6·0
Dungeness	7·4	4·1	11·5	7·7	2·1	4·8	8·4	8·1	*20·4*	11·8	4·3	7·1	2·1
Hurn	6·8	7·9	5·9	4·3	3·5	5·5	8·3	9·8	*14·8*	9·3	8·0	5·9	10·0
Ringway	2·5	4·7	6·1	6·8	5·4	8·2	*16·7*	10·3	7·2	9·9	8·9	4·0	9·2
Aberporth	5·8	4·1	4·6	4·4	4·5	9·6	*14·5*	10·8	10·6	10·3	8·4	6·6	5·6
Exeter	5·0	7·6	6·6	2·8	5·2	8·3	10·6	7·8	5·5	7·6	*13·3*	6·3	13·5
Scilly	6·2	6·2	6·2	7·0	6·3	4·4	7·0	9·9	10·4	*12·3*	11·2	7·0	5·9
Aldergrove	5·7	2·9	5·3	7·2	8·0	8·7	11·5	*13·2*	10·7	8·7	6·4	6·8	5·0
Belmullet	5·6	6·7	5·4	6·4	6·3	6·9	8·8	*13·5*	11·7	11·4	8·9	6·0	2·1
Shannon	6·2	4·8	3·3	5·8	9·8	9·9	7·8	7·6	11·7	*13·5*	10·8	6·6	2·3
Dublin	2·7	3·6	4·4	6·4	6·8	9·1	3·3	8·5	17·7	*18·0*	8·9	4·6	5·7
Rosslare	3·8	6·6	6·3	4·4	5·0	5·3	9·5	*17·1*	16·4	9·5	7·6	7·7	1·1
Valentia	6·5	5·1	7·2	6·5	4·4	7·5	*12·4*	10·7	7·4	9·6	7·7	7·4	7·1

*See Table 3.8 for station details; period 1957–71

data used being as shown in Table 3.3 for these stations. Looking down the columns the annual variation in the frequency of winds from each direction can be seen readily, the highest values appearing in italics. For some directions the variation is quite irregular but for others a seasonal pattern is easily discerned. Thus at Lerwick winds from all directions between 020°–040° and 140°–160° show in varying degree a spring or summer maximum, the variation being particularly regular for winds from 020°–040°, rising from a minimum of 2·9 per cent in October to a maximum of 12·0 per cent in May and then gradually falling again. Winds from between 170° and 220° are most common in the autumn while those from between 230° and 280° show quite a marked summer maximum. Winds from between 320° and 010° exhibit a winter maximum and a summer minimum at this station while those from 290°–310°, and also calms, show no very significant annual variations. At Tynemouth annual variations are most conspicuous for winds from 140°–160° and from the north, both with a spring to early summer maximum and a winter minimum, while calms have a distinct summer maximum and a winter minimum. Winds from between 230° and 280°, which overall are the most common directions at this station, have their greatest frequencies in autumn and winter and their lowest in spring or summer. As at Lerwick, there is evidence of a spring maximum for all winds having an easterly component. At Heathrow the directions with a most marked annual variation are 020°–040°, with a spring maximum, and those between 200° and 280°, all with summer maxima. There is also a tendency for spring maxima for winds from the north and from directions between 050° and 100°. Ringway too shows a spring maximum for winds from north to east, inclusive, while there is a distinct autumn maximum for southerly winds and summer ones for directions between 260° and 310°. At Scilly the spring maximum is seen clearly only for winds from 020°–040°. For directions between 050° and 100° the highest frequencies occur in winter and the lowest ones in summer. Winds from between 110° and 160° are most common in the autumn months and those from between 170° and 220° in the autumn and winter. Marked summer maxima occur at this station for directions from between 230° and 310°, and also for calms. Finally, at Aldergrove in Northern Ireland the spring maximum for easterlies is yet again in evidence, being particularly prominent for winds from 110°–130°. For directions from between 170° and 220° there is a marked winter maximum while for directions between 260° and 340° the summer maximum already noted at other stations is again apparent.

The results for these few representative stations give a good general indication of the variations in the direction of the surface airflow over the British Isles in the different months of the year. Information for other stations may be found in Meteorological Office (1968b).

Wind speed

The speed ranges used in Tables 3.1 and 3.2 and in Figs 3.1 to 3.3 were those corresponding to the wind forces 0 to 12 on the Beaufort scale. This was done because at one time the vast majority of observations of surface wind speed were estimated in accordance with the descriptive terms of the Beaufort scale for use on land and it has become customary therefore to express wind speed statistics in this way. Indeed a number of voluntary climatological stations still use the Beaufort scale, reporting the mean speed in knots which is equivalent to the estimated wind force. However, the above-mentioned tables and figures were in fact based almost entirely on measurements of wind speed obtained from anemometers. At the majority of the stations the instruments were of the recording type, known as anemographs, which provide continuous records of speed and direction on a chart.

Much of the detailed knowledge of wind structure is based on the records provided by anemographs. An early form of wind recording instrument used in Britain was the cup anemograph of Robinson (1850) and Beckley (1858), which recorded the direction and the run of the wind in miles, and which was in routine use at a few stations from 1868 onwards. Following the Tay Bridge disaster in 1879, however, special attention was directed to the measurement of gusts and the Robinson and Beckley instrument was superseded by the pressure-tube anemograph devised by W. H. Dines. The pressure-tube anemograph provides a continuous speed record consisting of a connected series of nearly vertical lines the tops of which indicate the gusts and the bottoms the lulls caused by the turbulent eddies which are always present in the natural wind. The speed shown by the pen at any instant is the mean value over a period of some 3 to 5 seconds, depending to some extent on the diameter and length of the pressure and suction pipes which lead from the head of the instrument to the recorder. For this reason the ½-inch pipes of the early models were later replaced by 1-inch diameter pipes. A full account of this anemograph has been given by Gold (1936), who included illustrations of records showing special characteristics. Details are also given in the *Handbook of Meteorological Instruments, Part 1* (Meteorological Office, 1956). The pressure-tube anemograph became the recognized standard and revolutionized the measurement of wind. After its introduction near the end of the nineteenth century, the numbers in operation in Britain increased steadily to about ten in 1910, thirty in 1930, fifty in 1950 and fifty-six in 1955. In that year the Meteorological Office electrical cup anemograph was introduced (Hartley, 1955) and has since then gradually replaced the pressure-tube type and has itself been improved in several ways (Else, 1974). Its chief advantages over its predecessor are its use of a strip chart lasting a month in place of separate daily charts, the fact that the recorder can be installed at a considerable distance from the head and its greater ease of maintenance. There is no reason to believe that this change of instrument has affected the homogeneity of the wind records, wind tunnel measurements having shown that mean values are in very good agreement and that the response times are about the same. It has been suggested that the maximum gust speeds recorded by the electrical cup anemograph may be a little higher than those recorded by the pressure-tube type, but sufficiently accurate comparisons, which

Table 3.5 *Monthly percentage frequency of winds from stated directions, plus calms, at six stations*

Station	Month	350°–010°	020°–040°	050°–070°	080°–100°	110°–130°	140°–160°	170°–190°	200°–220°	230°–250°	260°–280°	290°–310°	320°–340°	Calm
Lerwick	Jan.	10·1	4·4	3·6	4·0	6·9	9·5	11·1	10·1	11·2	8·7	7·5	8·8	4·3
	Feb.	*12·5*	4·7	3·7	3·1	6·8	9·0	9·7	9·8	10·6	8·7	6·8	10·2	4·6
	Mar.	6·4	4·0	3·3	3·4	6·9	*13·0*	14·2	8·8	12·4	7·7	6·1	9·8	4·1
	Apr.	8·4	7·2	*5·4*	4·4	6·3	8·4	13·2	8·6	12·1	7·2	5·5	9·8	3·4
	May	7·8	*12·0*	*5·4*	*6·4*	*8·8*	12·5	9·7	5·3	7·8	5·9	5·0	9·1	4·4
	June	5·4	11·4	4·9	2·7	5·0	11·8	13·8	5·6	14·1	9·5	6·1	5·5	4·2
	July	8·2	11·6	3·4	3·8	3·0	6·8	8·9	7·7	*14·3*	*11·1*	8·2	8·9	4·2
	Aug.	8·7	10·8	4·3	3·8	4·4	11·1	10·0	6·0	9·6	11·0	7·8	7·2	5·4
	Sept.	6·9	6·9	2·5	1·8	5·1	9·9	15·5	9·2	11·5	10·2	*8·4*	7·9	4·2
	Oct.	5·9	2·9	2·6	3·4	4·7	10·3	*15·9*	*14·3*	13·6	9·5	7·2	6·0	3·7
	Nov.	9·8	3·9	3·5	3·9	5·7	8·2	10·6	11·8	10·1	9·7	7·5	8·8	*6·6*
	Dec.	9·7	3·3	3·3	2·6	5·2	8·6	11·6	10·9	11·9	8·6	8·2	*11·4*	4·7
Tynemouth	Jan.	3·1	0·9	4·3	5·1	2·7	6·3	10·8	10·5	16·7	22·0	6·1	8·8	2·5
	Feb.	4·8	1·3	5·1	*7·3*	3·1	8·0	8·3	7·6	14·7	20·6	*7·0*	8·7	3·6
	Mar.	6·3	2·4	5·8	6·4	*5·2*	12·8	7·2	5·7	14·4	18·2	4·5	7·6	3·4
	Apr.	12·2	*4·5*	*6·0*	6·6	2·6	12·3	6·0	5·5	11·5	18·3	3·5	6·0	4·9
	May	*16·8*	3·4	5·5	4·3	2·7	*16·0*	6·9	5·5	12·2	14·8	2·3	4·4	5·2
	June	14·5	3·3	5·1	3·8	2·3	13·6	5·3	4·5	14·3	18·8	2·7	4·4	*7·5*
	July	13·3	2·2	3·6	3·6	2·0	12·3	4·9	4·4	13·3	22·6	4·9	5·7	7·1
	Aug.	14·5	2·7	3·9	3·7	1·9	10·6	6·4	5·2	15·2	19·0	3·9	6·2	6·8
	Sept.	6·6	2·3	3·5	4·0	2·0	11·2	10·3	10·4	15·9	19·0	3·5	5·5	5·9
	Oct.	2·8	1·4	2·7	3·4	1·6	9·4	*11·6*	*12·0*	*20·8*	20·6	3·7	5·1	4·8
	Nov.	3·4	2·1	5·3	6·0	2·2	4·7	9·9	10·6	16·6	22·1	5·1	9·2	2·8
	Dec.	2·2	1·1	4·1	5·0	1·9	2·8	9·0	11·8	18·9	*24·3*	6·7	*10·7*	1·5

Table 3.5 (continued)

Station	Month	350°–010°	020°–040°	050°–070°	080°–100°	110°–130°	140°–160°	170°–190°	200°–220°	230°–250°	260°–280°	290°–310°	320°–340°	Calm
Heathrow	Jan.	5·3	6·7	6·3	5·3	4·6	*7·1*	11·0	12·3	11·2	12·8	6·0	4·8	6·6
	Feb.	7·8	7·7	9·3	6·6	3·6	4·3	7·2	11·5	12·2	12·4	7·0	5·3	5·3
	Mar.	7·0	9·4	7·8	*8·1*	*5·0*	4·9	7·0	10·2	12·0	12·1	6·9	4·8	4·7
	Apr.	*9·3*	*13·4*	*8·2*	5·8	2·4	4·2	7·6	11·8	10·0	11·2	5·7	5·9	4·7
	May	7·6	10·0	7·2	5·7	3·1	4·7	9·3	15·7	11·4	9·7	5·1	5·8	4·9
	June	6·8	10·2	7·0	5·9	2·7	3·0	7·1	*16·0*	13·3	12·1	6·2	4·9	4·7
	July	7·0	8·0	6·4	5·8	3·3	4·7	8·6	13·6	12·7	11·9	6·5	5·3	6·0
	Aug.	6·9	5·6	3·8	5·5	2·4	3·1	7·5	*16·0*	*16·1*	*15·0*	6·9	5·7	5·5
	Sept.	6·7	7·5	6·3	7·8	3·3	5·0	8·3	14·2	13·4	10·2	5·4	4·2	7·7
	Oct.	4·8	5·3	5·6	5·9	4·0	6·7	*11·4*	15·0	12·8	9·8	6·4	3·6	*8·7*
	Nov.	8·8	7·4	6·1	5·5	3·4	5·6	9·9	12·2	13·2	9·5	5·7	*6·4*	6·3
	Dec.	7·7	7·6	6·5	3·5	2·9	5·0	9·1	13·0	11·6	12·4	*7·1*	5·8	7·8
Ringway	Jan.	2·3	4·1	5·2	7·2	6·1	*11·2*	20·6	9·1	6·2	7·1	6·3	3·4	*11·2*
	Feb.	2·9	4·3	*8·7*	7·4	7·1	8·6	15·0	9·6	6·4	8·6	8·2	4·3	9·0
	Mar.	2·7	4·8	8·1	*9·9*	*8·8*	8·8	11·9	9·8	6·6	9·2	9·2	3·5	6·8
	Apr.	*4·3*	*8·5*	8·6	7·5	4·8	6·0	12·4	9·0	5·8	10·4	10·7	4·5	7·6
	May	2·6	7·0	8·5	7·8	3·9	7·3	16·1	9·6	6·9	9·8	9·1	3·4	7·9
	June	2·0	5·4	7·3	6·8	3·5	5·6	13·9	10·8	9·0	12·0	11·5	3·9	8·2
	July	1·8	3·3	3·6	3·7	3·1	4·9	14·5	9·9	*9·1*	*16·3*	*14·5*	4·7	10·6
	Aug.	2·7	4·7	5·0	5·0	5·2	6·9	14·1	10·6	8·4	14·1	10·5	4·2	8·6
	Sept.	1·6	2·8	5·4	8·2	5·8	9·0	18·8	11·6	7·7	9·7	6·8	2·7	10·0
	Oct.	1·3	2·4	3·8	6·2	7·1	11·0	*21·8*	*13·4*	7·9	6·8	6·3	3·1	9·0
	Nov.	3·4	5·1	5·0	7·0	5·4	10·5	20·4	8·8	6·2	6·4	6·4	*5·3*	10·0
	Dec.	3·0	4·5	4·7	4·8	4·2	9·0	21·0	11·2	5·9	8·4	7·2	5·1	10·9

Table 3.5 (continued)

Station	Month	350°–010°	020°–040°	050°–070°	080°–100°	110°–130°	140°–160°	170°–190°	200°–220°	230°–250°–	260°–280°	290°–310°	320°–340°	Calm
Scilly	Jan.	4·4	3·9	5·1	*10·2*	8·1	6·0	9·1	11·9	9·8	11·3	10·1	6·4	3·8
	Feb.	4·0	4·4	*11·1*	9·3	8·6	5·1	8·2	9·9	9·7	10·6	8·1	5·7	5·3
	Mar.	6·9	7·1	6·8	7·0	9·2	6·8	6·9	7·7	9·0	10·6	9·8	7·6	4·7
	Apr.	6·8	*9·3*	7·7	6·7	6·1	3·8	7·3	9·7	8·9	10·7	12·1	6·6	4·3
	May	7·6	7·7	6·3	5·8	6·3	3·7	7·0	10·9	11·3	12·4	8·6	6·0	6·5
	June	7·1	7·5	6·3	5·3	4·2	3·4	5·8	10·0	11·1	13·0	13·5	5·2	7·6
	July	*7·9*	7·9	2·7	3·4	3·2	1·6	4·8	8·5	*12·2*	*15·8*	*14·4*	8·0	*9·5*
	Aug.	6·5	6·2	4·4	3·7	3·3	3·0	5·9	8·4	11·7	14·5	16·5	8·4	7·5
	Sept.	4·6	5·6	6·3	7·9	7·9	5·7	7·2	10·1	10·7	11·7	8·1	5·7	8·6
	Oct.	4·4	4·0	3·1	7·1	*9·4*	*7·2*	*9·3*	*12·2*	9·8	10·1	10·6	6·9	5·8
	Nov.	7·7	5·4	6·5	7·3	6·1	3·8	6·5	8·9	9·9	13·7	11·0	*8·7*	4·4
	Dec.	6·6	5·0	8·1	*10·2*	3·3	3·4	6·4	10·2	10·5	13·5	11·7	8·1	2·9
Aldergrove	Jan.	3·2	2·0	5·9	6·7	9·0	10·0	*16·5*	15·5	10·0	5·4	4·8	4·0	7·0
	Feb.	3·8	2·7	5·2	8·7	12·0	8·6	13·3	14·5	9·8	7·4	5·4	3·4	5·4
	Mar.	4·6	3·0	4·9	*9·7*	*13·3*	9·2	8·2	12·9	10·7	7·3	6·1	5·4	4·7
	Apr.	7·8	*4·0*	6·3	8·9	7·9	7·9	9·1	9·5	10·2	8·2	6·6	9·3	4·4
	May	*8·8*	2·8	*7·7*	8·7	8·1	9·9	9·4	8·9	8·4	8·6	6·2	8·6	3·8
	June	7·2	2·5	6·5	8·2	6·6	8·6	9·4	10·9	10·8	9·7	6·9	8·5	4·1
	July	8·3	2·4	4·7	5·5	3·6	6·2	7·4	10·2	11·5	*13·5*	*9·7*	*13·1*	4·0
	Aug.	8·2	3·6	5·7	6·4	5·5	6·6	9·4	9·7	10·5	11·9	8·1	9·5	4·8
	Sept.	3·7	2·5	3·7	6·3	8·2	*11·9*	12·9	14·1	9·7	9·0	6·4	6·5	5·1
	Oct.	3·3	2·2	3·0	4·4	9·1	10·6	14·1	17·8	*13·6*	8·4	5·5	4·4	3·8
	Nov.	5·7	3·6	4·9	7·1	7·6	6·9	12·9	15·4	11·3	7·4	5·6	4·6	*7·1*
	Dec.	3·7	3·2	5·4	5·9	5·5	7·8	15·6	*18·7*	12·2	7·4	4·9	4·3	5·5

would need to be carried out in the natural wind, have not yet been made to test this.

By 1965 there were about 100 anemographs providing records to the Meteorological Office, of which only about thirty were of the older pressure-tube type. Since then the total number has increased to about 120 and the number of the older type has fallen to a mere handful. This network of anemographs probably comprises the closest network of continuous and reliable wind records to be found anywhere in the world.

When siting an anemometer the meteorologist aims at getting a record that represents fairly the general airflow at the standard height of 10 m above open level ground. It is rarely possible to achieve this ideal exposure, which would only be met if the vane or cups were mounted at the top of a 10 m mast in the middle of a flat area with no obstacles whatever, i.e. no trees or buildings within a radius of 200 to 300 m. If there are local obstructions to the wind then it may be necessary to raise the anemometer head to a height greater than 10 m in order to minimize their effects. In such cases the so-called 'effective height' may exceed the standard height. The 'effective height' of an anemometer is defined as the height over open level terrain in the vicinity of the instrument, which, it is estimated, would have the same mean wind speeds as those actually recorded by the anemometer. A large majority of anemographs have effective heights of between 9 and 13 m, less than one in four having an effective height greater than 15 m. It is usually only in built-up areas that the effective height is greatly different from the actual height above the ground.

Because the speed of the wind usually increases with increasing height above the ground, higher wind speeds are generally recorded when an anemometer has to be mounted where it has an effective height greater than 10 m. It is customary, therefore, to reduce the speeds measured by such an anemometer to the standard height of 10 m, using an empirical formula, in order to make them directly comparable with those from other stations. For synoptic purposes the necessary correction is applied at the station, the corrected speed being inserted in the station report before transmission. The corrections range from +20 per cent for effective heights of 3 to 4 m to −20 per cent for effective heights of 23 to 42 m. For climatological purposes speeds are tabulated as recorded, i.e. without any correction, but when it is desired to compare statistics from different stations reduction to the standard height is effected by means of the power-law formulae:

$$V_{10} = V_Z \cdot \left(\frac{10}{Z}\right)^{0\cdot17}$$ for mean speeds (Carruthers, 1943) and

$$V_{10} = V_Z \cdot \left(\frac{10}{Z}\right)^{0\cdot085}$$ for gusts (Deacon, 1955)

where V_{10} is the speed at 10 m and V_Z is the speed as recorded at the effective height z m. In particular circumstances the power may need adjusting for variations in surface roughness.

Mean monthly and annual wind speeds

In Table 3.6 are given the monthly and annual averages of the hourly mean wind speeds recorded at 24 selected anemograph stations over the period 1961–70, reduced to the standard height of 10 m using the 0·17 power law formula.

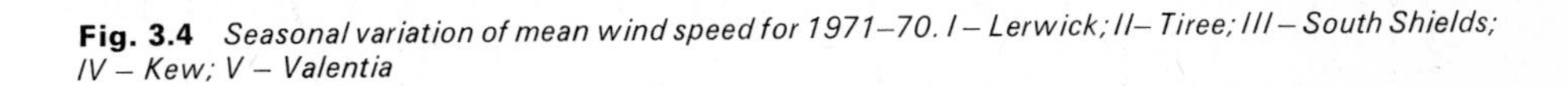

Fig. 3.4 *Seasonal variation of mean wind speed for 1971–70. I – Lerwick; II – Tiree; III – South Shields; IV – Kew; V – Valentia*

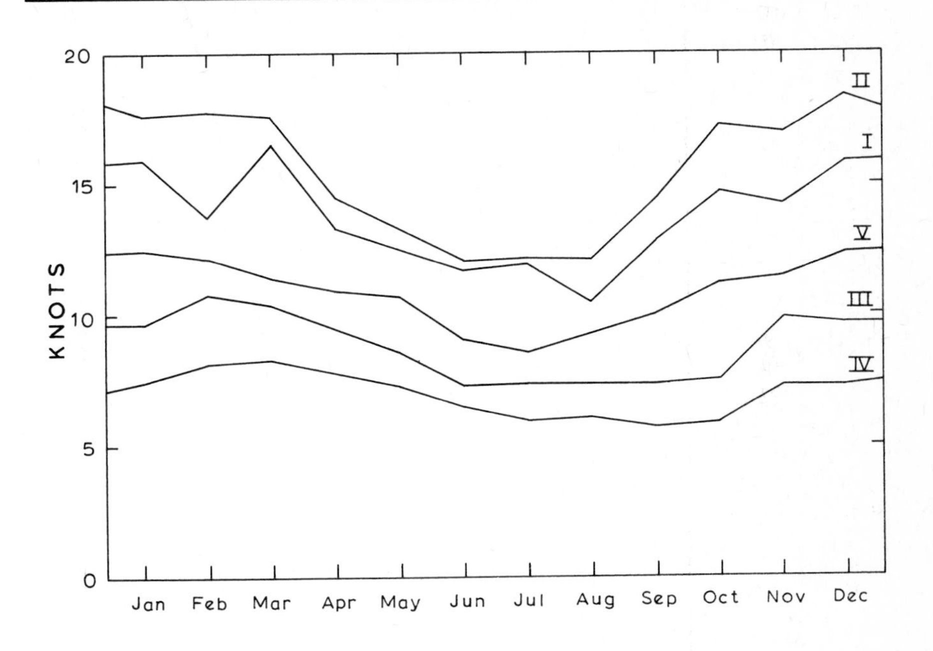

The monthly values for five of these stations, Lerwick, Tiree, South Shields, Kew and Valentia, are shown graphically in Fig. 3.4. The average speeds are highest at Tiree and lowest at Kew in all months of the year. The highest average speeds usually occur in one of the months November to March, most commonly in March, and the lowest in one of the months June to September, most commonly in July. The actual months of highest and lowest speed may depend to some extent on the averaging period used, but it may be noted that those for Kew and Valentia for the recent period 1961–70 are the same as they were for these stations over the much earlier 35-year period for which averages were given by Bilham (1938). The average speeds shown in Table 3.6 will to some extent reflect the degree of exposure to the wind of the individual stations and so they may have only local significance but the general form of the annual variation exhibited by the curves of Fig. 3.4 is a definite climatic characteristic.

Table 3.6 *Average values of mean wind speed in knots, for each month and the year, reduced to the standard height of 10 m,1961–70*

Station	J	F	M	A	M	J	J	A	S	O	N	D	Year
Lerwick	15·9	14·5	*16·5*	13·3	12·5	11·7	11·9	10·5	12·8	14·7	14·2	15·8	13·7
Dounreay*	15·2	15·4	*16·5*	14·2	13·9	12·1	12·2	11·9	12·9	15·0	14·6	15·3	14·2
Stornoway CG*	*16·7*	16·6	16·0	13·4	12·7	11·2	10·9	10·9	13·8	14·8	15·0	16·3	14·2
Dyce	9·6	10·1	*10·9*	9·1	9·0	8·1	8·4	8·2	8·5	9·2	9·3	9·7	9·2
Turnhouse	9·2	10·1	*11·0*	10·3	9·6	9·1	9·2	8·4	8·9	9·4	9·5	9·2	9·5
Tiree	17·6	17·8	17·6	14·5	13·4	12·1	12·2	12·1	14·4	17·2	16·9	*18·3*	15·4
Prestwick	10·0	10·1	*11·0*	9·9	9·3	8·9	8·9	8·5	8·5	9·6	9·8	9·3	9·5
Ronaldsway	16·2	*16·3*	15·4	13·1	11·7	10·3	10·1	11·0	12·2	14·1	15·4	15·7	13·5
South Shields*	9·7	*10·8*	10·4	9·5	8·6	7·3	7·4	7·4	7·4	7·5	9·9	9·7	8·8
Mildenhall	7·7	7·7	*8·1*	7·1	6·4	5·5	5·2	6·1	5·8	6·2	7·1	7·3	6·8
Elmdon	9·4	9·9	*10·2*	9·7	9·2	8·6	8·6	8·9	8·4	8·5	9·5	9·3	9·2
Heathrow	9·4	*10·1*	*10·1*	9·8	9·3	8·7	8·5	8·6	8·2	8·1	9·3	9·1	9·1
Kew*	7·5	8·2	*8·3*	7·8	7·3	6·5	6·0	6·1	5·7	5·9	7·3	7·3	7·0
Dover*	10·0	10·1	9·1	8·4	8·1	7·9	6·8	8·0	7·8	9·9	11·6	10·3	9·0
Hurn	9·1	*9·4*	*9·4*	*9·4*	8·6	8·0	7·7	7·9	7·7	7·8	8·9	9·0	8·6
Ringway	9·3	10·0	*10·5*	9·6	9·1	8·2	7·7	8·2	8·2	8·4	9·2	8·9	8·9
Aberporth	14·8	14·3	13·7	12·1	11·2	9·7	9·4	11·1	11·7	12·8	14·2	*15·7*	12·6
Scilly	15·4	14·7	13·2	13·0	11·6	9·4	9·0	9·5	10·6	12·5	15·0	*16·3*	12·5
Aldergrove	10·8	11·0	*11·3*	10·4	10·1	9·5	9·4	9·2	9·7	10·5	9·9	9·9	10·1
Belmullet	14·5	14·9	*15·0*	13·6	13·5	12·5	12·3	12·3	13·3	14·9	14·2	14·8	13·8
Dublin Apt.	11·5	*12·0*	11·7	10·6	9·3	8·3	8·3	8·6	8·9	10·0	11·0	11·6	10·2
Shannon	11·0	*11·8*	11·7	11·0	10·4	9·3	9·1	9·6	9·7	10·5	10·3	11·1	10·5
Rosslare	13·4	*14·2*	12·7	12·3	11·8	10·3	9·4	10·5	10·9	11·6	13·0	13·6	12·0
Valentia	*12·5*	12·2	11·4	10·9	10·7	9·1	8·6	9·3	10·1	11·2	11·5	12·4	10·8

*See Table 3.8 or 3.9 for station details

The frequency distribution of mean wind speeds

The average wind speeds discussed in the preceding section, like most averages, are of relatively minor interest and rather limited usefulness. In practice one often wants to know the frequency with which wind speeds lie within given limits or with which they exceed or fall below particular levels. Table 3.7 shows percentage frequencies

Fig. 3.5 *Annual percentage frequency of mean wind speed from all directions (for periods, see Table 3.3). The radius of the outer circle represents 100 per cent*

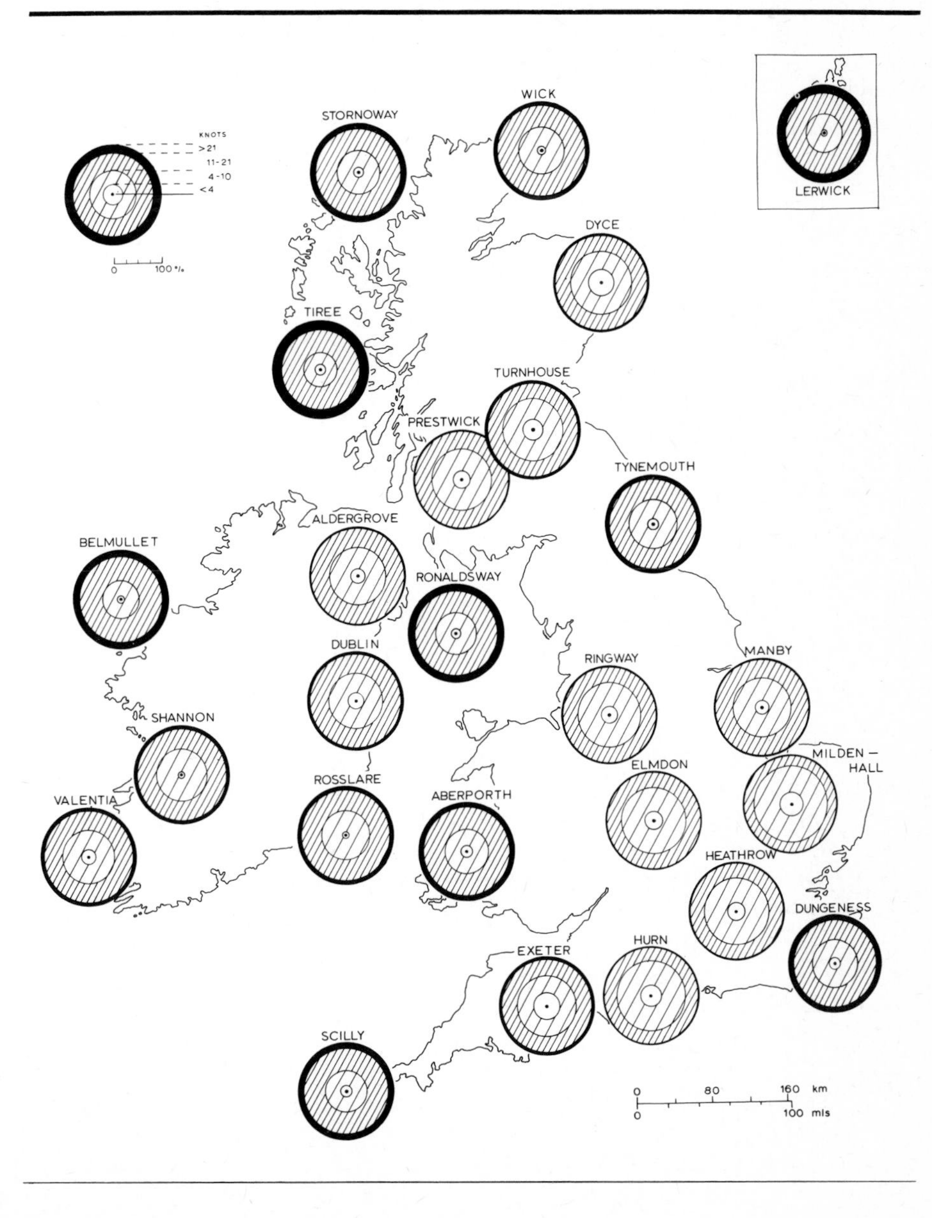

Table 3.7 *Percentage frequency of winds with stated speeds in January, July and the whole year, reduced to the standard height of 10 m*

Station	Speed ranges in knots	Less than 4	4–10	11–21	22–33	More than 33
Lerwick	*January*	6·8	23·4	42·5	23·2	4·1
	July	7·7	38·0	47·3	6·8	0·2
	Year	7·7	31·4	43·7	15·2	2·0
Wick	*January*	8·5	30·1	47·1	13·5	0·8
	July	11·1	46·1	39·1	3·6	0·1
	Year	9·7	37·1	43·9	8·8	0·5
Stornoway Apt	*January*	9·6	31·7	41·8	15·4	1·5
	July	7·8	39·2	47·2	5·8	<0·1
	Year	9·9	34·3	44·2	10·8	0·8
Dyce	*January*	24·3	32·4	38·0	5·0	0·3
	July	29·3	41·3	27·8	1·6	0·0
	Year	26·5	37·1	32·8	3·5	0·1
Turnhouse	*January*	26·1	37·0	30·4	5·8	0·7
	July	17·5	49·2	30·7	2·5	0·1
	Year	20·4	42·0	32·6	4·7	0·3
Tiree	*January*	8·4	19·0	40·5	28·3	3·8
	July	13·3	30·9	49·1	6·6	0·1
	Year	10·4	23·8	45·7	18·5	1·6
Prestwick	*January*	20·6	37·2	35·9	6·0	0·3
	July	16·7	48·0	34·4	0·9	0·0
	Year	18·0	43·0	35·6	3·3	0·1
Ronaldsway	*January*	8·1	24·3	42·3	22·2	3·1
	July	13·6	42·4	39·8	4·1	0·1
	Year	10·0	31·3	43·7	13·5	1·5
Tynemouth	*January*	8·2	30·5	47·2	12·3	1·8
	July	14·7	47·0	35·1	3·0	0·2
	Year	11·1	38·4	41·0	8·5	1·0
Manby	*January*	11·4	40·2	43·4	4·8	0·2
	July	17·2	53·4	28·5	0·9	0·0
	Year	14·5	45·5	37·2	2·8	<0·1
Mildenhall	*January*	18·9	50·7	28·4	2·0	<0·1
	July	30·5	56·4	13·0	0·1	0·0
	Year	24·6	53·1	21·3	1·0	<0·1
Elmdon	*January*	15·2	47·6	34·1	3·0	0·1
	July	18·2	53·6	27·5	0·7	0·0
	Year	17·0	49·6	31·3	2·1	<0·1

Station		Speed ranges in knots Less than 4	4–10	11–21	22–33	More than 33
Heathrow	*January*	16·6	45·9	34·7	2·7	0·1
	July	16·2	55·4	27·6	0·8	0·0
	Year	16·8	49·9	31·6	1·7	<0·1
Dungeness	*January*	8·4	33·0	40·5	15·9	2·2
	July	11·9	43·4	37·0	7·3	0·4
	Year	10·5	37·8	39·7	10·9	1·1
Hurn	*January*	18·4	47·3	31·6	2·7	0·0
	July	21·8	53·3	24·5	0·4	0·0
	Year	20·7	50·0	27·8	1·5	<0·1
Ringway	*January*	18·9	41·4	35·4	4·2	0·1
	July	20·9	53·3	25·4	0·4	0·0
	Year	17·6	47·5	32·5	2·4	<0·1
Aberporth	*January*	9·1	24·8	49·3	15·5	1·3
	July	13·7	42·3	39·9	4·0	0·1
	Year	10·7	33·9	45·2	9·7	0·5
Exeter	*January*	24·9	31·7	35·5	7·6	0·3
	July	27·7	38·9	30·8	2·6	0·0
	Year	27·5	34·2	33·0	5·1	0·2
Scilly	*January*	7·7	20·8	46·1	23·2	2·2
	July	16·7	44·1	36·3	2·8	0·1
	Year	11·3	31·4	44·4	12·0	0·9
Aldergrove	*January*	17·7	36·0	40·5	5·5	0·3
	July	15·1	49·6	34·1	1·2	0·0
	Year	15·2	43·7	37·6	3·4	0·1
Belmullet	*January*	8·9	29·0	43·6	16·4	2·1
	July	5·4	37·6	50·0	6·9	0·1
	Year	6·4	31·8	47·7	13·2	0·9
Shannon	*January*	10·8	40·3	39·2	9·0	0·7
	July	9·2	51·7	36·8	2·3	0·0
	Year	8·5	45·0	39·7	6·5	0·3
Dublin Apt	*January*	12·7	35·7	42·6	8·6	0·4
	July	20·1	50·9	27·9	1·1	0·0
	Year	15·3	42·1	37·2	5·2	0·2
Rosslare	*January*	5·8	36·8	43·8	13·1	0·6
	July	11·9	53·1	33·4	1·6	0·0
	Year	7·1	42·2	42·3	8·1	0·4
Valentia	*January*	13·9	29·6	43·5	12·3	0·7
	July	18·0	51·0	30·2	0·8	0·0
	Year	14·5	39·3	40·0	6·0	0·2

of winds located within stated speed ranges in the months of January and July and over the year as a whole, reduced where necessary to the standard height of 10 m, using the same data as were used in preparing Figs 3.1 to 3.3. The speed ranges used are those corresponding to the Beaufort forces 8 or more (gales), 6 and 7 (strong winds), 4 and 5 (moderate or fresh winds), 2 and 3 (light winds) and 0 and 1 (calms or light airs). The annual frequencies are also shown diagrammatically in Fig. 3.5. Here the radius of the central circle round each station gives the percentage frequency of winds under 4 kt and the widths of the three successive outer zones give the percentages falling within the ranges 4–10, 11–21 and 22 or more knots, the frequency of gales being too small at most stations to be shown in such diagrams. Tiree has the highest annual frequency of strong winds and gales with over 20 per cent (over 30 per cent in January) followed by Lerwick, Ronaldsway, Belmullet and Scilly and then, rather surprisingly, by Dungeness. Although winds of 22 kt and above at Scilly have a frequency of over 25 per cent in January, their frequency in July is less than three per cent, a figure which is exceeded in that month at such east coast stations as Wick and Tynemouth as well as at the more exposed west coast stations such as Tiree, Belmullet, Ronaldsway and Aberporth. Stations having the fewest strong winds and gales are Mildenhall, Hurn and Heathrow, all with an annual frequency of less than two per cent. Stations having a more than 20 per cent annual frequency of calms and light airs are Exeter, Dyce, Mildenhall (which has over 30 per cent in July), Hurn and Turnhouse, while those having the lowest frequencies of these quiet conditions are Belmullet, Rosslare, Lerwick and Shannon.

Gusts

The data considered so far have been concerned with the 'mean wind'. However, as stated earlier, the natural wind is always made up of a series of gusts and lulls caused by turbulent eddies. It is the higher gusts of stormy weather which are responsible for a great deal of the damage which then occurs, blowing down trees, removing roof tiles and sometimes causing loss of life and structural damage to buildings, bridges, masts, etc. Thus the gustiness of the wind is a matter of considerable practical importance. There are a number of measures of gustiness. One of these is the ratio of the range of the oscillations of the wind velocity to the mean velocity, usually called the gustiness ratio. At open coastal sites such as Scilly it has a value of about 0·5. Over the open sea it is probably only about 0·3, while in the centre of a city it may approach 2·0. Thus a mean wind of 30 kt would be accompanied by gusts averaging only about 35 kt over the open sea, about 37 kt on the coast, about 45 kt at a place such as Kew Observatory where the gustiness ratio is about 1·0, and as much as 60 kt in a city centre. A force 8 wind (mean speed 34–40 kt) at a well-exposed coastal station would be accompanied by gusts averaging about 47 kt. As already noted, it is unusual for the mean wind speed to reach force 8 inland in the British Isles, but gusts of over 47 kt are by no means uncommon. Figures given by Meteorological Office (1968b) show that most inland anemograph stations recorded an average of over ten hours per annum with gusts greater than 47 kt over the ten-year period 1950–9. At the ten inland stations for which values were included the averages in hours were: Renfrew 32, Eskdalemuir 53, Cranwell 20, Mildenhall 16, Edgbaston 14, Kew 7, South Farnborough 6, Larkhill 25, Ringway 32 and Aldergrove 43. These figures are in most cases of the same order as those recorded at coastal stations, e.g. Leuchars 19, Prestwick 40, South Shields 27, Gorleston 22, Thorney Island 15, Southport 58 and Plymouth Hoe 27. Thus, so far as gusts are concerned, there is much less difference between inland and coastal stations than might have been expected from statistics based on mean wind speeds. Even at Kingsway in central London, where the mean wind speed is generally low because of the frictional resistance of the surrounding buildings, an average of eight hours per year with gusts exceeding 47 kt occurred during the period 1945–54.

Another measure of gustiness is the ratio of the maximum gust in a particular period to the mean wind speed over that period, often called the gust factor and denoted by $G(N,n)$ where N is usually either 10 or 60 minutes and n is a few seconds. For $N=60$ minutes and n about 3 seconds, the minimum duration which a gust must have to be fully recorded by a standard anemometer, G has values around 1·35 over the sea, 1·5 at coastal stations, 1·6 over more or less flat country with few obstructions, 1·8 in well-wooded country, small towns and open suburbs and about 2·1 in the centres of large towns and in cities.

Extreme wind speeds recorded by anemographs

Bilham (1938) included a list of occasions between 1909 and 1936 when a gust of 90 mph (78 kt) or more was recorded by at least one anemograph. There were forty-three such occasions in the 28 years considered and the list occupied a full page. Bearing in mind that another 37 years of records are now available and that there are now several times as many anemographs as there were when Bilham was writing, it is not practicable to bring his list up to date here. Instead, in Table 3.8 are presented the highest hourly mean speeds and the highest gust speeds so far recorded at each of forty-five selected anemograph stations having 20 or more years of record, while in Table 3.9 are given some exceptionally high wind speeds which have been recorded at a number of shorter period stations, including some at high levels. Station details, i.e. position, height of anemometer above msl and above ground, and effective height, are included in both these tables.

It will be seen that at low-level stations in the British Isles, into which category nearly all the stations in Table 3.8 and all but one of those in Table 3.9 may be considered to fall, the highest hourly mean speeds so far recorded are 69 kt at Dounreay in September 1969 and at St Ann's Head in November 1938 and 68 kt at Stornoway in January 1968 and at Jersey in October 1964. Other low-level stations which have recorded 60 kt or more include Tiree, with 67 in January 1968 and 65 in September 1961, Scilly with 66 in March 1922 and 60 in November 1954, Lerwick with 65 in January 1961 and 64 in January 1953, Bell Rock with 65 in December 1951 and 60 in January 1953, Lizard with 64 in November 1954, Southport with 61 in October 1927 and Edinburgh with 60 in December 1964. Much higher mean speeds have occurred over high ground, however, as is shown (Table 3.9) by the values of 86 kt (99 mph) recorded at Lowther Hill in January 1963 and at Great Dun Fell on 15 January 1968, when Lowther Hill recorded 76 kt.

All the speeds in Tables 3.8 and 3.9 are shown as recorded, i.e. they have not been reduced to the standard height of 10 m. After reduction to the standard height, using the 0·17 power law formula, it is found that the highest hourly mean speed so far recorded at a low level station in the British Isles is 70 kt at Dounreay in Caithness in September 1969.

Looking now at the highest recorded gust speeds entered in Tables 3.8 and 3.9, it is seen that the highest values in knots at low level stations are 118 at Kirkwall in February 1969, 102 at Tiree in January 1968, 99 at Malin Head in Eire, over 98 at St Ann's Head in January 1945, 98 at Stornoway in February 1962, over 97 at Dounreay in September 1969, 96 at Benbecula in December 1956, at Bell Rock in January 1968 and at Scilly in December 1929 and 95 at Dunfanaghy Road, Eire, in January 1927 and also at Lerwick in January 1961. There have been many others of 90 kt or more and it interesting to note that Bilham's 1938 table contained only five entries of 104 mph (90 kt) or more. The speeds of over 98 kt at St Ann's Head and over 97 kt at Dounreay, quoted above, indicate that the recording pen went off the chart, so that the actual peak gust speeds are not known. The highest gusts so far recorded at high-level stations in Britain are 124 kt at Cairngorm in March 1967, 116 kt at Great Dun Fell in January 1968 and 108 kt at Lowther Hill in the same month. Further details of some very high wind speeds recorded in Britain and of the gales in which they occurred have been given by Harris (1970).

Apart from the gust of more than 97 kt at Quilty, Co. Cork in January 1920 noted by Bilham as being of doubtful authenticity, the highest gust speed quoted by him was 111 mph (96 kt) at Scilly in December 1929, and his table included only ten entries of 100 mph (87 kt) or more. Also a map of extreme gust speeds reduced to standard height and based on data for the period 1925 to 1947, which was published in 1952 in the *Climatological Atlas of the British Isles* (Meteorological Office, 1952), showed that gusts of 100 mph or more had occurred at only eight stations, the highest being the 110 mph (95 kt) recorded at Dunfanaghy Road in 1927. Clearly, the use of the highest gust speeds actually recorded over these earlier periods as a basis for the design of structures in the British Isles would not have been very sound, particularly when it is borne in mind that the force exerted by the wind is proportional to the square of its speed. Moreover there is every reason to believe that further new records will continue to be established in the future. It was for these reasons that in the late 1950s Shellard (1958) applied the statistical theory of extreme values due to Gumbel (1954) to all the available annual extreme wind speeds, and so obtained estimates of the hourly mean

Table 3.8 *Details of forty-five selected long-period anemograph records and their highest recorded wind speeds (in kt)*

Station	Lat N	Long	Height of anemometer		Effective height	No. of years and last year of record	Highest gust speed	Month of occurrence (most recent)	Highest hourly mean speed	Month of occurrence (most recent)
			above msl	above ground						
	° ′	° ′	m.	m.	m.					
Lerwick	60 08	1 11W.	93	10	10	41 1971	95	Jan. 1961	65	Jan. 1961
Stornoway CG	58 12	6 22W.	37	12	11	33 1969	98	Feb. 1962	68	Jan. 1968
Dyce	57 12	2 12W.	72	10	10	20 1971	88	Jan. 1953	55	Jan. 1953
Balmakewan	56 48	2 33W.	43	8	6	21 1935	73	Dec. 1920	41	Dec. 1920
Bell Rock	56 26	2 24W.	39	40	38	41 1971	96	Jan. 1968	65	Dec. 1951
Leuchars	56 23	2 52W.	25	13	13	23 1971	92	Jan. 1968	56	Jan. 1968
Edinburgh	55 55	3 11W.	148	12	7	50 1966	88	Dec. 1964	60	Dec. 1964
Tiree	56 30	6 53W.	24	16	12	45 1971	102	Jan. 1968	67	Jan. 1968
Renfrew	55 52	4 24W.	22	13	11	20 1965	84	Jan. 1954	45	Dec. 1951
Prestwick	55 30	4 35W.	21	10	10	28 1971	90	Jan. 1968	51	Jan. 1958
Eskdalemuir	55 19	3 12W.	259	10	10	48 1971	87	Jan. 1968	49	Jan. 1930
Point of Ayre	54 25	4 22W.	20	10	10	35 1971	78	Dec. 1966	56	Sept. 1945
Durham	54 46	1 35W.	119	16	10	34 1971	90	Jan. 1968	47	Nov. 1970
South Shields	55 00	1 26W.	22	17	13	38 1971	76	Jan. 1945	56	Jan. 1945
Spurn Head	53 35	0 07E.	20	13	10	32 1958	79	Apr. 1943	51	Jan. 1928
Cranwell	53 02	0 31W.	68	13	13	42 1971	86	Dec. 1952	45	Apr. 1943
Gorleston	52 35	1 43E.	16	13	13	45 1971	71	Mar. 1947	48	Mar. 1948
Mildenhall	52 22	0 28E.	30	25	18	32 1969	85	Mar. 1947	54	Mar. 1947
Cardington	52 07	0 25W.	69	41	41	40 1971	81	Mar. 1947	55	Nov. 1957
Felixstowe	51 57	1 20E.	26	23	20	22 1952	78	Feb. 1938	44	Nov. 1952
Shoeburyness	51 33	0 50E.	36	32	28	46 1971	75	Jan. 1962	55	Jan. 1962
Edgbaston	52 29	1 56W.	196	36	22	48 1971	73	Nov. 1938	38	Nov. 1954

Table 3.8 (continued)

Station	Lat N	Long.	Height of anemometer above msl	above ground	Effective height	No. of years and last year of record	Highest gust speed	Month of occurrence (most recent)	Highest hourly mean speed	Month of occurrence (most recent)
	° ′	° ′	m.	m.	m.					
Kew	51 28	0 19W.	28	23	15	41 1971	63	Mar. 1947	34	Jan. 1962
Croydon	51 21	0 07W.	95	32	21	27 1958	71	Nov. 1928	44	Jan. 1930
Dover	51 07	1 19E.	21	12	19	36 1970	80	Nov. 1957	50	Dec. 1966
Abingdon	51 41	1 19W.	90	12	12	25 1971	68	Dec. 1956	36	Jan. 1962
Calshot	50 49	1 18W.	18	15	13	24 1952	76	Jan. 1938	45	June 1938
S. Farnborough	51 17	0 45W.	97	21	11	24 1968	69	Mar. 1947	43	Mar. 1947
Larkhill	51 11	1 48W.	145	13	10	40 1970	70	Dec. 1971	41	Jan. 1943
Sellafield	54 25	3 30W.	25	12	11	22 1971	76	Dec. 1952	45	Jan. 1965
Southport	53 39	2 59W.	19	14	11	57 1971	83	Oct. 1927	61	Oct. 1927
Speke	53 21	2 53W.	30	10	10	23 1971	82	Mar. 1966	49	Nov. 1960
Ringway	53 21	2 16W.	80	10	10	27 1971	79	Apr. 1943	49	Apr. 1943
Valley	53 15	4 32W.	26	16	12	20 1971	84	Jan. 1965	56	Dec. 1966
Aberporth	52 08	4 34W.	147	12	12	25 1969	81	Nov. 1954	53	Dec. 1966
Port Talbot	51 34	3 45W.	28	10	11	20 1971	72	Jan. 1962	46	Jan. 1962
Plymouth Hoe	50 22	4 08W.	58	27	20	41 1965	83	Mar. 1922	56	Dec. 1929
Scilly	49 56	6 18W.	70	20	17	45 1971	96	Dec. 1929	66	Mar. 1922
Lizard	49 57	5 12W.	96	23	18	34 1971	85	Nov. 1954	64	Nov. 1954
Aldergrove	54 39	6 13W.	80	10	10	43 1971	79	Nov. 1928	49	Sept. 1961
Belmullet	54 14	10 00W.	9	12	9	15 1970	94	Jan. 1957	—	—
Dublin Apt.	53 26	6 14W.	65	12	10	25 1970	73	—	—	—
Shannon	52 41	8 55W.	8	12	10	31 1970	93	Sept. 1961	—	—
Rosslare	52 18	6 24W.	24	12	9	14 1970	87	Oct. 1961	—	—
Valentia	51 56	10 15W.	17	12	10	39 1970	88	Sept. 1961	—	—

Table 3.9 *Details of some additional anemograph stations which have recorded some exceptionally high wind speeds (in kt)*

Station	Lat N.	Long.	Height of anemometer above msl	Height of anemometer above ground	Effective height	No. of years and last year of record	Highest gust speed	Month of occurrence (most recent)	Highest hourly wind speed	Month of occurrence (most recent)
	° ′	° ′	m	m	m					
Kirkwall	58 58	2 54W.	41	15	10	15 1971	118	Feb. 1969	59	Jan. 1961
Dounreay	58 35	3 45W.	34	12	9	13 1971	>97	Sept. 1969	69	Sept. 1969
Benbecula	57 28	7 22W.	16	10	10	15 1971	96	Dec. 1956	58	Dec. 1961
Cairngorm	57 08	3 39W.	1075	10	10	5 1971	124	Mar. 1967	66	Oct. 1971
Lowther Hill	55 23	3 45W.	736	13	10	8 1968	108	Jan. 1968	86	Jan. 1963
Gt Dun Fell	54 41	2 27W.	857	10	10	10 1971	116	Jan. 1968	86	Jan. 1968
St Ann's Head	51 41	5 11W.	65	21	65	14 1949	>98	Jan. 1945	69	Nov. 1938
Jersey	49 12	2 11W.	98	18	12	14 1971	94	Oct. 1964	68	Oct. 1964
Malin Head	55 23	7 24W.	24	—	18	15 1970	99	Sept. 1961	—	—
Dunfanaghy Rd	55 11	7 58W.	55	14	9		95	Jan. 1927	—	—

and gust speeds having average recurrence periods, sometimes called return periods, of 10, 20, 50 and 100 years at each station for which a sufficiently long record was available. He also mapped the speeds having an average recurrence period of 50 years, i.e. a probability of 0·02 of being exceeded in any one year. These estimates of probable extreme wind speeds over the United Kingdom were updated by Shellard (1962, 1965a), and more recently by Hardman *et al.* (1973)

Fig. 3.6 *The 'once in 50 years' gust speed (ms^{-1}) at 10 m above the ground over open level country*

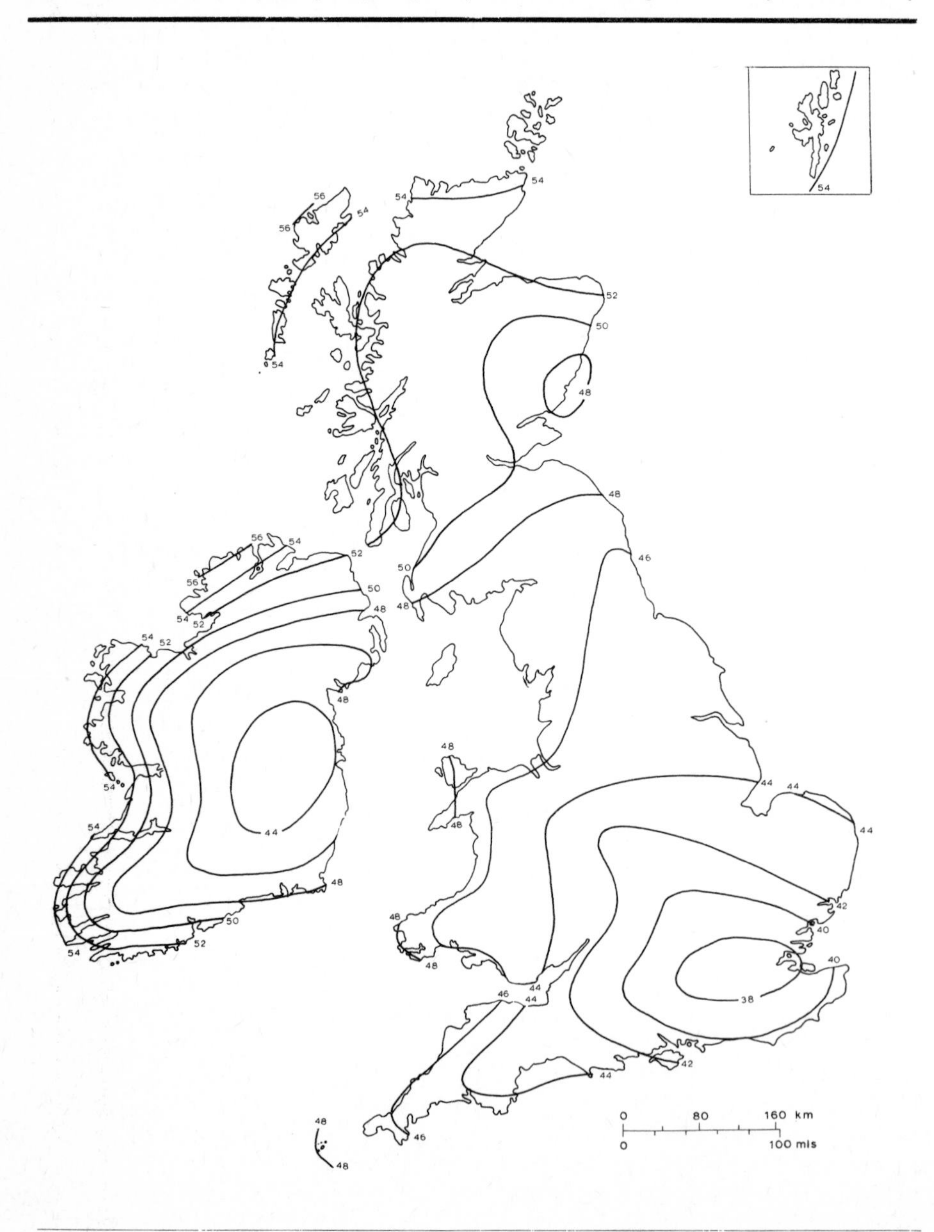

using all available anemograph data up to and including those for 1971. The maps of once-in-50-year gust speeds presented in these papers have been adopted by the British Standards Institution (1972) as its recommended source of basic design wind speeds for the United Kingdom in its code of practice dealing with the wind loading of structures. Similar statistical analyses of data from thirteen anemograph stations in Eire have been carried out by Logue (1971) who used data up to and including those for 1970. Table 3.10 presents, for the stations whose details are given in Table 3.8, the results of these statistical analyses after reducing all the recorded speeds to the standard height of 10 m above the ground, using the appropriate power law formula. Table 3.10 gives both estimated hourly mean speeds and estimated gust speeds in knots for average recurrence periods of 10, 20, 50 and 100 years, and also average annual maximum speeds, i.e. the speeds that on average are exceeded one year in two. The data were taken mainly from Hardman *et al.* (1973) or from Logue (1971).

It must be emphasized that the estimated extreme speeds for individual stations should be used with caution, however, because the exposure of the anemometers is rarely perfect and because due allowance has to be made for the fact that the 50- and 100-year values have in many cases been computed from only 20 to 30 years of record. Comparisons of values for neighbouring stations sometimes show inconsistencies due to the effects of different exposures or to different periods of record. Such inconsistencies can be substantially reduced by mapping the estimates and, in so doing, giving less weight to the shorter period and more suspect stations than to the longer period, more reliable and better exposed ones. Thus the best way of presenting the data for practical use is in the form of maps. As an example, Fig. 3.6 shows the distribution of the once-in-50-year gust speed in metres per second (1 ms^{-1} = 1·9426 kt) from Hardman *et al.* (1973) and extended over Eire following the corresponding map prepared for that country by Logue (1971). It will be noted that the map refers to 'open level country'. This may be taken to imply a more or less open and level area some 200 to 300 m in radius situated at an altitude approximately equal to that of the general surrounding terrain. In practice a particular site may depart considerably from this description, however, and if it does then certain corrections may have to be applied to the wind speed taken from the map, mainly to allow for topographical effects. Further details may be found in Shellard (1965b) and in Hardman *et al.* (1973).

Figure 3.6 indicates that maximum gust speeds ranging from about 37 ms^{-1} (72 kt) in the London area to about 56 ms^{-1} (109 kt) at places along the coasts of north-western Ireland and the extreme north-west of Scotland can be expected to be exceeded on the average only once in 50 years at the standard height of 10 m. These compare with individual station values, taken from Table 3.10, which range from 66 kt at Kew, reflecting its rather sheltered situation, to 107 kt at Stornoway and also at Belmullet. The range of once in 50-year hourly mean speeds at individual stations is from 32 kt at Kew to 71 kt at Stornoway and 72 kt at Lerwick.

Table 3.10 *Estimated extreme wind speeds in knots at 10 m above the ground at forty-five long-period anemograph stations*

Station	Maximum gust speeds					Maximum hourly mean speeds				
	Average annual maximum	Speeds likely to be exceeded only once in stated number of years				Average annual maximum	Speeds likely to be exceeded only once in stated number of years			
		10	20	50	100		10	20	50	100
Lerwick	76·5	91	96	104	110	50·5	62	66	72	77
Stornoway CG	78·1	93	99	107	113	52·1	62	66	71	75
Dyce	62·3	77	83	92	98	37·8	46	49	54	57
Balmakewan	54·5	66	71	77	82	31·9	39	42	46	49
Bell Rock	65·9	77	82	88	92	42·1	48	51	54	57
Leuchars	62·2	76	84	91	97	39·5	47	49	52	56
Edinburgh	76·0	86	90	95	99	44·1	51	54	58	60
Tiree	72·3	88	95	104	110	46·4	56	61	66	70
Renfrew*	65·8	81	87	95	101	36·6	45	49	53	57
Prestwick	63·5	78	84	91	97	37·9	45	48	52	55
Eskdalemuir	65·2	75	80	86	90	39·2	46	48	52	54
Point of Ayre	66·6	76	80	85	89	43·1	50	53	57	60
Durham	68·4	80	85	92	97	36·7	43	46	50	53
South Shields	59·5	70	75	81	85	37·4	45	48	52	55
Spurn Head	63·5	74	78	84	88	44·3	48	50	53	55
Cranwell	56·1	66	71	76	80	33·3	39	42	45	48
Gorleston	57·2	66	69	74	78	38·1	43	46	49	51
Mildenhall	60·0	72	77	84	89	31·8	39	42	46	49
Cardington	55·1	66	70	76	80	31·9	39	41	45	48
Felixstowe	57·7	69	75	81	85	33·9	39	42	44	46
Shoeburyness	56·3	65	69	74	78	35·4	42	44	48	50
Edgbaston	53·1	63	67	73	77	26·2	31	33	36	38
Kew Observatory	52·9	60	63	66	69	25·2	29	30	32	33
Croydon	55·5	64	68	73	76	30·1	36	37	40	43
Dover	57·2	68	73	79	84	33·6	39	41	44	46
Abingdon	50·6	61	65	70	74	28·1	33	35	37	39
Calshot	58·4	69	74	80	84	37·3	44	46	50	52
S. Farnborough	57·2	65	69	73	77	31·8	39	42	46	49
Larkhill	61·0	69	72	76	79	33·8	39	41	43	45
Sellafield	60·1	71	76	82	87	37·4	44	47	50	53
Southport	65·0	76	80	86	90	43·3	50	53	57	60
Speke	63·5	75	80	87	92	38·0	43	46	49	51
Ringway	59·8	70	76	84	89	35·7	43	47	51	54
Valley	68·9	80	85	91	95	45·8	54	57	61	64
Aberporth	64·9	76	80	86	91	41·1	49	52	56	59
Port Talbot	61·9	71	75	80	84	37·6	45	48	53	56
Plymouth Hoe	57·9	68	72	78	82	39·5	46	49	52	55
Scilly	71·8	82	87	93	97	45·1	53	56	61	64
Lizard	71·3	80	84	89	93	46·5	53	56	59	62
Aldergrove	61·6	72	77	82	87	35·1	42	45	49	51
Belmullet	—	93	99	107	115	—	—	—	—	—
Dublin Apt	—	74	78	82	86	—	—	—	—	—
Shannon	—	84	87	95	99	—	—	—	—	—
Rosslare	—	82	86	93	99	—	—	—	—	—
Valentia	—	82	86	91	95	—	—	—	—	—

* Includes data from Abbotsinch (55° 52′N., 4° 26′W.) for 1967–71.

Gales

Average numbers of days with gale at nineteen selected stations are presented in Table 3.11. A day with gale is defined as one on which the mean wind speed at the standard height of 10 m attains a value of 34 kt or more over any period of 10 consecutive minutes during the day. Values are given for the months of January, which usually has the most gales; July, which usually has the least; and for the year as a whole. As far as possible, the averages refer to the 30-year period 1941 to 1970. The annual averages vary from 0·1 at Kew to 51·8 at Lerwick. Other stations with frequent gales include Stornoway Airport with 32·5, Tiree with 29·1 and Ronaldsway with 26·3 days per year.

Table 3.11 *Average number of days with gale at nineteen stations*

Station	Average number of days with gale			Period of data	Highest in a year	Years of occurrence
	Jan.	July	Year		(days)	
Lerwick	9·1	0·5	51·8	1941–70	86	1949
Wick	3·8	0·1	17·9	1941–70	38	1949
Stornoway Apt	6·2	0·3	32·5	1941–70	74	1943
Dyce	1·3	0+	6·8	1943,4,6–70	17	1950
Turnhouse	2·0	0·1	10·1	1951–70	23	1954
Tiree	5·5	0·2	29·1	1941–70	55	1962
Prestwick	1·0	0·0	5·5	1941–70	21	1944
Ronaldsway	4·6	0·3	26·3	1941–70	55	1963
Tynemouth	2·5	0·2	15·2	1941–70	34	1941
Manby	0·9	0·0	2·8	1952–70	11	1954
Mildenhall	no data		1·3	1941–68	4	1947
Elmdon	0·5	0·0	3·0	1951–70	7	1952
Kew Obsy	0+	0·0	0·1	1941–70	1	1957,8,62
Dungeness	2·7	0·2	17·5	1961–70	27	1961
Ringway	0·5	0·1	3·3	1941–70	11	1954
Aberporth	2·7	0·1	13·2	1941–70	30	1967
Plymouth Hoe	1·7	0·2	12·9	1941–70	31	1962
Scilly	3·4	0·3	19·9	1941–70	53	1962
Aldergrove	0·9	0·0	3·4	1941–70	10	1954,63

On a January day the odds against a gale range from over 900 to 1 at Kew to only a little over 3 to 1 at Lerwick. In July over the periods of these averages no gales whatever have been registered at Prestwick, Manby, Elmdon, Kew or Aldergrove, while the highest July frequencies are 0·5 at Lerwick and 0·3 at Stornoway, Ronaldsway and Scilly. In the last two columns of Table 3.11 are given the highest numbers of days with gale in any year during the period under review together with the year in which they occurred. The variation from year to year is quite considerable. At Lerwick, for example, there were no fewer than 86 days with gale in 1949, while Stornoway had 74 in 1943, Tiree 55 in 1962 and Scilly 53 in the same year. In most cases these maximum numbers are more than twice the average numbers of days with gale and in a few cases they are well over three times the average, e.g. at Prestwick, Manby, Mildenhall, Ringway and Kew.

Generally speaking, gales occur on from 20 to 40 days per annum on our northern and western seaboard, while on our eastern coasts and in the English Channel they have a frequency of about 10 to 20 days per year. Inland, gales are much less frequent, usually occurring on only a few days per annum.

Diurnal variations

Variations in the frequency of winds from different directions at various hours of the day are generally small at inland stations, but at coastal stations they are often quite marked, particularly in quiet weather in the warmer months of the year. These variations are usually referred to as land and sea breezes and are brought about by the unequal radiational heating and/or cooling of the adjacent land and sea surfaces. On a sunny day the land becomes warmer than the sea and a sea breeze blows onshore; on a clear night the land becomes cooler than the sea and a land breeze blows in the reverse direction. The sea breeze is the more pronounced phenomenon, reaching its maximum strength and extending furthest inland during the afternoon. The land breeze may set in around midnight or in the early hours. Both are considerably influenced by the local topography and may vary in direction and strength from one part of the coast to another. In the British Isles a sea breeze rarely exceeds force 3 and does not often extend inland more than about 40 km. When well developed the line of greatest penetration of the sea breeze overland is marked by a line of convective activity which is known as a seabreeze front. This theme is discussed and illustrated more fully in Chapter 11.

Another local wind which is of a diurnal nature is the katabatic or downslope wind. On a quiet, cloudless night the air nearest to sloping ground is cooled rapidly and so becomes denser than air at the same level some distance away. It therefore begins to flow downhill, its direction depending on the local topography. The resulting gravitational flow is known as a katabatic wind. Such winds usually attain a speed of only 3 kt at most but when they flow into an open valley they may feed a down-valley wind whose speed may reach 5 kt or more. The night-time climate of a valley depends very much on the degree to which the flow along it is obstructed. If the topography, whether natural or man-made, is such that the downslope flow of air is blocked then very low night-minimum temperatures may be experienced. Katabatic winds and their associated 'frost-hollows' are thus of considerable importance to agriculture. The local wind which sometimes blows up a slope heated by the sun is the converse of the katabatic wind and is known as an anabatic wind but it is decidedly less common and is of much less practical importance. Again, further discussion of this theme follows later in Chapter 13.

Although the diurnal variation of wind direction at inland stations

Fig. 3.7 *Diurnal variation of wind speed at Kew in winter (December to February) and summer (June to August) for 1959–68:* (a) *for all days;* (b) *for the twelve sunniest days from each month over the 10-year period;* (c) *for the twelve least sunny days from each month over the 10-year period*

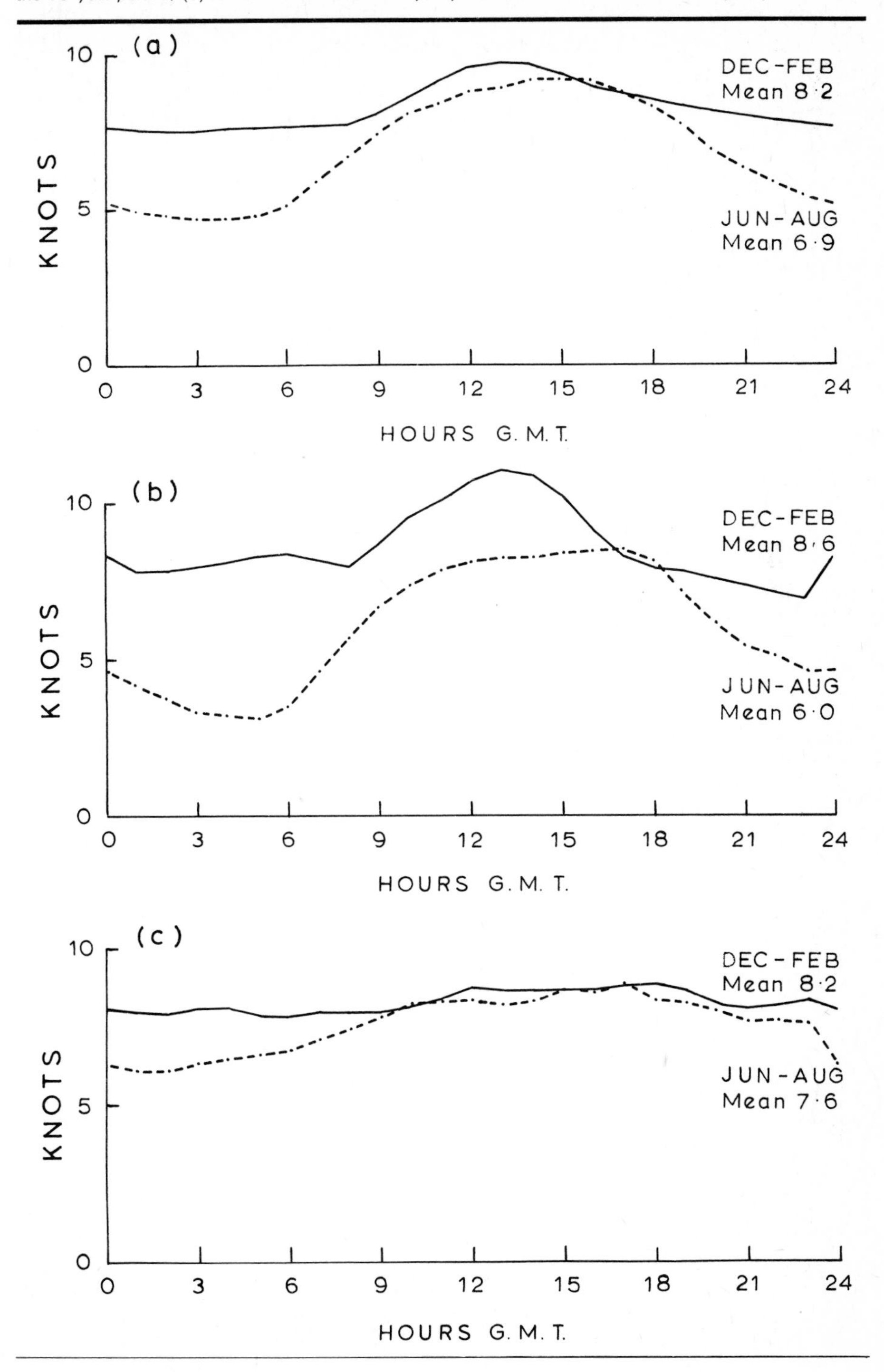

is generally small, that of wind speed is usually well marked with a maximum in the early afternoon and a minimum shortly before sunrise, the actual times and also the amplitude of the average variation depending on the time of year. Average curves for Kew Observatory in summer (June to August inclusive) and in winter (December to February inclusive) are shown in Fig. 3.7*a*, based on data for the 10-year period 1959–68. It will be noted that the mean speed is greater in the winter months (8·2 kt) than in the summer ones (6·9 kt) while the amplitude in summer (4·5 kt) is more than twice that in winter (2·2 kt). It is also of interest that although the maximum speed is reached some two to three hours earlier in the winter than in the summer, the highest values reached by the two curves do not differ by very much, at least at Kew Observatory.

Fig. 3.8 *Diurnal variation of wind speed at Ocean Weather Station 'J' (52½°N. 20°W.) in winter, summer and the whole year, for 1948–51 and 1953–6*

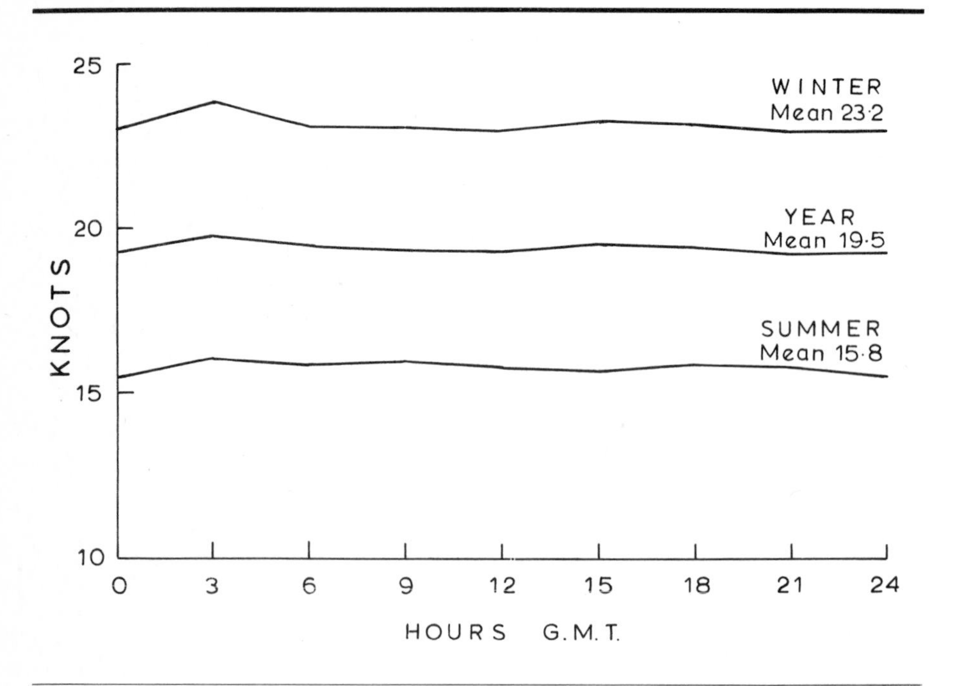

These diurnal variations in wind speed are a direct result of the increased mixing in the atmospheric boundary layer brought about by solar radiation. The resulting exchange of momentum, which speeds up the flow of air near the ground and reduces that of the flow at higher levels, reaches its greatest development soon after midday. Hence the average wind speed near the ground attains its maximum value at that time, as also does the air temperature. To illustrate more clearly this relationship between the diurnal variation of wind speed and solar radiation, average winter and summer curves are presented in Fig. 3.7*b* for sunny days and in Fig. 3.7*c* for dull days at Kew. The twelve sunniest days in each of the three winter and in each of the three summer months were selected from the 10 years' data to construct Fig. 3.7*b*, while the curves in Fig. 3.7*c* were similarly prepared using data for the 12 days from each summer and winter month having the lowest sunshine totals.

It is clear that the amplitudes of the diurnal variations are greatest on sunny days, being about 4·1 kt in winter and 5·4 kt in summer, compared with only about 1·1 kt in winter and 2·7 kt in summer on dull days. It may also be noted that the mean speed on dull days at Kew is not much less in summer (7·6 kt) than it is in winter (8·2 kt) but that on sunny days the mean speed in summer (6·0 kt) is appreciably less than it is in winter (8·6 kt).

Finally, since there is very little diurnal variation of temperature over the sea it is to be expected that the diurnal variation of wind speed there will also be small. In Fig. 3.8 are shown average curves of mean wind speed in knots at the Ocean Weather Station J (position 52½°N, 20°W) for summer, winter and the year based on three-hourly observations over the eight years 1948–51 and 1953–6. They are plotted on the same scale as those in Fig. 3.7. The amplitudes of all three curves are less than 1 kt and the variations shown are probably largely accidental, especially bearing in mind the very much higher speeds that are experienced over the North Atlantic than over the British Isles (see Table 3.6).

Solar radiation is the only significant external source of energy available to the earth and plays a vital role in determining our weather. The systematic variation with latitude in the amount of solar radiation which is absorbed at the surface provides the main driving force behind the general circulation, but solar radiation is also important on much smaller scales; for instance, it often controls the dissipation of fog and low cloud.

Terminology and units

Solar radiation is the electromagnetic radiation emitted by the sun and intercepted by the earth. The span of wavelengths is extremely wide, from the extremely short wavelength γ rays (10^{-10}m) to radio waves whose wave length is measured in tens of metres. However, as far as the radiation reaching the surface of the earth is concerned, 98 per cent of the total energy received from the sun is contained in wavelengths in the range 0·3 micrometres to 4 micrometres (1 micrometre, or μm, is 10^{-6} m or 10^{-3} mm).

Outside the atmosphere solar radiation travels essentially in straight lines and is thus only received from the direction of the sun itself, which is a circular source subtending an angle of 32 minutes of arc at the earth. Once the radiation enters the atmosphere however it encounters numerous 'particles', and a proportion is scattered into different directions. Subsequently, a further proportion of the scattered radiation is itself scattered and so on. A considerable part of the solar radiation intercepted by the earth and atmosphere is reflected back to space; this fraction is known as the planetary albedo and measurements from satellites have shown that the annual average over the whole globe is about 30 per cent. At the surface of the earth therefore, incoming solar radiation is no longer essentially a parallel beam but has contributions from all directions of the hemisphere. Solar radiation striking the earth's surface is in turn partially absorbed and partially reflected back.

There are three main scattering agencies in the atmosphere. Firstly, there are the molecules of the atmospheric gases (nitrogen, oxygen, water vapour, etc.); the scattering from these is strongly wavelength dependent (shorter wavelengths are scattered more than longer wavelengths) and this accounts for the blue colour of the sky. Secondly, there are natural and artificial aerosol particles which are much larger than molecules; scattering from these is much less wavelength dependent and these particles are responsible for haze. Thirdly, there are cloud particles (water drops or ice crystals) which are much larger in size again and the scattering from these is practically independent of wavelength.

The rate at which radiation energy falls on unit area of a plane surface is known as the 'irradiance'. It will be seen easily that in general this irradiance at any point in the atmosphere will vary with the orientation of the receiving surface and this orientation must therefore always be specified or clearly implied. Global solar radiation (often symbolized by G) is defined as the solar irradiance on a plane surface coming from the whole hemisphere. If not otherwise specified the plane surface is understood to be horizontal. Global solar radiation will in general be made up of two main components; firstly a component coming directly from the sun and secondly the diffuse component from the remainder of the hemisphere. If, however, there is significant cloud between the observation point and the sun then the direct component is effectively zero and there is only a diffuse component. The diffuse component (D) of the global solar radiation is often measured separately.

Direct solar radiation (I), on the other hand, is the solar irradiance on a surface perpendicular to the direction of the sun and contained in a narrow solid angle centred on the sun's direction. It can be seen that G, D and I are related by the equation

$$G - D = I \sin h$$

where h is the solar elevation.

It is necessary to distinguish carefully between the irradiances discussed above, which are the rates at which energy falls on a unit area, and what is often called the corresponding 'irradiation'. This latter is the time integral of the irradiance or the amount of energy falling on unit area in some stated time interval (e.g. hour, day, month, etc.).

The SI unit for the irradiance is 'watts per square metre' (Wm^{-2}) and this is a convenient size for general use; the 'solar constant', which is the mean value of the direct solar radiation outside the earth's atmosphere, is about 1 353 $W\ m^{-2}$. The unit 'calorie per minute' (cal min^{-1}) has often been used in the past in place of the watt but it is not now recommended. For reference 1 cal min^{-1} = 0·0698 watts; 1 cal cm^{-2} min^{-1} = 698 Wm^{-2}). The SI unit for amount of energy is the Joule (an energy rate of 1 watt lasting for 1 second gives rise to 1 Joule) and the corresponding unit of irradiation is Joules per square metre (Jm^{-2}). In practice, this unit is very small and a more convenient unit is megajoules per square metre ($MJ\ m^{-2}$).

This unit is not, however, at present in widespread use and the conversion factors between it and more common units are set down here for reference.

$$1\ MJ\ m^{-2} = 278\ Wm^{-2} = 23{\cdot}9\ cal\ cm^{-2}$$

$$1\ Wm^{-2} = 0{\cdot}0036\ MJ\ m^{-2} = 0{\cdot}0860\ cal\ cm^{-2}$$

$$1\ cal\ cm^{-2} = 0{\cdot}0419\ MJ\ m^{-2} = 11{\cdot}6\ Wm^{-2}.$$

It can also be noted that the unit calorie per square centimetre is, in American usage, often called a Langley (Ly).

Radiation is usually measured with a thermopile instrument in which the equilibrium temperature difference between a black receiving surface, and a reference surface, which is either white coloured or shielded in some way from the radiation to be measured, is converted into an electrical output and measured. This temperature difference is directly related to the radiation intensity. In instruments for measuring global solar radiation the receiving surface is usually covered by one or two clear glass hemispheres; good quality clear glass is a very convenient material for this purpose since it is highly transparent to radiation of wavelengths between about 0·33 and 2·8 μm which covers most of the solar radiation wavelength range. The small amounts of radiation outside this wavelength range are in fact included in the measurement because the calibration constants are determined by comparison with instruments measuring direct solar radiation without any filter.

Solar radiation

Solar radiation outside the Earth's atmosphere

The global solar radiation on a horizontal surface outside the earth's atmosphere (I_0) varies systematically on two distinct time scales. Because of the rotation of the earth, the elevation of the sun has a characteristic diurnal variation and this in turn means that G increases from zero during the night to a maximum at local apparent noon, when the sun is due south, and falls to zero again at sunset. The maximum intensity is related directly to the solar elevation h_0 at midday by the equation

$$G = I_0 \sin h_0$$

However, because of the motion of the earth in its orbit around the sun, h_0 varies systematically from a minimum in mid-winter to a maximum in mid-summer. If δ is the sun's declination then

$$h_0 = (90 - L) + \delta$$

where L is the latitude. δ varies from $+23°27'$ to $-23°27'$. There is also a small systematic variation in I_0 because the earth's orbit is slightly elliptical. I_0 varies by $\pm 3{\cdot}3$ per cent from its mean value, with a minimum in early July and a maximum in early January.

Figure 4.1 is a diagram showing the variation in solar azimuth and elevation for a selection of dates throughout the year for a latitude of 51 °N. The total amount of radiation received on a horizontal surface in a day outside the atmosphere can easily be calculated and Fig. 4.2 shows this on a world wide basis.

Solar radiation at the surface

The attenuation and scattering of solar radiation caused by the atmosphere is another source of variability which is superimposed on the diurnal and annual variations caused by the rotation of the earth about its axis and its movement in its orbit around the sun. By far the most important factor is the amount and type of cloud but the effect of atmospheric aerosol can also be clearly detected.

Diurnal variations of global and diffuse solar radiation Figure 4.3*a* shows the diurnal variation of global solar radiation at Kew Observatory for March based on 10 years' data (1959–68) as well as the corresponding diurnal variation of diffuse solar radiation for the same selection of days. Curve A can be taken as representative of clear days and curve C as representative of heavily overcast days. Figures 4.3*b* to 4.3*d* give similar data for June, September and December. It will be seen that, except in December, the average diffuse radiation on clear days is, in the middle of the day, less than the average for all days and that on the heavily overcast days the global and diffuse radiation are practically equal.

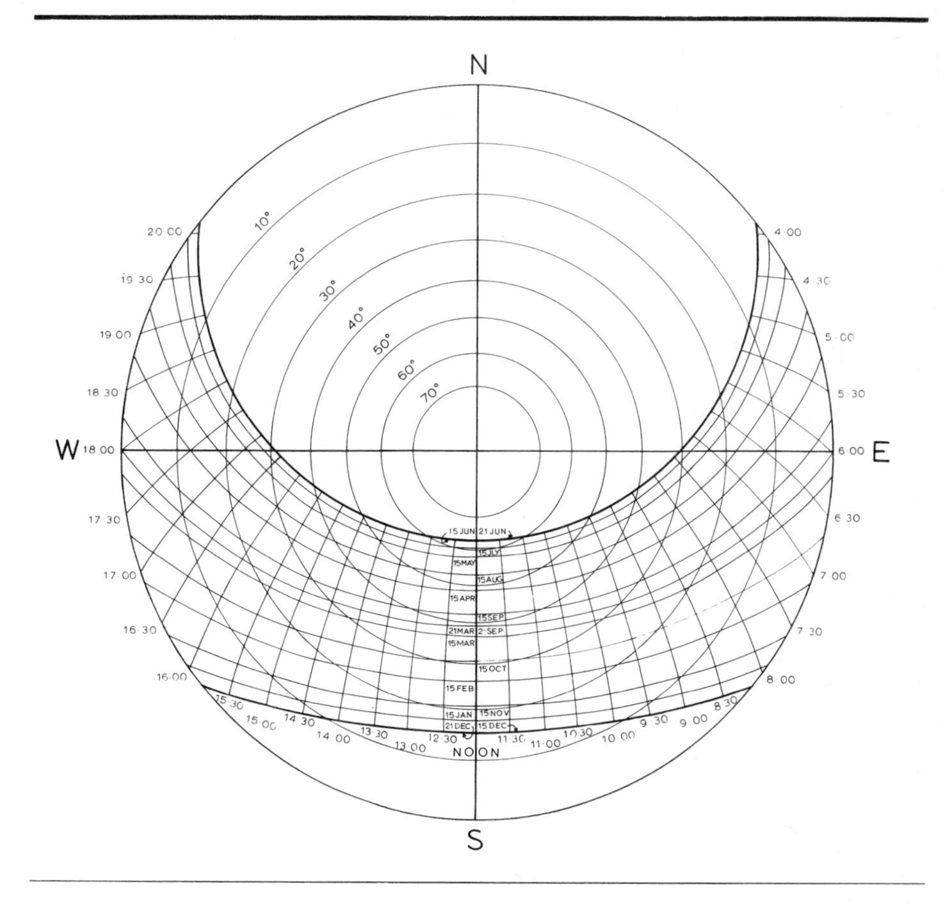

Fig. 4.1 *Stereographic sunpaths for 51° N. (after Petherbridge, 1969)*

On any given day, of course, the actual solar radiation intensities may fluctuate wildly, according to the cloud conditions. This variability, which may mean the occurrence of opposite extreme values on the same day within perhaps a period of an hour or so, makes the summary of radiation measurements difficult.

It should be noticed that the observations in Fig. 4.3 are mean hourly values. The radiation also varies however on a much shorter time scale and, in particular, on days with broken clouds (especially with cumulus type clouds) much higher values than those in the curves marked A can be experienced for periods of a few minutes at a

Fig. 4.2 *Daily totals of global solar radiation outside the atmosphere (MJ m^{-2})*

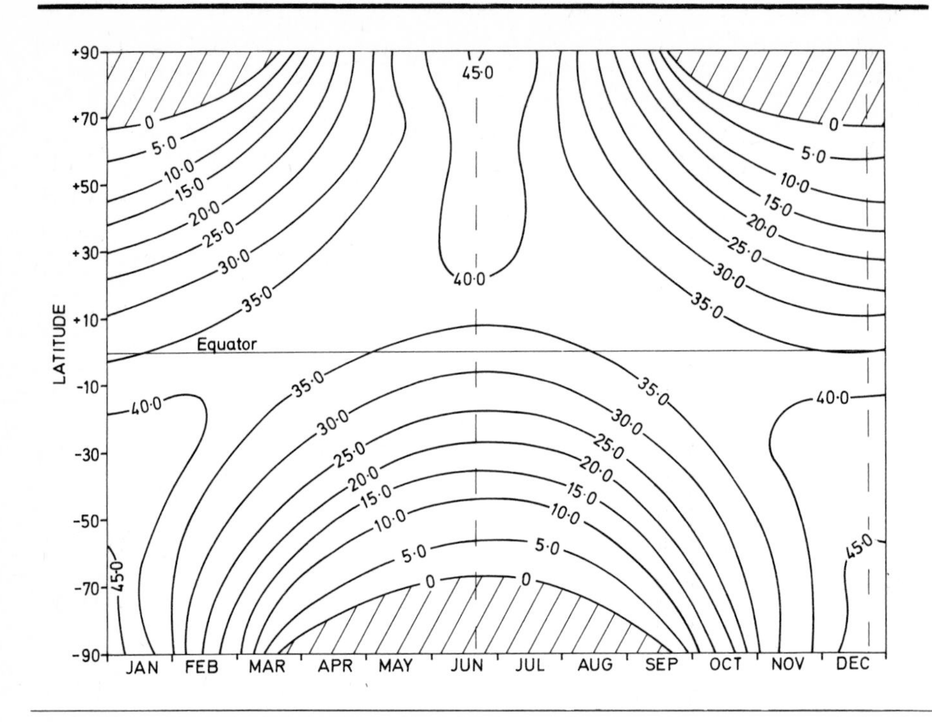

time. That this is possible can be readily understood; on days with broken cloud the direct component on the horizontal surface can reach the values appropriate to the clear days when the sun shines through a patch of blue sky. However, the diffuse component is still at the high value, appropriate to curve B or even higher, and the global solar radiation (equal to the sum of the diffuse and direct components) is thus much higher than average. In extreme cases, the global solar radiation can exceed the solar constant. On the other hand, of course, when the sun is obscured then the global solar radiation falls to a very low value so that the mean hourly value is not excessive.

Annual variation of global and diffuse radiation The description of the annual variation of solar radiation is best carried out by considering daily totals. In Fig. 4.4*a* is plotted the average values for each calendar month of the daily totals of global radiation at four stations with long periods of measurements: Kew, Eskdalemuir, Lerwick and Valentia. The standard deviation of a single observation of a monthly mean value in the United Kingdom varies from about 18 per cent of the mean value in January–February to about 10 per cent in mid-summer.

It will be seen that the variations from summer to winter are large and the variation from north to south across the United Kingdom, when expressed as a fraction, is much less in summer than it is in winter. Both these factors would be expected from the corresponding variations outside the atmosphere. The differences between Valentia and Kew are mainly due to the greater aerosol content in the atmosphere above Kew

The annual variation of diffuse radiation at the three Meteorological Office stations can be deduced from Fig. 4.4*b* which gives the annual variation of the diffuse component expressed as a fraction of the corresponding global solar radiation. It will be seen that the diffuse component is an exceedingly important part of the average radiation income of the British Isles.

Fig. 4.3 *Diurnal variation of global and diffuse solar radiation at Kew.* (a) *March;* (b) *June;* (c) *September;* (d) *December. In each graph: A—the average of the 2 per cent of days with the highest total of global radiation B—the average for all days; C—the average of the 2 per cent of days with the lowest total of global radiation*

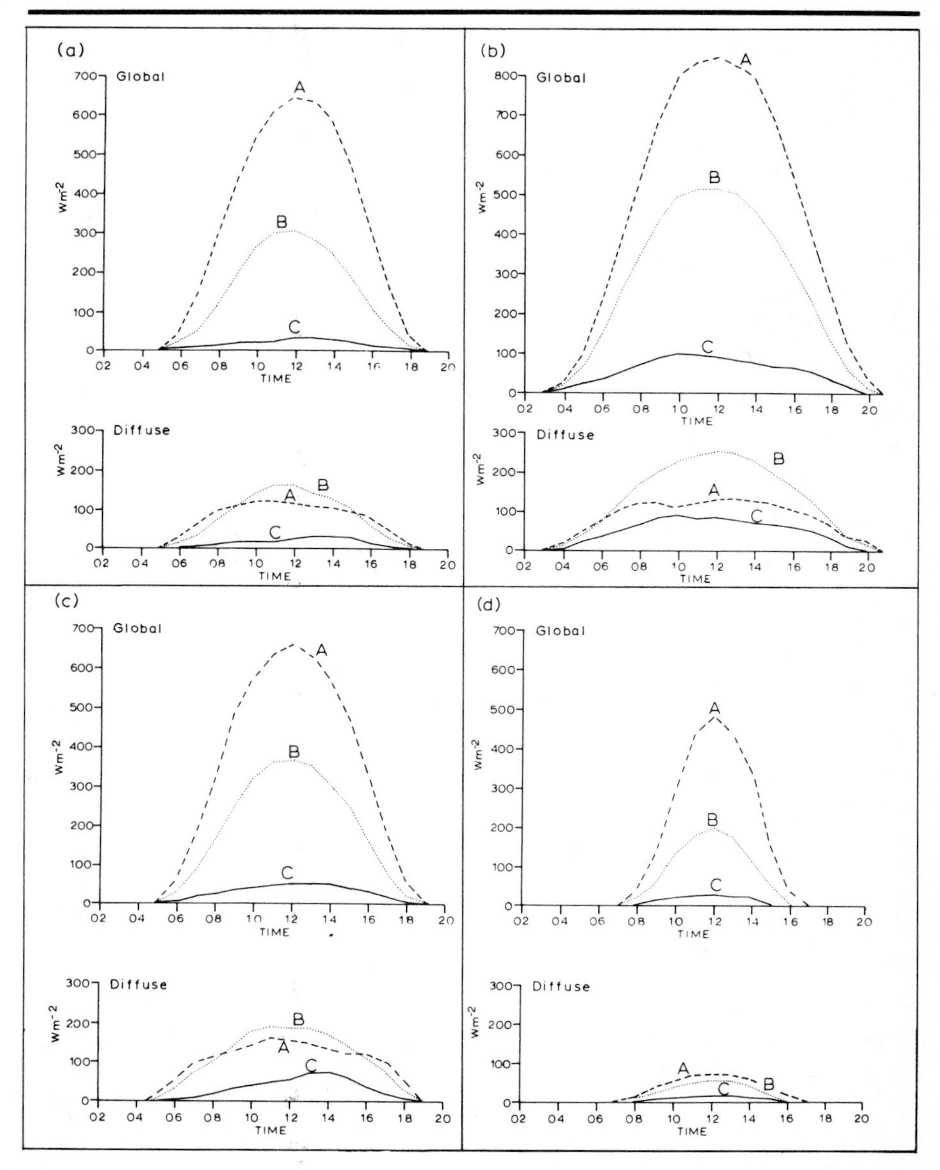

Fig. 4.4 (a) *Annual variation of mean daily totals of global solar radiation;* (b) *Ratio of mean daily diffuse solar radiation to mean daily global solar radiation*

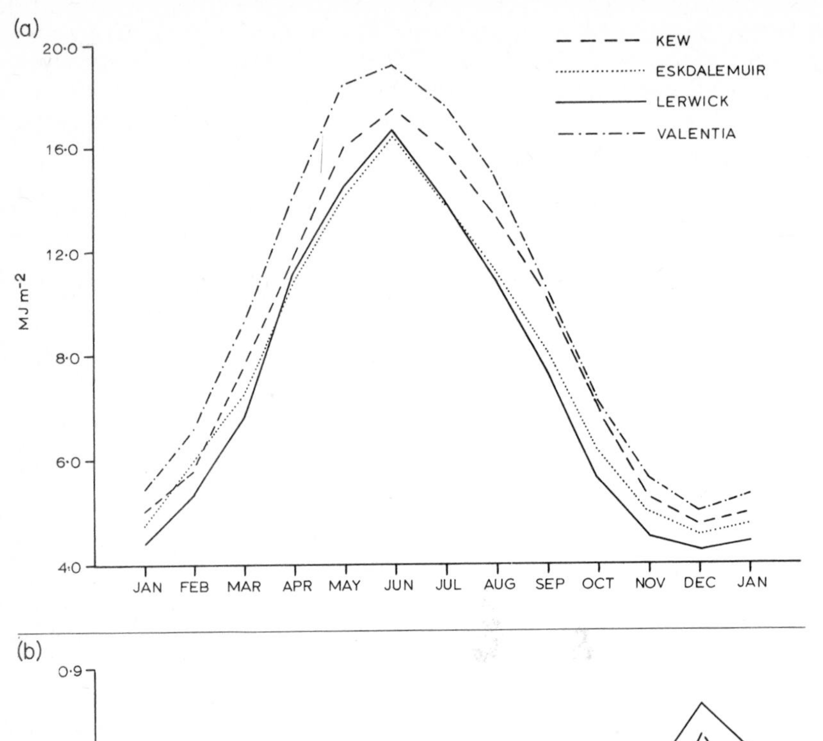

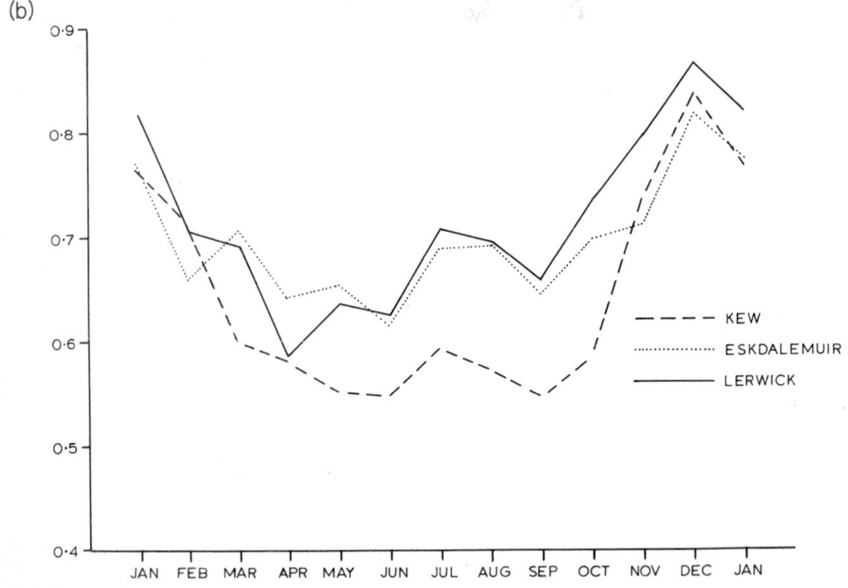

Geographical variation of global radiation over the British Isles Solar radiation has not yet been measured at sufficient stations and for a long enough period to enable a description to be given in the detail that would be desired, but using such data as are now available, maps

(Figs 4.5 and 4.6) have been drawn showing the major features in the distribution of the average daily total of global solar radiation over the British Isles in the months March, June, September and December, and for the whole year. These maps are based mainly on the results from stations with more than 10 years of data; the measurements from other stations have been used with caution.

Fig. 4.5 *Average daily totals of global solar radiation (MJm⁻²).* (a) *March;* (b) *June;* (c) *September;* (d) *December*

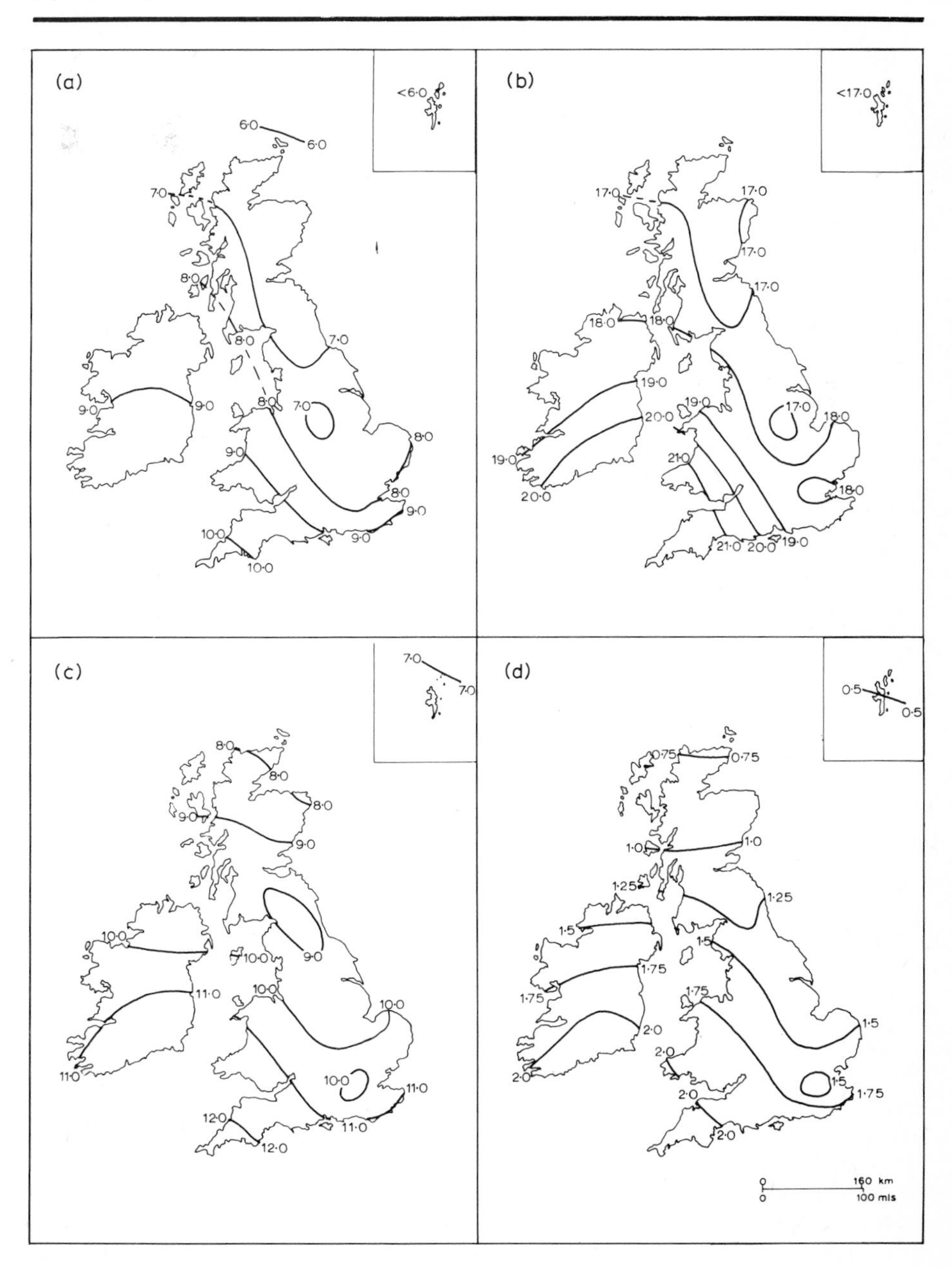

Fig. 4.6 *Average daily total of global solar radiation ($MJ\ m^{-2}$) for the year*

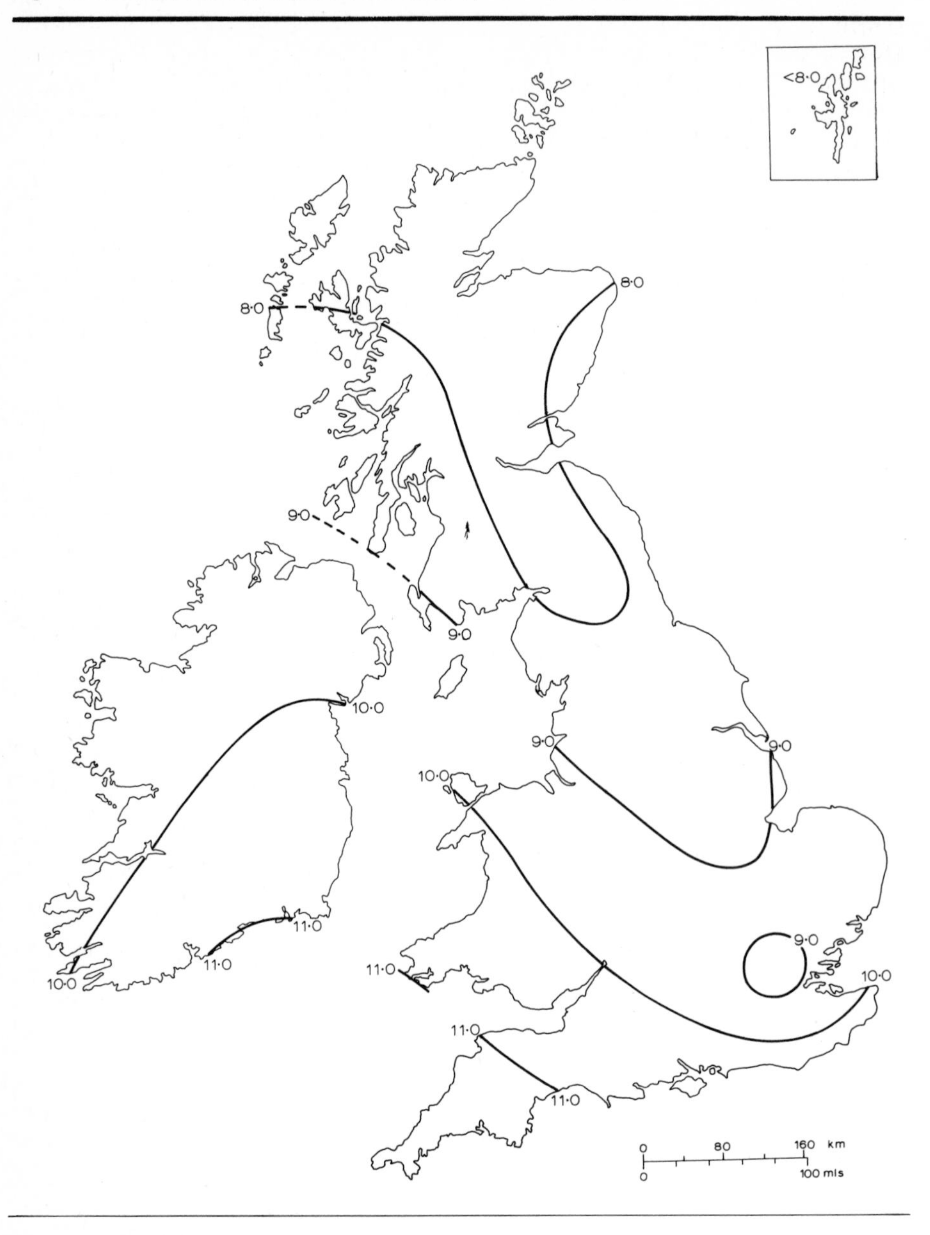

It will be seen that, in fact, the patterns of the isopleths on each of the maps have a number of features in common. Firstly, the amounts decrease from south to north but, for any given latitude, the amounts are higher in the west (particularly in areas bordering the Irish Sea) than they are in the east where there are areas of below average radiation around London and in the Midlands, Yorkshire and southern lowlands of Scotland.

It must be emphasized that these maps are based on radiation measurements made at stations established in general for other purposes, and the stations are thus usually at a low level. While the maps are considered to show reasonably correctly the main features of the distribution of solar radiation over the British Isles there are undoubtedly local systematic deviations from the simple patterns shown, particularly over high ground. Data are also sparse for the south-west of the United Kingdom.

Fig. 4.7 *Cumulative frequency distribution of daily totals of global solar radiation as a ratio of the median value.* (a) *Kew;* (b) *Eskdalemuir*

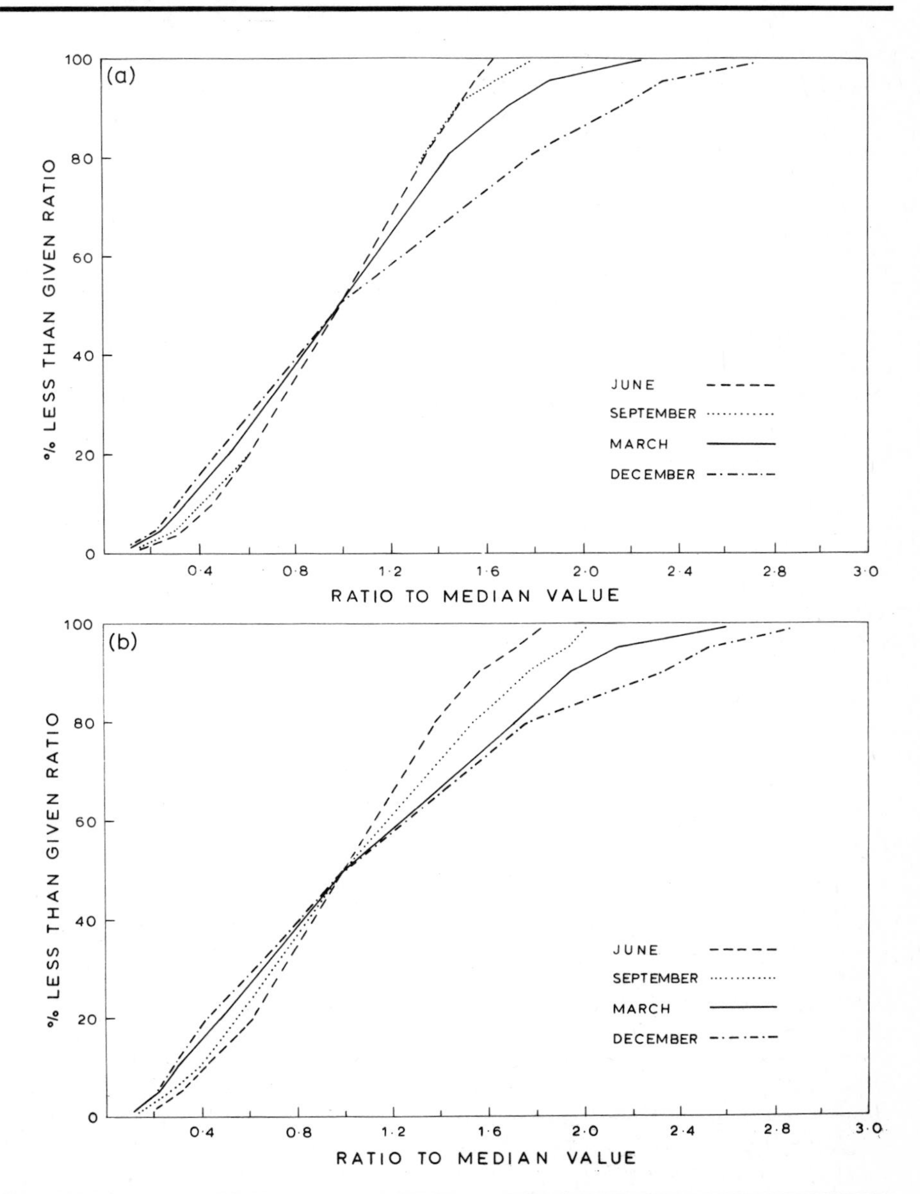

Variability of daily totals of global radiation The day to day variability of solar radiation can be assessed from the cumulative frequency distribution of daily totals of global solar radiation presented in Fig. 4.7. In Fig. 4.7*a* the distribution for Kew is given of the ratio of the actual daily total to the median daily total (and the median total is very close to the mean value); data are for March, June, September and December. In Fig. 4.7*b* are the corresponding data for Eskdalemuir. These diagrams can be used to estimate the frequency with which a given daily total is exceeded in any given month. It will be seen that the relative variability is greatest in December and least in June, and that March is more variable than September.

Secular changes It was explained at the beginning of this chapter that atmospheric aerosol was one of the factors determining the attenuation of solar radiation in its path through the atmosphere. It is thus natural to ask whether the changes in atmospheric pollution, brought about both by the 1956 Clean Air Act and the general realization that air pollution should be controlled, have led to a general increase in radiation received at the surface. It has been noted by Jenkins (1969) that the duration of sunshine in the London area has increased significantly, especially in the winter months, and Monteith (1966) has noted that there has been a slight but noticeable increase in the ratio between the annual mean direct components of solar radiation and the mean duration of sunshine for the year. This has been confirmed by a more detailed analysis of hourly radiation data (J. P. Cowley, to be published). As part of this analysis for each month the observed hourly values of ($G/I_0 \sin h_0$) have been correlated with the duration of sunshine in that hour (S) by fitting a linear equation in the form

$$G/I_0 \sin h_0 = a + bS$$

where I_0 is the direct solar radiation outside the earth's atmosphere and h_0 is the solar elevation; a and b are regression coefficients. The predicted value of ($G/I_0 \sin h_0$) for a value of S equal to 1 (i.e. complete sunshine) is thus ($a+b$).

Figure 4.8a shows a plot of the smoothed values of ($a+b$) for December and for June for the period 1950–73. It will be seen that in both months there is a significant increase in the later years but that this is much more marked in December. In fact, there is some evidence for a fall in June since 1965. The radiation received for a complete hour of sunshine has apparently increased by up to 20 per cent in December. In Fig. 4.8a are also plotted the corresponding curves for Lerwick in March and June. It will be seen that not only is there no significant variation with time but the value of ($a+b$) at Lerwick is significantly higher than at Kew, as would be expected by the much cleaner atmosphere at Lerwick.

Unsworth and Monteith (1972) have also described a method of interpreting the monthly mean values of global and diffuse radiation and sunshine hours which enables them to derive a mean turbidity coefficient to describe the mean attenuation due to aerosol, which enables different sites to be compared. Unsworth (1974) applied this method to data from four sites in the United Kingdom for the period 1956–72 for the period between April and September, and the results are shown in Fig. 4.8b. It will be seen that the turbidity at Kew and Aberporth has tended to increase since 1964 (and this confirms the results for Kew in Fig. 4.8a) whereas there is little change at Cambridge and Lerwick (again confirming the results in Fig. 4.8a). The changes at Kew and Aberporth may well be linked with an increase in the frequency of easterly winds in these later years.

Fig. 4.8 (a) *Secular variation of global solar radiation, as a proportion of the extra-terrestrial, for a complete hour of sunshine.* (b) *Five-year running means of turbidity (April to September) at four sites*

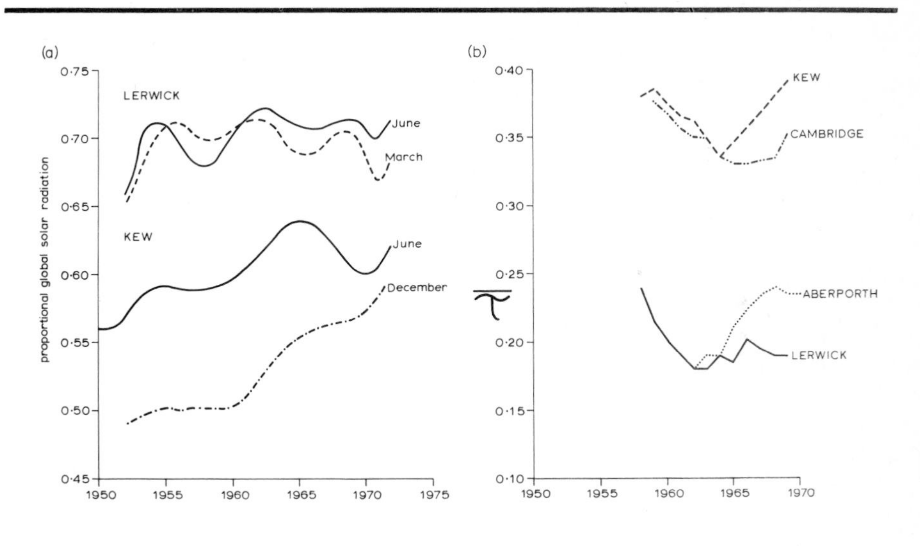

Radiation on vertical surfaces The amount of radiation falling on a sloping or vertical surface is important for some purposes and presented in Fig. 4.9a–d are the mean values of the daily total of global solar radiation on vertical surfaces facing north, south, east and west at Bracknell based on seven years' data. Also included in the figures are the median daily totals and the 95 per cent and 5 per cent cumulative frequency totals (i.e. the daily totals exceeded on 5 per cent and 95 per cent of the days considered). It will be seen that the distribution of the daily totals on the vertical surface facing south is very skew in the winter months with a few days having very high totals. This is due to the geometry; in winter, with low solar elevations, the direct radiation component is often very nearly perpendicular to the receiving surface.

It should be noted that these measurements were made without the component of global radiation reflected from the ground in front being allowed to fall on the measuring instrument. This standard condition was selected because any other form of ground surface would have been an arbitrary choice.

Fig. 4.9 *Daily totals of global solar radiation on vertical surface at Bracknell:* (a) *north facing;* (b) *south-facing;* (c) *east-facing;* (d) *west-facing*

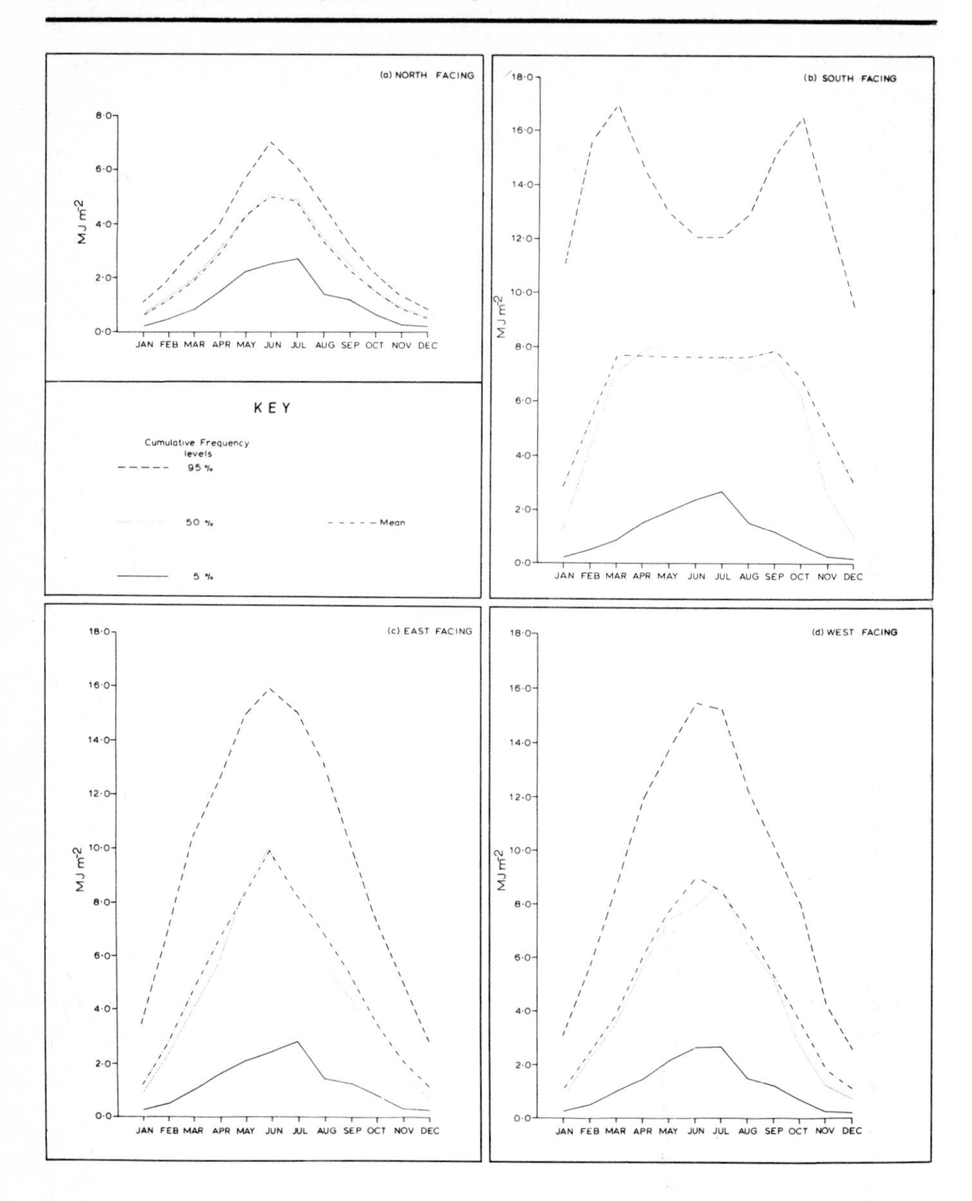

Direct solar radiation The direct solar radiation has been recorded continuously in the United Kingdom only at Kew Observatory. The first 14 years of data (1932–46) have been discussed by Stagg (1950) and reference should be made to this for fuller information. It is informative, however, to reproduce from this publication Fig. 4.10*a* which shows, for pairs of months, the intensity of direct solar radiation on the clearest occasions as a function of the solar elevation. It will be seen that there is very little difference between the peak values for a given solar elevation in different months of the year; in fact the intensities in the solar elevation range 10°–40° are slightly higher in the winter half year than in the summer half year, and this difference is in the right direction and of about the right amount to be accounted for by the ellipticity of the earth's orbit around the sun.

Fig. 4.10 (a) *Relation of intensity of radiation (I_0) to sun's altitude (h) on the clearest occasions in pairs of months (after Stagg, 1950)*; (b) *Solar radiation and cloud (after Lumb, 1964)*

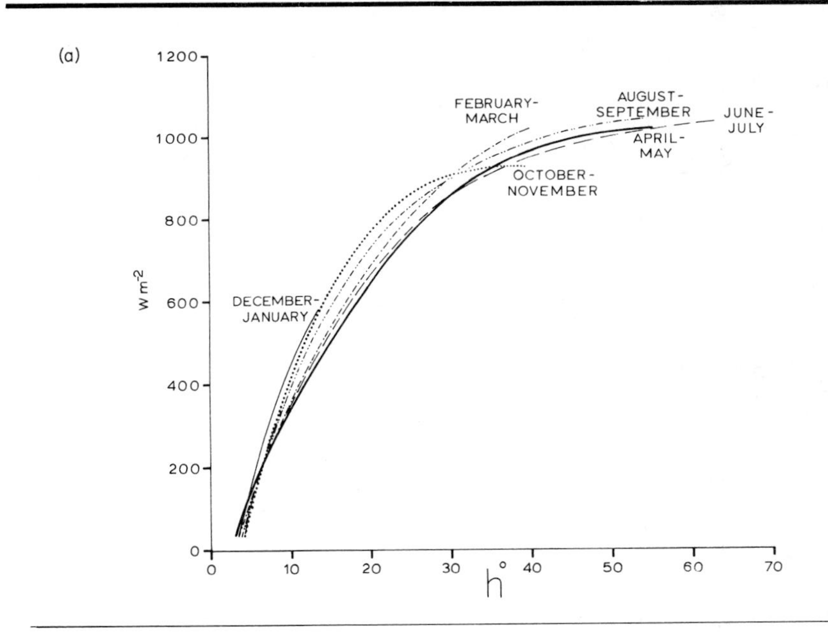

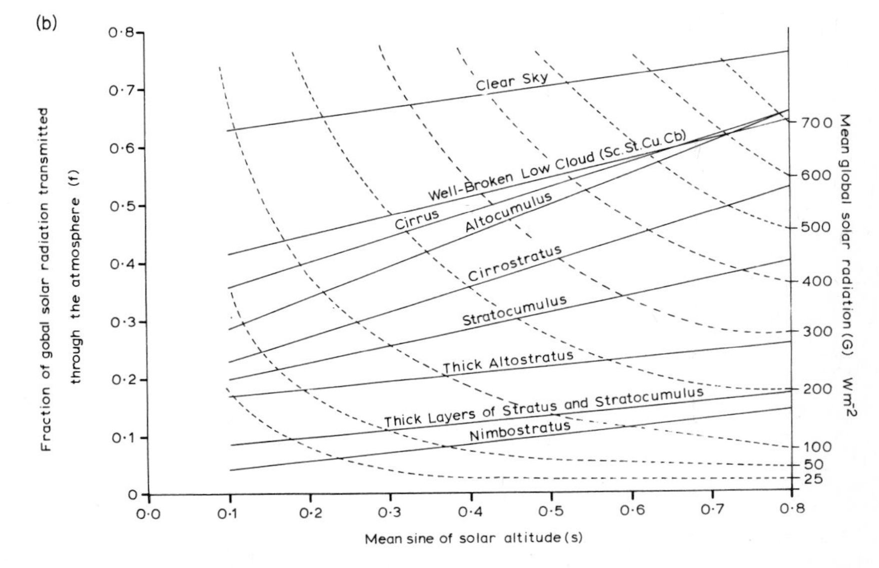

Variation of global solar radiation with cloud type It has been mentioned that clouds are the most important influence on the transmission of solar radiation by the atmosphere. The effect of different kinds of clouds has been studied by Lumb (1964) using data from the records of solar radiation on Ocean Weather Ships. He classified the cloud observations into nine groups and then studied the relation in each group between the global radiation (G) and the sine of the solar elevation s using a relationship of the form

$$G/I_0s = a + bs$$

Figure 4.10b shows the results obtained. These results are probably applicable to the clear conditions on the western coasts of the British Isles but less so to the more polluted inland regions.

Reflected solar radiation Part of the global solar radiation is reflected by the ground surface and the rest is absorbed. The fraction that is reflected (the albedo) varies very much with the nature of the surface and some typical values, expressed as percentages, are given in Table 4.1.

Table 4.1 *Albedo of various natural surfaces (after Kondratyev, 1969)*

Vegetation	Albedo (per cent)	Soil	Albedo (per cent)
		Black earth, dry	14
Summer wheat	10–25	Black earth, moist	8
Winter wheat	16–23	Grey earth, dry	25–30
Winter rye	18–23	Grey earth, moist	10–12
		Blue clay, dry	23
		Blue clay, moist	16
Grass cover:		Fallow field, dry surface	8–12
High standing grass	18–20	Fallow field, wet surface	5–7
Green grass	26	Ploughed field, moist	14
Dry grass wizened in the sun	19	Surface of clayey desert	29–31
		Yellow sand	35
Forest vegetation:		White sand	34–40
Tops of oak	18	Gray sand	18–23
Tops of pine	14	River sand	43
Tops of fir	10	Fine light sand	37
Different small plants:			
Cotton	20–22		
Lucerne (early florescence)	23–32		
Rice field	12		
Lettuce	22		
Beet	18		
Potatoes	19		
Heather wasteland	10		

The albedo of a water surface is rather variable. In smooth conditions the albedo for direct solar radiation depends greatly on the solar elevation; at moderate or high elevations (say 30° or more) the albedo is low (6 per cent or less) but it increases rapidly at lower elevations reaching 35 per cent at an angle of 10° and higher values at still lower elevations. However, the values for diffuse radiation are much less dependent on solar elevation (as would be expected) and average 6–10 per cent.

Fig. 4.11 *Annual variation of the mean daily total of net radiation*

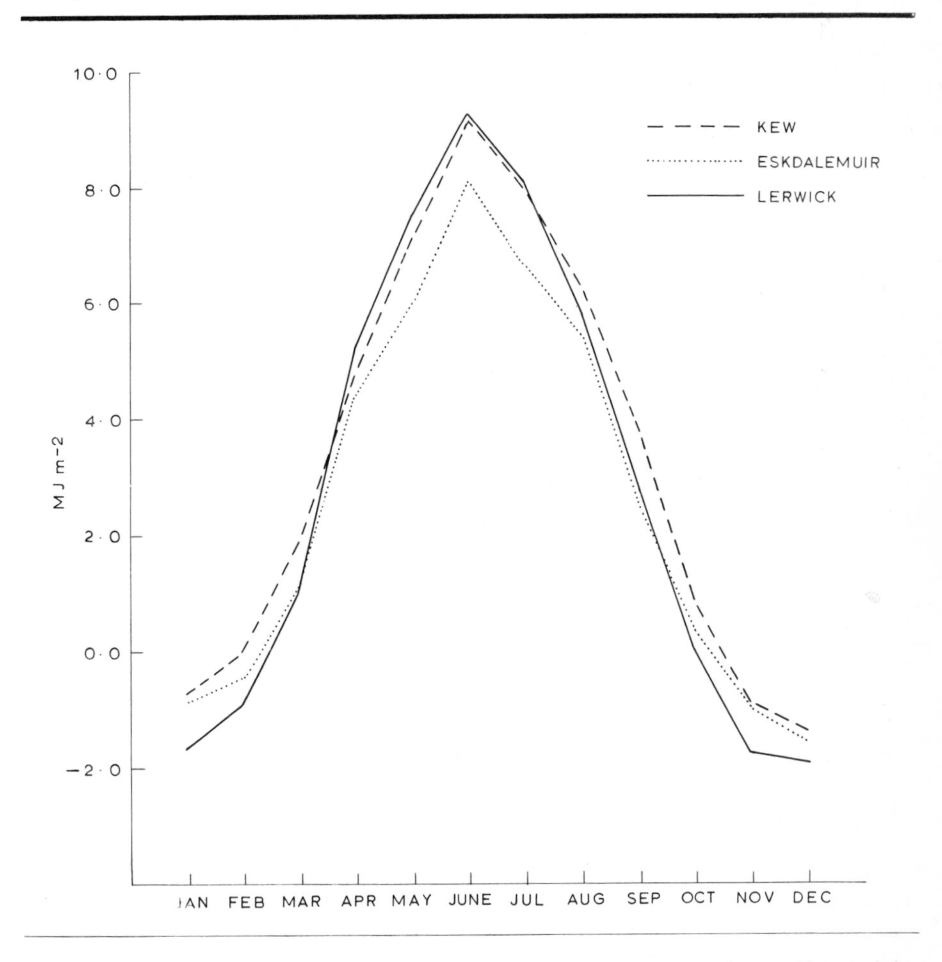

The values for direct solar radiation are however also affected by the surface roughness. At low solar elevations the albedo of a rough sea is much less than that of a smooth sea but at higher solar elevations there is an increase in albedo with an increase in surface roughness. The reader is referred to Kondratyev (1969) for a fuller discussion.

Net radiation The previous sections of this chapter have dealt with solar radiation. The thermal radiation from the ground and from the atmosphere, however, also forms a vital part of the heat budget of the earth's surface. Thermal radiation from a 'black body' at surface and atmospheric temperatures covers a wavelength range from about 5 μm to 50–100 μm and meteorologists often call this radiation 'long-wave radiation', as opposed to 'shortwave' (solar) radiation.

Unsworth and Monteith (1975) have recently described methods of estimating the down-coming longwave radiation (L) from the atmosphere both on the horizontal surface and on sloping surfaces. The net radiation (Q^*) is defined as the net difference between the down-coming streams of radiation of all wavelengths ($Q\downarrow$) and the corresponding upward ($Q\uparrow$) stream of radiation

$$Q^* = Q\downarrow - Q\uparrow$$

In general, both $Q\downarrow$ and $Q\uparrow$ will have two components (short-waves and longwave) but at night, of course, the shortwave components are both zero.

Net radiation is measured on a continuous basis at only a few stations in the United Kingdom but the monthly mean values for some of these are plotted in Fig. 4.11. It will be seen that the balance is negative in the winter months and positive in the rest of the year.

Sunshine

The duration of 'bright sunshine' is measured by a simple instrument which uses a glass sphere to focus direct solar radiation on to a specially treated card. If the intensity is sufficient to cause a perceptible discolouration of the card (i.e. the card is burnt) the sun is said to be 'shining'. The rotation of the earth on its axis causes the spot to move along the card and thus it is possible to measure the length of time during which the sun was shining. These instruments are simple to use and are widely distributed across the British Isles.

An investigation, using the records of direct solar radiation at Kew Observatory, has shown that the minimum direct solar radiation required to produce a burn varies between about 90 and 260 W m^{-2} with an average value of about 130 W m^{-2}; this rather wide range of values can be caused by differences in the moisture content of the card, or by the presence of dew or rime on the sphere. This means that, in clear conditions, the sunshine recorder does not start to record until the sun has reached an elevation of about 3°; below this elevation the direct solar radiation is not sufficient to cause a burn.

There is, of course, a marked annual variation in the average daily sunshine duration, caused by the annual variation in day length associated with the changes in declination of the sun. The maps in Fig. 4.12 and 4.13 show the distribution of the average duration of sunshine over the British Isles for the months March, June, September and December together with the complete year. As an example of the variability of sunshine, Fig. 4.14*a* gives the cumulative frequency distribution for daily sunshine durations for Kew for the months March, June, September and December.

Relationship between solar radiation and sunshine There must obviously be a relationship between the duration of sunshine (S) and

Fig. 4.12 *Average daily bright sunshine in hours, 1941–70.* (a) *March;* (b) *June;* (c) *September;* (d) *December; (based mainly on Meteorological Office, 1974, and Eire Meteorological Service, 1971)*

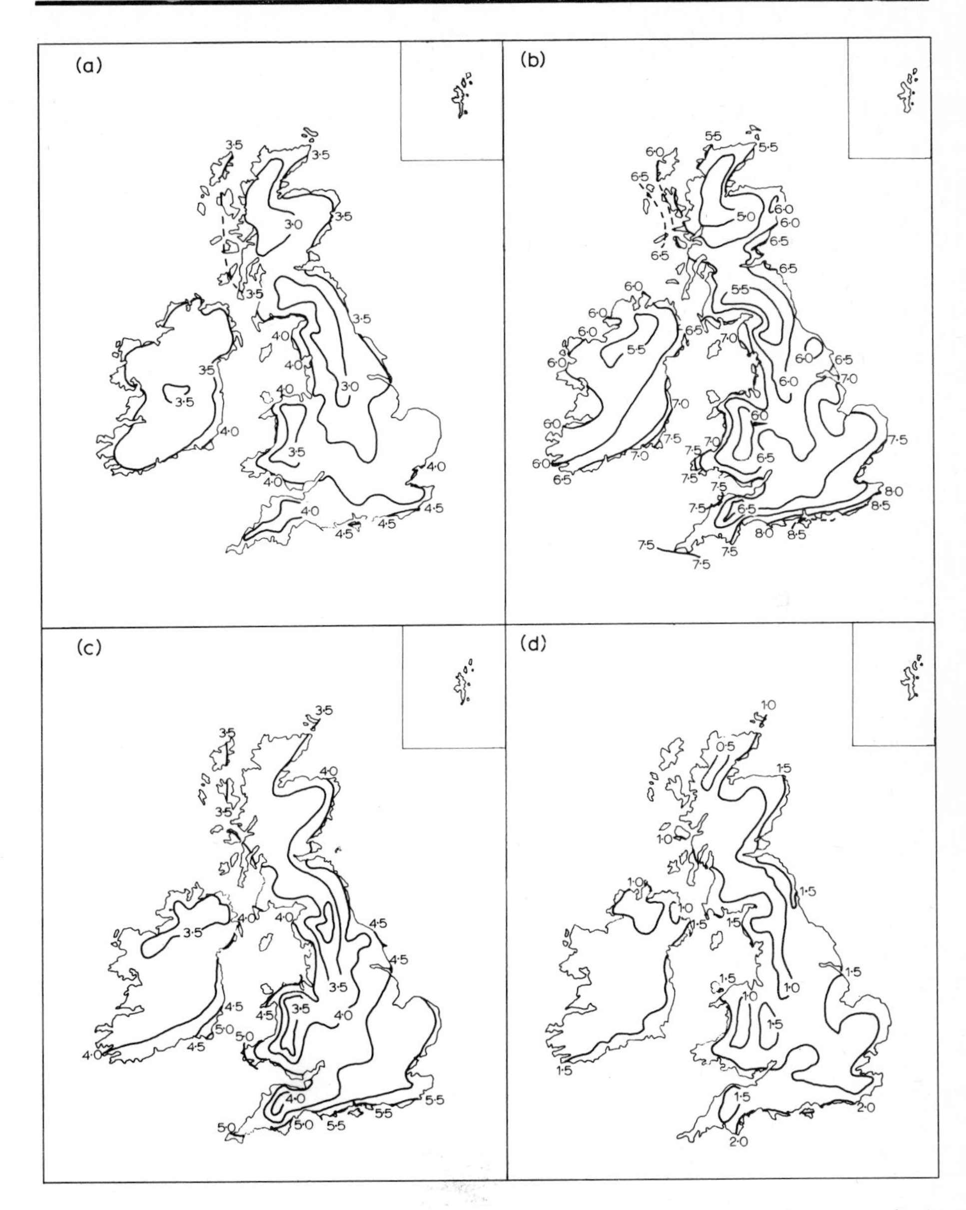

the global solar radiation (G) and a major investigation was carried out by Angström (1924). There have been various forms in which the general relationship proposed by Angström has been expressed but one of the most useful is that between (G/G_0) and (S/S_0), where G_0 is the daily total of solar radiation on a horizontal surface outside the atmosphere at the place in question (see Fig. 4.2) and S_0 is the day length. Figure 4.14*b* shows a plot of the daily data for the month

Fig. 4.13 *Average daily bright sunshine in hours, 1941–70, for the year (based mainly on Meteorological Office, 1974)*

Fig. 4.14 (a) *Cumulative frequency distribution of daily totals of sunshine at Kew;* (b) *Fractional global solar radiation and fractional bright sunshine (daily totals) for June, 1952–72, at Lerwick*

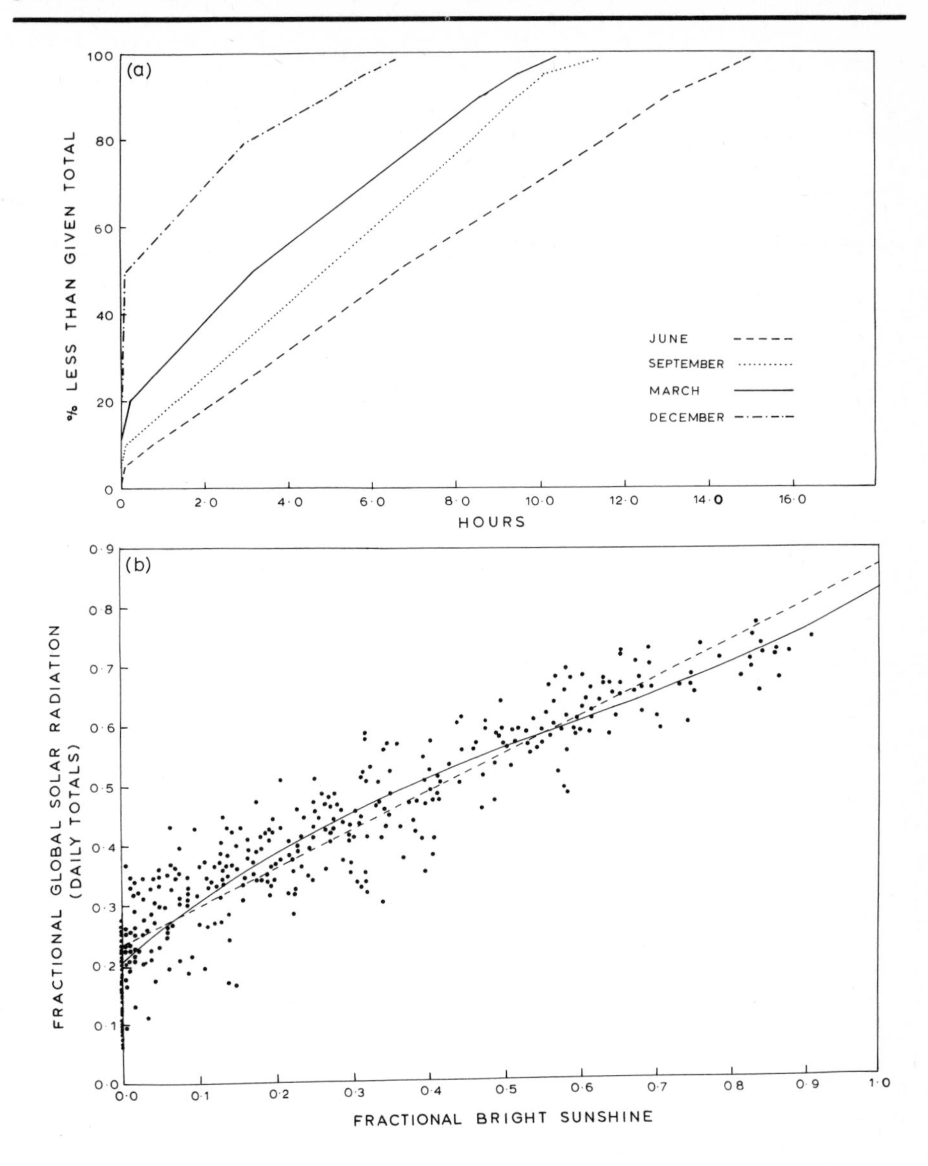

of June for Lerwick (21 years of data). Also plotted is the best fitting straight line, using the duration of sunshine as the independent variable, and also the best fitting third order polynomial. It is clear that there is a very significant correlation between the two sets of data. The standard deviation of the individual daily observation about the linear regression line is 0·065, and 0·061 about the polynomial. Thus for, say, a mean value of G/G_0 equal to 0·6, a measurement of the daily total of sunshine would enable a prediction of the daily total of solar radiation with a standard deviation of about 10–11 per cent. It is clear that the polynomial curve does not add useful extra precision when compared with the linear regression line.

The equation linking (G/G_0) and (S/S_0) can be written

$$G/G_0 = a + b\ (S/S_0)$$

Fig. 4.15 *Annual variation of regression constants* a *and* b *linking global radiation and sunshine duration (for daily totals) at Lerwick (L) and Kew (K)*

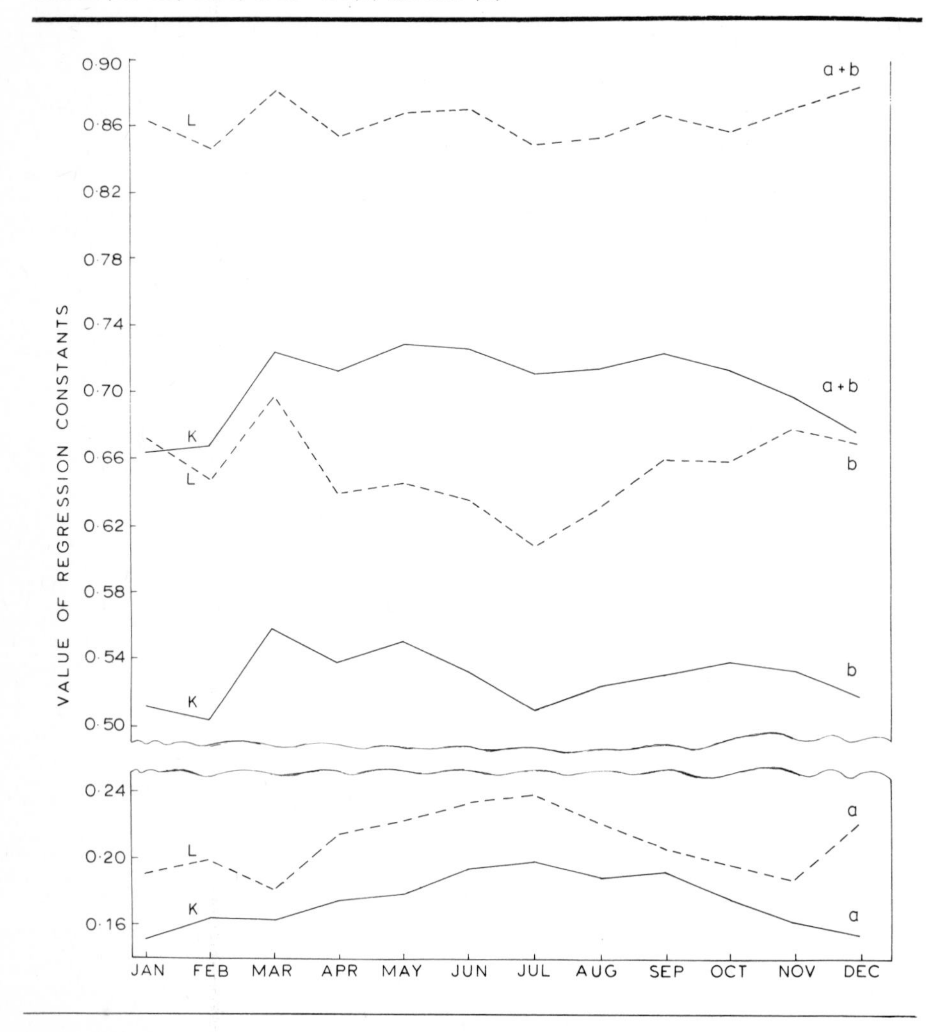

Figure 4.15 shows the annual variation of a and b for Lerwick and Kew, together with the annual variation of $(a+b)$, which is the fractional global radiation for a hypothetical day with complete sunshine $(S/S_0=1)$. Several interesting features are apparent. Firstly, the value of $(a+b)$ for Lerwick is much higher than the value for Kew. This is almost certainly due to the air pollution at Kew which affects the

radiation proportionally more than the sunshine. Secondly, there is an annual variation in $(a+b)$ at Kew which is not seen in the Lerwick data. This would also be expected if there were greater attenuation in the atmosphere at Kew, which would be most marked for the lower solar elevations in winter. Thirdly, the annual variation of b at the two stations shows an interesting number of similarities; each has a maximum in March and minimum in July. The reasons for these changes are probably quite complex but they suggest clearer air on the average in spring and thinner more transparent clouds in the summer. The annual variation at Kew has also, of course, been affected by the secular changes in the transmission of the atmosphere discussed above.

The comparison of these two stations shows that it is by no means allowable in general to use a relationship determined at one station to estimate radiation at another station. It is far better to rely on radiation observations themselves wherever possible.

Introduction

The temperature regime is one of the most distinctive features of the climate of the British Isles. It is generally recognized that these islands, situated as they are between approximately 50°N and 60°N latitude, are very favoured with regard to their range of temperature, experiencing winters which are exceptionally mild for such a northerly position and summers which are never unbearably warm. Unknown in the British Isles are the oppressively hot, humid spells of areas such as the north-eastern United States or the prolonged below freezing-point winter temperatures of countries further east on the European mainland. For most of the year the temperature at sea-level lies between approximately 5°C and 20°C. Prolonged spells in winter when the temperature fails to reach 5°C are uncommon, and likewise the summer 'heatwave' is a rare event, but exceptional months and even entire seasons do occur. The winter of 1962–3, with spells of continuously low temperatures which, in terms of duration, had not been experienced in the British Isles in living memory, will long be remembered by those who endured it.

The range of temperature is controlled to a large extent by the fact that the British Isles lie in the mid-latitude westerly wind belt, are exposed to the full effects of the comparatively warm oceanic North Atlantic Drift and are surrounded by water whose temperature varies very slowly from month to month. Near the Shetland Islands, for example, the sea surface temperature ranges from a mean of about 7°C in March to a mean of about 12°C in August (Höhn, 1973). Areas directly exposed to the Atlantic Ocean experience mean temperatures which never rise much above nor fall far below the sea surface temperature. Inland, away from the maritime influence, greater extremes are possible. Occasionally, in time periods ranging from a day to a whole season, the normal westerly circulation is interrupted and Continental influences become important; it is at these times that the extremes of temperature are recorded.

Within the British Isles it is customary to make a division between Highland and Lowland Britain and the contrasting relief features of these two sub-divisions give rise to somewhat different temperature characteristics, especially in the more mountainous areas where, as Manley (1945a) has shown, temperature falls quite rapidly with height in this exposed temperate climate (see Chapter 12).

The measurement of temperature

Air temperature is normally measured by a mercury-in-glass dry bulb thermometer exposed in a white-painted, louvred Stevenson screen with the bulb at a height of 1·25 m above a short grass surface. A wet bulb thermometer, necessary for measuring the humidity conditions of the air, and a maximum and a minimum thermometer to record the extremes of temperature, are also housed in the screen. The thermometers are mounted inside a screen so that the bulbs are shielded from direct and reflected short-wave radiation from the sun and also from long-wave terrestrial radiation but with the louvres allowing through ventilation.

Readings of dry and wet bulb temperature may be taken hourly, as at major airports, or possibly only once daily, usually at 09.00 hrs, at less important meteorological and climatological stations. Observations of the maximum and minimum temperature are also taken with the minimum credited to the day of observation, as the lowest temperature is likely to have occurred in the early hours of the morning, and the maximum to the previous day, as the highest temperature normally occurs a few hours after midday. At some stations the maximum refers to the 'day' period and the minimum to the 'night' period while at others both maximum and minimum refer to 24-hour periods. In winter the average night minimum may be nearly 1 °C higher at an inland station than the average 24-hour minimum, particularly when the 24-hour minimum is observed at 09.00 hrs (Meteorological Office, 1963). The requirements for the accurate observation of air temperature and the errors, both instrumental and human, which may occur are covered in detail elsewhere (Meteorological Office, 1956, 1969).

Crowe (1962, 1971) and Kalma (1968) have discussed, at some length, the various methods of obtaining the daily mean air temperature and it is not proposed to elaborate on this point. Suffice it to say that in this chapter the following definitions are applicable: The *mean daily maximum* and the *mean daily minimum* are the means of the individual daily maximum and minimum temperatures for the period in question; the *mean temperature* is half the sum of the mean daily maximum and the mean daily minimum and the *mean annual temperature* is the average of the mean temperature for each month; the *mean daily range* is the difference between the mean daily maximum and the mean daily minimum temperatures; the *annual range of mean temperature* refers to the difference between the mean temperatures of the warmest and coolest months; the *absolute extremes* of temperature refer to the highest and lowest values on record for a particular month.

The presentation of the temperature data

In the presentation of the temperature data, it was decided to concentrate on the latest period for which reasonably complete figures are available, namely the 30-year period from 1941 to 1970. It is now recognized that means based on a recent, relatively short, period are more likely to be a guide to conditions to be expected in the near future than means based on a much longer time period, which may encompass several climatic epochs (Wright, 1970). The differences in temperature conditions between 1911 and 1940 and 1941 to 1970 are discussed in Chapter 10.

Data from 217 meteorological and climatological stations in the United Kingdom were used in the preparation of the original maps.

The records of these stations were either complete for the 1941–70 period or had been weighted against adjacent stations with complete records in instances where breaks occurred in the 30-year record. The lack of adequate coverage of temperature conditions at even moderate altitudes is shown by the fact that 77 per cent of these stations are at less than 100 m and 94 per cent are less than 200 m. Only one station, Malham Tarn, is at an altitude of more than 350 m. Oliver (1964) has discussed this problem and outlined the upland temperature data which are available. The statistics used to extend the isotherms over southern Ireland are not of comparable standardization but represent the most recent records available. For this reason the isotherms over Ireland should be interpreted with caution.

In order to eliminate the altitude factor in the discussion of the spatial distribution of temperature over the British Isles, the actual, station-level, temperatures have been reduced to mean sea-level values by taking the standard correction factor of 0·6 °C per 100 m. Without this correction the temperature maps would appear as complicated as orographic maps. To reduce mean daily maximum temperatures to sea-level a correction of 0·7 °C per 100 m was used and for mean daily minimum temperatures a value of 0·5 °C per 100 m, in accordance with the fact that maximum and minimum temperatures decrease with height at different rates (Harrison, 1974). For many purposes uncorrected temperatures are required and these are given in Table 5.7 at the end of this chapter, for twenty-five stations which have been chosen to provide as adequate an areal coverage as possible.

The standard months of January, April, July and October, representing winter, spring, summer and autumn conditions respectively, were chosen to analyse the variation of temperature through the year and maps of mean temperature, mean daily maximum, mean daily minimum and mean daily range of temperature in these four months are presented, as well as a map of the mean annual temperature. Isotherms, to a certain extent generalized, have been drawn at intervals of 1 °C on all these maps. In areas where significant differences were seen to exist, as in the London urban area, for example, isotherms have not been generalized. Where isotherms have been extrapolated and the position is uncertain, the isopleths have been drawn as broken rather than as continuous lines. The values for Baltasound and Lerwick in the Shetland Islands have been plotted on the inset but the isotherms have not been extended north of the Orkney Islands.

Individual observations of temperature cannot, of course, be reduced to sea-level values in a similar manner to that described above and the map of conditions at a particular observation hour (Fig. 5·8*a*) and the two maps of maximum temperatures on a particular day (Figs 5·8*b* and 5·8*c*) include details of station temperature within the station circle, also wind speed and direction, at the hour of observation or at 12.00 hrs in the case of maximum temperatures. Isotherms on these three maps are at 5 °C intervals.

In the discussion which follows an attempt has been made to include not only the mean conditions but also the temperature variation experienced over a recent 30-year period. As important as the spatial distribution of mean values is information on the extent to which individual years fluctuate above and below these mean figures. Temperature values over a sufficiently long period approximate to a normal distribution, and to test for normality, the January, April, July, October and annual mean temperatures for the 1941–70 period a one sample Kolmogorov–Smirnov test was used. All distributions tested were found to be normal so standard deviations were calculated and probabilities determined on this basis.

Oceanicity/continentality

In order to demonstrate the extreme oceanic regime of the north-west of the British Isles and the increasingly continental temperature regime of the south-east, Conrad's (1946) Continentality Index was calculated for over 200 stations, using the formula

Continentality Index $(k) = (1{\cdot}7A/\sin(\emptyset + 10^\circ)) - 14$

where A is the annual range of mean temperature in °C and $\emptyset$ is the latitude angle. With this type of index it is the relative magnitude which is of interest and in this case values range from 1·3 at Cape Wrath on the very exposed north-west tip of the Scottish mainland, to 12·5 at Heathrow (London Airport). The limits of Conrad's formula give about zero for the oceanic regime of Thorshavn in

Fig. 5.1 (a) *Conrad's continentality index, 1941–70;* (b) *Mean annual temperature 1941–70, in °C, reduced to sea-level*

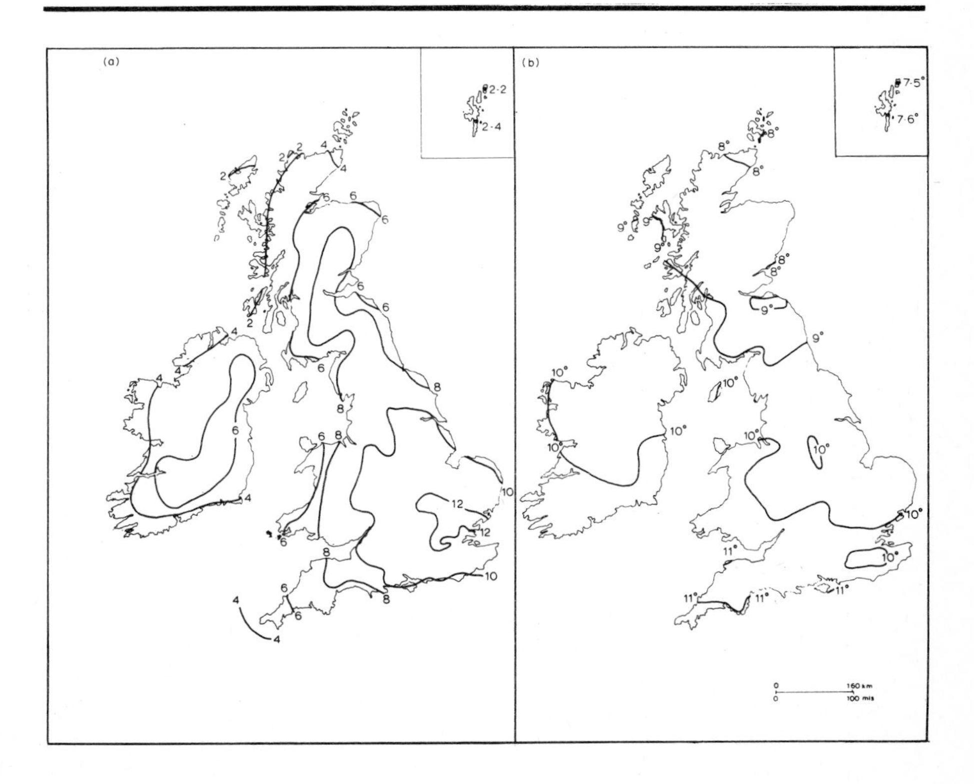

the Faeroes and about 100 for the extreme continental climate of Verkhoyansk in Siberia. Several points of interest emerge from Fig. 5.1*a*: (i) the extreme oceanicity of stations on the margins of the Atlantic Ocean, e.g. Tiree (Inner Hebrides) 2·1, Valentia (south-west Ireland) 2·6, Scilly Isles 3·7; (ii) the 'maritimeness' of the whole of Ireland with a maximum value of 6·7 at Armagh; (iii) the comparative continentality of parts of east and south-east England.

It is, however, important to bear in mind that although an increased trend of continentality is evident in a south-easterly direction over England the values are still very low compared with the distribution of continentality over, for example, North America, where values rise to more than 60 and the area less than 12 is restricted to the tip of the Florida peninsula and the Pacific coastlands (Trewartha, 1966).

Mean annual temperature

Figure 5.1*b* shows the mean annual temperature over the British Isles for 1941–70, reduced to mean sea-level. Values range from 7·5 °C at Baltasound on Unst in the Shetland Islands to 11·9 °C in the Isles of Scilly. Mean annual temperatures decrease from south-west to north-east, modified by a northward bending of the isotherms over the sea and a southward bending over the land. The areas which experience the highest mean annual temperature are those in a windward position exposed to the full influence of the Atlantic. It is the mildness of the resulting winter which gives rise to the high mean annual temperature of the extreme south-west of England.

The effect of the Atlantic and the Irish Sea can be seen in the position of the 10 °C isotherm which bends southwards over Ireland, northwards again to include parts of the Isle of Man, thence southwards and eastwards to include much of Wales and southern England. The only stations where the annual temperature is in excess of 11 °C are situated in the extreme south-west of England. The effect of the Atlantic is also seen in the position of the 9 °C isotherm which runs approximately east to west in the region of the Scottish border but then bends northwards towards the Scottish islands.

Fluctuations of the mean annual temperature

As might be expected, the mean annual temperature fluctuates within rather narrow limits. To quote three examples from different parts of the British Isles, during the 1941–70 period the range of the annual temperature was 2·2 °C at Plymouth (Mt Batten), 1·8 °C at Aldergrove in Northern Ireland and 1·7 °C at Eskdalemuir in the Southern Uplands of Scotland. In Fig. 5.2*a* the annual temperature, uncorrected for altitude, for each year of the 30-year period for these three stations has been plotted and it can be seen that there is a clear correlation between the fluctuations of annual temperature above and below the mean in these three contrasting regions.

The standard deviation of the annual temperature is about ±0·5 °C (Table 5.1) so that a figure more than 1 °C above or below the mean would only be expected one year in thirty. The actual range of the annual temperature of approximately 2 °C over the 30-year period is very close to the theoretical value.

Table 5.1 *Mean temperature and its standard deviation in °C, for 1941–70*

	Mean	S.D.	Year in which > ±2 S.D.
January			
Lerwick	3·0	1·3	1959+
Stornoway	4·1	1·3	1945—
Aldergrove	3·5	1·5	1963—
Kew	4·2	1·8	1963—
Plymouth	5·9	1·7	1963—
April			
Lerwick	5·4	0·8	1951—
Stornoway	6·9	1·0	
Aldergrove	7·9	1·0	
Kew	9·5	1·2	
Plymouth	9·2	0·9	
July			
Lerwick	11·7	0·8	1947+ 1965—
Stornoway	12·9	0·7	1955+
Aldergrove	14·4	0·8	1955+ 1965—
Kew	17·5	1·0	
Plymouth	15·9	0·8	1954— 1955+
October			
Lerwick	8·5	0·9	
Stornoway	9·5	0·9	1959+
Aldergrove	10·1	1·0	1969+
Kew	11·6	1·1	1968+
Plymouth	12·1	0·9	
Annual			
Lerwick	7·1	0·5	
Stornoway	8·3	0·4	1953+
Aldergrove	8·9	0·4	1949+
Kew	10·6	0·5	1959+
Plymouth	10·7	0·5	1949+ 1963—
Eskdalemuir	7·2	0·5	1959+
Gorleston	9·8	0·6	1963—

Fig. 5.2 (a) *Fluctuations and frequency distribution of mean annual temperature, 1941–70, for Plymouth Aldergrove and Eskdalemuir;* (b) *Fluctuations of mean temperature for January and July at Plymouth (Mount Batten), 1941–70;* (c) *Frequency distribution of mean temperature for January and July at Plymouth (Mount Batten), 1941–70*

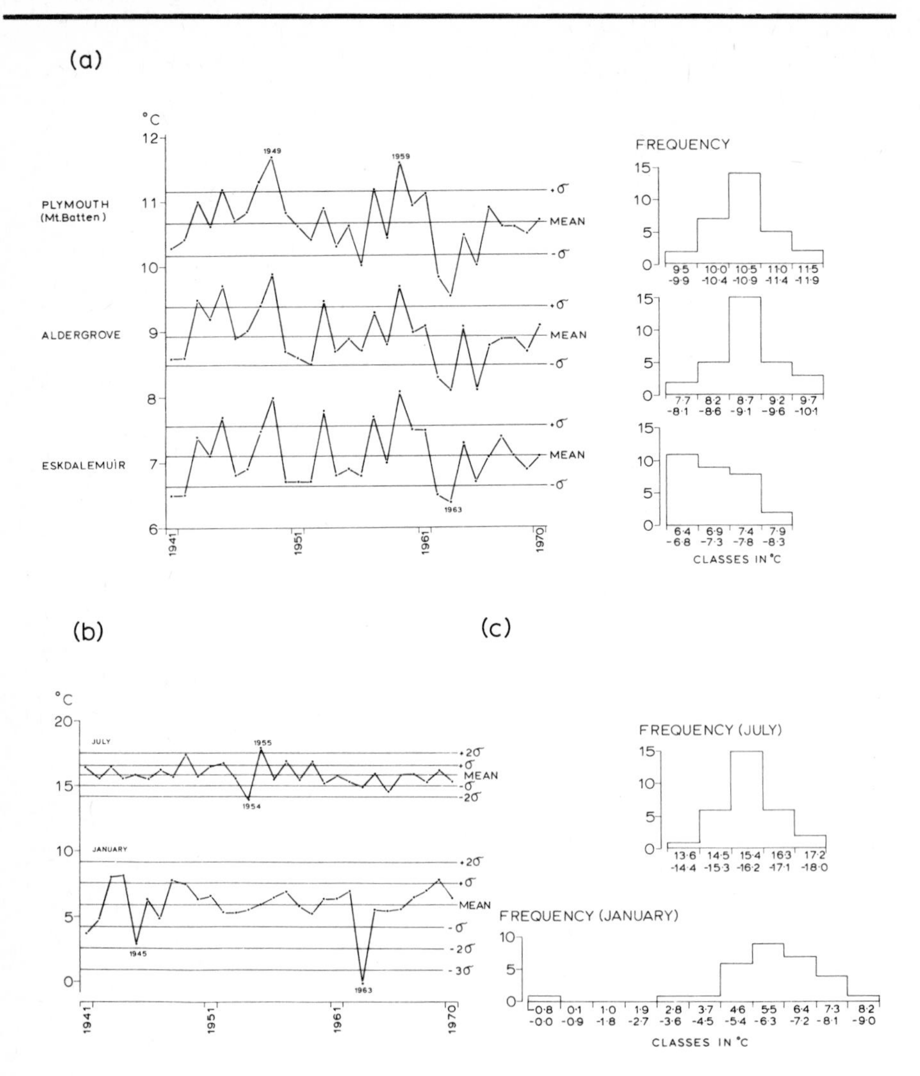

Seasonal temperature characteristics

Winter

As indicated above, the months of January, April, July and October were chosen to represent the temperature conditions in the four main seasons. The January mean temperature is shown in Fig. 5.3*a*, with sea-level values ranging from 2·4 °C at several stations in Scotland to as high as 8·0 °C in the Scilly Isles. The area with a mean temperature of 6 °C or above is restricted to coastal areas of southern

Ireland, Dyfed and the south-west of England. If 6 °C is taken as the threshold temperature above which accelerated growth of vegetation occurs, this means that there is little, if any, check to continuous growth in these favoured localities and this fact is exploited to the full by the early vegetable and flower industries of regions such as Cornwall and the Scilly Isles (Hogg, 1967b). The exposure of these areas

Fig. 5.3 *January temperatures, 1941–70, in °C, reduced to sea-level.* (a) *Mean;* (b) *Mean daily maximum;* (c) *Mean daily minimum;* (d) *Mean daily range*

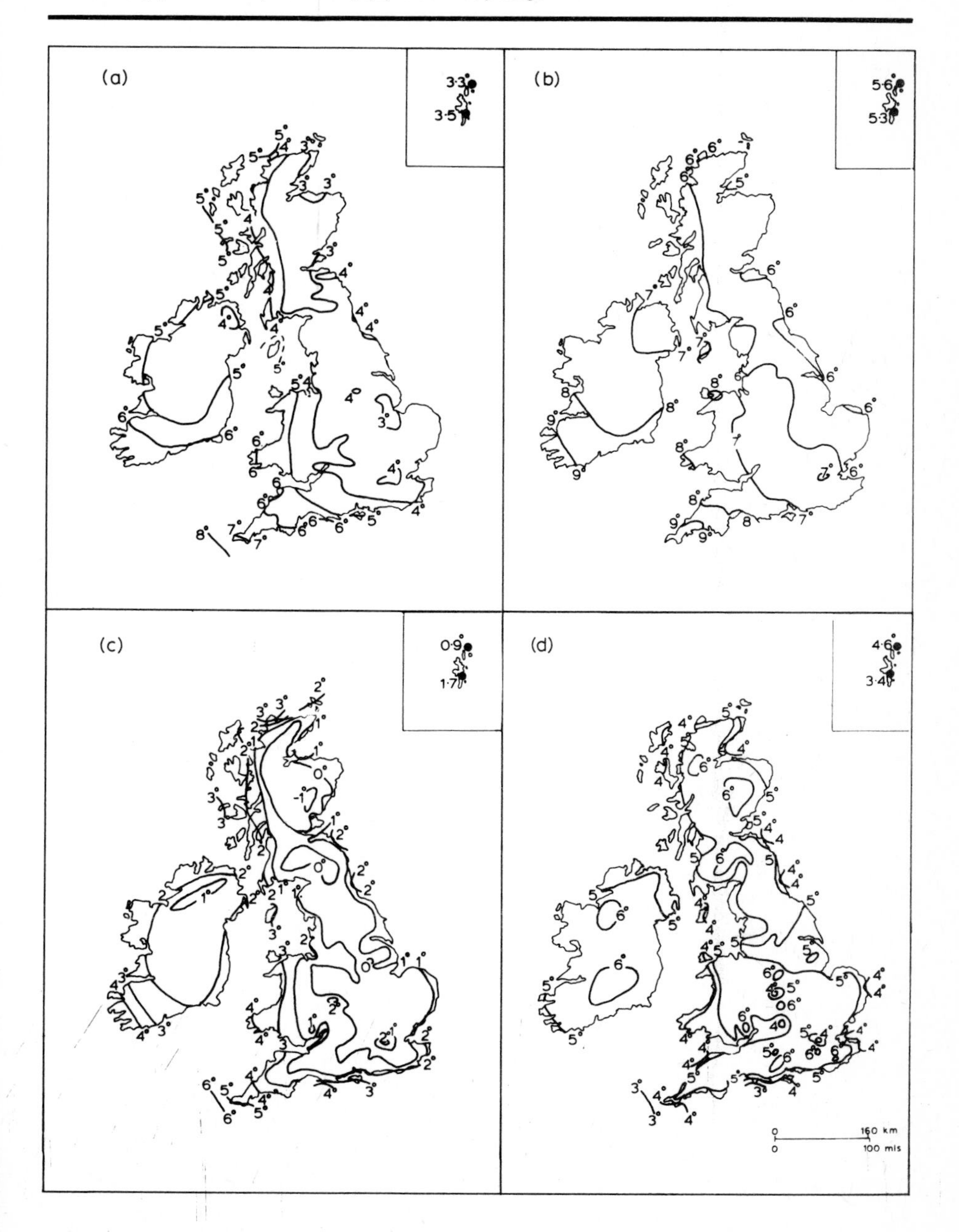

to the full moderating effect of the Atlantic gives rise to January temperatures which compare very favourably with sea-level stations in the Mediterranean, e.g. Nice (Airport), latitude 43°39′N, altitude 10 m, January (CLINO) mean temperature 7·5 °C (E.S.S.A., 1966).

The dominant Atlantic maritime influence in January is seen in the general north to south alignment of the isotherms, with the 4 °C line separating a warmer Scottish Islands, Ireland, Wales, south-western and extreme southern England from colder areas, more exposed to Continental influences, to the east. Over Ireland and the Irish Sea the isotherms bend northwards over the sea and southwards over the land. From Fig. 5.3a it can be seen that the west coast of Scotland is as warm as the south coast of England, eastwards from the Isle of Wight. The moderating effect of a large urban area on winter temperatures results in an outlier of the 4 °C isotherm enclosing the Greater London area (see Chapter 14).

The fluctuation of winter mean temperatures is much greater than in other seasons of the year. This point is illustrated in Fig. 5.2*b*, which shows the spread of January and July mean temperatures for 1941–70 for Plymouth (Mt Batten). The standard deviation of ±1·7 °C for January is more than twice as great as the July figure. In January there is a much greater contrast between air masses than in July and the mean temperature of a month with an easterly airflow and predominantly cP air may be much lower than a more normal month of vigorous cyclonic activity with strong west or south-westerly winds and mP/mT air.

In summer the two contrasting airflows result in a smaller range of mean temperature. January mean temperatures at Plymouth have ranged (1941–70) from a high of 8·2 °C in 1944 to a remarkable low of −0·2 °C in 1963. The degree of abnormality of this latter figure is seen on the histogram of Fig. 5·2*c* where no other January was below +2·9 °C in the 30-year period. From Fig. 5.2b it can be seen that −0·2 °C is more than three standard deviations below the mean of 5·9 °C, a figure which would be expected only once in 740 years if we accept the probabilities based on a normal distribution.

The extremely abnormal winter of 1962–63, over central England the coldest winter since 1739–40, has been analysed by many authors. Clarke (1964) has looked at the synoptic situation as it developed in the days prior to 22 December when air from Europe reached England to mark the start of the long cold spell. During this the British Isles were subject to continental influences which to a large extent replaced the usual maritime influences as high pressure to the north maintained airstreams with an easterly component. The cold spell finally ended on 4–6 March when a mild south-westerly airstream spread in from the Atlantic. In places, temperatures remained below freezing point for over a week in mid-January and it was the length of the cold spell which was its outstanding feature. Hosking (1968) states that even at Ryde (Isle of Wight) in January and February 1963, 55 out of 59 minimum temperatures were 0 °C or below and on the other four days the minimum was only just above freezing point. The maximum temperature did not exceed 5 °C in either month. Murray (1966) investigated the large-scale features of the winter and found exceptional developments even in the Pacific sector. The Icelandic Low was replaced by a high pressure region for much of the winter. The North Atlantic and Western Europe were subject to an abnormal frequency of blocking activity, especially in January 1963. Booth (1968) has compared it with other cold winters. Shellard (1968b) has analysed the winter in terms of the actual temperatures recorded and concluded, on the basis of a statistical examination of the 268 winter mean temperatures provided by the central England series (Manley, 1953), that 1962–63 was well within the population of other cold winters experienced in this country and needed no exceptional explanation. This isolated example of a severe winter illustrated very effectively the extent to which we are normally excluded from continental influences in the winter months.

The pattern of isotherms of mean daily maximum temperature in January (Fig. 5.3b) is very similar to that of mean temperature with a north-north-west to south-south-east trend but with the North Sea coastal areas having slightly higher values than places inland. South-west Ireland and extreme south-west England have mean daily maxima in excess of 9 °C.

During the cooler part of the year maximum temperatures well above average may be recorded when mT air with its source region in the sub-tropics covers parts of the British Isles. This situation can occur even in mid-winter and a recent example of this is illustrated in Fig. 5.8a. At 06.00 hrs a cold front separated very mild mT air over England and Wales from cooler mP air over Ireland and Scotland. There was a uniformity of conditions within the warm sector with temperatures as high as 13 °C even at this early hour and indeed the maximum can occur at any hour of the day or night when such a warm sector crosses the British Isles. On other occasions even higher temperatures are recorded but tend to be more localized and occur particularly to the lee of high ground. Lawrence (1953) describes the synoptic situation under which a temperature of at least 15 °C was recorded in north-east Scotland on 18 February 1945 at 03.00 hrs, a time which precludes insolation as an operative factor.

The distribution of mean daily minimum temperature (Fig. 5.3c) reveals a far more complicated picture, with distance from the sea assuming an increased importance. The range over the British Isles is nearly twice as great as in the case of mean daily maximum temperature. Areas with mean daily minima in excess of 4 °C are restricted to south-west Ireland, south-west Wales and parts of the coast of south-west England. Only the Scilly Isles, with 6·4 °C, have a minimum above the growth threshold. In the Highlands of Scotland values, even reduced to sea-level, drop below −1 °C. The uncorrected figure for Braemar, at an altitude of 339 m, is −2·7 °C and January 1963 gave −8·6 °C. The effect of a large conurbation in increasing the minimum temperature is seen in the London area where the temperature is nearly 2 °C higher than in the surrounding countryside.

The mean daily range of temperature in January (Fig. 5.3d) away from the immediate sea coast is 5 °–6 °C, with a few localized exceptions. At coastal stations the range falls below 4 °C in some areas and even

Fig. 5.4 *April temperatures, 1941–70, in °C, reduced to sea-level.* (a) *Mean;* (b) *Mean daily maximum;* (c) *Mean daily minimum;* (d) *Mean daily range*

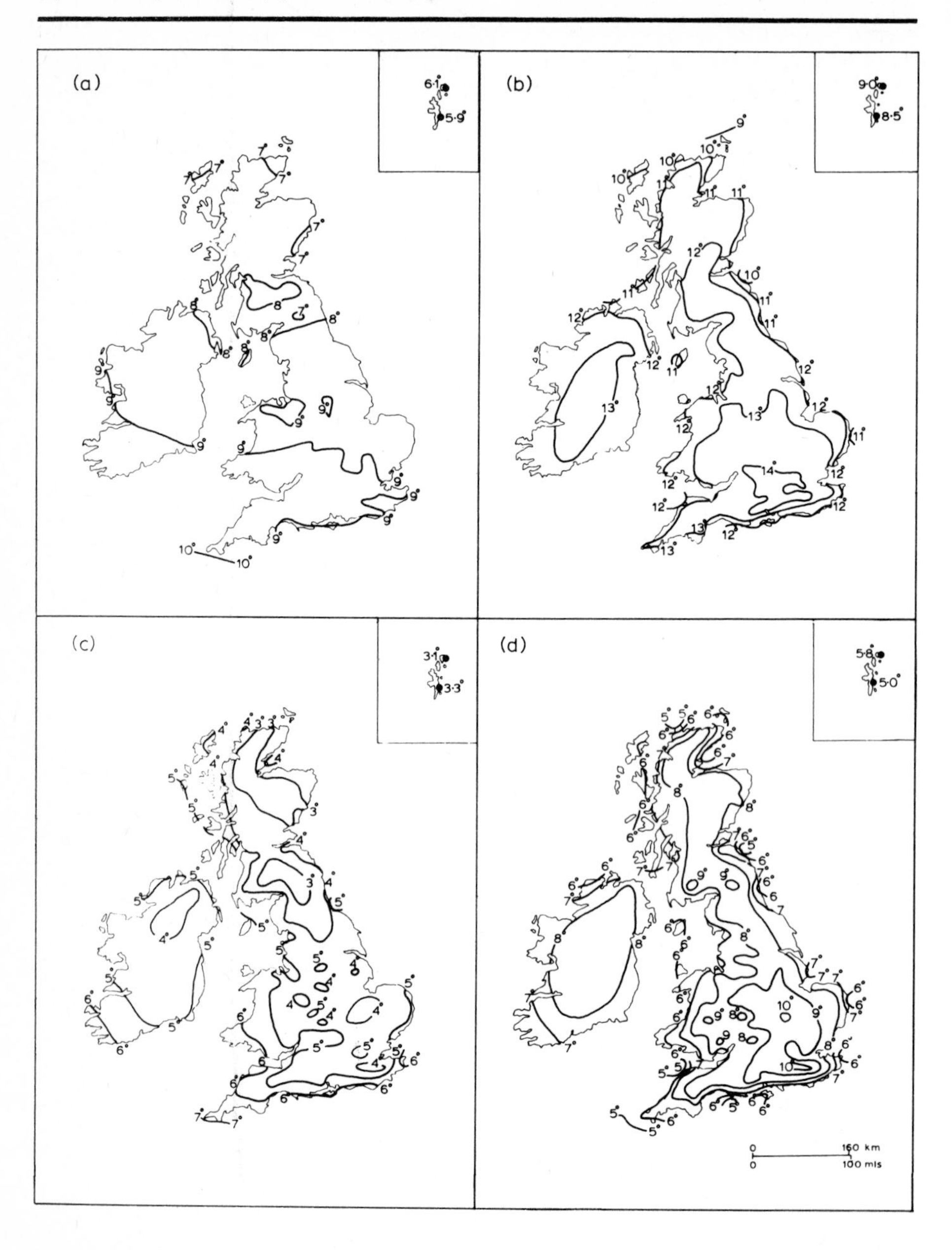

below 3 °C in extreme maritime situations in south-west England.

Due to the differences in the thermal properties of land and sea surfaces the seas surrounding the British Isles do not reach their lowest temperature until February or even March in the case of the northern North Sea and coastal areas in the south and west exposed to the Atlantic Ocean. This fact is reflected in the lowest mean temperature at

Fig. 5.5 (a) *Maximum temperatures recorded at Tynemouth and Blackpool during 2–10 April 1974.* (b) *Fluctuations of mean monthly temperature for April and October at Kew and Lerwick, 1941–70*

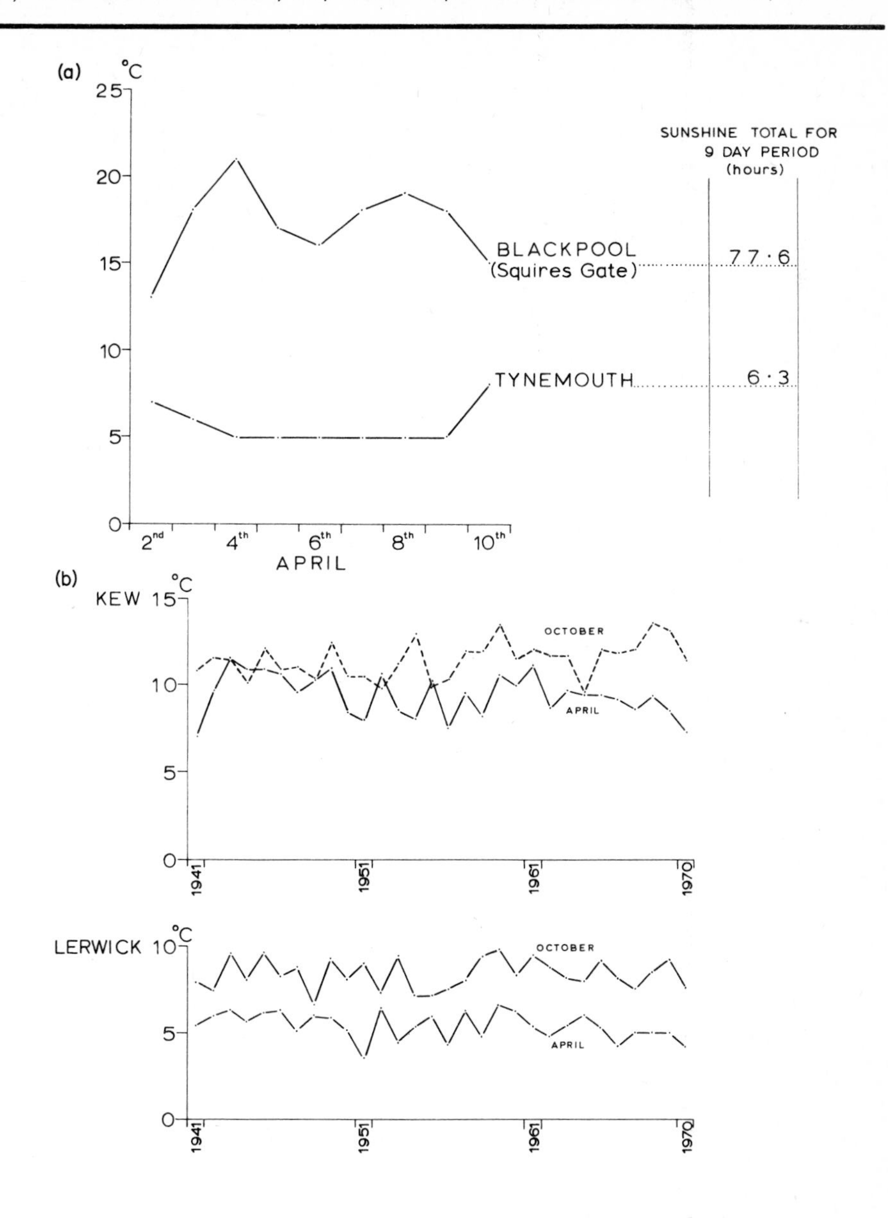

some stations, especially at coastal sites fully exposed to Atlantic maritime influences, being recorded in February rather than January. The Scilly Isles, for example, are 0·4 °C colder in February than in January. Stations east of the Isle of Wight as far as Eastbourne are equally cold in January and February and this applies also to many inland Irish stations, another indicator of the maritime nature of the

Irish climate. The North Sea coasts are not affected to the same extent as they tend to be on the leeward side of the country. Skegness, for example, is on average 0·3 °C warmer in February than in January but at St Abbs Head February is 0·1 °C cooler than January.

Spring

By April the pattern of isotherms of mean temperature (Fig. 5.4a) has changed fundamentally from the January picture. The isotherms now run generally from west to east and values range from 5·9 °C in Lerwick to 10·2 °C in the Scilly Isles. The latter is still the warmest station in the British Isles, although the temperature has risen only 2·6 °C since February, compared with 5·0 °C in the case of Kew Observatory over the same time period. In Fig. 5.5b April and October mean temperatures for 1941–70 have been plotted for Lerwick and Kew. At the maritime station, October was warmer than April in each year, but at the inland site this was not always the case, especially in the earlier part of the period. Standard deviations (Table 5.1) are about ±1 °C for April, indicating a less variable month than January, as air mass contrasts decrease.

Compared with January, the April pattern of isotherms of mean daily maximum temperature (Fig. 5.4b) has also changed fundamentally and no longer are the highest maxima found in the exposed western locations. Latitude and distance from the sea are now of greater significance and the coasts are in all cases cooler than inland areas. In central southern England maxima just exceed 14 °C in a few places.

Figure 5.8b, showing the distribution of maximum temperatures over the British Isles on 4 April 1974, illustrates the cooling effect of the North Sea in spring when a persistent onshore airflow can keep temperatures very low on the east coast. In this particular example there is a temperature gradient of 16 °C from a maximum of 5 °C at Tynemouth to one of 21 °C at Blackpool. The synoptic situation, with a high pressure block persisting over the North Sea, gave dull, cool conditions on the North Sea coast for much of April 1974 while further inland continuous sunshine was recorded on many days, with considerably higher temperatures. In Fig. 5.5*a* maximum temperatures for 2–10 April 1974 have been plotted for Tynemouth on the North Sea coast and Blackpool to the lee of the Pennines. The maximum temperature of 5 °C, recorded on each of 6 consecutive days at Tynemouth, is close to the average sea surface temperature of 6·5 °C for this part of the North Sea in April (Höhn, 1973). The sunshine total of 6·3 hours at Tynemouth was recorded on 1 single day while Blackpool averaged 8·6 hours of bright sunshine over the 9-day period.

On April nights (Fig. 5.4c) coastal areas are still warmer than inland, as in January, and once again the highest minima are found in the extreme south-west of England, with the Scilly Isles recording 7·7 °C. Large urban areas again stand out as regions of slightly higher minima. In parts of southern England the mean daily range is twice as great in April (Fig. 5.4d), as in January, with 10 °C being exceeded in a few places. Over inland areas the range is generally greater than 8 °C but over some exposed coastal areas this falls below 5 °C. Unusually large ranges of temperature can occur in spring. On 29 March 1965 at Birmingham Airport the temperature climbed from −2·8 °C in the early morning to a maximum of 23·7 °C in the afternoon, a range of 26·5 °C.

Summer

In July the isotherms bend northwards over land areas and coastal areas are generally cooler than inland areas in similar latitudes (Fig. 5.6a). Most of England south of a line from the Wirral to the Humber has a mean temperature above 16 °C with some areas exceeding 17 °C, notably in the Greater London area. Wales, apart from the south-east, Ireland and Scotland have temperatures below 16 °C, falling to 12·2 °C in the cloudy, northerly latitudes of the Shetland Islands where long hours of daylight fail to compensate for their unfavourable position. As indicated above, July temperatures fluctuate within a much narrower range than those of January (Fig. 5.2c) and there is no instance of a July which gave an uninterrupted flow of continental air sufficient to elevate the temperature well above the normal.

At Kew, July mean temperatures ranged from a low of 15·6 °C in 1954 to a high of 19·1 °C in 1959 during the 30-year period. This latter year was the classic example of a summer dominated by an extension of the Azores high far to the north-east and over a time period in excess of one single month. Johnson (1960), in his analysis of the 1959 summer, shows that blocking highs were centred mainly over and to the east of Britain. July produced a temperature of 36 °C in Lincolnshire. By way of contrast the summer of 1954 was cyclonic in character and the persistent westerly winds gave rise to cool, dull weather with below average temperatures. Lamb and Johnson (1960) compare the sea-level pressure maps for July 1954 and July 1955, another month when the Azores high pressure area spread north-eastwards over the British Isles and Scandinavia.

Murray and Ratcliffe (1969) have drawn attention to the anomalous summer of 1968 when the north-west seaboard experienced long fine spells and east England was unusually cool, dull and wet. Blocking activity was very frequent with the centres north of about 55° and frequent easterly and northerly winds over the British Isles. To quote Manley (1970a) : 'The character of the season can vary considerably within England alone. Around London, the very poor weather of the 1968 summer holiday season is fresh in the memory, but at Manchester it was agreeably dry and sunny, and in western Scotland superb.' The position of the blocking high is thus seen to be very important in terms of the actual cloud conditions and temperature levels experienced in different parts of the British Isles.

The temperature of the ocean surfaces surrounding the British Isles reaches its highest value in August and this in turn affects the month of highest mean temperature at coastal stations. Compared to the February situation far more coastal stations are affected since the phenomenon also appears on the south coast east of Eastbourne and at stations on the North Sea coast as far north as Scarborough. The

Scilly Isles are 0·3 °C warmer in August than in July and the exposed Cape Wrath is as much as 0·5 °C warmer.

Fig. 5.6 *July temperatures, 1941–70, in °C, reduced to sea-level:* (a) *Mean;* (b) *Mean daily maximum;* (c) *Mean daily minimum;* (d) *Mean daily range*

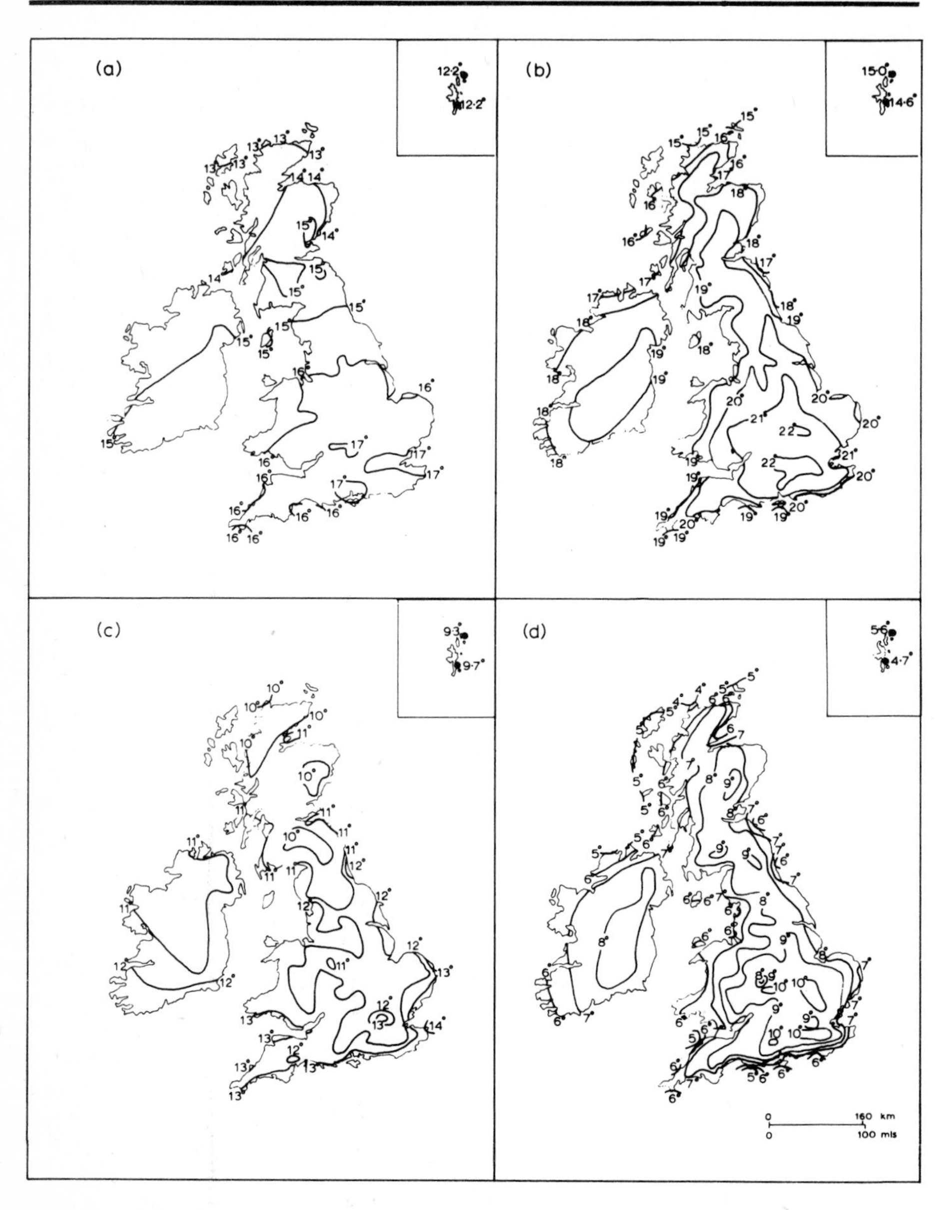

The isotherms of July mean daily maximum temperature (Fig. 5.6b) are almost identical in pattern to those of April but have increased in value by about 6 °C in the far north and 8 °C in the south. Highest values, in excess of 22 °C, are found in central southern

England. At Lerwick the maximum does not reach 15 °C. Irvine (1968) points out that a maximum of 20 °C is considered as a very warm summer day in Shetland and even a maximum of 15 °C is considered warm.

The highest summer temperatures recorded in the British Isles are usually the result of an inflow of cT air from its source region over North Africa–Iberia or south-east Europe. This air mass reaches south-east England over a predominantly land track, often over the intensely heated landscape of Europe, and can result in temperatures in excess of 30 °C and occasionally above 35 °C. An example of such an occasion was 1 July 1968 when the maximum temperature in the London area reached 32 °C. The source region of the air near the surface was Europe and the Mediterranean while the layer between about 4,300 m and 5,200 m originated in the region of the southern Sahara Desert (Stevenson, 1969). Dust was carried northwards from the desert to be deposited over England and Wales and severe storms broke out at the boundary of the cT air and the much cooler, much more moist air from the Atlantic.

Figure 5.8c illustrates the characteristic distribution of maximum temperatures over England and Wales on a hot summer day, 29 June 1957. Temperatures exceeded 30 °C in central southern England with 34 °C at Heathrow (London Airport). Around the coasts of England and Wales there was evidence of a sea-breeze effect which kept the maximum temperature below 20 °C in some cases. The synoptic situation was classified by Lamb (1972) as ASW with a cold front separating the warm air over England and Wales from slightly cooler air over Ireland and Scotland.

Perry (1968), in his analysis of the occurrence of 'summer days' in the British Isles, found that at the majority of the sixteen stations which he considered for the period 1947–66, Lamb's 'Anticyclonic' weather type provided the largest percentage of summer days. The westerly type is important along the east coast and at western stations the easterly type becomes increasingly important. In 1962 the temperature did not reach 25 °C on a single day. An interesting point brought out by Perry is that by averaging 'summer days' frequencies at Kew 'odd' years average 19·7 days and 'even' years only 8·6 days.

Glasspoole (1944) has shown that the distribution of the mean number of days with a maximum temperature above 25 °C ranges from an area which can expect more than 10 days in inland southern England to one or even none over Ireland, western coastal areas of England, Wales and Scotland and also extreme northern Scotland.

Perry (1967) has also investigated the relationship between days on which the maximum temperature at two or more of the stations quoted in the Daily Weather Report reaches or exceeds 32 °C and the associated synoptic situation. Between 1900 and 1966 this occurred only fifty times, of which 1911 contributed seven. In about two-thirds of the years, no such hot days were observed. Spatially, the south-east of England and the Midlands are the most favoured areas. Perry found that on 57 per cent of such hot days a surface high was centred in the rectangle between 50 °N. and 60 °N. and 5 °E. and 20 °E. He emphasized that the trajectory of the air mass and the state of the ground over which the air travels play a large part in determining the maximum temperatures in Britain during quiet summer weather and that deep subsidence is probably a pre-requisite for very high temperatures.

The variation of mean daily minimum temperature over the British Isles in July is far less than for the mean daily maximum, ranging from 9·3 °C at Baltasound to 14·2 °C at Margate (Fig. 5.6c). Highest temperatures are found along the coasts of southern England and in the London urban area. The pattern is, in general, the inverse of maximum temperature distribution with isotherms bending southwards over land areas and northwards over coastal areas, resulting in inland areas being cooler than coasts in the same latitude.

The map of mean daily range of temperature in July (Fig. 5.6d) is remarkably similar in pattern and isoline values to its April counterpart, with a steep gradient along the south coast of England and highest ranges in parts of inland southern England.

Autumn

The pattern of mean temperature in October (Fig. 5.7a) resembles that of January as the land and maritime influences again make themselves felt and isotherms bend southwards over land areas and north wards over the sea. This is illustrated in the position of the 11 °C isotherm which bends southwards over Ireland then northwards to include the Isle of Man, then southwards almost to the Bristol Channel and the London area, then northwards again along the North Sea coast as far as Redcar. Coastal areas of south-west Wales, southern England and also a small area in central London exceed 12 °C with the Scilly Isles and Portland Bill topping 13 °C. It has already been pointed out that October temperature levels exceed those of April. In the case of Lerwick, October is 2·9 °C warmer than April and even 0·9 °C warmer than May. At Kew, October is 2 °C warmer than April. Standard deviations in October are approximately ±1 °C (Table 5.1), similar to the April figure but higher than the July range.

Shellard (1959) has investigated the spatial distribution of the standard deviation of each monthly mean for the 1921–50 period for fifty-two stations in the United Kingdom. His results, presented in a series of twelve maps, show that in all months the standard deviation is smaller at coastal stations than inland, indicating the maritime effect of reducing the fluctuation of monthly temperatures about their means. As confirmed above, Shellard found that deviations were twice as great in winter as in summer, with January and February the most variable months. In general, May showed the least variability. The greatest variability noted was in February for a station in central southern England with a standard deviation of ±2·3 °C and the least variability at Tiree in May with a figure of ±0·4 °C.

In October the isotherms of mean daily maximum temperature (Fig. 5.7b) still bend northwards over land areas and southwards over the sea. This is particularly marked on the North Sea coasts where all the isotherms dip steeply towards the south. Much of southern England has a maximum in excess of 15 °C.

At night (Fig. 5.7c) the familiar pattern of warm coasts, especially

Fig. 5.7 *October temperatures, 1941–70, in °C, reduced to sea-level.* (a) *Mean;* (b) *Mean daily maximum;* (c) *Mean daily minimum;* (d) *Mean daily range*

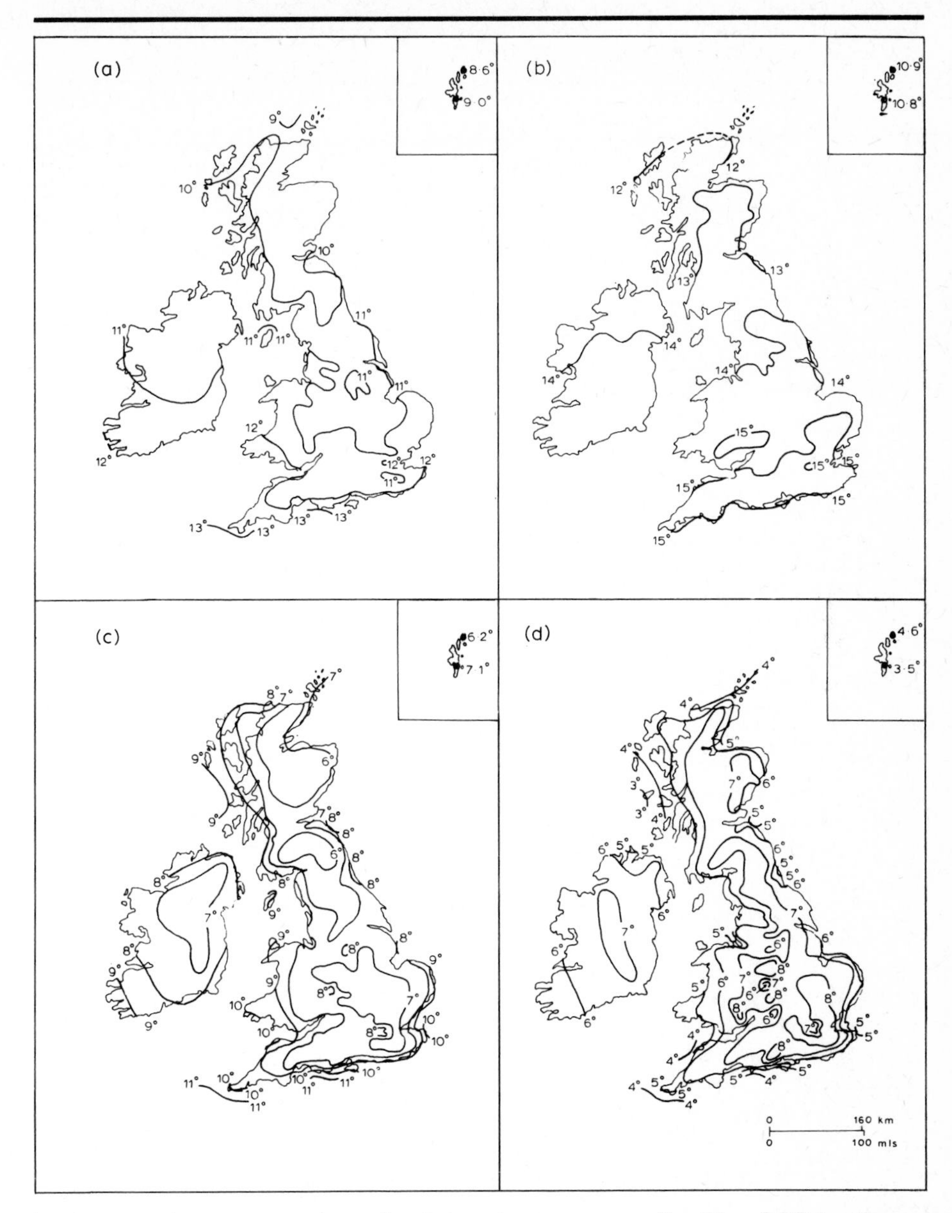

in the south-west, and cooler inland areas prevails. The 8 °C isotherm runs from north-west Scotland down the west coast, through central Wales to near the south coast of England before turning eastwards and again turning northwards up the coast of East Anglia. The October mean daily range of temperature (Fig. 5.7d) closely resembles the July pattern although ranges have decreased generally by a few degrees, mainly due to a greater decrease in maximum temperatures than in minimum temperatures.

Fig. 5.8 (a) *Temperatures in °C and wind data at 06.00 hrs on 15 January 1974 (data from Daily Weather Report, No. 40962)*; (b) *Maximum temperatures in °C recorded on 4 April 1974, with wind data for 12.00 hrs (data from Daily Weather Report, No. 41042)*; (c) *Maximum temperatures in °C recorded on 29 June 1957, with wind data for 12.00 hrs (data from Daily Weather Report, No. 34919)*; (d) *Annual range of mean temperature, 1941–70, in °C*

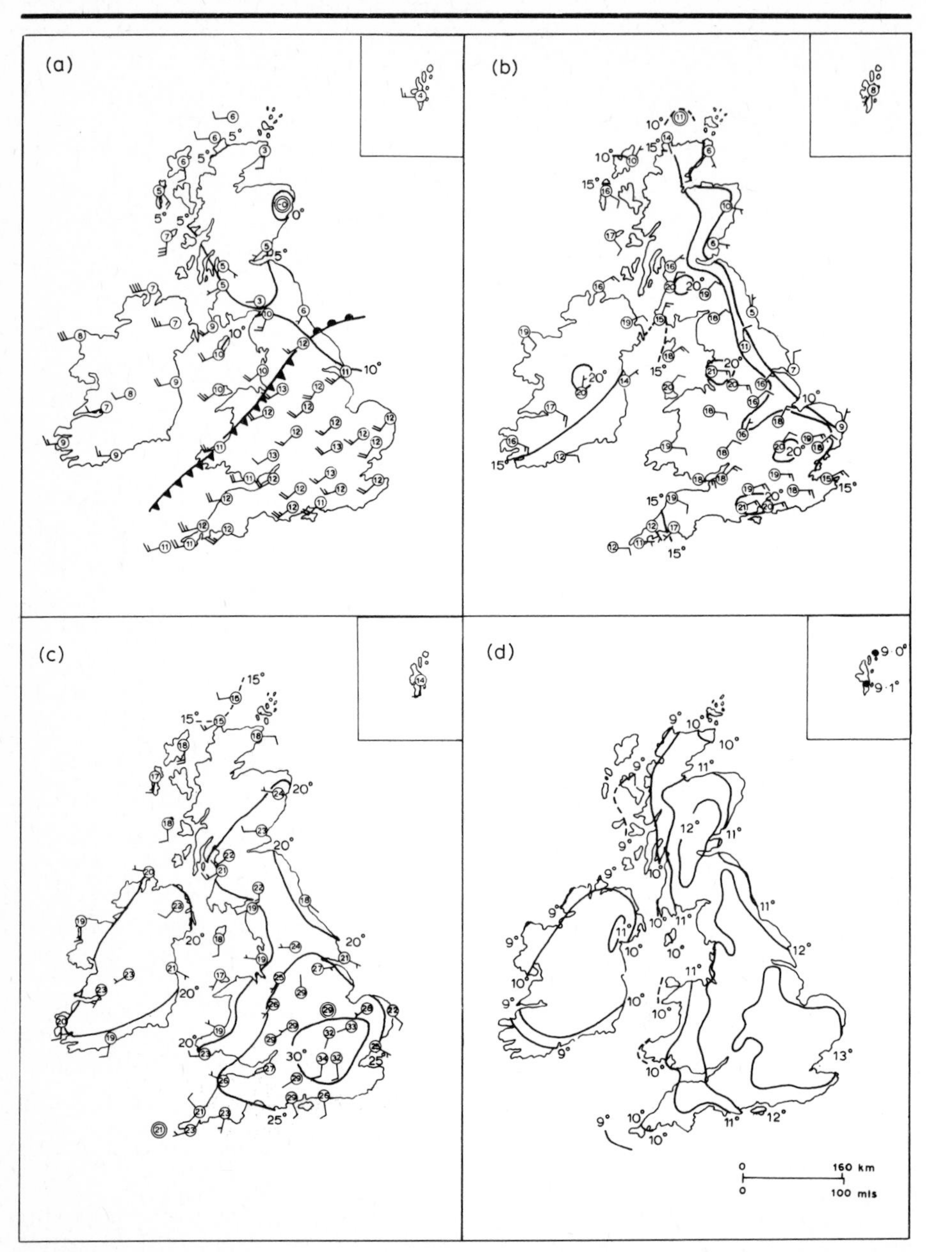

The actual mean temperature of each month for three contrasting stations has been plotted in Fig. 5.9. In the November to February period, inclusive, Lerwick in the far north of the British Isles is not much colder than Kew in the more continental south-east of England but during the rest of the year the difference is much greater so that it is clear that it is the contrast in summer temperature which gives the south-east of England its higher annual temperature rather than any appreciable winter temperature difference. The mildness of the Scilly Isles in winter is very evident from the figure and only in the May to August period is Kew warmer.

Fig. 5.9 *Mean monthly temperatures, 1941–70, for Scilly, Kew and Lerwick*

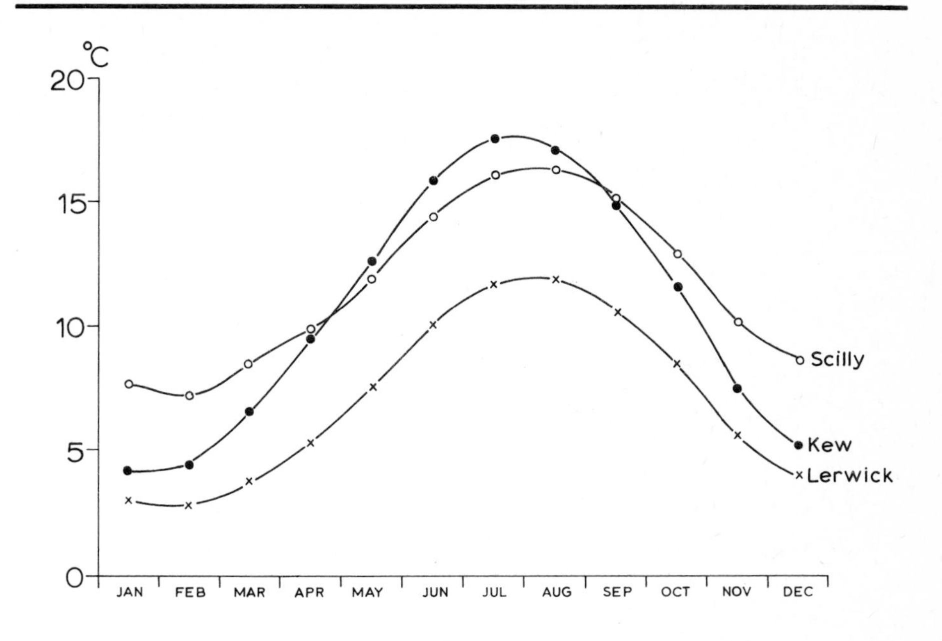

One of the characteristics of a maritime situation is a small annual range of mean temperature, the difference between the highest and lowest mean temperatures, compared to more continental areas. Figure 5.8d shows that areas in the extreme west have ranges less than 9 °C, e.g. Cape Wrath 8·4 °C, Valentia 8·6 °C. Ranges increase towards the east and south with much of east and south-east England having ranges in excess of 13 °C with a maximum of 13·7 °C at Heathrow (London Airport). North of the River Humber the effect of the North Sea results in lower ranges compared with inland areas to the west. As indicated above it is the higher summer temperatures of the south-east which give rise to the larger annual range of temperature experienced there compared with stations in the far north.

Extremes of temperature

A total of eighty-three stations with a complete 1941–70 record were used to prepare a working map (not reproduced) of the absolute maximum temperature recorded in the British Isles over the 30-year period. Similarly 141 such stations were used to obtain a working map of the absolute minimum temperature. All temperature values quoted in this section refer to *actual* station temperatures.

At most of the stations in England and Wales, whose records were used, 30 °C was exceeded. Gorleston, on the East Anglian coast, failed to reach this figure, as did much of the extreme south-west, e.g. the Scilly Isles recorded an absolute maximum of only 27·8 °C. Over the Isle of Man the temperature also failed to reach 28 °C. In Ireland the maximum temperature only barely exceeded 30 °C (e.g. Shannon Airport 30·3 °C), while Dublin Airport was remarkably cool with a maximum of only 26·2 °C. A few stations in Scotland exceeded 30 °C, e.g. Dumfries 30·6 °C, but the extreme north is crossed by the 25 °C isotherm of absolute maximum temperature and Lerwick only managed 23·3 °C.

Table 5.2 *Absolute maximum temperature recorded in the United Kingdom for each month, corrected to December 1971 (data from the Meteorological Office)*

Month	Date	Place	°C
January	10, 1971	Llandudno and Aber	18·3
February	28, 1891	Barnstaple	19·4
March	29, 1929	Wakefield	25·0
	29, 1965	Wakefield	25·0*
	29, 1968	Santon Downham and Cromer	25·0*
April	16, 1949	Camden Square	29·4
May	22, 1922	Camden Square	32·8
	24, 1922	Greenwich	32·8
	29, 1944	Camden Square, Regent's Park, Tunbridge Wells and Horsham	32·8
June	29, 1957	Camden Square	35·6
July	22, 1868	Tonbridge	38·1
August	9, 1911	Greenwich	37·8
September	2, 1906	Bawtry	35·6
October	5, 1921	Camden Square and Kensington	28·9
	6, 1921	Greenwich, Hampstead, Kensington and Westminster	28·9
November	4, 1946	Prestatyn	21·7
December	2, 1948	Achnashellach	18·3

* For a description of the synoptic situation on these occasions and a general discussion of the circumstances favourable to high temperatures in the early part of the year see Lyall (1971 b).

Table 5.2 gives the officially accepted absolute maximum temperature recorded in each month since records began, corrected to December 1971. The highest temperature recorded in the British Isles is 38·1 °C on 22 July 1868 at Tonbridge, Kent (altitude 23 m) (Lamb, 1958). In fact Lamb and Johnson (1960) consider that the summer of 1868 was probably the longest and hottest on record in England and temperatures above 32 °C were recorded in south-east England every month from May to September inclusive. In July the temperature exceeded 32 °C on nine days and the soil had dried out to an abnormal degree. On the day on which the record occurred temperatures above 35 °C were observed over a wide area around London. Lamb and Johnson's map of the July 1868 mean sea-level pressure distribution shows a north-eastward extension of the Azores high over the British Isles and Scandinavia. From 7–23 July, anticyclones centred south-east of Lyon, France, were continuously in evidence and gave rise to airflows which travelled over a very dry surface before reaching the British Isles.

The record for August is only slightly lower, 37·8 °C on 9 August 1911. Lamb (1958) states that prolonged warm weather beforehand was probably also important in this instance. The highest temperature recorded in January is 18·3 C on the 10th in 1971 (Lyall, 1971a) at two stations in Gwynedd. Very mild mT air with its origin south of 30 °N. arrived over the British Isles on the 9th. The 500 mb contour chart for 00.00 hrs on 11 January clearly indicated the development of a 'warm high' and the synoptic situation of 9–11 January closely resembled those favourable for the development of early warm spells (Lyall, 1971b).

The range of the absolute minimum temperature recorded over the British Isles between 1941 and 1970 is much greater than in the case of the absolute maximum. At most coastal locations around the entire British Isles approximately −10 °C was the absolute minimum, e.g. Lerwick, Swanage and Tynemouth all −9 °C. Absolute minima were even higher at stations fully exposed to the west and south-west, e.g. Scilly Isles −5 °C; Malin Head −6 °C; Valentia −7 °C. Inland the absolute minimum fell to values below −20 °C even in southern England, e.g. Woburn −21 °C, and in Scotland −25 °C was recorded at Braemar. Over Ireland approximately −12 °C was the absolute minimum although stations in the north-west recorded lower temperatures, e.g. Claremorris −17 °C; Markree Castle −18 °C.

Lyall (1973) has described the conditions necessary for widespread low temperatures (below −10 °C) in southern Britain. Cold air stagnating in an easterly weather type over freshly-fallen snow and with clear sky conditions gives rise to low minima generally, but additional factors such as below average sea-surface temperatures over the North Sea and unusually severe conditions over the Continent are necessary for the exceptionally cold spells such as those which occurred in January 1963.

Table 5.3 *Absolute range of temperature for each month in °C*

J	F	M	A	M	J	J	A	S	O	N	D
45·0	46·6	47·8	44·4	42·8	41·2	41·4	40·6	43·4	40·6	45·0	45·0

The absolute minimum temperature officially accepted for the British Isles is −27·2 °C, recorded at Braemar on 11 February 1895, (in comparison with the lowest temperature recorded in the 1962–63 winter of −22·2 °C also at Braemar on 18 January 1963). This means that the British Isles have experienced an absolute range of temperature of 65·3 °C. On a monthly basis, Table 5.3 shows that the absolute range is in excess of 40 °C in each case.

Table 5.4 *Absolute minimum temperature recorded in the United Kingdom for each month, corrected to December 1971 (data from the Meteorological Office)*

Month	Date	Place	°C
January	17, 1881	Kelso	—26·7
February	11, 1895	Braemar	—27·2
March	14, 1958	Logie Coldstone	—22·8
April	2, 1917	Newton Rigg	—15·0
May	17, 1891	Ben Nevis	—10·0
June	9, 1955	Dalwhinnie	—5·6
	1, 1962	Santon Downham	—5·6
	3, 1962	Santon Downham	—5·6
July	10, 1888	Ben Nevis	—3·3
August	13, 1887	Ben Nevis	—2·8
	29, 1959	Alwen	—2·8
	30, 1885	Alston	—2·8
	26, 1952	Glenlivet	—2·8
September	23, 1893	Ben Nevis	—7·8
October	28, 1948	Dalwhinnie	—11·7
	20, 1880	Braemar	—11·7
November	14, 1919	Braemar	—23·3
December	3, 1879	Kelso	—26·7

A temperature below freezing point has occurred in each month of the year (Table 5.4). With one exception, Newton Rigg in Cumbria, the station which has recorded the absolute minimum in each month is located in Scotland, not surprisingly as high level stations such as the summit of Ben Nevis (record from December 1883 to September 1904) are included. But in June 1962, Santon Downham (altitude 24 m) in the Breckland of west Norfolk equalled the June low of −5·6 °C recorded at Dalwhinnie in 1955. Oliver (1966) analysed the Santon Downham temperature record which is available from 1958 and concluded that a particular feature of the site, in a clearing in Thetford Forest, is the low night minima recorded from June to August. Air frosts have been observed there in every month of the year. On 4 May 1941 a minimum of −9·4 °C was recorded near Thetford, close to Santon Downham, this being only 0·6 °C different from the lowest temperature on record in May in the British Isles. It is important to bear in mind that local site characteristics, as exemplified in the case of Santon Downham, can make a station unrepresentative of the larger area in which it is situated.

Frost

An 'air frost' is recorded when the temperature in the Stevenson screen falls below 0 °C. From 1906 to 1960 a 'ground frost' referred to a grass minimum temperature of 30·4 °F or below. 'From 1 January 1961, the statistics issued have referred to the "number of days with grass minimum temperature below 0 °C" and no statistics have referred to "ground frost" ' (Meteorological Office, 1972).

The maps of mean daily maximum and minimum temperatures show that there is a much greater spatial variation of the latter and local topographical features become of paramount importance in a discussion of frost frequency, especially for ground frost where even the vegetation cover and soil type are important factors which can increase or decrease the incidence of grass minima below 0 °C (see Chapter 13).

Table 5.5 *Mean number of days of air frost, 1946–60 (Meteorological Offices, 1972)*

Station	J	F	M	A	M	J	J	A	S	O	N	D	Year
Lerwick	10·8	11·5	7·9	4·9	1·3	0·1	—	—	—	0·8	2·7	5·9	z5·9
Stornoway	9·1	10·0	6·7	3·3	0·6	—	—	—	—	1·0	3·9	6·1	40·7
Eskdalemuir	19·2	19·0	15·7	13·0	4·9	0·9	0·2	0·1	1·3	4·8	10·4	15·3	104·8
Douglas	6·3	8·7	3·5	0·3	—	—	—	—	—	0·1	0·7	2·9	22·5
Gorleston	7·8	9·3	5·0	0·7	0·1	—	—	—	—	0·3	1·7	5·0	29·9
Cambridge	13·7	14·0	10·2	4·5	0·7	—	—	—	0·1	2·5	6·1	10·5	62·3
Birmingham (Edgbaston)	11·1	12·5	8·3	0·9	—	—	—	—	—	0·6	2·5	6·2	42·1
Kew	10·5	10·3	7·2	1·1	—	—	—	—	—	1·3	3·5	7·0	40·9
Southampton	12·1	10·8	7·3	1·1	—	—	—	—	—	1·1	3·5	7·3	43·2
Valley	6·3	7·7	4·7	1·0	—	—	—	—	—	—	0·8	2·4	22·9
Falmouth	4·0	5·7	2·4	1·0	—	—	—	—	—	—	0·2	1·3	14·6
Aldergrove	11·5	12·5	7·6	4·3	1·3	0·2	—	—	0·3	1·1	5·1	7·2	51·1

The annual number of days of air frost over the British Isles varies from less than 5 in the Scilly Isles to over 100 in certain areas of Highland Britain. Table 5.5 gives the mean number of days of air frost for twelve stations in the United Kingdom for 1946–60. The annual totals of 15 for Falmouth and 23 for Valley contrast greatly with the 105 total for Eskdalemuir.

The variation from year to year can be great, as shown in Table 5.6, where 1957, with a mild winter and early spring, and 1963, with a very cold winter, were chosen to represent calendar years with below and above average numbers of days of frost respectively. In 1957 the Scilly Isles did not record a single air frost and the minimum temperature for the year was +1 °C and even in 1963 only nine air frosts were recorded.

Manley (1970b) has demonstrated the very large differences in the incidence of air frost which can be found within a few kilometres of a particular location. Using Cambridge as an example, Manley quotes Santon Downham (45 km to the north-east) as having a mean of 102 days of air frost between 1956 and 1965 and Mildenhall (35 km to the north-east) as having only 57. For the same period Moorhouse at 550 m in the Pennines experienced 131 days of air frost.

The mean date of the first air frost ranges from mid-August in the highest areas of inland Scotland to mid-December in exposed coastal regions in the west of Wales and the extreme south-west of England. The last frost may be expected in early March in these latter regions whereas air frosts usually occur until the beginning of June in upland areas of Scotland. However, the variation from one year to another of the length of the frost-free period and the dates of the first and last frosts is considerable.

Table 5.6 *Number of days of frost in 1957 and 1963 (Meteorological Office 1972 and* Monthly Weather Reports*)*

Station	Air frosts			Ground frosts	
	1957	1963	1946–60 mean	1957	1963
Lerwick	42	50	46	50	93
Stornoway	33	51	41	60	100
Eskdalemuir	86	104	105	106	144
Douglas	10	39	23	40	77
Gorleston	17	61	30	33	88
Cambridge	43	81	62	90	140
Birmingham (Edgbaston)	19	69	42	40	93
Kew	13	61	41	78	100
Southampton	22	68	43	35	99
Valley	13	56	23	38	87
Falmouth	7	33	15	35	74
Aldergrove	36	70	51	72	111

Human comfort

In some parts of the world, the dry-bulb temperature alone is a poor guide to the actual conditions experienced by the population and biometeorological indices, which include other meteorological and also physiological parameters, have been developed to describe the meteorological environment more adequately. In the British Isles, conditions can hardly be described as dangerous to human survival, at least at sea-level, but they are frequently below the comfort zone for temperate latitudes. This comfort zone can be defined in terms of effective temperature (ET), a biometeorological index mainly used to express conditions in tropical and sub-tropical climates. In England, during the summer months, the comfort zone can be taken as lying between ET 15·5 °C and 19 °C and in winter between ET 14 °C and 17 °C (Air Ministry, 1959).

Foord (1968) has analysed the climate of London by calculating

Table 5.7 *Temperature data for twenty-five stations in the British Isles, 1941–70 (see Station Index)*

(i) Mean daily maximum temperature; (ii) Mean daily minimum temperature; (iii) Mean temperature; (iv) Highest mean daily maximum; (v) Lowest mean daily maximum; (vi) Highest mean daily minimum; (vii) Lowest mean daily minimum.

Lerwick 60° 08′ N.; 01° 11′ W.; 82 m (see also Irvine, 1968)

Month	*(i)*	*(ii)*	*(iii)*	*(iv)*	*(v)*	*(vi)*	*(vii)*
Jan	4·7	1·3	3·0	7·0	2·7	3·8	−1·5
Feb	4·7	0·9	2·8	7·4	2·0	3·4	−1·9
Mar	5·8	1·8	3·8	8·4	2·7	4·3	−2·1
Apr	7·9	2·9	5·4	9·3	6·2	4·4	0·9
May	10·1	5·1	7·6	11·7	8·2	6·4	3·2
June	12·5	7·6	10·1	13·8	11·3	9·1	5·9
July	14·0	9·3	11·7	16·1	12·5	10·7	6·9
Aug	14·1	9·6	11·9	17·1	12·3	11·6	7·7
Sept	12·7	8·5	10·6	14·1	10·6	10·2	6·1
Oct	10·2	6·7	8·5	11·6	8·6	8·3	5·1
Nov	7·3	3·9	5·6	8·9	4·6	5·7	0·5
Dec	5·6	2·4	4·0	7·6	3·5	4·8	−0·2
Annual	9·1	5·0	7·1	10·1	8·4	5·8	4·3

Aberdeen 57° 08′ N.; 02° 08′ W.; 52 m

Month	*(i)*	*(ii)*	*(iii)*	*(iv)*	*(v)*	*(vi)*	*(vii)*
Jan	5·2	−0·2	2·5	7·9	2·4	1·8	−3·4
Feb	5·6	−0·2	2·7	9·4	0·4	2·4	−3·6
Mar	7·5	1·6	4·5	12·3	1·9	4·6	−1·8
Apr	10·5	3·1	6·8	12·9	7·4	4·8	1·2
May	12·5	5·5	9·0	15·1	10·6	7·1	4·1
June	15·8	8·5	12·1	18·0	14·3	10·1	7·1
July	17·3	10·2	13·7	19·5	14·7	11·1	8·4
Aug	16·8	9·9	13·3	20·8	14·6	11·6	8·3
Sept	15·3	8·4	11·9	17·3	12·6	10·7	6·3
Oct	12·3	6·2	9·3	14·6	10·5	8·6	4·7
Nov	8·2	2·5	5·3	10·2	5·8	4·8	−0·8
Dec	6·3	1·1	3·7	9·2	3·6	2·8	−1·7
Annual	11·1	4·7	7·9	12·3	10·2	5·5	4·2

Wick 58° 27′ N.; 03° 05′ W.; 36 m

Month	*(i)*	*(ii)*	*(iii)*	*(iv)*	*(v)*	*(vi)*	*(vii)*
Jan	5·1	1·2	3·1	7·2	3·0	3·3	−1·6
Feb	5·6	0·9	3·3	8·3	2·1	3·1	−2·2
Mar	7·3	2·1	4·7	10·3	4·1	4·5	−1·9
Apr	9·5	3·3	6·4	11·1	6·9	5·0	1·2
May	11·2	5·4	8·3	12·9	9·3	7·1	3·6
June	14·1	7·9	11·0	16·0	12·8	9·5	6·6
July	15·4	9·6	12·5	17·7	13·7	11·1	7·7
Aug	15·5	9·7	12·6	18·2	13·8	11·2	7·7
Sept	14·2	8·7	11·5	15·8	11·8	10·5	6·2
Oct	11·7	6·9	9·3	13·4	10·2	8·9	5·3
Nov	8·1	3·7	5·9	10·1	5·2	6·4	0·3
Dec	6·0	2·3	4·1	7·8	3·4	4·6	−0·3
Annual	10·3	5·1	7·7	11·4	9·5	6·1	4·5

Braemar 57° 00′ N.; 03° 24′ W.; 339 m

Month	*(i)*	*(ii)*	*(iii)*	*(iv)*	*(v)*	*(vi)*	*(vii)*
Jan	3·5	−2·7	0·4	6·5	−0·7	0·3	−8·6
Feb	3·8	−2·8	0·5	7·1	−1·4	1·7	−8·4
Mar	6·4	−0·9	2·7	10·8	2·1	2·5	−4·9
Apr	9·4	1·0	5·2	12·1	6·2	3·4	−1·2
May	12·9	3·5	8·2	15·6	9·5	6·0	1·8
June	16·3	6·7	11·5	18·8	14·4	8·6	5·2
July	17·1	8·3	12·7	22·7	15·0	10·1	6·5
Aug	16·6	8·0	12·3	22·8	13·6	10·1	6·1
Sept	14·3	6·4	10·3	16·9	12·1	8·3	4·4
Oct	11·0	4·0	7·5	14·2	8·8	6·9	1·4
Nov	6·5	0·3	3·4	9·0	3·7	2·6	−3·3
Dec	4·5	−1·1	1·7	6·7	1·1	2·0	−4·7
Annual	10·2	2·6	6·4	11·4	9·3	3·7	1·5

Table 5.7 continued

(i) Mean daily maximum temperature; (ii) Mean daily minimum temperature; (iii) Mean temperature; (iv) Highest mean daily maximum; (v) Lowest mean daily maximum; (vi) Highest mean daily minimum; (vii) Lowest mean daily minimum.

Tiree 56° 30′ N.; 06° 53′ W.; 9 m

Month	(i)	(ii)	(iii)	(iv)	(v)	(vi)	(vii)
Jan	6·9	3·4	5·1	8·6	4·3	5·7	0·4
Feb	6·9	2·8	4·9	8·9	3·3	5·6	−1·7
Mar	8·5	3·9	6·2	10·4	6·3	6·7	0·0
Apr	10·4	4·9	7·7	11·7	8·0	6·8	3·2
May	12·8	7·0	9·9	15·3	11·0	8·4	5·8
June	14·9	9·5	12·2	16·2	13·4	10·9	8·6
July	15·9	10·9	13·4	17·8	14·2	11·8	9·7
Aug	16·1	11·1	13·6	19·5	14·6	12·4	9·7
Sept	14·8	10·3	12·5	16·6	13·3	11·8	7·5
Oct	12·5	9·6	10·5	14·4	11·4	10·6	7·1
Nov	9·5	5·7	7·6	11·4	7·3	8·4	3·3
Dec	7·9	4·5	6·2	9·6	5·2	6·8	1·8
Annual	11·4	6·9	9·1	12·2	10·7	7·8	6·1

Aldergrove 54° 39′ N.; 06° 13′ W.; 68 m

Month	(i)	(ii)	(iii)	(iv)	(v)	(vi)	(vii)
Jan	5·8	1·3	3·5	8·3	1·9	4·4	−2·9
Feb	6·5	1·1	3·8	9·6	1·6	4·6	−1·9
Mar	9·0	2·4	5·7	12·4	6·2	6·0	−1·3
Apr	11·8	3·9	7·9	13·8	9·1	6·1	2·3
May	14·7	6·1	10·4	16·7	12·5	8·3	4·6
June	17·4	9·2	13·3	19·9	15·8	11·3	8·2
July	18·1	10·7	14·4	21·3	16·2	11·8	9·4
Aug	18·0	10·5	14·3	22·4	16·1	12·1	8·8
Sept	16·0	9·3	12·7	18·6	13·6	10·9	6·0
Oct	12·9	7·2	10·1	15·4	11·7	9·8	5·5
Nov	8·9	3·9	6·4	10·2	7·4	5·6	1·4
Dec	6·7	2·5	4·6	8·9	3·6	5·2	−0·9
Annual	12·1	5·7	8·9	13·3	11·3	6·3	4·9

Eskdalemuir 55° 19′ N.; 03° 12′ W.; 242 m

Month	(i)	(ii)	(iii)	(iv)	(v)	(vi)	(vii)
Jan	3·7	−0·9	1·4	6·2	0·5	1·7	−4·3
Feb	4·4	−1·3	1·5	7·2	−0·1	2·2	−6·3
Mar	7·2	0·1	3·7	11·3	3·2	3·6	−3·2
Apr	10·4	1·8	6·1	12·6	7·8	4·2	−1·2
May	13·8	4·1	8·9	16·6	11·1	6·1	2·5
June	16·6	7·2	11·9	19·0	14·7	9·1	5·4
July	17·4	8·8	13·1	21·9	15·1	10·2	6·9
Aug	17·1	8·8	12·9	22·8	14·6	10·5	6·3
Sept	14·9	7·3	11·1	17·5	11·7	9·8	5·2
Oct	11·6	5·1	8·3	14·6	9·7	7·9	2·7
Nov	7·2	1·9	4·5	8·6	4·5	4·3	−0·5
Dec	4·8	0·3	2·5	6·9	1·7	2·9	−2·8
Annual	10·8	3·6	7·2	12·1	9·9	4·7	2·9

Mullingar 53° 31′ N.; 07° 21′ W.; 108 m (1944 to 1970)

Month	(i)	(ii)	(iii)
Jan	6·6	1·2	3·9
Feb	7·2	1·2	4·2
Mar	9·9	2·6	6·3
Apr	12·5	3·9	8·2
May	15·3	6·3	10·8
June	18·0	8·9	13·5
July	18·9	10·5	14·7
Aug	18·9	10·3	14·6
Sept	16·7	8·7	12·7
Oct	13·5	6·5	10·0
Nov	9·3	3·4	6·3
Dec	7·4	1·9	4·7
Annual	12·9	5·5	9·1

Table 5.7 continued

(i) Mean daily maximum temperature; (ii) Mean daily minimum temperature; (iii) Mean temperature; (iv) Highest mean daily maximum; (v) Lowest mean daily maximum; (vi) Highest mean daily minimum; (vii) Lowest mean daily minimum.

Dublin Airport 53° 26′ N.; 06° 15′ W.; 81 m

Month	*(i)*	*(ii)*	*(iii)*
Jan	7·2	2·1	4·7
Feb	7·4	2·0	4·7
Mar	9·5	3·2	6·3
Apr	11·9	4·6	8·3
May	14·3	6·8	10·5
June	17·4	9·6	13·5
July	18·6	11·2	14·9
Aug	18·4	11·1	14·7
Sept	16·6	9·7	13·1
Oct	13·7	7·5	10·6
Nov	9·8	4·4	7·1
Dec	8·2	3·0	5·6
Annual	12·8	6·3	9·5

Valley 53° 15′ N.; 04° 32′ W.; 10 m

Month	*(i)*	*(ii)*	*(iii)*	*(iv)*	*(v)*	*(vi)*	*(vii)*
Jan	7·2	3·2	5·2	9·2	3·1	6·0	−2·2
Feb	7·3	2·7	5·0	9·4	1·9	5·7	−2·8
Mar	9·5	3·6	6·5	12·3	7·1	7·1	0·8
Apr	11·5	5·4	8·5	13·8	9·2	7·3	3·7
May	14·4	7·6	11·0	16·9	12·1	9·4	6·3
June	16·8	10·4	13·6	19·8	15·2	11·9	9·1
July	18·1	12·0	15·1	21·6	16·1	13·0	11·0
Aug	18·3	12·2	15·3	23·7	16·4	13·1	10·6
Sept	16·9	11·2	14·1	19·8	14·8	12·5	8·7
Oct	14·2	8·9	11·5	16·8	12·7	11·3	7·4
Nov	10·5	6·1	8·3	12·0	8·4	8·7	3·9
Dec	8·5	4·6	6·5	11·0	5·7	7·0	2·1
Annual	12·8	7·3	10·1	14·1	11·8	8·1	6·2

Valentia Observatory 51° 56′ N.; 10° 15′ W.; 14 m

Month	*(i)*	*(ii)*	*(iii)*
Jan	9·2	4·2	6·7
Feb	9·1	4·0	6·5
Mar	11·1	5·2	8·1
Apr	12·5	6·3	9·4
May	14·7	8·0	11·3
June	16·7	10·5	13·6
July	17·6	12·1	14·9
Aug	18·0	12·1	15·1
Sept	16·7	11·0	13·9
Oct	14·5	9·1	11·8
Nov	11·4	6·3	8·9
Dec	10·1	5·3	7·7
Annual	13·5	7·8	10·7

Llandrindod Wells 52° 14′ N.; 03° 22′ W.; 235 m

Month	*(i)*	*(ii)*	*(iii)*	*(iv)*	*(v)*	*(vi)*	*(vii)*
Jan	5·4	−0·0	2·7	8·6	−0·7	2·2	−6·5
Feb	5·9	−0·1	2·9	9·7	−0·9	3·5	−5·9
Mar	9·1	1·2	5·1	12·9	6·0	4·9	−3·1
Apr	12·3	3·2	7·7	15·0	9·2	5·2	1·0
May	15·5	5·8	10·7	17·6	13·3	7·8	3·6
June	18·5	8·7	13·6	21·1	16·1	10·4	7·2
July	19·5	10·4	14·9	22·7	16·6	11·6	8·9
Aug	19·1	10·5	14·8	23·7	16·8	11·7	8·9
Sept	16·8	8·5	12·7	20·1	13·9	10·4	6·2
Oct	13·3	5·9	9·6	15·7	11·7	8·9	3·2
Nov	8·6	2·8	5·7	10·6	6·4	5·2	0·7
Dec	6·5	1·0	3·7	8·8	3·2	4·3	−2·1
Annual	12·5	4·8	8·7	14·3	11·1	5·7	3·6

Table 5.7 continued

(i) Mean daily maximum temperature; (ii) Mean daily minimum temperature; (iii) Mean temperature; (iv) Highest mean daily maximum; (v) Lowest mean daily maximum; (vi) Highest mean daily minimum; (vii) Lowest mean daily minimum.

Swansea 51° 37′ N.; 03° 55′ W.; 8 m

Month	*(i)*	*(ii)*	*(iii)*	*(iv)*	*(v)*	*(vi)*	*(vii)*
Jan	7·5	2·6	5·1	9·8	2·2	5·0	−2·4
Feb	7·7	2·6	5·1	10·1	2·3	5·9	−1·8
Mar	10·0	3·9	6·9	12·4	7·5	7·3	0·8
Apr	12·9	6·1	9·5	15·1	10·4	8·0	4·3
May	15·8	8·8	12·3	17·4	14·0	10·6	6·6
June	18·6	11·8	15·2	21·2	16·3	13·3	10·4
July	19·8	13·3	16·5	23·5	17·1	15·2	11·9
Aug	19·7	13·3	16·5	23·9	17·8	16·2	12·1
Sept	17·9	11·9	14·9	20·8	15·9	14·6	9·2
Oct	14·9	9·3	12·1	17·0	13·6	11·7	7·0
Nov	10·9	5·9	8·4	12·4	8·8	7·8	3·7
Dec	8·7	3·9	6·3	11·2	5·2	7·1	1·6
Annual	13·7	7·8	10·7	15·0	12·4	8·7	6·9

Douglas 54° 10′ N.; 04° 29′ W.; 87 m

Month	*(i)*	*(ii)*	*(iii)*	*(iv)*	*(v)*	*(vi)*	*(vii)*
Jan	6·7	2·5	4·6	8·8	3·1	4·7	−0·5
Feb	6·6	2·1	4·3	9·2	1·3	4·9	−1·7
Mar	8·2	3·2	5·7	10·4	5·1	5·9	0·6
Apr	10·5	4·8	7·7	12·3	8·8	6·4	3·0
May	13·4	7·1	10·3	16·0	11·2	9·0	5·7
June	16·2	9·8	13·0	18·8	14·7	11·9	8·6
July	17·2	11·3	14·3	20·8	15·7	12·7	10·1
Aug	17·1	11·5	14·3	21·6	14·9	14·1	9·7
Sept	15·5	10·5	13·0	17·4	13·5	12·5	8·1
Oct	12·9	8·5	10·7	14·9	11·4	10·6	6·3
Nov	9·6	5·4	7·5	11·1	7·7	7·2	3·2
Dec	8·1	3·8	5·9	10·2	5·2	6·1	1·1
Annual	11·8	6·7	9·3	12·9	11·0	7·8	6·2

Tynemouth 55° 01′ N.; 01° 25′ W.; 29 m

Month	*(i)*	*(ii)*	*(iii)*	*(iv)*	*(v)*	*(vi)*	*(vii)*
Jan	5·6	2·2	3·9	8·4	2·6	4·4	−0·3
Feb	6·0	2·1	4·1	9·9	1·2	4·8	−1·7
Mar	7·8	3·1	5·5	11·8	4·2	5·9	0·2
Apr	10·4	4·9	7·7	13·4	7·7	6·9	3·1
May	12·2	7·1	9·7	14·5	9·6	8·7	5·6
June	15·7	10·1	12·9	18·3	13·4	11·3	9·0
July	17·4	12·0	14·7	19·0	14·4	12·9	10·1
Aug	17·2	12·0	14·6	19·8	14·6	13·8	10·0
Sept	15·9	10·6	13·3	17·6	13·1	12·7	8·9
Oct	13·0	8·4	10·7	15·0	11·3	10·3	6·7
Nov	8·8	5·1	6·9	10·8	7·2	7·2	2·9
Dec	6·6	3·2	4·9	8·6	3·6	5·4	0·6
Annual	11·4	6·7	9·1	12·8	10·3	7·5	5·9

Malham Tarn 54° 06′ N.; 02° 10′ W.; 395 m. (More information on the temperature regime at this high level station can be found in Manley, 1955).

Month	*(i)*	*(ii)*	*(iii)*	*(iv)*	*(v)*	*(vi)*	*(vii)*
Jan	3·1	−1·2	0·9	6·2	−0·8	0·6	−4·4
Feb	3·5	−1·4	1·1	7·2	−2·2	1·8	−5·8
Mar	6·0	−0·1	2·9	10·1	2·2	3·7	−3·3
Apr	9·4	2·0	5·7	11·9	5·6	3·9	0·4
May	12·7	4·6	8·7	15·4	10·0	7·0	2·8
June	15·6	7·7	11·7	18·1	13·5	9·6	6·6
July	16·5	9·2	12·9	20·9	14·4	10·6	7·9
Aug	16·2	9·3	12·7	21·0	13·7	11·2	7·6
Sept	14·0	7·8	10·9	17·2	11·1	10·2	5·3
Oct	10·6	5·3	7·9	13·6	8·8	7·7	3·5
Nov	6·4	1·9	4·1	8·0	4·7	4·4	−0·4
Dec	4·4	−0·1	2·1	6·6	0·8	2·8	−2·9
Annual	9·9	3·7	6·8	11·4	8·7	4·7	3·0

Table 5.7 continued

(i) Mean daily maximum temperature; (ii) Mean daily minimum temperature; (iii) Mean temperature; (iv) Highest mean daily maximum; (v) Lowest mean daily maximum; (vi) Highest mean daily minimum; (vii) Lowest mean daily minimum.

Manchester (Ringway Airport) 53° 21′ N.; 02° 16′ W.; 75 m

Month	(i)	(ii)	(iii)	(iv)	(v)	(vi)	(vii)
Jan	5·5	1·2	3·3	8·3	0·8	3·9	−3·6
Feb	6·3	1·0	3·7	9·8	0·5	4·2	−3·4
Mar	9·2	2·3	5·7	13·7	5·9	6·2	−0·7
Apr	12·3	4·4	8·3	14·5	7·7	6·9	2·2
May	15·7	6·9	11·3	18·6	13·4	9·5	4·6
June	18·8	9·9	14·3	21·8	16·6	11·9	8·1
July	19·6	11·7	15·7	23·6	16·7	12·6	10·3
Aug	19·4	11·5	15·5	24·6	16·9	13·1	9·7
Sept	17·3	10·0	13·7	20·9	14·1	11·8	7·0
Oct	13·7	7·3	10·5	17·4	11·9	10·4	5·1
Nov	8·9	4·0	6·5	10·8	6·7	6·3	1·2
Dec	6·4	2·2	4·3	8·7	2·8	4·9	−1·8
Annual	12·8	6·0	9·4	14·6	11·6	6·9	5·4

Cromer 52° 56′ N.; 01° 17′ E.; 54 m

Month	(i)	(ii)	(iii)	(iv)	(v)	(vi)	(vii)
Jan	5·9	0·9	3·4	9·2	1·3	2·8	−3·1
Feb	6·3	0·9	3·6	11·6	−0·1	3·2	−3·8
Mar	8·7	2·4	5·5	12·7	4·8	5·1	−0·4
Apr	11·9	4·6	8·3	15·7	9·5	6·6	3·0
May	14·9	7·2	11·1	17·8	12·5	8·7	5·7
June	18·0	10·2	14·1	21·4	15·8	12·2	8·7
July	19·8	12·1	15·9	22·8	16·4	13·7	10·6
Aug	19·7	12·1	15·9	22·7	17·0	14·5	9·9
Sept	17·9	11·0	14·5	21·1	14·8	14·3	9·1
Oct	14·4	8·2	11·3	16·8	12·3	11·6	6·7
Nov	9·6	4·4	7·0	10·8	7·4	6·4	2·9
Dec	7·3	2·2	4·7	9·1	4·0	5·4	−0·3
Annual	12·9	6·3	9·6	14·4	11·3	7·3	5·3

Waddington 53° 10′ N.; 00° 31′ W.; 68 m

Month	(i)	(ii)	(iii)	(iv)	(v)	(vi)	(vii)
Jan	4·9	0·8	2·9	7·8	0·1	3·3	−4·0
Feb	5·9	0·7	3·3	10·0	−0·5	4·8	−4·0
Mar	9·0	1·8	5·4	13·0	5·1	4·9	−1·2
Apr	12·4	4·1	8·3	15·7	9·4	6·2	2·6
May	15·7	6·6	11·1	19·8	13·1	8·0	5·0
June	18·9	9·6	14·3	21·7	16·5	11·0	8·6
July	20·3	11·6	15·9	22·9	16·9	13·5	9·8
Aug	20·0	11·3	15·7	26·1	16·9	13·1	9·6
Sept	17·8	9·8	13·8	20·7	14·7	12·2	7·6
Oct	13·9	7·1	10·5	17·1	12·1	9·7	5·2
Nov	8·6	3·8	6·2	10·6	6·3	5·4	1·4
Dec	6·0	1·9	3·9	8·1	2·6	5·3	−1·4
Annual	12·8	5·8	9·3	14·3	11·5	6·7	4·8

Rugby 52° 22′ N.; 01° 15′ W.; 117 m

Month	(i)	(ii)	(iii)	(iv)	(v)	(vi)	(vii)
Jan	5·5	0·2	2·9	9·0	−0·2	2·8	−5·4
Feb	6·4	0·2	3·3	10·4	−0·1	3·8	−4·9
Mar	9·5	1·6	5·5	13·1	5·8	5·2	−0·8
Apr	13·0	3·8	8·4	16·1	10·4	6·2	1·5
May	16·3	6·4	11·3	18·9	14·0	8·4	3·9
June	19·6	9·6	14·6	22·0	17·6	11·1	8·3
July	21·0	11·3	16·1	24·2	17·6	12·7	9·8
Aug	20·5	11·0	15·7	24·8	17·4	13·1	9·9
Sept	18·1	9·4	13·7	21·6	15·1	11·9	6·7
Oct	14·1	6·4	10·3	16·7	12·3	9·7	4·1
Nov	9·1	3·0	6·1	10·8	6·1	4·9	0·5
Dec	6·7	1·0	3·9	8·8	2·9	4·3	−2·0
Annual	13·3	5·3	9·3	14·9	11·7	6·2	4·4

Table 5.7 continued

(i) Mean daily maximum temperature; (ii) Mean daily minimum temperature; (iii) Mean temperature; (iv) Highest mean daily maximum; (v) Lowest mean daily maximum; (vi) Highest mean daily minimum; (vii) Lowest mean daily minimum.

Oxford 51° 46′ N.; 01° 16′ W.; 63 m

Month	(i)	(ii)	(iii)	(iv)	(v)	(vi)	(vii)
Jan	6·2	0·9	3·5	9·2	−0·2	3·7	−5·8
Feb	6·9	1·1	4·0	10·9	−0·2	4·9	−4·4
Mar	10·1	2·3	6·2	14·1	7·2	5·9	−0·7
Apr	13·5	4·8	9·1	16·2	11·1	7·0	2·6
May	16·7	7·4	12·1	19·2	14·2	9·4	5·0
June	20·1	10·4	15·3	22·3	17·9	11·9	9·3
July	21·5	12·3	16·9	24·7	18·5	13·6	11·0
Aug	21·0	12·1	16·5	25·8	18·3	13·6	10·6
Sept	18·6	10·2	14·4	22·1	15·7	12·5	7·2
Oct	14·7	7·2	10·9	17·3	13·3	10·8	4·9
Nov	9·8	3·9	6·9	11·7	7·1	6·3	1·3
Dec	7·3	1·9	4·6	9·6	3·7	4·9	−1·6
Annual	13·9	6·2	10·1	15·6	12·3	6·8	5·3

Bexhill-on-Sea 50° 50′ N.; 00° 28′ E.; 4 m

Month	(i)	(ii)	(iii)	(iv)	(v)	(vi)	(vii)
Jan	6·5	1·9	4·2	9·2	0·7	4·5	−3·5
Feb	6·7	1·9	4·3	9·6	1·1	5·2	−3·3
Mar	8·9	3·3	6·1	11·8	6·4	7·0	0·6
Apr	11·9	5·6	8·7	14·1	9·6	8·6	3·4
May	15·0	8·5	11·7	16·9	12·5	10·5	6·2
June	17·9	11·5	14·7	19·8	16·2	13·4	9·4
July	19·5	13·4	16·5	21·3	17·3	14·7	11·8
Aug	19·6	13·5	16·5	24·4	17·9	15·7	12·2
Sept	18·1	12·1	15·1	21·3	16·2	14·9	8·8
Oct	14·9	9·3	12·1	16·7	13·2	11·7	6·2
Nov	10·6	5·5	8·1	12·3	8·1	7·9	2·8
Dec	8·0	3·1	5·5	10·6	4·4	6·4	0·0
Annual	13·1	7·5	10·3	14·5	11·6	8·3	6·3

Kew Observatory 51° 28′ N.; 00° 19′ W.; 5 m

Month	(i)	(ii)	(iii)	(iv)	(v)	(vi)	(vii)
Jan	6·1	2·3	4·2	8·8	0·6	5·0	−2·7
Feb	6·8	2·3	4·5	10·3	0·7	6·1	−2·2
Mar	9·8	3·4	6·6	13·3	6·7	6·8	0·8
Apr	13·3	5·7	9·5	15·8	10·6	7·8	3·4
May	16·8	8·4	12·6	19·2	13·8	10·4	5·6
June	20·2	11·5	15·9	23·2	17·8	13·1	10·2
July	21·6	13·4	17·5	24·3	18·7	14·8	12·3
Aug	21·0	13·1	17·1	25·3	18·2	14·8	11·7
Sept	18·5	11·4	14·9	22·2	15·9	14·1	8·6
Oct	14·7	8·5	11·6	17·7	12·7	11·0	5·9
Nov	9·8	5·3	7·5	11·8	7·4	7·2	2·7
Dec	7·2	3·4	5·3	9·7	3·8	6·5	0·6
Annual	13·8	7·4	10·6	15·4	12·5	8·1	6·6

Shanklin 50° 37′ N.; 01° 11′ W.; 55 m

Month	(i)	(ii)	(iii)	(iv)	(v)	(vi)	(vii)
Jan	6·9	2·4	4·7	9·4	0·8	4·8	−3·4
Feb	6·9	2·1	4·5	9·6	0·9	5·7	−2·7
Mar	9·0	3·3	6·1	11·8	6·9	7·1	0·3
Apr	11·8	5·3	8·5	14·1	9·8	7·7	3·4
May	14·8	7·9	11·3	16·6	12·0	9·7	6·1
June	17·8	10·8	14·3	20·0	15·7	12·4	9·6
July	19·1	12·8	15·9	22·4	17·4	14·2	11·9
Aug	19·1	13·0	16·1	22·8	17·1	14·4	11·7
Sept	17·4	11·8	14·6	20·3	15·9	14·4	8·9
Oct	14·5	9·3	11·9	16·6	12·7	11·9	6·7
Nov	10·6	5·7	8·1	12·2	8·3	7·9	3·1
Dec	8·3	3·6	5·9	10·5	4·7	6·5	0·0
Annual	13·0	7·3	10·1	14·6	11·6	8·2	6·3

Table 5.7 continued

(i) Mean daily maximum temperature; (ii) Mean daily minimum temperature; (iii) Mean temperature; (iv) Highest mean daily maximum; (v) Lowest mean daily maximum; (vi) Highest mean daily minimum; (vii) Lowest mean daily minimum.

Scilly Isles 49° 56′ N.; 06° 18′ W.; 48 m

Month	*(i)*	*(ii)*	*(iii)*	*(iv)*	*(v)*	*(vi)*	*(vii)*
Jan	9·1	6·2	7·7	10·8	4·9	7·9	2·1
Feb	8·9	5·6	7·3	11·0	5·7	8·3	1·6
Mar	10·5	6·5	8·5	12·7	8·0	9·1	4·1
Apr	12·2	7·5	9·9	14·0	10·4	9·2	5·7
May	14·4	9·3	11·9	16·2	12·8	10·8	7·5
June	17·1	11·7	14·4	18·9	15·6	12·8	10·7
July	18·7	13·3	16·0	22·0	16·8	14·7	12·3
Aug	19·0	13·6	16·3	22·7	17·2	15·8	12·2
Sept	17·5	12·8	15·1	20·7	16·8	14·8	11·0
Oct	14·9	11·0	12·9	16·7	13·4	12·6	9·7
Nov	11·8	8·6	10·2	13·1	10·5	10·4	7·4
Dec	10·1	7·3	8·7	12·1	7·4	9·2	4·8
Annual	13·7	9·5	11·6	15·0	12·6	10·4	8·6

the ET using individual observations, not mean values, of the dry-bulb and wet-bulb temperature and the wind speed for Kew Observatory and the London Weather Centre. He found that effective temperatures above 19 °C occur for periods of a few hours only rather than in prolonged spells, and on 9·5 days per year on average. During the 1947–66 period which he examined, only one day had an ET above the comfort zone of a tropical area (19 °–24·5 °C). The residents of London have to endure far longer periods when the ET is below the lower limit of 14 °C. Effective temperatures during the months of November to February inclusive were constantly below this figure during the 20-year study period and an ET below 14 °C can in fact occur in any month of the year.

At low temperatures, the speed of the wind is important in accelerating the loss of heat from the body and the 'windchill' index was developed by Siple and Passel (1945) to express more adequately conditions of discomfort in areas which experience a cold season. This index is derived as follows:

$$H = (\sqrt{(100v)} + 10{\cdot}45 - v)\ (36 - t)$$

where H = windchill, expressed in kilogram calories per square metre of exposed surface per hour
v = wind speed in metres per second
t = air temperature in °C.

The limitations of this biometeorological index include the fact that it does not take into account solar radiation, which would mitigate the windchill effect (Court, 1948; Falconer, 1968). Nevertheless, although there are other avenues of heat loss, it is thought that windchill represents 75 to 80 per cent of the total heat loss from the body at these low temperatures. It should be remembered, however, that other variables such as the amount and type of clothing worn, whether the head is covered and the state of health of the person are also important in an assessment of personal comfort levels.

This index is of undoubted value in the study of the physiological winter climate of Canada, but Howe has emphasized that conditions can be quite hazardous at times even in these islands when low temperatures combined with strong winds give high windchill values. Howe (1962) has prepared a map of mean windchill in the United Kingdom in January based on mean values of temperature and wind speed for 1956–61 and using the Siple–Passel formula. The area with the least severe windchill, less than 750 kg cals m^{-2} hr^{-1}, is found in the area south of London and extending to the coast. The highest values occur in the Orkneys and Shetlands, the Outer Hebrides and the extreme north of Scotland where values exceed 900 kg cals m^{-2} hr^{-1}.

Observations of windchill at specific hours naturally give much higher values than these means and Howe (1962) quotes a value of 1 100 kg cals m^{-2} hr^{-1} at London Airport at 18.00 hrs on 23 December 1961. It has been established that windchill values in excess of 1 400 kg cals m^{-2} hr^{-1} correlate quite well with the incidence of frostbite of exposed flesh (Wilson, 1967), and Howe (1972) states that this figure was reached in Newcastle-upon-Tyne in the early evening of 31 December 1962. It is fair to assume that in the winter months windchill values greater than 1 400 occur with some regularity in areas such as the Welsh and Scottish mountains. In spite of such hazards most people would consider the normal range of temperature conditions experienced in the British Isles as favourable, although rarely ideal for outdoor activities.

6 Precipitation

B. W. Atkinson and P. A. Smithson

Any climatological study of precipitation in the British Isles has the advantage, bordering on an embarrassment, of a tremendous wealth of data and conceptual development on which to draw. The long-recognized importance of precipitation, not only as a meteorological curiosity but also as being essential to life on our planet, has meant that observations have been taken in the British Isles for over 200 years. By the mid-nineteenth century the enthusiasm and dedication of G. W. Symons led to the initiation of the British Rainfall Organization whose main objective was to establish a country-wide network of rain gauges where measurements were made daily. Over the succeeding century the number of gauges has grown to over seven thousand, providing the densest routine meteorological observation network in the world. The information from the daily, weekly and monthly instruments is complemented by that from several hundred autographic gauges throughout the British Isles.

Although the bulk of the information in the many *British Rainfall* volumes is indeed concerned with rainfall, snowfall has not been ignored. The major snowfall events from 1876 to 1965 have been documented by Bonacina (1927, 1936, 1966) and since 1945 the annual figures of the snow survey of Great Britain have appeared successively in the *Journal of Glaciology*, the *Meteorological Magazine* and ultimately back in the *British Rainfall* volumes. In contrast, climatological analyses of hail in the British Isles are rare, no doubt because of the local and comparatively infrequent occurrence of this form of precipitation. The short duration of hail falls also puts them beyond the conventional climatological pale as most analysis is perforce based upon the day as its time unit.

With such a long history of data collection it is not surprising that the main elements of the precipitation climatology of the British Isles were identified in the early decades of this century (Glasspoole, 1938). As a result of the mapping programme instituted by Mill the spatial distributions of monthly and annual amounts were soon described and slightly more complicated analyses, such as mapping monthly amounts as percentages of the annual amounts (i.e. isomeric maps as in Figs 6.8 to 6.11), were available by 1915 (Mill and Salter, 1915). More detailed spatial resolution and a more explicit treatment of temporal changes, including variability in general and sequences of wet or dry days in particular, are to be found in the many papers produced by Glasspoole in the period from the mid-1920s to about 1950.

The essentially statistical manipulation of data involved in these analyses of daily, monthly and annual precipitation amounts prior to 1950 is the most familiar approach employed in climatological investigations. Being based upon the 24-hour rainfall-day, the result of any such climatological synthesis can recognize only the direct effects of atmospheric circulations which have lifetimes greater than one day, or indirectly register the cumulative effect of circulations of duration less than one day. Figure 6.1 illustrates the spectrum of significant precipitation events that can occur and emphasizes the important change of approach which is forced upon the analyst as he crosses the rain-day time averaging threshold. For time periods less than one day it is possible, indeed most profitable, to analyse precipitation distributions in terms of the 'cloud-scale' motions that cause them. For time periods greater than one day it becomes increasingly difficult and ultimately impossible to do this. Up to about 3 days, synoptic scale systems can perhaps be used as a valid circulationary context to explain precipitation distributions; but as such systems rarely stay over an area such as the British Isles for as long as three days this approach has practical limitations.

Fig. 6.1 *Spectrum of precipitation-producing systems (after Mason, 1970)*

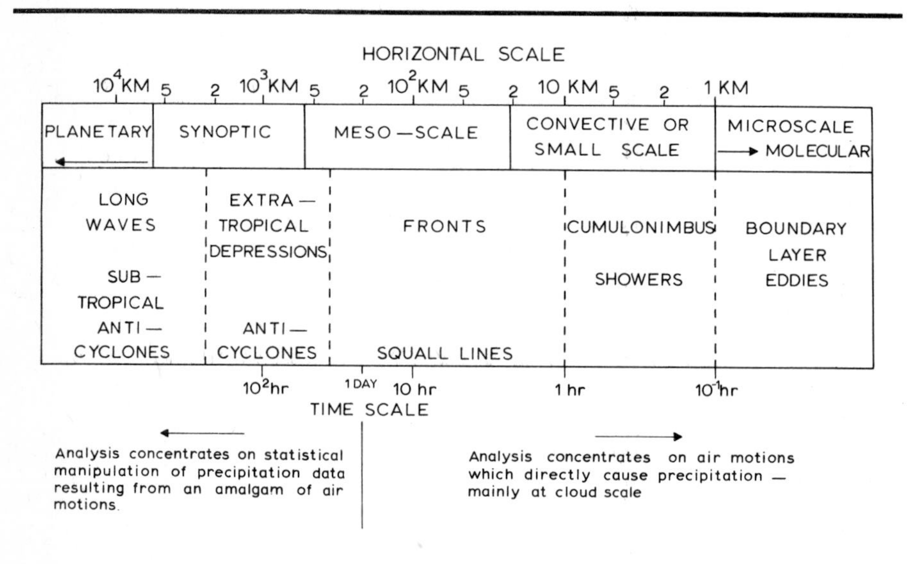

Despite these difficulties in reconciling the two approaches, any treatment of the precipitation climatology of the British Isles which ignored the underlying mechanics would lead to a shortfall of understanding. Fortunately the period since 1950 has seen much progress in the analysis of air motion on all scales ranging from that of the single cloud to that of the synoptic-scale system. The importance of these air motions, as opposed to the micro-physics of clouds, is emphasized by Mason (1969), who notes that they govern the formation, growth, dimensions, shape, organization, lifetime and vigour of clouds. Similarly, they govern the rate of the micro-physical processes of condensation and aggregation, the period over which they operate and thus the maximum size which the particles can attain. If precipitation should result, it is again the air motion which determines its distribution, intensity and duration. Here lies the essence of a genetic precipitation climatology.

Precipitation mechanisms in the British Isles

It is widely recognized that the uplift required to produce precipitation may occur as a result of one or more of three basic mechanisms: first, low level convergence, usually at fronts in cyclones, providing widespread uplift with magnitudes of the order of 5–10 cm s^{-1}; secondly, buoyancy, resulting from local atmospheric static instability giving upward motion of up to 30 m s^{-1} but extending over fairly restricted areas; and thirdly, forced uplift by orography, an effect which may trigger or intensify the other two mechanisms. This threefold division was recognized both in the British Isles and Norway in the early decades of this century (Salter, 1918; Bjerknes and Solberg, 1921) and has not been invalidated in the subsequent period. The important point to note, of course, is that all three mechanisms frequently operate simultaneously.

Convergent, dynamic or cyclonic precipitation

Although most cyclones last for 5–7 days, they usually do not take more than 2 days to cross the British Isles. The resultant precipitation is primarily explicable in terms of the advection of this synoptic-scale system of dynamically-induced uplift over the country (Sawyer, 1952)

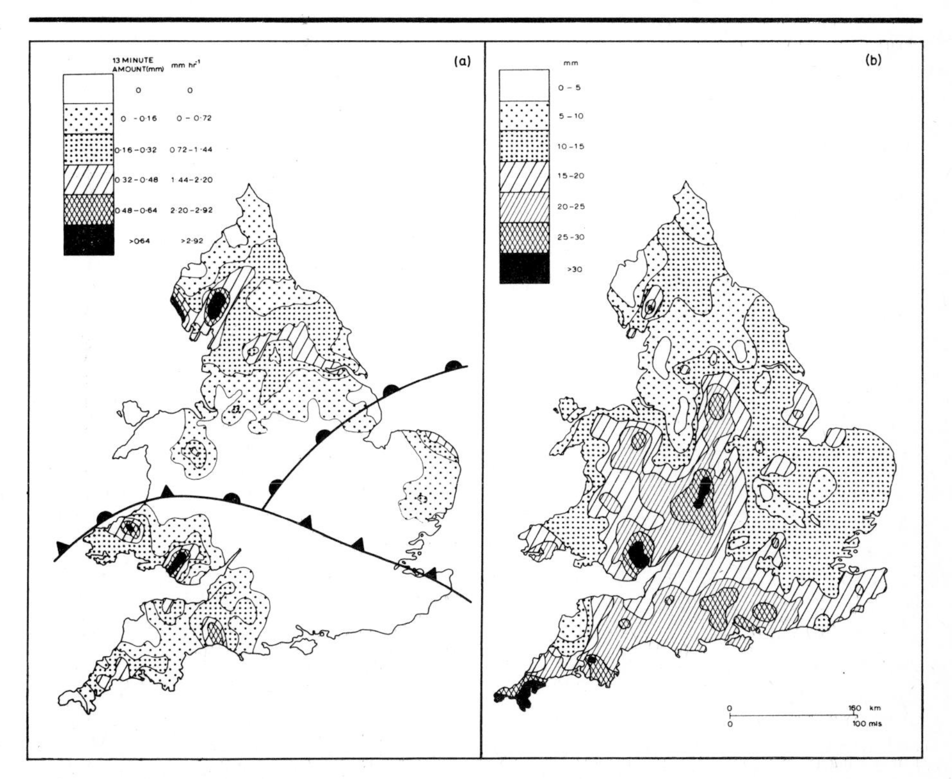

Fig. 6.2 *Rainfall patterns associated with an occluding depression (after Atkinson and Smithson, 1974).* (a) *precipitation from 10.32 to 10.45 hrs, 9 March 1967;* (b) *total precipitation from 15.00 hrs, 8 March to 17.00 hrs, 9 March 1967*

Fig. 6.3 *Mean rainfall relative to depression tracks (after Sawyer, 1956a)*

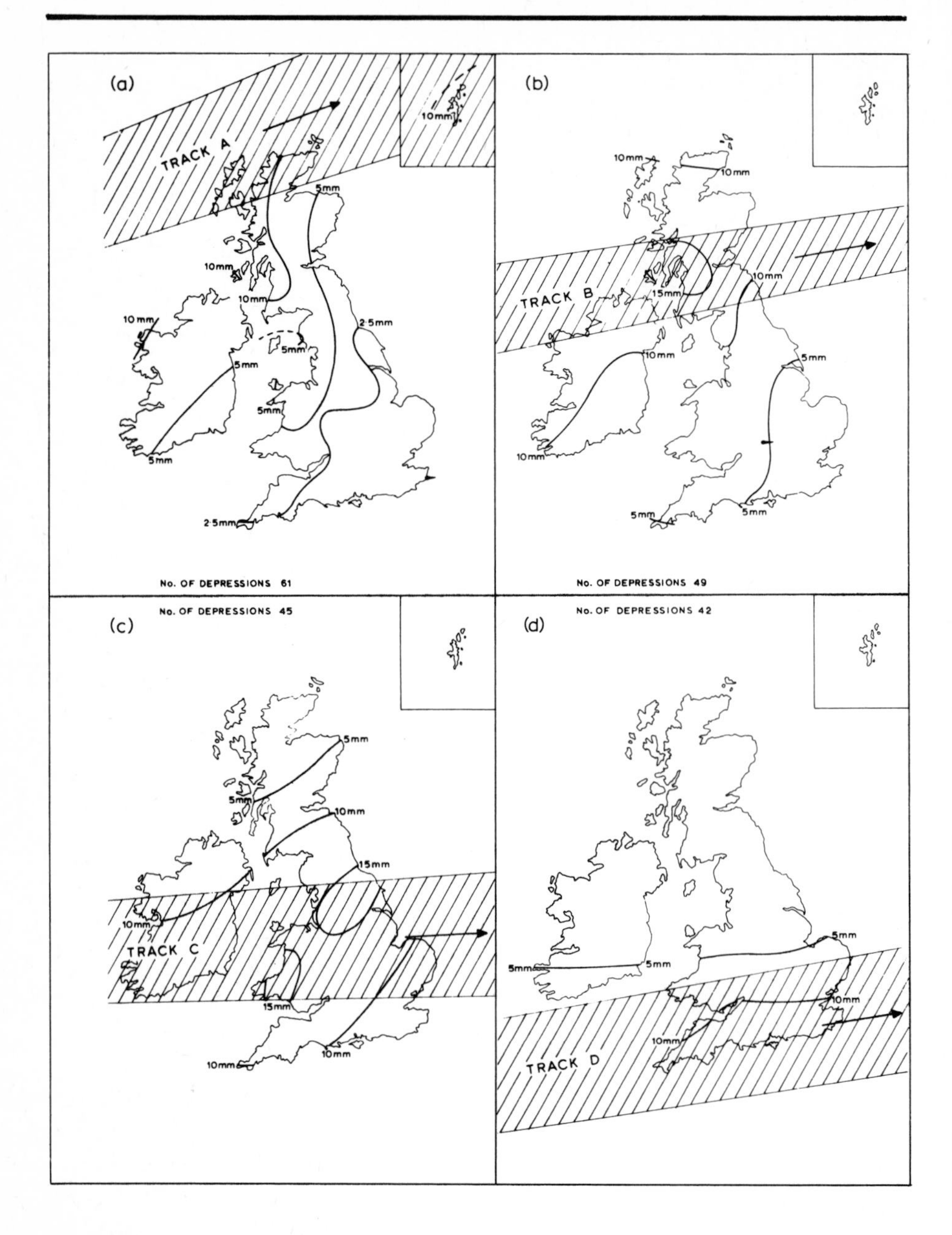

but orographic effects are also frequently significant. Under similar conditions of uplift a slow-moving system will obviously have a greater chance to precipitate more over the British Isles than a faster moving one. A typical vertical velocity and associated rainfall distribution at an occluding system is shown in Fig. 6.2*a*. The effect of sinking air in countering the orographic uplift in North Wales is strikingly illustrated. In this particular case, the dynamic component of the rainfall overshadowed the orographic component giving maximum amounts in the Midlands and south-central England (Fig. 6.2*b*) (Atkinson and Smithson, 1974). In contrast, the cases analysed by Sawyer (1952) and Atkinson and Smithson (1972) showed a strong orographic effect within a more uniform, widespread uplift at frontal zones. It is clear that the dynamic and orographic causes of precipitation interact in space and time in intricate fashion and that these interactions are identified only by spatial analysis for short time periods.

The nearest approach to a climatology of the 'purely' dynamic or convergent precipitation associated with cyclones is provided by Sawyer (1956a) in his analysis of precipitation within numerous depressions which passed along four tracks eastward over the British Isles. Figure 6.3 shows that all types of cyclone give 10–15 mm near the centres with values falling off to north and south. Typical intensities are shown in Fig. 6.3. South of the track of the depression western areas received more rain than corresponding eastern areas, but this is not so noticeable north of the track where onshore winds maintain high moisture values and convergence is greater in the front left quadrant of the depression. South of the track of the depression, the high-level stations received up to three times more precipitation than adjacent lowland stations. North of the depression track the orographic effect is smaller, possibly because the reversal of wind with height is unfavourable for the development of systematic uplift over the hills.

Convective, buoyancy precipitation

In contrast to the uplift associated with cyclones, penetrative convective motion is most frequently localized, comparatively transient and yet may have magnitudes which are capable of providing the most damaging forms of precipitation such as giant hail. Commonly, however, precipitation ensues from weak convective clouds giving the familiar short-lived shower (tens of minutes duration at most).

The basic cause of convection in most shower clouds and in several so-called 'air mass' thunderstorms is frequently solar heating. This tends to be strongest and most persistent in the south-eastern parts of our islands and the resultant instability in the lower atmosphere in those areas may be triggered by convergence zones, occasionally at sea-breeze fronts (Findlater, 1964; Wallington, 1961b) or by added heating due to urban areas (Atkinson, 1968, 1969, 1970, 1971). A moisture supply is also vital to the existence of storms and it is probably the proximity of a long fetch over the South-western Approaches together with orographic effects which results in some of the heaviest daily falls measured in the British Isles occurring over the south-west peninsula (Bleasdale, 1973).

It is difficult to summarize in a climatological sense the convective mechanisms of precipitation but one comparatively simple way is to gain an appreciation of isomeric maps for summer (Bleasdale, 1971) which show that up to 42 per cent of the annual precipitation in lowland Britain occurs in the five months May to September, whereas in the same period in highland Scotland only 35 per cent of the annual precipitation occurs (Fig. 6.11).

Orographic precipitation

The increase of precipitation with altitude (for all heights in the British Isles) has long been recognized and is an effect which dominates the distributions portrayed on monthly and annual maps and not infrequently on daily maps. Salter (1918) quoted values of increases of 1·5–2·0 inches per 100 ft (38·1–50·8 mm per 30 m) in south-east England and 2–3 inches per 100 ft (50·8–76·2 mm per 30 m) in west Wales but it is only recently that a more comprehensive measure of orographic effects in Britain has been made.

Using data from over 6 500 stations, Bleasdale and Chan (1972) established the following relationship between average annual precipitation and altitude in Britain:

$$R = 714 + 2{\cdot}42\, H$$

where R is average annual precipitation in millimetres and H is height in metres. Mapping the 6 500 plus anomalies from this regression line reveals positive anomalies exceeding 600 mm in western mainland Scotland from Cape Wrath to Glasgow and up to about 80 km inland in the south. To the east, the Cairngorms have a negative anomaly exceeding 600 mm, strikingly illustrating the west to east 'rain shadow' effect. The zero anomaly line identified by Bleasdale and Chan closely follows the main west to east water parting in Scotland, northern England and Wales with 600 mm positive anomalies occurring in the Lake District, and north and south Wales. To the east of the line, with the exception of the Cairngorms, negative anomalies are less than 600 mm.

There are two main reasons why precipitation values are greater over upland than lowland areas. First, the hills act as a barrier to moist airstreams which are therefore forced to rise; and secondly, the hills act as high-level heat sources on sunny days so that convective clouds tend to form preferentially over them giving showery precipitation of the type described in the previous section. Most of the studies this century concerned with orographic precipitation in Britain have concentrated on an evaluation of the barrier effect. A useful conceptual framework was provided by Sawyer (1956b) who showed that any explanation of orographic precipitation must take account of: first the synoptic-scale factors which determine the temperature, humidity, stability, wind speed and direction of the air crossing the hills; secondly, the dynamics of the air motion over and around the hills as they determine to what depth and through what layers the air mass is lifted; and thirdly the micro-physics of the cloud and rain, which determine the spectrum of droplets and thus whether the water which is condensed as cloud will reach the ground as rain or snow, or whether it will be re-evaporated on the leeward side. The first and third of these factors are applicable to all types of precipitation but their inclusion here is also important in seeing the orographic effect in a proper light.

In the post-war period, most studies of orographic precipitation in the British Isles have, either implicitly or explicitly, employed Sawyer's approach to the problem. One of the first such studies, by Douglas and Glasspoole (1947), showed that falls of 2·5 inches (63·5 mm) per day over western upland areas were most frequently associated with warm sectors which had a deep ($1\frac{1}{2}$ km) moist layer of air at low levels. As expected, the role of instability in these situations was considered to be minimal, despite Brunt's (1945) belief that heavy precipitation could occur only as a result of instability. More detailed studies have been made in north Wales and Scotland by Pedgley (1967, 1970), but in both cases simplifying assumptions were made about the airflow over the hills and about the micro-physics involved. Nevertheless useful results were forthcoming.

In Snowdonia, the mean annual precipitation at Cwm Dyli is 3 500 mm in a mean duration of 1 300 hours. At Holyhead the mean annual precipitation is 1 000 mm and the mean duration 713 hours. Hence the average rates of fall at Cwm Dyli are about twice as great as at Holyhead, i.e. 2·7 mm hr^{-1} compared with 1·4 mm hr^{-1}. Pedgley argues that part of this difference is due to heavy falls which themselves are a result of an augmentation of rains already being produced in association with cyclones. This increase in precipitation rate is produced by a scouring of the orographic cloud as rain drops fall through it. In the absence of these scouring drops, the orographic cloud is unable to precipitate unless winds are light and even then only if large cloud droplets are already present in a cloudy airstream as it approaches the mountains. This result, which is in accord with previous opinions that it must already be precipitating, or very nearly precipitating, before orography has any marked effect, is further supported by Pedgley's (1967) reappraisal of one of the two-day rainfalls of greatest areal extent in the British Isles, which occurred in Scotland on 16–18 December 1966. Of a total of approximately 75 mm in 2 days over most of western mainland Scotland, about 25 mm could be attributed to frontal systems; the remainder is attributed to orographic lifting in a persistent, moist airstream flowing at high speed.

It is clear from these examples that orographic effects are by no means simple in their mechanisms. The important inter-related factors listed by Sawyer allow an almost infinite number of combinations which change with time and which therefore result in a correspondingly near-infinite number of different precipitation distributions. It is only when these distributions are aggregated on a daily, monthly or annual basis that a simpler spatial distribution establishes itself.

Whichever of the preceding mechanisms or combination of mechanisms produce the precipitation, the total recorded at a point on the ground is a function of the duration and intensity. Tables 6.1 and 6.2 show the mean monthly and annual duration of precipitation at various stations in the British Isles. The dryness of the lowland south-east is well illustrated by the 407 hours in a year at Ipswich, which contrasts markedly with the 1 422 hours at Loch Sloy, which although the gauge is only 12·5 m above sea level faithfully records the wetness of the surrounding mountains. This three-fold increase in duration of measurable rain as one moves upward, northward and westward must be primarily due to the combined effects of more active and more frequent frontal systems crossing Scotland, a marked orographic influence and a rain-shadow effect in south-east England. It is discussed more fully in a later section.

Table 6.1 *Mean monthly and annual duration of rainfall, in hours, 1951–60, for selected stations in the United Kingdom (British Rainfall, 1959–60)*

Station	Height above sea-level (m)	J	F	M	A	M	J	J	A	S	O	N	D	Year
Camden Square	33·5	49	54	44	33	35	33	33	35	38	48	50	54	506
Croydon	61·6	54	52	44	32	35	33	31	32	39	51	57	59	519
Kew	5·5	45	42	36	27	30	28	27	31	33	40	43	45	427
Folkestone	39·0	66	59	42	30	29	29	31	35	35	49	57	65	527
Strood	34·1	60	54	46	29	35	32	32	39	39	44	51	60	521
Worthing	7·6	60	50	38	27	29	28	29	35	42	44	53	62	497
Farnborough South	69·5	58	56	45	31	37	32	30	38	45	49	61	57	539
Rothamsted	128·0	63	62	51	38	37	42	38	47	45	57	61	68	609
Peterborough	2·7	52	45	41	26	30	30	21	28	28	35	41	40	417
Ipswich	38·1	44	39	35	24	24	27	26	30	29	39	43	47	407
Marham	23·7	61	57	47	33	32	32	29	39	35	46	50	52	513
Boscombe Down	127·4	65	48	49	36	37	32	32	44	43	55	63	69	573
Plymouth	27·1	75	53	58	37	41	33	36	42	47	52	69	73	616
Exeter	32·3	62	52	51	35	41	30	28	35	39	49	64	67	553
Chivenor	6·1	72	61	52	47	45	38	43	47	56	63	79	82	685
St Mawgan	102·7	78	58	57	38	41	33	36	44	50	53	69	80	637
Bridgwater	7·0	57	49	44	33	33	32	32	34	42	48	67	66	537
Little Rissington	225·6	81	63	66	43	51	48	42	59	49	63	81	82	728
Ross-on-Wye	68·3	52	41	40	31	36	29	22	29	34	38	62	57	471
Shawbury	71·6	61	47	47	31	40	47	36	48	41	46	58	57	559
Elmdon (Birmingham)	98·1	59	47	56	35	38	40	34	45	41	47	62	61	565
Loughborough	82·0	58	44	53	32	39	44	26	40	35	44	54	59	528
Cranwell	62·8	60	57	49	32	37	43	33	40	32	40	53	50	526
Watnall	117·4	66	50	55	32	42	49	35	46	39	49	57	65	585
Ringway (Manchester)	75·6	64	48	48	41	35	47	52	55	50	54	58	71	623
Speke (Liverpool)	24·7	63	48	50	39	38	47	51	55	53	57	62	69	632
Squires Gate (Blackpool)	10·1	77	58	53	47	51	52	50	63	60	69	78	91	749
Leeds (Weetwood)	100·0	55	51	45	31	35	40	40	51	39	47	58	60	552
Skipton	111·0	71	58	48	39	42	52	55	71	57	65	74	93	725
Felixkirk	164·6	50	48	38	30	35	42	43	57	37	48	57	49	534
Acklington	42·7	61	47	46	34	39	45	41	56	39	45	54	54	561

Table 6.1 (continued)

Station	Height above sea-level (m)	J	F	M	A	M	J	J	A	S	O	N	D	Year
Silloth	8·2	65	45	41	42	43	49	56	61	58	56	74	77	667
Swansea	7·9	75	50	58	45	44	43	48	52	58	64	77	82	696
Aberporth	115·2	84	56	69	45	53	51	49	55	65	73	91	90	781
Swansea Waterworks	318·0	132	104	93	69	74	72	76	85	92	101	121	138	1157
Holyhead	8·5	76	61	53	41	49	44	46	52	59	61	80	81	703
Ronaldsway (Isle of Man)	17·4	61	50	41	34	39	40	40	44	54	55	68	74	600
St Peter (Jersey)	83·8	66	56	42	30	29	24	30	35	38	43	49	73	515
West Freugh	15·8	80	61	53	41	44	52	60	60	59	57	75	94	736
Eskdalemuir	242·1	120	89	79	74	67	76	91	90	86	93	116	149	1130
Swinton House	61·0	55	43	35	27	34	38	42	53	37	43	52	59	518
Turnhouse	35·4	49	47	44	33	44	51	64	62	42	43	57	70	606
Prestwick	9·1	69	56	46	43	41	53	63	57	53	62	70	89	702
Greenock	60·7	84	69	65	61	50	55	69	65	65	79	88	126	876
Loch Sloy	12·5	144	107	105	95	77	88	104	108	105	139	143	207	1422
Loch Fyne (Lephinmore)	9·1	110	71	64	68	44	57	79	70	81	98	113	150	1005
Tiree	9·5	102	70	64	57	46	58	68	62	71	89	87	116	890
Leuchars	10·1	49	47	51	39	41	42	53	51	37	47	60	70	587
Dyce	56·4	58	57	50	43	40	39	61	51	38	54	61	69	621
Dalcross	11·3	51	42	32	37	36	46	56	66	35	51	46	61	559
Benbecula	5·5	104	70	63	55	50	54	58	61	61	93	88	99	856
Stornoway	3·7	105	62	58	60	52	55	59	57	56	85	83	101	833
Wick	36·9	80	62	55	42	44	49	53	57	44	62	72	77	697
Grimsetter	25·6	93	78	69	50	49	56	54	62	57	88	93	104	853
Lerwick	82·0	98	76	57	56	50	51	59	55	60	78	92	106	838
Aldergrove	67·1	80	59	51	45	43	55	59	62	53	54	66	95	722

Table 6.2 *Mean monthly and annual duration of rainfall, in hours, for periods within 1948–73, for selected stations in the Republic of Ireland*

Station	Height (m)	J	F	M	A	M	J	J	A	S	O	N	D	Year
*Dublin	81	56·4	45·7	43·3	42·6	45·5	38·1	36·1	44·0	42·9	44·2	53·3	61·8	557·0
†Middleton	9	62·4	56·4	55·5	44·0	49·1	41·3	36·5	52·9	61·5	62·8	72·1	85·1	696·2
*Shannon	7	71·1	48·0	49·6	46·3	44·1	42·2	44·4	47·1	53·2	54·9	66·0	76·8	649·7
*Valentia	14	100·2	70·2	67·4	56·2	57·3	53·2	58·5	60·5	65·6	75·6	88·8	99·9	857·7
‡Claremorris	69	103·9	77·5	79·8	59·2	57·8	70·4	77·9	71·7	84·4	84·4	90·1	118·6	975·5
§Clones	89	83·9	67·5	56·6	42·8	46·8	58·3	65·1	66·2	62·2	66·5	75·6	105·0	796·5
§Mullingar	110	70·1	54·0	50·0	41·4	40·6	41·8	56·8	56·5	60·1	61·2	65·2	89·6	686·7
‖Malin Head	25	77·0	55·8	59·4	55·9	51·1	53·4	54·8	56·8	61·9	66·6	76·8	81·2	754·5
‖‖Belmullet	9	77·1	54·3	58·5	53·8	51·7	48·8	45·5	53·2	59·0	67·6	85·3	77·4	738·8

Source: Department of Transport and Power, Meteorological Service, Dublin. *Monthly Weather Report*, Part III, Annual Summaries, 1948–73.
Averaged over: *26 years; †8 years; ‡11 years; §10 years; ‖18 years; ‖‖13 years.

Table 6.3 *Average number of rainfalls of any duration, with precipitation rate exceeding specified values at Kew, 1933–50, and the maximum rate recorded (after Best, 1951)*

Rate (mm hr $^{-1}$)	J	F	M	A	M	J	J	A	S	O	N	D
>10	9·1	6·3	4·6	7·2	11·4	16·1	18·3	22·3	15·1	15·7	10·3	7·2
>25	2·3	1·5	0·7	1·9	3·7	6·7	8·1	9·4	6·6	5·3	3·2	1·7
>50	0·7	0·4	0·1	0·4	1·1	2·3	3·4	3·6	2·3	1·2	0·9	0·3
Maximum rate of rainfall recorded (*mm hr* $^{-1}$)												
	128	69	147	111	137	129	200	171	160	148	115	79

Table 6.4 *Number of rainfall periods during 6 months, each of which lasted for more than t minutes, during which the rate of precipitation exceeded p mm hr* $^{-1}$ *(after Best, 1951)*

t	p (mm hr $^{-1}$)					
min	6	10	25	50	75	100
0	159	52	21	11	4	2
2	100	27	11	3	1	—
5	42	10	2	—	—	—
10	14	1	—	—	—	—
20	2	—	—	—	—	—

True rates of precipitation are difficult to measure and the usual form of analysis of intensity is to present the amounts of precipitation falling in certain defined time periods. By far the easiest time period to take is the 24-hour rainfall-day which, despite drawbacks to be outlined later, has allowed a substantial increase in our knowledge of the spatial distribution of so-called intensities. For time periods less than a day data are sparse but Best (1951), sacrificing spatial variation for more detail at one station, produced some of the few figures available on measured rates of rainfall. Using data from the Jardi rate-of-rainfall recorder at Kew for the period 1933 to 1950, Best clearly showed the marked summer increase in the number of falls for three given intensities (Table 6.3) and that maximum intensities could reach as high as 200 mm hr $^{-1}$, a striking contrast to the average figures in Table 6.17. Further analysis provided data on the frequency of rainfalls of a given intensity showing that in a given 6-month period for which data were available nearly all falls with an intensity of 6 mm hr $^{-1}$ existed for less than 20 minutes and extremely high intensities occurred in less than 5 minutes (Table 6.4). Hail falls are also usually short-lived, 70 per cent of them lasting for less than 5 minutes and almost all falls lasting for less than 30 minutes.

In the absence of more instantaneous measurements of intensity (incorporating all their own peculiar errors) the analyst is forced to use amounts in finite time periods – data which are more readily available. Figure 6.4 shows the maximum intensities (strictly maximum magnitude–duration relationships) at Grendon Underwood and Loch Grennoch. For a 5-minute period the maximum amount recorded was from 5–6 mm, for 1 hour 15–18 mm, and so on. The maximum daily falls were 65 mm at Loch Grennoch and 48 mm at Grendon. Rodda (1970) provides a further diagram (see his Fig. 2) showing maximum magnitude–duration relationships up to durations of 2 years. The nearest approach to a spatial pattern of intensities lies in the maps of precipitation amounts in 13-minute periods over England and Wales prepared by Atkinson and Smithson (1972, 1974) in their studies of frontal precipitation. Figure 6.2 is a typical example of such maps which, although not as accurate as Best's figures, do give an indication of the complexity of precipitation distributions when viewed on a short time scale.

Fig. 6.4 *Maximum rainfall amount for given durations at two gauges (after Rodda, 1970)*

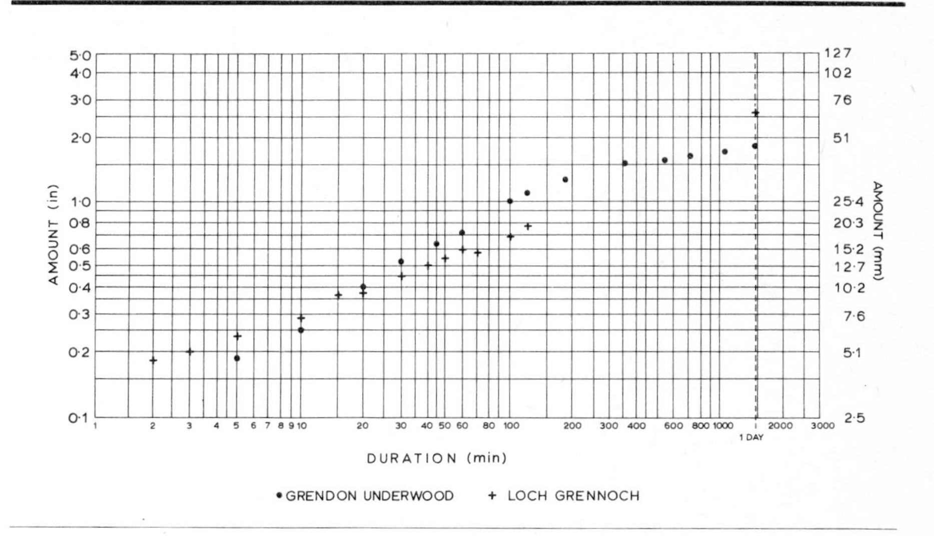

A climatologically informative compromise between the once-only maximum amount–duration analysis in Figs 6.5 and 6.6, and the nearly instantaneous picture in Fig. 6.2, is reached in Tables 6.5 and 6.6. These show the mean frequency of days in ten years on which given amounts of precipitation fell in different time intervals of less than one day. The data for Scotland are taken from Plant (1971), and in view of the sparseness of autographic data in Scotland we must be content with only a few stations despite their probably limited spatial representativeness. Although there are more stations in England and Wales, most of them are in the lowlands and records are far from being as complete as could be desired. Nevertheless the cautious amalgamation of records to give means for six areas (Fig. 6.6*b*) in England and Wales gives some useful data. The records for three individual upland stations in the Lake District, Wales and Dartmoor respectively (Table 6.6) show that some smoothing of the orographic effect has accompanied the amalgamation. The Irish data are presented in a slightly different way (Table 6.7).

Several points emerge from a study of Tables 6.5 to 6.7. Within Great Britain falls of 5–10 mm in periods up to 1 hour, and falls of 15–25 mm in periods of 2–4 hours, are markedly more frequent in

Fig. 6.5 *Frequency of specified daily rainfall totals for January and July (from the Meteorological Office series* Hydrological Memoranda for 1916–50*).* (a) *Average number of days in January with more than 0·2 inches;* (b) *Average number of days in July with more than 0·2 inches;* (c) *Average number of days in January with more than 0·6 inches;* (d) *Average number of days in July with more than 0·6 inches*

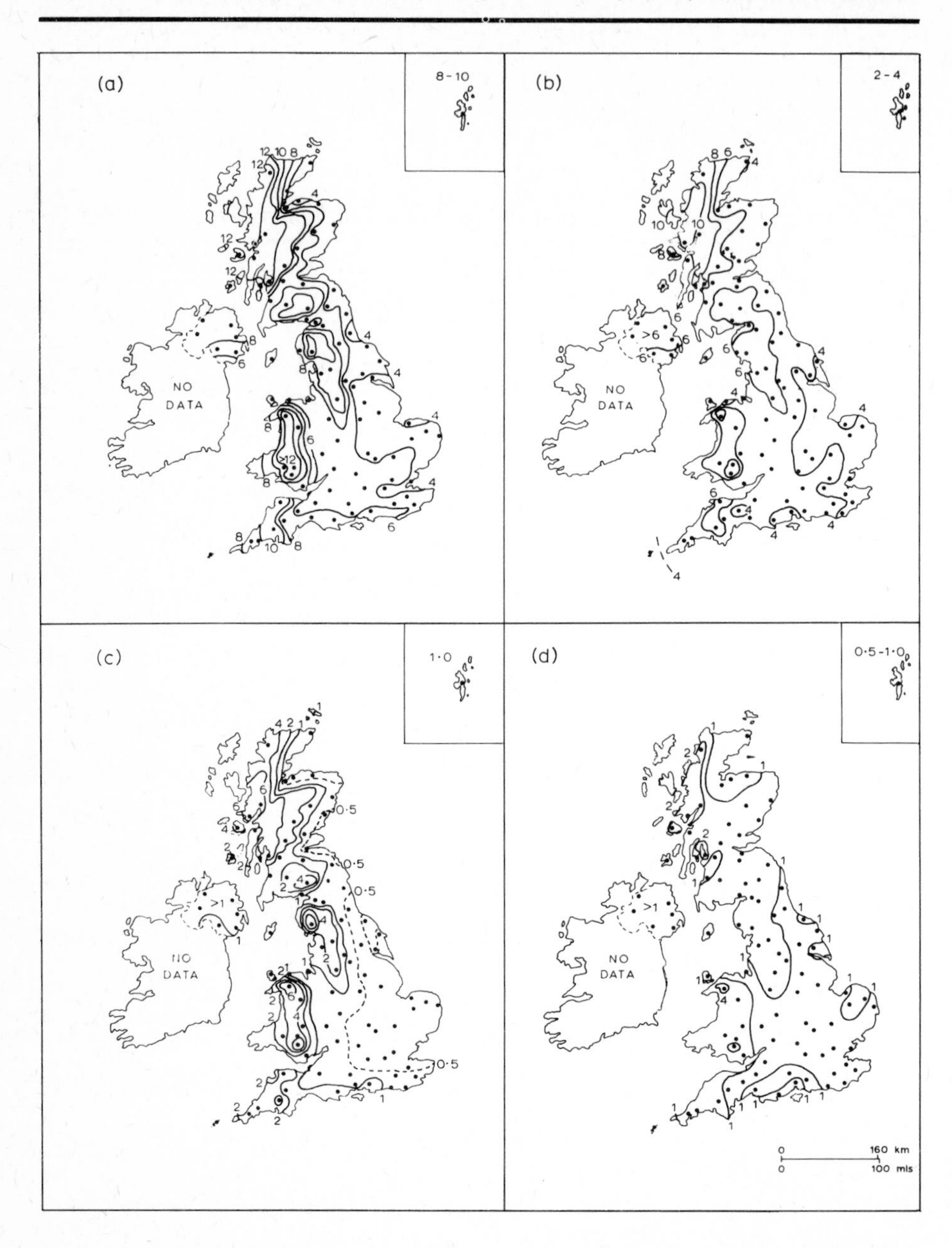

lowland England and Wales than in Scotland. For heavy falls with durations over 4 hours however, frequencies in Scotland are far greater. The high frequencies of short duration–heavy intensities in south-eastern Great Britain are no doubt due to the greater summer convective activity mentioned earlier in the chapter. Areas A and G (Fig. 6.6*b*) have the highest frequencies of heavy falls with durations up to 30 minutes. Heavy falls over 2 hours' duration are most frequent over the south-west peninsula, particularly the upland of Dartmoor. The frequency distributions of all three 'upland' stations are quite similar in form to those of Scotland if somewhat muted in their extreme values. It is quite evident that the areas with large annual precipitation amounts receive a substantial proportion of those amounts from prolonged (over 4 hours) heavy falls, and that these falls usually occur in upland areas for the reasons outlined earlier in the chapter. Plant (1971) further suggests that hills in Scotland are also responsible for the higher frequencies of short duration heavy falls at Eskdalemuir and Rhum, for example. If such is the case we see again that the marked orographic effect of the Scottish mountains is not as significant as convective uplift in determining the climatology of short-period heavy falls. In Ireland falls of 5 mm in 15 minutes occur in 3 out of 4 years, and the longer falls – 15 mm in 4 hours, 2 mm in 6 hours and 25 mm in 12 hours – have similar frequencies (Table 6.7). The heavier falls in shorter periods occur much less frequently, in accord with the popular view of gentle Irish rain.

Fig. 6.6 *Average number of days with more than 1 inch of rainfall (from the Meteorological Office series* Hydrological Memoranda for 1916–50*).* (a) *January;* (b) *July. Figure 6.6*(b) *also shows, in broken lines and capital letters A–G, the regions referred to in the text (p. 141)*

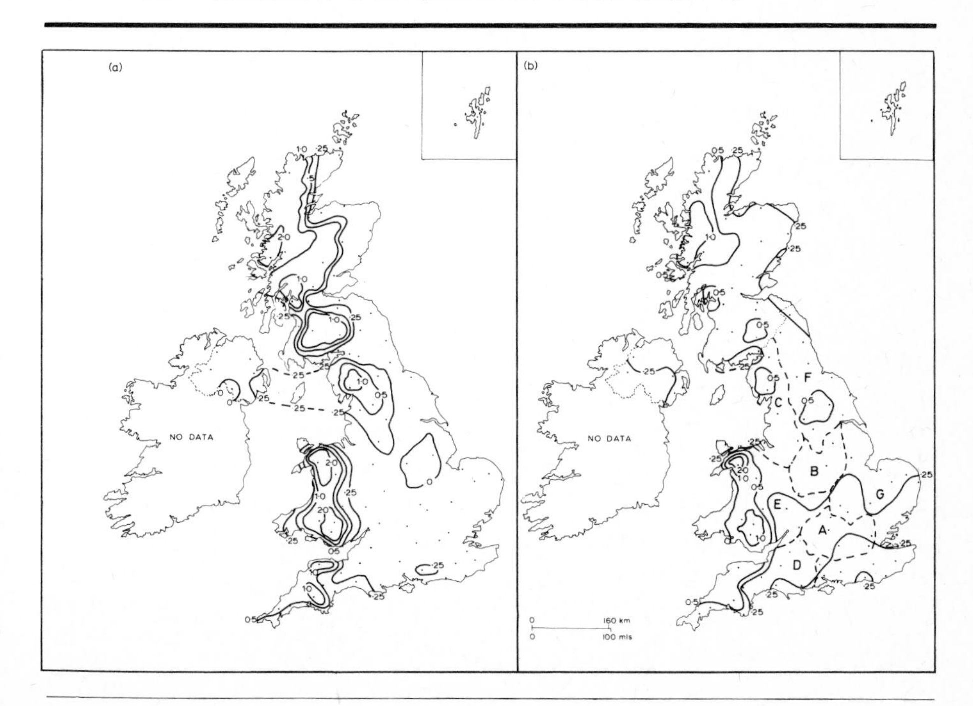

Table 6.5 *Observed frequency of the number of days in 10 years with specified amounts of rain falling in a specified time, for selected Scottish stations (after Plant, 1971)*

Station	5 mm within			10 mm within				15 mm within					20 mm within					25 mm within				
	5 min	10 min	15 min	5 min	15 min	30 min	1 hr	15 min	30 min	1 hr	2 hr	4 hr	30 min	1 hr	2 hr	4 hr	8 hr	1 hr	2 hr	4 hr	8 hr	16 hr
Eskdalemuir	2·7	8·0	17·3	0·3	2·4	6·1	15·8	0·6	0·9	3·6	13·3	71·2	0·3	0·3	2·7	20·9	78·2	0·3	0·6	7·3	37·3	84·8
Sloy Power Station	1·0	6·0	14·0	0·0	2·0	2·0	27·0	1·0	1·0	4·0	36·0	185·0	1·0	1·0	8·0	78·0	202·0	1·0	3·0	29·0	120·0	202·0
Machrihanish	0·0	1·7	3·3	0·0	0·0	0·0	1·7	0·0	0·0	0·0	0·0	28·3	0·0	0·0	0·0	10·0	23·3	0·0	0·0	0·0	11·7	23·3
Isle of Rhum	1·1	11·1	15·5	0·0	0·0	5·5	44·4	0·0	0·0	4·4	51·1	192·2	0·0	2·2	12·2	76·7	192·0	0·0	1·1	30·0	106·7	181·1
Stornoway	0·0	0·7	2·0	0·0	0·0	0·6	2·3	0·0	0·0	0·0	1·8	10·0	0·0	0·0	0·0	1·8	13·5	0·0	0·0	0·6	2·9	11·8
Kirkwall Apt.	0·0	1·3	3·3	0·0	0·0	0·8	3·3	0·0	0·0	0·4	1·7	11·7	0·0	0·0	0·4	1·7	12·1	0·0	0·0	0·4	4·6	12·9
Mullardoch Dam	0·0	2·5	3·7	0·0	0·0	1·3	16·3	0·0	0·0	1·3	7·5	53·3	0·0	0·0	3·7	18·7	63·7	0·0	0·0	7·5	37·5	75·0
Leuchars	2·0	6·0	10·0	0·0	2·1	3·2	7·6	0·3	0·9	1·8	5·6	16·8	0·0	0·6	1·5	6·5	17·9	0·0	0·3	2·3	7·3	17·3

Table 6.6 *Observed frequency of the number of days in 10 years with specified amounts of rain falling in a specified time, for selected areas and stations in England and Wales (data from the Meteorological Office)*

Area (see Fig. 6.6b)	5 mm within 5 min	10 min	15 min	10 mm within 5 min	15 min	30 min	1 hr	15 mm within 15 min	30 min	1 hr	2 hr	4 hr	20 mm within 30 min	1 hr	2 hr	4 hr	8 hr	25 mm within 1 hr	2 hr	4 hr	8 hr	16 hr
A	5·9	15·8	18·8	0·4	3·1	7·4	13·0	0·7	2·4	4·4	8·6	20·0	1·0	2·0	3·2	7·6	16·3	1·0	1·6	3·9	7·7	14·3
B	5·7	14·2	19·8	0·1	3·8	7·6	12·3	1·0	2·1	4·7	7·2	18·3	0·9	1·7	3·0	6·8	13·3	0·9	1·3	3·1	7·5	12·7
C	5·9	12·3	22·2	0·0	3·3	6·5	14·2	1·2	2·5	4·0	9·1	22·3	0·5	1·9	4·0	8·0	20·3	0·7	1·8	3·4	9·7	18·4
D	5·6	12·9	21·9	0·0	2·9	6·7	14·2	0·5	1·5	4·2	9·9	30·4	0·5	1·9	3·5	10·8	26·4	0·7	1·6	4·2	9·3	21·4
E	3·7	11·7	20·0	0·0	2·4	6·4	12·9	0·3	1·4	3·7	8·0	23·2	0·5	1·3	2·7	7·6	22·7	0·8	1·1	2·7	8·4	20·5
F	4·5	10·8	15·6	0·4	2·4	5·3	10·4	0·5	1·8	3·2	6·9	15·6	0·4	1·2	3·1	6·7	14·0	0·7	1·1	2·5	5·7	11·9
G	5·5	16·2	23·2	0·1	3·2	7·5	14·4	0·5	2·2	4·5	9·0	19·5	0·5	1·5	3·8	7·5	16·5	0·7	1·2	3·2	7·4	13·6
Station (with area)																						
Keswick (C)	6·7	10·0	16·7	0·0	2·7	7·3	17·3	1·8	2·7	3·6	17·3	71·8	0·9	1·8	6·4	25·5	79·8	1·8	2·7	14·5	41·8	76·4
Princetown (D)	14·8	31·9	51·4	0·0	4·3	14·3	52·4	0·5	1·9	10·9	49·5	—	0·0	0·9	12·9	70·5	—	0·5	3·3	31·9	—	—
Upper Lliw (E)	1·0	11·0	31·0	0·0	1·0	1·0	21·0	0·0	1·0	4·0	20·0	99·0	0·0	1·0	4·0	31·0	104·0	0·0	1·0	10·0	44·0	94·0

Table 6.7 *Ireland – number of years in which specified amounts of rain fell within specified times (data from Eire Meteorological Service)*

Station name	Number of years of record	5 mm 15 min	10 mm within 15 min	30 min	1 hr	15 mm within 15 min	30 min	1 hr	2 hr	4 hr	20 mm within 30 min	1 hr	2 hr	4 hr	6 hr	25 mm within 1 hr	2 hr	4 hr	6 hr	12 hr
Belmullet	17	10	1	5	9	1	2	4	7	17	1	3	5	9	14	0	2	4	8	15
Birr	18	15	5	10	13	1	4	5	9	15	1	1	4	7	12	1	2	2	4	12
Claremorris	30	23	3	9	19	0	2	3	17	30	1	2	2	17	28	1	1	4	12	25
Clones	24	20	5	11	17	1	3	6	14	20	1	3	6	11	17	2	3	7	10	17
Dublin Airport	33	23	5	11	20	2	3	7	14	29	2	5	5	21	24	2	3	5	16	29
Kilkenny	16	13	2	5	8	1	2	2	6	12	0	1	1	8	9	0	1	1	4	11
Malin Head	17	12	0	2	9	0	0	1	4	15	0	0	1	6	12	0	0	1	6	12
Mullingar	30	25	5	11	21	2	4	4	12	26	1	2	4	12	24	1	2	6	11	25
Roches Point	18	15	2	10	14	0	0	5	9	17	0	1	4	12	16	0	3	5	9	17
Rosslare	16	16	3	8	14	1	3	7	12	15	0	2	4	10	15	2	2	5	10	16
Shannon Airport	35	28	9	11	25	2	5	7	23	33	2	3	5	17	26	3	3	8	14	26
Valentia Observatory	34	26	3	13	28	0	3	10	29	34	0	4	9	32	34	0	4	20	27	34

Daily amounts

If we further extend our time-base in this analysis of precipitation 'intensities', by far the most convenient, and thus widely used, time period is the 24 hour rainfall-day, starting at 09.00 hrs. The reason for this popularity lies of course in the observational practice which requires daily observations to be taken at this hour. Analyses of frequencies of daily amounts at one station were undertaken as early as 1902 (Sowerby Wallis, 1902) and began to appear in the monthly supplement of the *Daily Weather Report* in 1917, but it was not until 1932 that Bilham and Lloyd (1932) provided frequency tables on a uniform basis for twenty-four stations. This first major effort was not followed up until the production of the *Hydrological Memoranda* of the British Meteorological Office in the early 1960s. Figures 6.5 and 6.6 are based on data extracted from these *Memoranda*, and attempt to illustrate the broad outlines of the spatial and seasonal variation of the frequencies of days receiving (in inches (mm)) more than 0·2(5), 0·6(15) and 1(25) of precipitation. Readers requiring monthly data on a greater range of precipitation amounts should consult the original *Memoranda*.

Perhaps the most striking feature of Figs 6.5 and 6.6 is the close correspondence of the isolines with the major upland areas in the British Isles. Further inspection reveals that the magnitude of the orographic effect varies with both 'intensity' and season. For 'light intensities' (0·2 inch (5 mm) per day), frequencies in upland Scotland and Wales range only from 10 to 7 days per month between January and July (a 30 per cent decrease); with intensities of the order of 0·6 inch (15 mm) per day the seasonal variation in the same upland areas extends from 4 days per month in January to 2 days per month in July (50 per cent); and with intensities of 1 inch (25 mm) per day the frequencies range from 1 day per month in January to 0·4 day per month in July in Scotland (60 per cent) but show little seasonal change in Wales and south-west England.

The relationship between upland and lowland frequencies also changes with 'intensity' and season. For the lighter falls, lowland frequencies do not appear to change significantly with season, so any winter increase in the ratio of upland to lowland frequencies is due to the increased frequencies in upland areas in winter. In the two cases of heavier daily falls this argument no longer remains true: the lowland frequencies alter significantly with season, increasing in July due to the convective activity in summer. This is well illustrated by the values recorded on Fig. 6.6.

Heavy falls

Most of the analysis of rainfall data in this century has been based upon the rainfall-day as the basic time-unit, and the investigation of heavy falls is no exception. The disadvantages of being forced into this analytical straitjacket are implicit in the material already presented in this chapter and Bleasdale (1963, 1970) is at pains to stress them before presenting data on exceptionally heavy daily falls in the United Kingdom over the last century or so. The most important qualification to such analyses are first that they give no recognition of differences in heavy falls which occurred in periods of less than 24 hours and secondly that a heavy fall within a period of 24 hours may have been split into two smaller amounts 'by the arbitrary (though necessary) convention of the rainfall day' (Bleasdale, 1963). Bleasdale (1963) also stresses the difficultues of establishing the areas affected to give an assessment of the overall 'magnitude' of a storm. Some rainfalls are extremely heavy but very localized; others do not reach record point values but deposit very large amounts of water over very large areas and are thus possibly of greater danger (Glasspoole, 1929–30).

As 4 inches (approximately 100 mm) of precipitation in one rain-fall-day have at one time or other been recorded in virtually all districts of the United Kingdom, a daily fall of 5 inches (126 mm) was chosen by Bleasdale (1963) as a threshold for exceptionality. As this daily total is more frequent in areas of heavy orographic rainfall, to be classed as an 'exceptional fall' in volumes of *British Rainfall* it must be at least 15 per cent of the annual average. With this criterion, 142 occasions of heavy precipitation were recorded between 1863 and 1960, with the possibility that several such falls in the early part of this period occurred unobserved. Since 1960, the terminal date of Bleasdale's analysis, at least twelve further occasions have been recorded, three of them in the notable year 1968 (Grindley, 1969; Bleasdale, 1970). Tables 6.8 and 6.9 summarize these heavy falls, implicitly indicating the strong orographic effect in the winter half of the year (October to March) on intensities and durations of precipitation within the rainfall day. If all occasions of heavy precipitation within 24 hours (not necessarily a rainfall-day) could be identified and quantitatively analysed, the number of occurrences in this group would be appreciably increased, possibly by 50 per cent or more. Away from the upland areas the heavy falls were very scattered but were remarkably absent from an area stretching from southern Lancashire to Kent and from the Welsh Border and the Thames to the Suffolk–Cambridgeshire–Derbyshire area. Table 6.8 clearly shows that these heavy falls are summer occurrences, most no doubt due to thunderstorms. As thunderstorm occurrence shows a marked diurnal variation with a maximum in late afternoon and early evening it is quite probable that most of the convective precipitation is adequately represented by observations taken the following morning at 09.00 hrs. It is perhaps worth noting that the top nine daily falls ever measured in this country have been at gauges in the three counties of Devon, Dorset and Somerset. These occurred on only 4 days – four on 18 July 1955, three on 28 June 1917, and one each on 15 August 1952 and 18 August 1924. This fact could reflect three characteristics of the area which in combination are not found elsewhere: a long sea fetch to the west; several upland areas; and, perhaps the most critical, a location well within the reach of much thundery activity travelling from the south. Table 6.9 clearly shows how rare are heavy falls in Ireland, due to the general lack of relief and of severe convective activity.

Data on heavy falls for periods greater than 1 or 2 days are few in the literature. Hawke (1942) remarked on this when he compiled a table based on the stations at Seathwaite, Ben Nevis, Kinlochquoich and Blaenau Festiniog. Whilst the data are far from being fully comprehensive, they indicate that for periods of 2, 3 . . . 7 days values (in inches (mm)) of 12·42 (315·5), 15·52 394·2), 16·63 (422·4), 17·91 (454·9), 19·12 (485·7) and 20·57 (522·5) have been recorded, all at Seathwaite, except the last one which was at Ben Nevis. These values require mean daily amounts of from 3 inches (76·2 mm) to 6 in (152·4 mm) and contributed from 9 to 15 per cent of the annual average precipitation at these stations.

Table 6.8 *Occasions with 5 inches (c. 125 mm) of rain in the rainfall day, grouped according to district and season (after Bleasdale, 1963)*

Areas	Summer half-year	Winter half-year
Dartmoor	2	2
South Wales	3	8
Snowdonia	10	15
Lake District	8	28
Western Highlands	4	23
South-west Peninsula (excluding Dartmoor)	8	0
Northern Ireland	2	1
Other areas	26	2

Table 6.9 *Values of rainfall amounts exceeding 125 mm in one rainfall day, based on 330 stations in Ireland, 1941–70 (data from Eire Meteorological Service)*

Station	Latitude ° ′ N.	Longitude ° ′ W.	Height (m)	Annual average rain (mm)	Amount (mm)	Period of record used
Dunmanway (Inchanadreen)	51 44	9 10	125	1 800	131·3	1951–70
Lough Eske (Edergole)	54 44	8 03	94	2 108	126·3	1954–70
Beaufort	52 04	9 38	35	1 358	141·1	1941–70
Cloone Lake	51 57	9 52	122	2 560	126·8 126·9	1950–70
Kenmare (Sheen Falls)	51 52	9 34	15	1 721	139·2	1941–70
Killarney (Muckross For. Stn.)	52 02	9 30	32	1 474	167·1	1941–70

Periods of duration greater than one day but less than one month

Deficiency and excess of precipitation

Most studies of precipitation climatology on the basis of a time interval less than one month but greater than one day have used the 'rain-day' as their basic building block. These blocks were themselves first fashioned in 1865 as a 24-hour period starting at 09.00 hrs in which 0·01 inch (0·2 mm) of precipitation was recorded. Although not as widely used today, the notion of the rain-day proved invaluable for the marshalling of our knowledge of precipitation distribution in the early decades of this century. The idea was extended in 1919 when a 'wet-day' was defined as a similar 24-hour period to the 'rain-day' in which 0·04 inch (1·0 mm) or more of precipitation was recorded. From the above definitions sprang the ideas of 'rain spells', 'wet spells', 'dry spells' and 'absolute droughts'. The former two are periods of at least 15 consecutive days to each of which is credited 0·01 inch (0·2 mm) or more and 0·04 inch (1·0 mm) or more respectively: the latter two were defined respectively in 1919 and 1887 as periods of at least 15 consecutive days to none of which is credited 0·04 inch (1 mm) or more and 0·01 inch (0.2 mm) or more.

These definitions were used in the volumes of *British Rainfall* until 1960 and consequently the bulk of British rainfall records have been analysed within the framework they impose. Tables 6.10 and 6.11 give a broad indication of the frequency, duration and other characteristics for rain and wet spells at eighty-two stations in the United Kingdom between 1940 and 1960 and for nine stations in Ireland from 1948 to 1966. The year-to-year variations reflect, of course, the different circulation characteristics of each calendar year. Tables 6.12 and 6.13 give similar kinds of data for dry spells and absolute droughts. Lewis (1939) attempted to sketch the seasonal and geographical distribution of absolute drought in England, using six stations with continuous records of at least 80 years. Her data, together with a map presented by Glasspoole in the discussion of the paper, suggest frequencies of just over one absolute drought per year south-east of a line from the Wash to Portland, falling to only one in 5 years in north Wales and extreme north-western England. Lewis found no very obvious seasonality in drought occurrence except for a general tendency at all six stations for any one day between March and September to have a higher chance of being in a drought than any day between October and February.

Since 1961 a different method of describing the occurrences of spells has been possible with the aid of data processing by computer. For any station, cumulative amounts of precipitation at intervals (usually pentads) throughout the year are first expressed as percentages of the average annual precipitation, and then as departures from the percentage amounts which would accumulate at a uniform rate corresponding to the average. When plotted on a graph, approximately horizontal sections correspond to precipitation at more or less the average rate for the station. Significantly descending sections represent precipitation deficiency spells, with a limiting slope corresponding to no rain at all. Ascending sections, for which there is no accurately assignable limiting slope, represent periods of excess. Figure 6.7 shows the results of such an analysis applied to the data at four stations over the period 1916–50. In general, stations in western locations experience precipitation excess from October to February and a deficiency for the remainder of the year apart from a spell of zero deficiency or excess for about a month in the late summer. Eastern areas do not have such a marked annual variation but generally experience rainfall excess in the autumn and deficiency in the spring with little or no excess or deficiency in the summer and winter.

Fig. 6.7 *Cumulative amounts of rainfall at 5-day intervals expressed as percentage departures from the amounts which would accumulate at a uniform rate corresponding to the average, all values being given as percentages of the average annual rainfall (1916–50) for the station (after Stephenson, 1971). Months are defined by vertical broken lines, and pentads by vertical full lines*

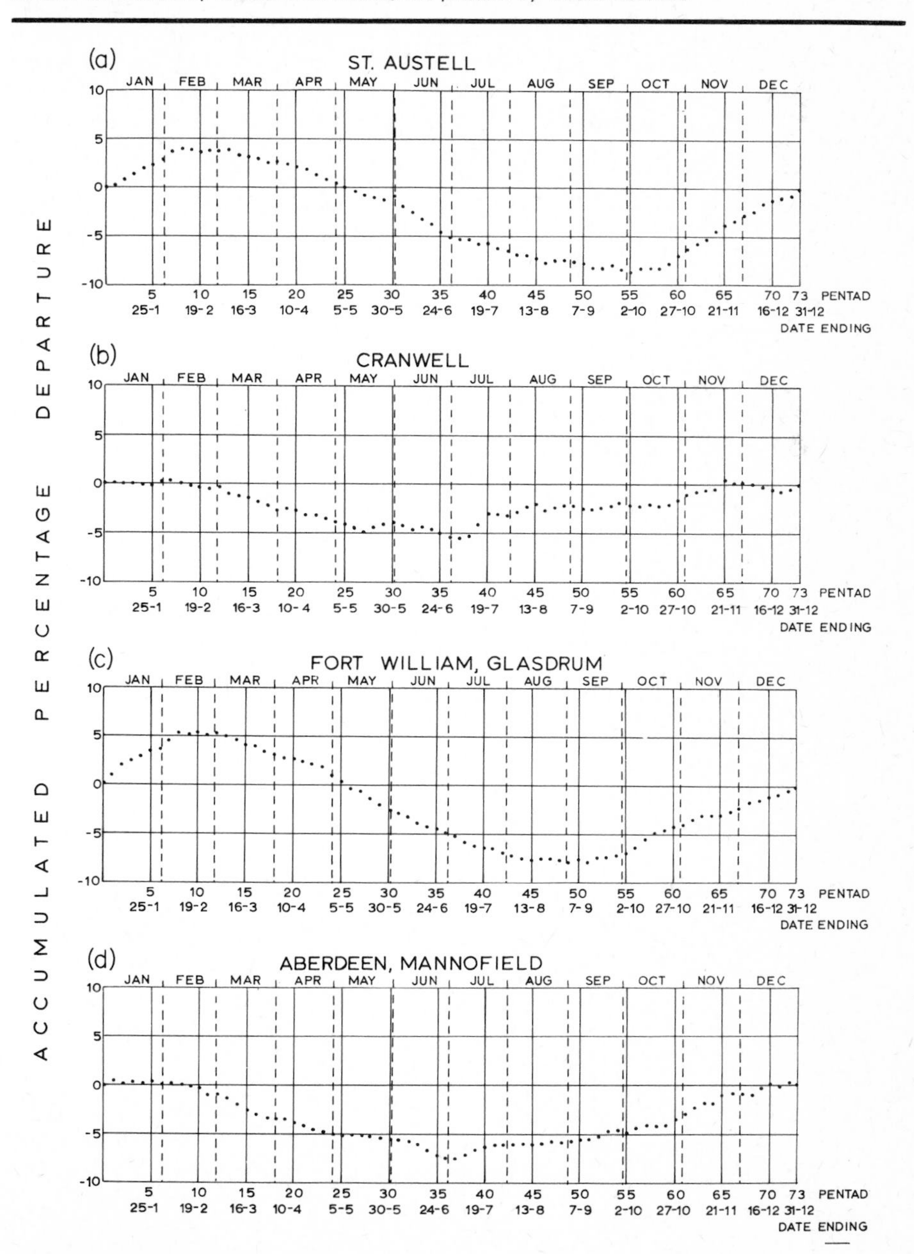

Table 6.10 *Rain spells and wet spells at eighty-two stations, 1940–60 (from* British Rainfall, *1959–60)*

	Rain Spells*						Wet spells†					
Year	Number at 82 stations	Mean duration (days)	Mean rain per day (inches)	Longest duration (days)	Mean rain per day (inches)	Stations without one	Number at 82 stations	Mean duration (days)	Mean rain per day (inches)	Longest duration (days)	Mean rain per day (inches)	Stations without one
1940	67	19	0·27	47	0·44	45	4	20	0·34	24	0·44	79
1941	43	19	0·25	34	0·26	54	4	16	0·36	18	0·34	78
1942	94	20	0·27	47	0·15	32	22	17	0·36	24	0·54	63
1943	98	22	0·27	42	0·28	42	15	21	0·37	26	0·63	71
1944	95	21	0·26	67	0·19	35	21	18	0·63	25	0·49	66
1945	86	19	0·26	40	0·24	35	22	16	0·35	20	0·41	65
1946	134	21	0·28	42	0·18	22	42	19	0·38	37	0·40	57
1947	82	20	0·26	34	0·20	26	21	17	0·35	25	0·25	64
1948	107	21	0·30	54	0·37	37	28	19	0·38	31	0·28	63
1949	109	22	0·30	41	0·44	48	32	20	0·40	34	0·54	68
1950	139	21	0·30	50	0·35	35	38	19	0·41	38	0·64	61
1951	112	19	0·28	49	0·47	23	23	18	0·42	32	0·39	66
1952	115	20	0·24	40	0·34	36	24	18	0·38	26	0·31	66
1953	65	20	0·29	31	0·18	50	16	19	0·44	29	0·56	67
1954	151	21	0·30	116	0·35	33	42	18	0·41	37	0·38	59
1955	51	18	0·20	27	0·25	45	6	17	0·34	21	0·24	76
1956	62	20	0·29	32	0·36	45	10	20	0·45	29	0·25	74
1957	107	20	0·27	37	0·16	27	29	19	0·37	24	0·47	59
1958	77	19	0·23	31	0·22	43	9	18	0·30	23	0·27	75
1959	107	19	0·28	37	0·15	33	30	17	0·37	23	0·18	58
1960	117	20	0·24	39	0·25	19	26	17	0·33	28	0·42	60

* At least 15 days, each credited with 0·01 inches or more

† At least 15 days, each credited with 0·04 inches or more.

Table 6.11 *Rain spells and wet spells at Irish stations, 1948–66 (from the Irish* Monthly Weather Report—*Annual Summaries, 1948–66)*

	1948	**49**	**50**	**51**	**52**	**53**	**54**	**55**	**56**	**57**	**58**	**59**	**60**	**61**	**62**	**63**	**64**	**65**	**66**
Rain spells																			
Dublin	0	0	0	0	0	0	0	0	0	0	0	0	0	1	0	0	1	0	0
Midleton	2	1	1	0	0	0	0	1	—	—	—	—	—	—	—	—	—	—	—
Shannon	1	3	2	1	1	0	1	1	0	2	2	2	1	4	1	2	1	1	0
Valentia	4	4	7	3	2	2	7	1	2	3	4	4	5	3	4	4	2	3	2
Claremorris	—	—	5	2	1	1	4	1	2	2	4	4	3	—	—	—	—	—	—
Mullingar	—	—	0	2	1	1	2	1	0	1	2	2	2	—	—	—	—	—	—
Clones	—	—	—	1	0	0	2	1	0	3	0	1	3	—	—	—	—	—	—
Malin Head	—	—	—	—	—	—	—	—	1	3	2	3	3	3	4	1	2	2	2
Belmullet	—	—	—	—	—	—	—	—	—	—	—	—	—	4	3	4	1	3	1
Wet spells																			
Dublin	0	0	0	0	0	0	0	0	0	0	0	0	0	0	0	0	0	0	0
Midleton	0	0	0	0	0	0	0	0	—	—	—	—	—	—	—	—	—	—	—
Shannon	0	0	0	0	0	0	0	0	0	1	0	1	0	1	0	0	0	1	0
Valentia	3	1	3	0	1	0	1	1	0	1	1	1	1	1	1	0	0	2	0
Claremorris	—	—	0	0	0	0	1	0	0	0	0	1	1	—	—	—	—	—	—
Mullingar	—	—	0	0	0	0	0	0	0	0	0	0	0	—	—	—	—	—	—
Clones	—	—	—	0	0	0	1	0	0	0	0	0	0	—	—	—	—	—	—
Malin Head	—	—	—	—	—	—	—	—	0	2	0	0	1	2	0	0	1	2	0
Belmullet	—	—	—	—	—	—	—	—	—	—	—	—	—	2	0	0	0	1	0

Table 6.12 *Droughts and dry spells at eighty-two stations, 1940–60 (from* British Rainfall, *1959–60)*

Year	Absolute droughts*				Dry spells†			
	Number at 82 stations	Mean duration (days)	Longest (days)	Stations without one	Number at 82 stations	Mean duration (days)	Longest (days)	Stations without one
1940	40	20	31	53	83	23	48	39
1941	82	21	31	25	196	21	37	1
1942	96	20	30	26	206	22	36	9
1943	61	21	41	47	129	22	47	28
1944	25	17	21	62	96	19	37	41
1945	33	19	24	55	125	20	35	14
1946	57	17	25	36	132	18	29	10
1947	114	25	44	3	179	25	53	1
1948	29	17	23	57	107	19	32	18
1949	75	20	37	27	162	22	55	12
1950	41	16	21	49	106	17	27	24
1951	27	16	23	56	94	18	32	12
1952	32	17	23	61	119	18	23	24
1953	71	24	36	20	163	23	43	7
1954	44	20	27	43	97	21	29	15
1955	111	20	36	15	224	22	44	2
1956	12	17	21	72	69	19	36	43
1957	46	17	40	47	159	18	42	19
1958	18	18	33	64	71	18	33	29
1959	152	21	57	15	253	22	59	5
1960	27	17	25	58	98	18	31	17

* 15 days without a rain-day   † 15 days without a wet day

Table 6.13 *Droughts and dry spells at Irish stations, 1948–66 (from the Irish* Monthly Weather Report – *Annual Summaries, 1948–66)*

	1948	49	50	51	52	53	54	55	56	57	58	59	60	61	62	63	64	65	66
Absolute droughts																			
Dublin	0	1	0	0	1	1	1	2	0	0	1	0	0	2	1	1	0	0	0
Midleton	0	0	0	0	0	1	0	4	—	—	—	—	—	—	—	—	—	—	—
Shannon	0	0	0	0	0	1	1	2	0	1	0	0	0	0	0	0	1	1	0
Valentia	0	0	0	0	0	1	0	2	0	0	0	0	1	0	1	0	0	1	0
Claremorris	—	—	0	0	0	1	0	2	0	0	0	0	0	—	—	—	—	—	—
Mullingar	—	—	0	0	0	1	1	2	0	1	0	0	0	—	—	—	—	—	—
Clones	—	—	—	0	0	1	1	2	0	1	0	0	0	—	—	—	—	—	—
Malin Head	—	—	—	—	—	—	—	—	0	0	0	0	0	0	0	0	0	1	0
Belmullet	—	—	—	—	—	—	—	—	—	—	—	—	—	0	0	0	0	1	0
Dry spells																			
Dublin	1	3	1	0	2	1	1	3	0	3	1	2	0	3	1	1	3	0	1
Midleton	0	2	2	1	3	3	1	4	—	—	—	—	—	—	—	—	—	—	—
Shannon	1	1	1	0	2	1	1	3	0	2	1	1	0	3	5	1	2	2	0
Valentia	1	0	1	0	1	1	1	2	0	1	0	0	0	0	2	0	0	1	0
Claremorris	—	—	1	1	0	1	1	3	0	2	1	0	1	—	—	—	—	—	—
Mullingar	—	—	3	0	2	1	1	3	0	2	0	1	1	—	—	—	—	—	—
Clones	—	—	—	0	2	1	1	3	0	1	1	0	1	—	—	—	—	—	—
Malin Head	—	—	—	—	—	—	—	—	0	1	0	0	1	0	0	1	0	1	1
Belmullet	—	—	—	—	—	—	—	—	—	—	—	—	—	1	1	0	0	1	0

Monthly rainfall

The tradition of totalling rainfall into monthly periods stems from administrative convenience rather than having any climatological significance. Monthly totals provide more detail than seasonal or annual values about the variations in rainfall throughout the year for periods of time which are nationally known and understood. To a certain extent they can also be equated with parts of the seasons, but in the British Isles at least the monthly rainfall total is simply a convenient period of time within which the precipitation has been summed. In spite of calendar months being artificial divisions, monthly rainfall data are easily obtainable and consequently often form the basis for analysis. Early attempts are summarized by Mill (1908) in *British Rainfall*, and further developments by the previously mentioned work of Mill and Salter in 1915. The latter paper used the idea of isomers to illustrate monthly rainfall. By this method, mean monthly rainfall is depicted as a percentage of the mean annual total and isolines indicate the areas experiencing similar percentages. Mill and Salter took the 35-year period 1875–1909 on which to base their annual and monthly means and the isomeric maps were produced accordingly. Such maps have two main advantages over maps of actual totals; first, they remove the orographic effects which normally dominate rainfall distribution and secondly the smoother patterns enable an easier interpretation of the map. However the apparent simplicity disguises two serious disadvantages. The marked variability of monthly averages means that whenever the standard period is changed, different percentages will result from identical monthly rainfall totals. In addition, because the means for successive months are rarely the same it is impossible to determine the percentage for a longer period simply by summing the percentages for the appropriate months. This variability is clearly shown in Table 6.14 where monthly totals for the record at Oxford between 1815 and 1970 are summarized (Smith, C. G., 1974). As long as these limitations are understood, monthly percentage maps are useful as a guide to the way in which mean rainfall varies throughout the year for the period specified. To reduce this problem, more recent isomeric maps base their values on the more stable *annual* mean rather than on monthly values.

The isomeric maps for 1916–50 are shown in Figs 6.8 to 6.11. Bearing in mind that, for an even distribution throughout the year, months with 31 days would have 8·5 per cent, months with 30 days 8·2 per cent and February 7·6 per cent, it is clear from the maps that there is appreciable monthly diversity. The pattern in December and January is very similar with high values on the western side of the country, but only slightly above 9 per cent in the English Midlands and eastern areas. This reflects the frequency of depressions crossing from the south-west and especially the contribution from warm sectors which produce a marked orographic increase and are slightly more common at this time of year. The distribution in February is similar, but the westerly factor is less pronounced with a much reduced percentage difference between areas of high and low mean annual rainfall.

Fig. 6.8 *Isomeric maps of monthly rainfall, 1916–50, with average monthly rainfall as a percentage of average annual rainfall (after Bleasdale, 1971);* (a) *January;* (b) *February;* (c) *March;* (d) *April*

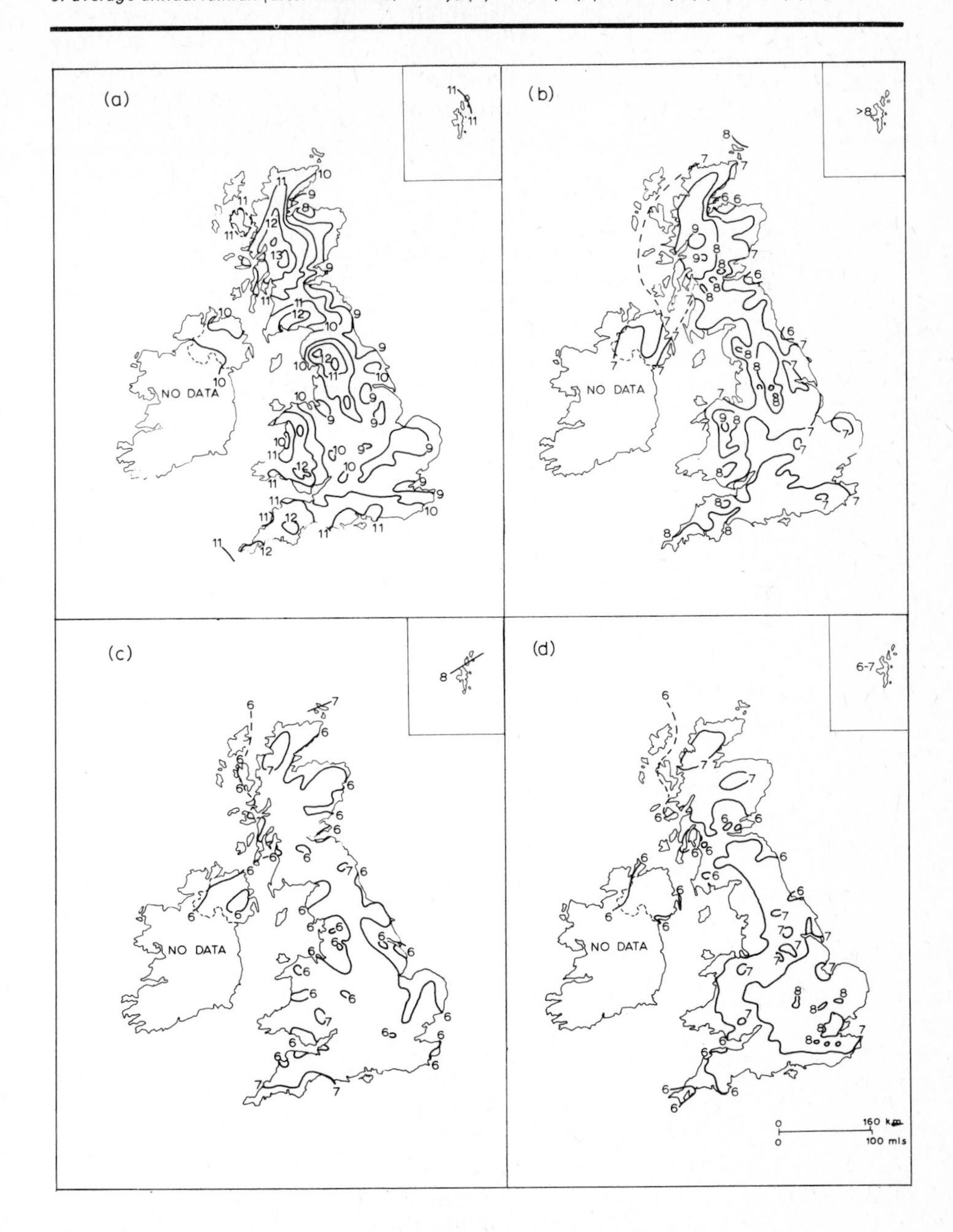

March and April are intermediate in character between the winter and summer situations with almost uniform values and a low percentage of the annual total everywhere. The low incidence of westerlies in May and June is confirmed in the isomeric maps by low percentage values in the west, reaching less than 5 per cent in the North-West Highlands of Scotland. By high summer the patterns change drastically with high values in the drier eastern areas, decreasing towards the west and being lowest over upland areas. This represents the continental element of the climate with a decreased frequency of mobile depression systems, lighter winds and stronger surface heating combining to give a greater incidence of convectional rain. For the period September to November the pattern becomes less distinct, but is marked by an increase in percentage values especially in the south of the country. Only the salient features of the isomeric maps have been outlined. They would be apparent for whichever time period was used but there would be changes in emphasis and particularly in the position of the isomers for periods other than 1916–50, with marked changes for individual years.

Table 6.14 *Variability of monthly rainfall at Oxford, 1815–1970 (after C. G. Smith, 1974)*

	Mean (mm)	Standard Deviation (mm)	Per cent of annual mean	Coefficient of variation (per cent)
Year	652·3	142·7	—	22
January	52·8	27·7	8·1	52
February	41·7	25·9	6·4	62
March	41·1	26·4	6·3	64
April	44·7	23·1	6·9	53
May	50·3	26·9	7·7	54
June	53·9	29·7	8·3	55
July	61·7	34·5	9·4	57
August	61·0	31·0	9·3	51
September	59·9	36·1	9·2	60
October	67·3	36·6	10·3	54
November	60·7	34·0	9·3	56
December	57·2	30·7	8·8	55

Considerable stress has been placed on the variations of monthly rainfall which result in average values that could be misleading. Table 6.14 shows that mean monthly values at Oxford have a coefficient of variation, without exception, above 50 per cent. Even 10-year running-means of the Oxford data (not shown) exhibit marked fluctuations being responses to the subtle variations of atmospheric circulation and rain-bearing influences from year to year. Spells of similar weather patterns lasting as long as 30 days are infrequent, so the monthly total indicates the summation of a series of weather sequences, some giving dry weather and some abundant rain. As the circulation is not dictated by calendar months it is relatively rare for the same circulation pattern to persist throughout a calendar month, though when it does, extremes are likely to occur. For example, in September 1959 parts of East Anglia recorded no measurable rainfall and in April 1974 Stornoway in the Outer Hebrides received 1·0 mm, 58 mm less than average. At the other extreme, November 1970 was

Fig. 6.9 *Isomeric maps of monthly rainfall, 1916–50, with average monthly rainfall as a percentage of average annual rainfall (after Bleasdale, 1971);* (a) *May;* (b) *June;* (c) *July;* (d) *August*

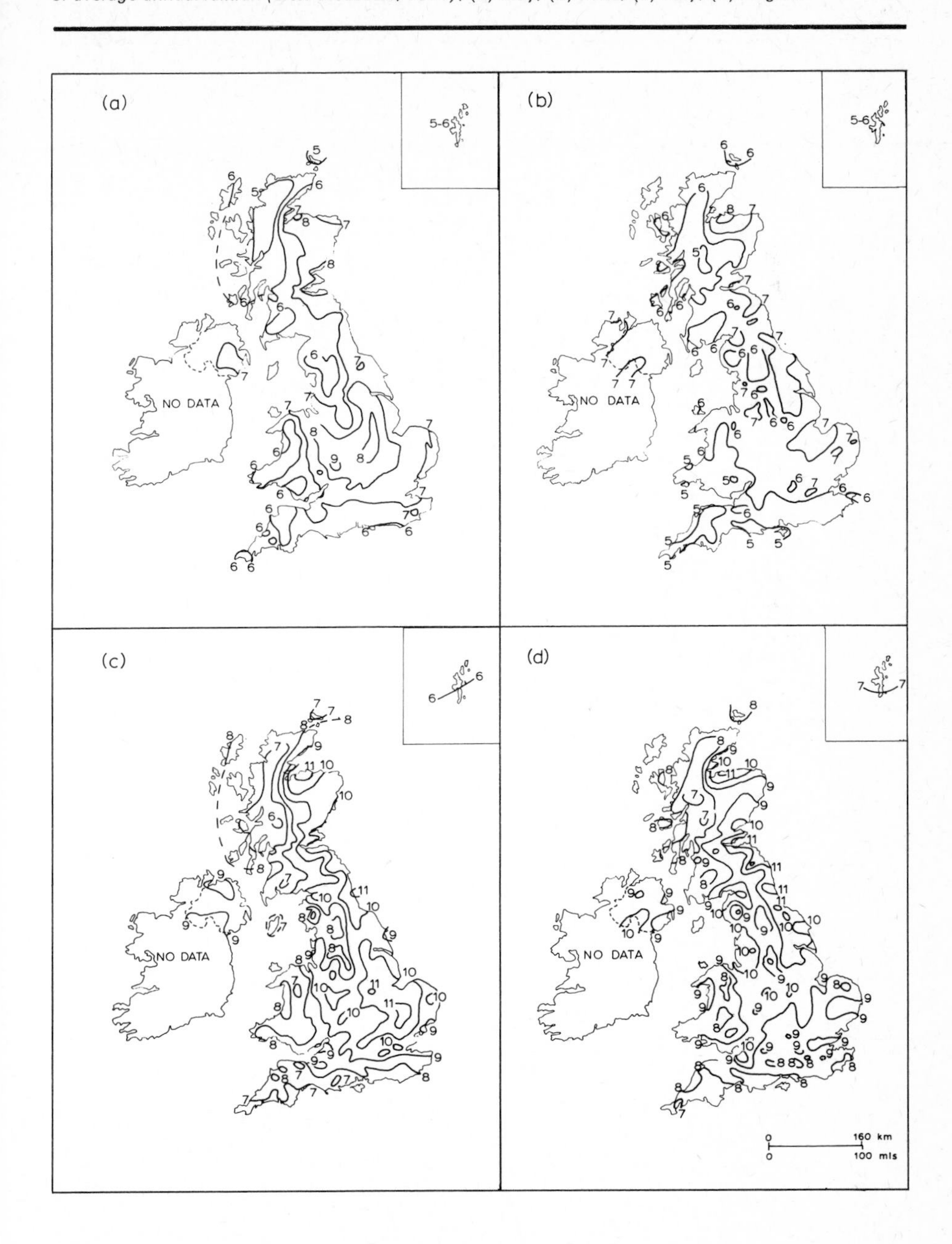

the wettest on record in many parts of south-east England with totals more than twice the average after a westerly and cyclonic month. In any month and in all parts of the country a monthly total of at least three times the average is possible. As would be expected, these are more frequent in the drier parts where anomalous depression

Fig. 6.10 *Isomeric maps of monthly rainfall, 1916–50, with average monthly rainfall as a percentage of average annual rainfall (after Bleasdale, 1971);* (a) *September;* (b) *October;* (c) *November;* (d) *December*

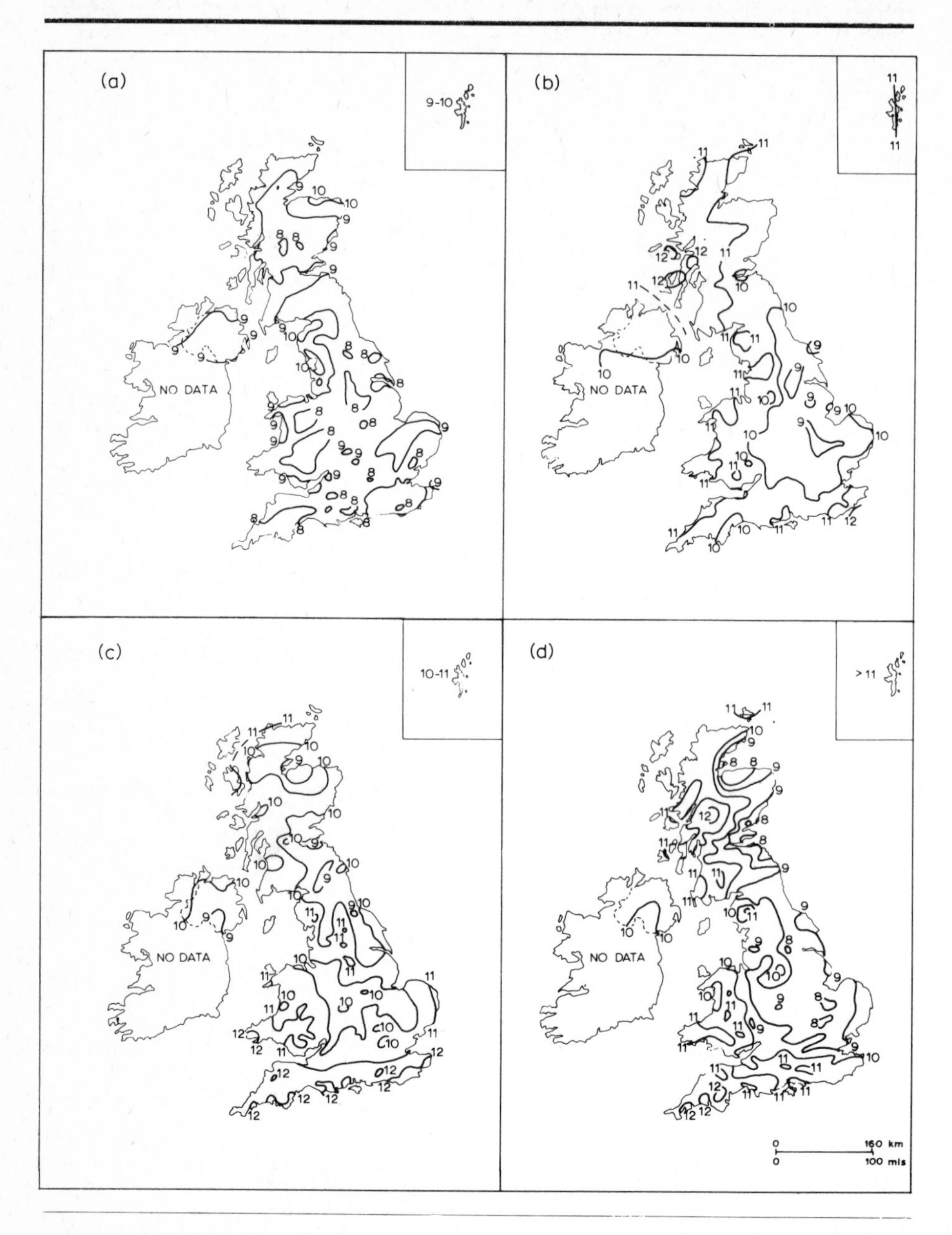

tracks or heavy thunderstorms can affect the values to give up to five or six times the monthly average. Because of the instability of these monthly averages, monthly totals have been expressed as a percentage of the annual average in volumes of British Rainfall since 1961 and so monthly extremes subjectively appear less impressive. It also seems probable that no measurable precipitation during a complete calendar month could be experienced anywhere in the British Isles. However, as its probability is low, the relatively short period of instrumental records means that it may not yet have been recorded. Surprisingly, in view of its low annual average, Kew has never recorded a monthly total of zero, but 4 months have this achievement for at least one gauge in the London area (Brazell, 1968). In both February and August 1947 no rain was recorded at Loch Quoich, where the mean annual fall is 3 711 mm. In February 1963 a normally very wet area north-west of Fort William was rainless and again in February 1965 the Aran Islands were similarly affected. April 1938 was the driest month so far this century in Ireland with parts of south Cork, Wexford, Wicklow and Tipperary rainless and parts of south-west England were also completely dry. It is rare for February to be completely dry, but for the period 1841–1949 it was the driest month most often in London. Also, taking a mean rainfall for England and Wales for the period 1915–64 40 per cent of the Februarys could be classed as dry, receiving less than 50 mm (Table 6.15). No rain for a period of 30 days is more common but is not included within the present criteria.

Maximum monthly totals inevitably occur in the mountainous areas of the west. Rain gauges are relatively sparse in these locations, especially at higher altitudes where totals would be expected to be greater. Consequently the highest totals recorded are partly a function of gauge siting and frequency rather than the maximum which has fallen at some spot in the British Isles. The largest totals are usually in the winter half of the year when depressions are more active and frequent, have a more southerly track across the Atlantic and are accompanied by strong moist winds blowing over the warm oceans. The highest recorded monthly fall was at Llyn Llydaw (Snowdonia) for October 1909 when 1 436·1 mm fell, but totals over 1 000 mm have been noted at several gauges in upper Borrowdale, Loch Quoich and Glen Garry, Snowdonia and Ben Nevis Observatory. Monthly rainfall in the summer half-year does not normally exceed 800 mm even in these very wet locations.

A summary of the range of values for monthly rainfall from 1727–1964 has been obtained by Bayliss *et al.* (1965). They found considerable differences in frequencies of wet and dry months between the different half-centuries. For their final period, 1915–64, it was found that July, August and November were most reliably wet (wet being defined as a monthly total above 100 mm for March to August and 125 mm for September to February), and March to June rarely receiving more than 100 mm, as shown in Table 6.15. For dry months (less than 50 mm in September to February and less than 37·5 mm March to August), March, June and February were most frequently dry and July had the lowest frequency of dry occasions.

These percentages reflect the modal circulation for any month. Large percentages of dry conditions imply that anticyclones have a higher probability of occurrence for that calendar month, or at least that depressions and westerly influences are at their weakest. Wet months will be associated with frequent and intense depressions in

Table 6.15 *Frequencies of monthly rainfall for England and Wales, 1915–64 (after Bayliss* et al., *1965)*

	Per cent dry (inches/mm)	Per cent wet (inches/mm)		Per cent dry (inches/mm)	Per cent wet (inches/mm)
	<1·5 (38) per month <6 (152) per season <13 (330) per half year	>4 (100) per month >11 (279) per season >19 (483) per half year		<2 (51) per month <7 (178) per season <15 (381) per half year	>5 (127) per month >12 (365) per season >23 (584) per half year
March	30	8	*September*	16	8
April	18	2	*October*	14	20
May	16	8	*November*	10	28
June	26	4	*December*	14	10
July	8	28	*January*	10	18
August	12	34	*February*	40	10
Spring **Summer**	22 16	4 14	**Autumn** **Winter**	8 12	34 24
Spring and Summer	16	12	**Autumn and Winter**	10	30

winter, but with stagnant thundery lows in summer with few anti-cyclones on average. Individual convectional storms giving large totals will be averaged out as the figures are means for England and Wales.

Throughout this section it has been stressed that as monthly rainfall is made up of a wide range of rainfall distributions and synoptic situations it is not surprising that fluctuations from month to month and year to year are large. However, examination of monthly figures based on an overall average for England and Wales has shown that certain monthly sequences exist (Murray, 1967b). All months from 1866 to 1965 were classed into terciles and designated as dry, average or wet for the lower, middle and upper terciles respectively. Murray then analysed sequences of months with similar rainfall (as indicated by the monthly total), to determine whether certain months were more likely to have strong persistence of similar rainfall conditions than others. An interesting result of the study was that all frequencies were similar to the geometric distribution with a very rapid fall off in persistence with time. The longest spells of dry weather lasted for 6 months, one starting in April and the other in September. For wet weather, July showed the longest persistence with one occasion of 7 months, but persistence from initial to the following month was least from August to September and greatest from September to October. Sequences at individual stations would show greater persistence as the area covered by England and Wales ensures that some areas will exhibit different characteristics and so modify the overall average.

Seasonal rainfall

As the time scale of investigation lengthens, so the underlying factors affecting the rainfall totals become of increasingly large scale. The atmospheric circulation belts have seasonal changes linked to the apparent motion of the overhead sun and also changes from year to year and even longer periods, for reasons which are not clear as yet.

Table 6.16 *Variability of seasonal rainfall at Oxford from 1815 until 1970 (after C. G. Smith 1974)*

	Mean (mm)	**Standard Deviation (mm)**	**Per cent of annual mean**	**Coefficient of variation (per cent)**
Winter	151·4	49·3	23·2	32·6
Spring	135·9	43·9	20·8	32·3
Summer	177·0	62·2	27·1	35·2
Winter	188·2	58·7	28·9	31·2

The average rainfall of a season will depend in general terms on the mean frequency, intensity and track of rain-bearing systems near the British Isles, but for a specific season the rainfall will be determined by deviations from this mean state. Persistent anomalous circulations can give either very large or very low totals depending upon their nature, whereas normal circulations would be expected to give average rainfall. As the global circulation varies over time, so we can expect seasonal rainfall averages to change from one 'standard' period to the next (Crowe, 1940). Fortunately, Murray's conclusions about monthly persistence suggest that seasonal variations should be somewhat less and this is confirmed by Smith's work on the Oxford record (Table 6.16) where the coefficients of variation for each season are all about 33 per cent.

Fig. 6.11 *Isomeric maps of seasonal rainfall, 1916–50, with average seasonal rainfall as a percentage of average annual rainfall (after Bleasdale, 1971)*; (a) *March and April*; (b) *May to August*; (c) *September and October*; (d) *November to February*

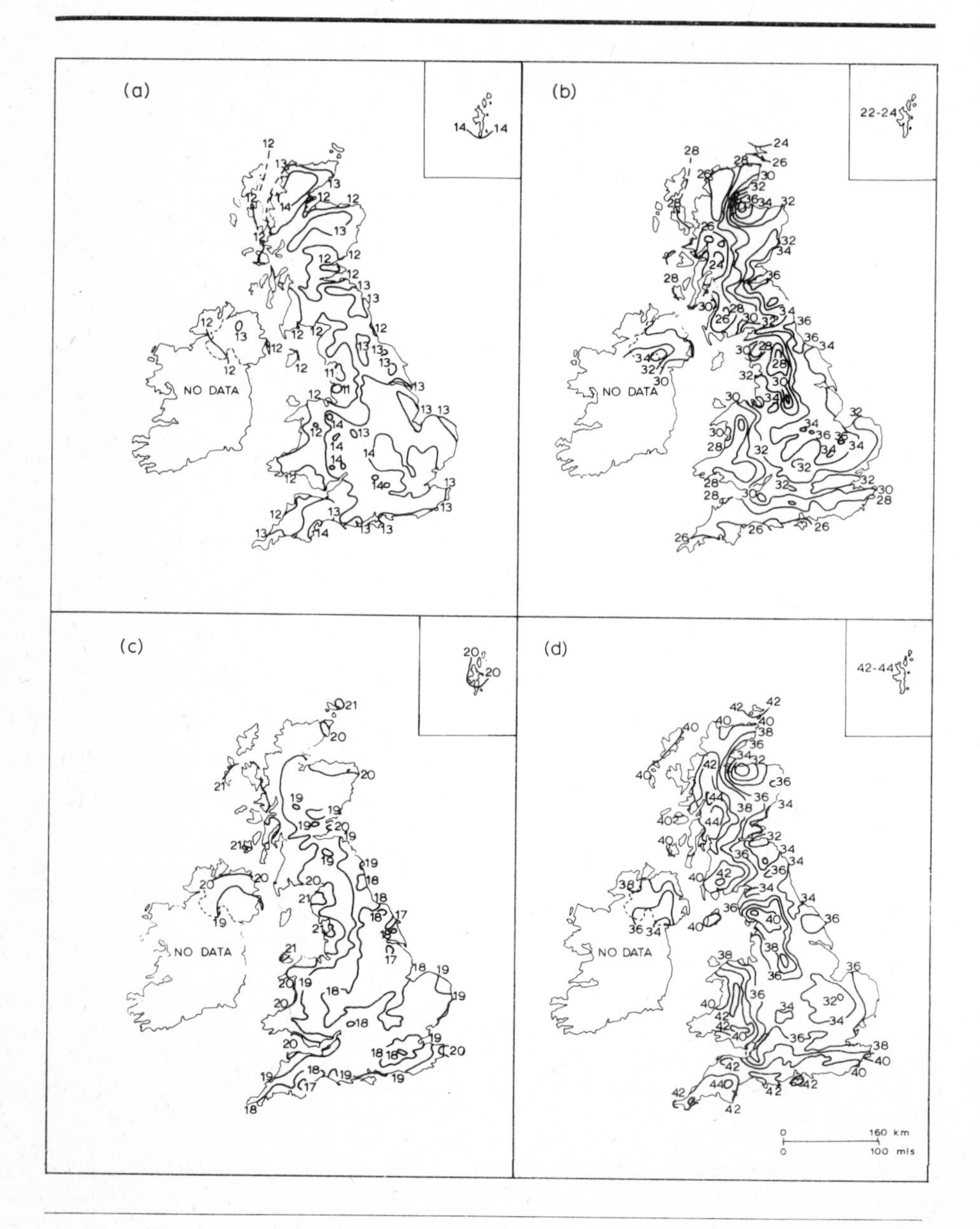

Lamb (1950) proposed that the year could be sub-divided into natural seasons on the basis of changes in the atmospheric circulation. Using these criteria, autumn would be from early September to mid-November, early winter from late November to mid-January, late winter from late January to late March, spring from April to early June and high summer from late June to the end of August (see Chapter 2). As there is nothing sacrosanct about months, this division could be appropriate, but from the rainfall point of view the record at Oxford suggests that variability in most months is such as to make a constant 'natural season' unrealistic and seasons based on calendar months are equally appropriate (Smith, C. G., 1974). The choice of months often depends upon the purpose of the analysis. From the agricultural view-point, summer may be considered as May to August when evapotranspiration is at a maximum and winter as from November to February when soils reach field capacity in most areas; spring and autumn are March and April and September to October respectively. With an even distribution of rainfall, spring would be expected to receive 16·7 per cent of the annual total. For the 1916–50 period, nowhere in the British Isles had such a high proportion (Fig. 6.11). The highest percentage of 14 was found in the Home Counties of England, a few points to the east of the Welsh uplands, the Torbay area, part of the north-west Highlands and the Shetlands. Lowest values are on Islay and parts of coastal Lancashire with only 11 per cent. The range from highest to lowest is relatively small. The summer period shows much greater variation. Areas receiving high percentages are around the Moray Firth with 38, then an area in general comprising southern and eastern Britain as shown in Fig. 6.11. This does not necessarily coincide with regions of summer maximum but is purely an area with a tendency for a greater proportion in summer than would be expected from an even distribution. The lowest values are the areas with high orographic components or more northern parts such as the Shetlands with only 24 per cent of the mean annual fall between May and August. This season shows the greatest effect of sheltering from oceanic influences. Wherever rain-shadowing from the autumnal and winter westerly depressions is marked then summer percentages are naturally greater.

For September and October, rainfall everywhere is above an even proportion. It reaches its maximum in western coastal areas, receiving the brunt of the equinoctial storms with a secondary maximum along the English Channel coast as far east as Kent. The percentage distribution of winter precipitation is, as would be expected, almost a complete reversal of the summer situation. Only in the Moray Firth area is it less than for an even distribution rising to a maximum of 44 per cent in the Rannoch Moor area of the Grampians, parts of Dartmoor and Cornwall and south-west Ireland. Thus the greater frequency of depressions and their more southerly tracks at this time make winter one of the wetter seasons of the year in almost all parts and make it appear to be even wetter by its lower mean intensities and hence longer durations. As short-period fluctuations in circulation continue, so the relative seasonal proportions will change from those of the 1916–50 period. For example, autumn has become much drier in recent years with rainfall in October being far less than average in England and Wales for 7 of the last 8 years (1968–75) and to a lesser extent in September associated with a greater frequency of anticyclones. This trend could, of course, change at any time.

Investigations into sequences in seasonal rainfall have come to similar conclusions to those devoted to monthly totals. Stephenson (1967) studied the period 1727–1964 in terms of whether the rainfall in the conventional seasonal periods – September to November, etc. – had any persistence or singularities. Of the sixteen contingency tables produced, only the one from autumn to winter gave a statistically significant result using the χ^2-test and that was only at the 6 per cent level of significance. This implied that a wet autumn was likely to be followed by a wet winter but the association was not strong enough for use in long-range forecasting. Stephenson concluded that although persistence may occur for certain periods when studied on a long-term basis, seasonal rainfall could be treated as independent between preceding and succeeding periods.

Annual rainfall

Mean annual rainfall over the British Isles varies widely from low totals about 500 mm in the English Fenland (Stretham) and the Thames Estuary (Isle of Grain) to a maximum of 5 000 mm above Sprinkling Tarn in the English Lake District for the standard period 1931–60 (Fig. 6.12). The area experiencing values less than 508 mm (20 in) has shrunk and changed slightly since Bilham (1938) described them for the 1881–1915 period. Only the small area to the south-west of Ely now lies within the 508 mm isohyet; the lowest values in the Thames Estuary are now higher than this value with the exception of a single gauge on the Isle of Grain.

Annual rainfall in most years shows a similar pattern with large totals in the higher areas of Ireland, Wales, Scotland and the English Lake District, gradually decreasing towards the south and east, though totals in the east of Ireland are lower than those of the windward coasts of England and Wales. This pattern is governed by the frontal depressions which develop within the northern hemispheric, westerly circulation reaching their maximum frequency and intensity in the area of the Icelandic Low. As this is located to the north-west of the British Isles, it is not surprising that sea-level rainfall totals are highest in the north-western parts and decrease towards the south-east. By chance, the mountain areas are also in the west and north which exaggerate the sea-level rainfall gradient and through orographic intensification give these areas the highest mean annual totals. Everywhere rainfall increases with altitude as explained earlier, so that even within south-east England totals reaching 1 000 mm are found on the South Downs with the isohyets approximately following the contours within any specific area. Further modifications of this pattern are achieved through topography where funnelling and carry-over effects can produce meso-scale changes which do not accord with the usual regular extrapolation between mean annual rainfall and altitude. Because of the scarcity of gauges in upland areas, many such topoclimatic effects will be unrecorded. In spite of the decrease in westerly influences in recent years (Chapter 2), the overall mean annual rainfall distribution remains similar though the differences between north and south are decreasing as the frequency of stagnating depressions over England and Wales increases (Lawrence, 1973a).

Fig. 6.12 *Mean annual rainfall in mm, 1931–60 (after Meteorological Office, 1968a, and Eire Meteorological Service, 1973)*

The variations of rainfall intensity tend to mask the raininess of a climate as exhibited by its annual total. It has already been pointed out that the annual mean at Cwm Dyli is 3·5 times that at Holyhead, but because of the higher mean intensity of rain at the former site its mean duration is less than double. This can be complicated by different synoptic systems which can produce widely different rainfall intensities and so make it difficult to generalize. For example, in mountainous areas frontal rainfall is usually intensified, but if the front is very weak and giving rain only over the mountains then intensity there may be low. Similarly within a warm sector, prolonged drizzle by orographic scouring (Pedgley, 1970) would also have a low intensity. The net result is that mean rainfall intensities are not always higher. For the period 1931–60, mean intensity at Eskdalemuir (242 m) is 1·30 mm hr^{-1}, whereas at Greenock (61 m) only 130 km to the north-west, the mean value is 1·63 mm hr^{-1} (Table 6.17).

The mean annual rainfall at the two sites is almost identical (1 582 mm and 1 587 mm respectively) but with the lower mean intensity at the former site, rainfall duration is about 200 hours longer or an excess of almost 18 per cent. The mean intensity at sea-level varies little throughout the British Isles from 1·3 mm hr^{-1}. In the eastern counties of England it is somewhat lower, about 1·1 mm hr^{-1}, as the low intensities of frontal rainfall in winter counteract the higher intensities from summer convectional rain. Mean values increase slightly towards the west with 1·5 mm hr^{-1} at Falmouth and 1·4 mm hr^{-1} at Aberporth. Northwards there is a decrease to 1·4 mm hr^{-1} at Tiree and 1·2 mm hr^{-1} at Lerwick. This is complicated by topographic modifications. As mentioned above, altitude is not necessarily the cause of higher intensities, but wherever hills induce sudden lifting of air an intensification of pre-existing rainfall is probable with resulting higher intensities. Thus we might expect higher rates at Greenock where hills are numerous in an upwind direction compared with Eskdalemuir which is on a plateau surface where initial uplift will have taken place further west. Only the slightly lower rate of evaporation in the shorter distance between cloud base and surface at Eskdalemuir should affect rainfall apart from any standing waves which the Southern Uplands may induce. Where we find examples of topographic intensification, mean intensities increase to above 2 mm hr^{-1} as in the Lake District, Snowdonia and parts of western Scotland and Ireland.

Three major points may be drawn from Table 6.17. First, the annual rates of precipitation are very small at about 1·3 mm hr^{-1}. Secondly, the summer increase in rates due to the increased convective activity is clearly shown by the majority of stations in the south and east but decreases in importance elsewhere, partly as a result of more intense winter rain and partly by a reduction in summer intensity. Thirdly, there is a hint in the Swansea Waterworks records that rates of precipitation increase with altitude. Although this was not true for Eskdalemuir for the reasons stated, other data could be used to support this premise.

The scarcity of autographic raingauges and problems of analysis make it impossible to produce a map of mean annual rainfall duration

Table 6.17 *Mean monthly and annual rates of rainfall, in mm hr⁻¹, for selected stations (UK data 1951–60; Irish data 1961–70)*

	Height above sea level (m)	Month												Annual mean
		J	F	M	A	M	J	J	A	S	O	N	D	
Camden Square	33·5	1·04	0·91	0·97	0·99	1·37	1·68	1·96	2·08	1·65	1·40	1·27	1·02	1·30
Boscombe Down	127·4	1·12	1·09	1·04	1·02	1·37	1·63	1·50	1·68	1·52	1·47	1·40	1·22	1·32
St Mawgan	102·7	1·27	1·30	1·27	1·24	1·45	1·50	1·85	2·13	1·93	1·78	1·55	1·52	1·55
Ross-on-Wye	68·3	1·30	1·27	1·27	1·37	1·65	1·93	2·11	2·16	2·13	1·73	1·60	1·40	1·60
Cranwell	62·8	0·91	0·76	0·89	0·89	1·35	1·40	1·55	1·65	1·35	1·22	1·04	0·99	1·14
Squires Gate	10·1	1·02	0·91	0·81	0·86	1·14	1·07	1·63	1·60	1·42	1·22	1·22	1·07	1·17
Felixkirk	164·6	1·27	1·12	1·07	1·12	1·47	1·45	1·80	1·73	1·65	1·45	1·32	1·22	1·40
Swansea WW	317·9	1·70	1·52	1·70	1·68	1·73	1·78	1·78	1·91	2·13	1·96	2·01	1·78	1·80
Holyhead	8·6	1·12	1·12	0·91	1·02	1·27	1·22	1·37	1·68	1·63	1·50	1·35	1·24	1·30
Eskdalemuir	242·0	1·30	1·09	1·09	1·10	1·37	1·30	1·50	1·40	1·45	1·40	1·35	1·22	1·30
Greenock	60·7	1·73	1·42	1·37	1·42	1·55	1·42	1·70	1·65	1·93	1·80	1·65	1·73	1·63
Dyce	58·4	1·10	1·07	1·14	1·07	1·35	1·45	1·70	1·47	1·60	1·57	1·52	1·37	1·37
Stornoway	3·7	1·24	1·22	1·09	1·19	1·14	1·19	1·27	1·40	1·50	1·35	1·37	1·42	1·30
Lerwick	82·0	1·12	1·17	1·12	1·17	1·09	1·17	1·24	1·42	1·50	1·37	1·24	1·32	1·24
Aldergrove	67·1	0·91	0·89	0·91	0·81	1·07	1·22	1·50	1·37	1·37	1·37	1·12	1·04	1·12
Claremorris	89	1·15	1·15	1·06	1·10	1·16	1·43	1·30	1·43	1·57	1·49	1·22	1·24	1·27
Belmullet	9	1·42	1·41	1·37	1·19	1·33	1·49	1·45	1·67	1·79	1·76	1·58	1·51	1·51
Birr	70	1·10	1·09	1·02	1·18	1·23	1·31	1·41	1·68	1·48	1·54	1·18	1·18	1·27
Clones	69	1·04	1·06	1·02	0·99	1·20	1·37	1·23	1·58	1·33	1·35	1·11	1·09	1·18
Dublin Airport	68	1·20	1·09	1·05	1·22	1·20	1·40	1·73	1·64	1·69	1·38	1·41	1·21	1·33
Kilkenny	61	1·11	1·08	1·04	1·11	1·08	1·27	1·46	1·50	1·59	1·45	1·19	1·09	1·22
Malin Head	25	1·26	1·34	1·12	1·12	1·09	1·33	1·29	1·55	1·44	1·58	1·48	1·29	1·34
Mullingar	110	0·99	0·95	0·90	1·03	1·04	1·11	1·27	1·31	1·34	1·28	1·14	0·99	1·10
Roches Point	40	1·40	1·27	1·37	1·35	1·28	1·37	1·73	1·91	1·74	1·72	1·62	1·31	1·48
Rosslare	18	1·35	1·26	1·19	1·33	1·22	1·34	1·54	1·70	1·69	1·90	1·70	1·32	1·45
Shannon Airport	7	1·36	1·39	1·29	1·31	1·46	1·56	1·44	1·81	1·67	1·63	1·40	1·42	1·48
Valentia	14	1·67	1·61	1·57	1·45	1·48	1·44	1·42	1·54	1·79	1·83	1·60	1·44	1·57

Fig. 6.13 *Mean annual duration of measurable precipitation, 1931–40, in hundreds of hours, (after Meteorological Office, 1952)*

with the same reliability as rainfall amounts. The area with the shortest duration of rain, having less than 500 hours per year, is found on low ground in East Anglia and the Fens, the lower Thames Valley and estuary and in the lee of the South Wales hills where Ross-on-Wye frequently has less than 500 hours per year of recorded rain (Fig. 6.13 and Tables 6.1 and 6.2). Most of England to the south and east of a line from Plymouth to the Tees has fewer than 600 hours, except areas above about 200 m, and an embayment of lower values occurs to the lee of the Welsh Mountains and on the Cheshire Plain. Coastal areas of north-east England, the extreme east of Scotland and the Inverness area also come within this drier zone. All Wales, the remainder of Scotland, and much of Ireland have mean annual durations above 700 hours reaching 850 hours in the Hebrides and northern isles with maxima of about 1 500 hours in the wetter parts around Fort William and Loch Lomond. This means that measurable rain is falling here, on average, about 17 per cent of the year.

Another index of wetness is the rain-day. This does not distinguish between the sudden heavy shower which may last for only a few minutes and prolonged drizzle which could continue for several hours, but it is at least some measure of the propensity for rainfall. Average values of the number of rain-days range from less than 150 days per year in the lower Thames Estuary to over 250 in the Outer Hebrides, the Shetlands and the hillier parts of south-west and western Ireland. The isoline for 200 days can be traced from County Wexford in a north-north-east direction to Belfast, through Galloway, the margins of the Lake District and Pennines, then following the east coast as far as Ross and Cromarty. Most of Wales, Cornwall, Dartmoor, Exmoor, the Isle of Man, the Yorkshire Moors and Wolds and, surprisingly, a few parts of upland Norfolk also experience a mean of more than 200 rain-days. The standard deviation of the number of rain-days is low, about 16 to 20, irrespective of the mean, so on the usual assumptions of normality, we could expect one year in a hundred in the dry south-east of England to have about 110 rain-days and in the north-west about 310 rain-days. At Kew in 1921, only 104 days with rain were recorded (and only 84 at Ramsgate) but this was the driest since records began in 1874 and confirms the earlier calculation. With the decrease of westerly systems affecting the British Isles, it is not surprising that there has been a significant decrease (at the 5 per cent level) in the average rainfall and number of days with rain between the periods 1916–50 and 1951–71 in north-western areas. At Kew and Falmouth, where the proportion of rain from active westerly depressions is less, the differences were non-significant for both annual totals and the number of rain-days, though there had been a slight fall in the number of days with rain at both sites.

Rainfall variability

The British Isles are renowned for rainfall reliability and, to the general public, for rainfall frequency. However, whilst the former may be true on a world scale, there is some variation from year to year as a response to the ever changing atmospheric circulation and in particular to the location, intensity and duration of the main centres of depression and anticyclone activity relative to the British Isles. To take two extremes, in 1954 much of the year and particularly the summer period was dominated by westerly cyclonic systems which had more southerly tracks than normal and which continually crossed the British Isles (Barnes *et al*., 1956). This had the effect of giving very large annual totals in the north and west of the country reaching a record of 6 527 mm at Sprinkling Tarn, the highest annual total in the British Isles since records began. The frequency of depressions meant that areas to the east of the main uplands also had higher than average totals, though because summer convective rain was less than usual they were only slightly above average. In contrast, 1921 was a year of unprecedently low rainfall in England and Wales with a general total of only 71 per cent of the 1881–1915 mean. In Ireland the general value was 89 per cent and for Scotland 97 per cent. The months of 1921 were not individually striking, apart from February, June and July. The unusual feature was that a period of sub-normal rainfall coincided with a calendar year, only January and August having rainfall above the monthly average in England and Wales, and the large area experiencing below average rainfall. Other periods, for example, July 1972 to March 1973, have been drier, but did not coincide with the calendar year. During 1921, pressure was above normal from February to October reaching the maximum positive anomaly off south-west Ireland. This was the result of a northward extension of the Azores anticyclone which prevented many depressions from following their normal routes near the British Isles. As the main centre of blocking was in the south-west, it was southern areas which experienced the greatest deficits and Scotland received only slightly less than normal (Brooks and Glasspoole, 1922). The lowest total was at Margate in Kent where only 236 mm were recorded, 44 per cent of the 1916–50 average, although the gauge may be suspect.

As the frequency distribution of annual rainfall totals approximates to normality, it is possible to assess the range of values which would be expected to occur in a given period and also the probability of occurrence of specified amounts. For example, at Kew the 1916–50 mean is 608·6 mm with a standard deviation of 110·5 mm; therefore we would expect values between 829·6 mm and 397·6 mm (±2 S.D.) in about 96 years out of any 100-year period. Between 1871 and 1973, only 4 years were outside these limits: 1879, 1903, 1915 and 1921 with 841·0 mm, 969·5 mm, 832·9 mm and 308·4 mm respectively. This can be applied to any gauge record although the use of the standard deviation demands that there is no underlying long-term trend in the annual mean.

An alternative method of assessing variability is to take the extreme percentage range of annual rainfall. This value will tend to increase as records lengthen as it is merely a measure of the difference between the highest and lowest falls relative to the long period average. In the British Isles the lowest values are found in the north and west with a range of approximately 70 per cent for a 100-year record, i.e. between 65 per cent of the average for the lowest total and 135 per cent for the highest. This increases towards the south and east with the highest values above 100 per cent for parts of the English Midlands and the Home Counties. The pattern is not quite as regular as this might suggest; the highest in Ireland is Dublin with a range of 87 per cent (154 per cent to 67 per cent) for the period 1837–1971, and several of the drier areas of Scotland have large ranges such as Fortrose with a range from 136 per cent to 53 per cent, the latter occurrence being the dry year of 1972. For the period 1901–30, the Isle of Man and east Ulster had the smallest range with below 40 per cent, but in the latter area, if the period is extended to 1872–1973, the range increases to 80 per cent, emphasizing the weakness of this method. If all data are based on the same period, however, it does give a useful indication of variability.

Fig. 6.14 *Coefficient of variability of annual rainfall, 1901–30 (after Gregory, 1955)*

A more accurate presentation of variability is the coefficient of variation, which relates the standard deviation of the data to the average, expressed as a percentage. For the period 1901 to 1930, the coefficient of variation over the British Isles is shown in Fig. 6.14 (Gregory, 1955). Even this index is affected by the period studied. At Kew, Fig. 6.14 shows a coefficient of 18 per cent; for 1916–50 it was 18·2 per cent and 1951–73 it fell to 16·3 per cent. Similar slight variations exist in other parts of the country although they are statistically non-significant. Some of these variations for other time periods are described by Senior (1969). In general, the areas of the north and west have the lowest coefficients and higher values occur in the south and east, although in recent years the wetter areas of Scotland have become rather more variable with coefficients above 16 per cent.

Fig. 6.15 (a) *Percentage probability of receiving an annual rainfall of less than 20 inches (508 mm) (after Gregory, 1957)*; (b) *Percentage probability of receiving an annual rainfall of more than 40 inches (1 016 mm) (after Gregory, 1957)*

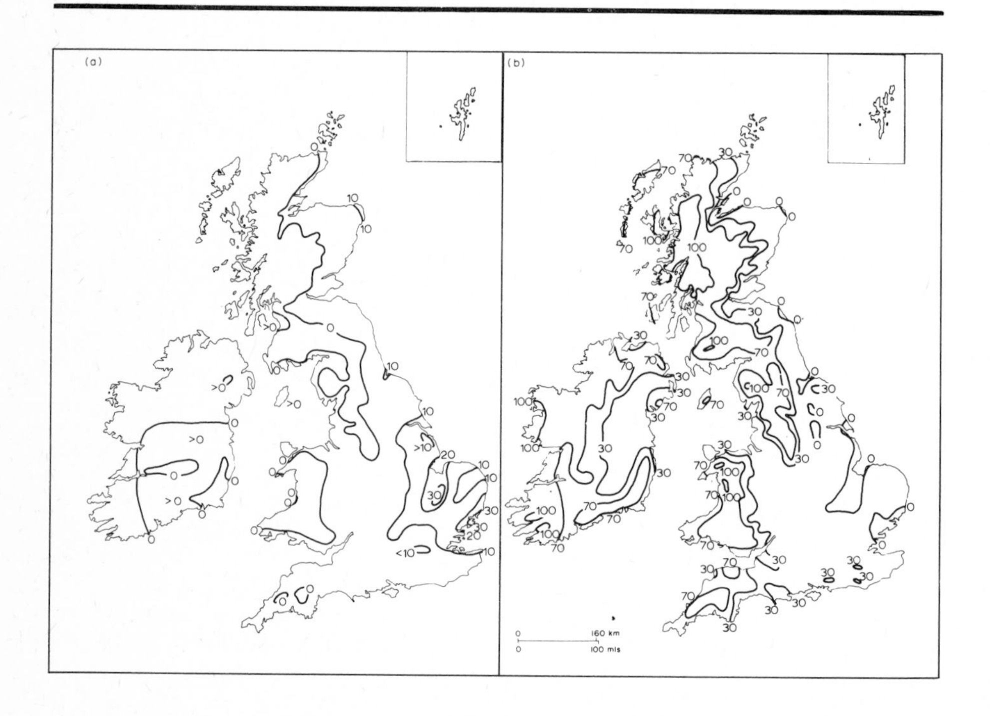

A third method of illustrating rainfall variability is by a statistical assessment of certain annual totals being reached. This has been investigated by Gregory (1957) for the period 1901–30 when the incidence of westerly conditions affecting the British Isles was reaching a peak. For that period no gauge had a mean fall below 508 mm (20 in), but with a normal frequency distribution a mean near 508 mm would imply that nearly half the years in any standard period would have an annual total below 500 mm. With an annual total below

500 mm and an average seasonal distribution, agriculture would experience drought conditions during part of the year with irrigation required for many crops, so a knowledge of the probability of this annual fall being exceeded is important. Figure 6.15 shows that for most of the British Isles there is a 90 per cent probability of 508 mm being exceeded. Only eastern England, especially the Fens and the Essex/Suffolk coastlands, has a probability of more than 1 year in 10 receiving less than 508 mm with the lowest values being near Ely where the probability falls to between 60 per cent and 70 per cent. This does not appear to change much by taking a longer period as for the years 1908–73 at Cambridge (Botanic Gardens) the percentage probability of receiving more than 508 mm was 69 per cent.

If we take 1 000 mm (40 in) as the boundary of the wetter parts of the British Isles it can be seen (Fig. 6.15*b*) that most of the population lives in areas with less than a 10 per cent probability of this amount being received. Only Clydeside, parts of central Lancashire and South Wales are significant centres of population inside this area. It is interesting to note that few areas have a value of zero per cent. Contrasts between Scotland and Ireland are marked. In the former there is a rapid transition from the east coast lowlands, where 1 000 mm is unusual, to the neighbouring hills where it is frequent. As most of the central area is upland, values remain high and even on the west coast 1 000 mm is reached in most years. In the wettest parts with low standard deviations, rainfall has never been less than 1 000 mm. In Ireland, much of the centre is lowland and so there is a more gradual transition from the drier east where 1 000 mm is rare to the west where in Connemara it is always exceeded even on the coast. Again there seems to be little modification by increasing the period. For Douglas, Isle of Man with a mean of 1 122·4 mm there is a 77 per cent probability of exceeding 1 000 mm based on the period 1909–73 which is in accord with Gregory's work for 1901–30.

The British Isles are often described as having an oceanic climate with the implication that the rainfall regime of the whole area is uniform. On detailed examination this is far from true. Glasspoole (1922) calculated correlation coefficients between annual rainfall at the Radcliffe Observatory, Oxford, for the period 1881–1915 and a selection of reliable raingauges throughout the British Isles. These coefficients are shown in map form in Fig. 6.16. If a single regime were operative we should expect a strong positive correlation with totals at Oxford. This positive correlation was found over England, Wales, Ireland and southern Scotland, but there was a very rapid decrease to the north-west of a line from Islay to Aberdeen. By the Outer Hebrides the coefficient was just negative at −0·01 though this merely means that there was no association whatever in annual rainfall totals. The high value for Crieff (+0·66) is surprising but suggests that rainfall changes there are comparable to those at Oxford. This work was amplified by Salter and Glasspoole (1923) where, from an examination of maps expressing annual rainfall, 1868–1921, as a percentage of average, they concluded that patterns fell into three types, (i) excess of orographic rain; (ii) deficiency of orographic rain; and (iii) excess of cyclonic rain. With an excess of orographic rain, the north and west would have totals above average, but with an excess of cyclonic rain the tracks of the depressions would determine the pattern. It was noted that in the 12 years from 1875 to 1886 only 1 year had an orographic pattern, whereas from 1911 to 1921 6 years had such a pattern. This is undoubtedly a reflection of the increase in vigour of the westerly circulation by the latter period. To see how this affected the correlation coefficients a preliminary investigation was made for the period 1916–72 and listed in Table 6.18. The correlation between Oxford and sites to the north and west has shown a significant decrease from which it may be presumed that either rainfall at Oxford now has less oceanic influences or that continental effects are less strong in the north and west whilst the westerly circulation built up to its peak in the 1930s. The correlation with Cambridge increased and with Kew (not shown) remained the same, suggesting a greater uniformity of regime in the southern and eastern parts of England. The high correlation with southern Scotland centred on the area around Crieff has now disappeared (as exemplified by Aberfeldy) which is more in keeping with the expected decrease of correlation with distance. At Inverie House (annual mean 2 136 mm) on the west coast of Scotland there is a negative tendency developing though it is not statistically significant. However it would appear that there is some support for reverse regimes between north and south which Perry (1972b) found for Ireland, as well as between west and east.

Fig. 6.16 *Annual rainfall correlation isolines, 1881–1915 (after Glasspoole, 1925);* (a) *based on Oxford;* (b) *based on Glenquoich*

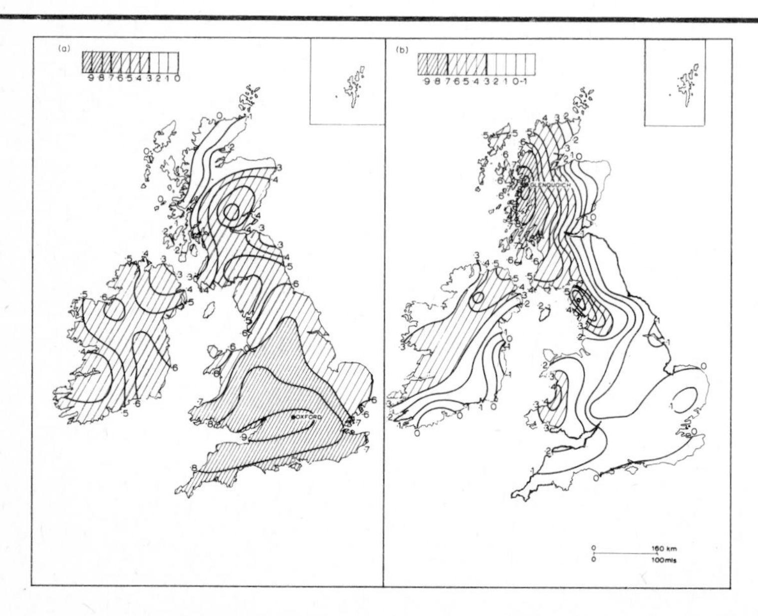

For correlation between other gauges, it can be seen that west to east contrasts across England are far less than across Scotland with its greater topographic influences. Falmouth and Swansea both have a

Table 6.18 *Correlation matrix for annual rainfall totals of selected stations, 1916–72*

	Oxford 1916–72	1881–1915	C	W	Fa	Sw	Sh	T	D	O	A	H	Ft	I	St
Oxford	1·000	1·00	—	—	—	—	—	—	—	—	—	—	—	—	—
Cambridge	0·830	0·70	1·000	—	—	—	—	—	—	—	—	—	—	—	—
Wye	0·797	0·75	0·803	1·000	—	—	—	—	—	—	—	—	—	—	—
Falmouth	0·575	0·76	0·593	0·630	1·000	—	—	—	—	—	—	—	—	—	—
Swansea	0·563	0·80	0·527	0·491	0·709	1·000	—	—	—	—	—	—	—	—	—
Sheffield (Rivelin)	0·566	0·75	0·551	0·618	0·606	0·496	1·000	—	—	—	—	—	—	—	—
Tynemouth	0·589	0·52	0·602	0·524	0·443	0·363	0·598	1·000	—	—	—	—	—	—	—
Douglas	0·436	0·60	0·538	0·404	0·491	0·663	0·436	0·438	1·000	—	—	—	—	—	—
Omagh	0·148	0·47	0·300	0·191	0·357	0·530	0·367	0·176	0·723	1·000	—	—	—	—	—
Aberfeldy	0·303	0·50	0·204	0·234	0·344	0·549	0·390	0·379	0·509	0·529	1·000	—	—	—	—
Helensburgh	0·039	0·35	0·030	—0·003	0·076	0·472	0·106	0·130	0·548	0·662	0·699	1·000	—	—	—
Fortrose	0·311	0·25	0·317	0·242	—0·029	0·108	0·217	0·284	0·418	0·399	0·373	0·387	1·000	—	—
Inverie (Mallaig)	—0·170	0·00	—0·123	—0·200	—0·176	0·209	—0·165	—0·200	0·283	0·546	0·398	0·729	0·327	1·000	—
Stornoway	—0·016	—0·01	0·006	—0·058	0·023	0·377	—0·130	—0·123	0·268	0·491	0·342	0·665	0·191	0·831	1·000

Significance levels 5 per cent=0·25 1 per cent=0·33 0·1 per cent=0·42 high correlation with all the other English sites whereas Stornoway has significant associations only with north-western locations. The regime at Fortrose is even more distinctive with no coefficient above +0·42. Distance does seem to be important in determining the resultant coefficient, but it is not the only factor. The distance from Sheffield to Helensburgh is almost the same as to Wye but there is no significant correlation with the western oceanic and orographic regime of Helensburgh whereas the correlation with Wye is high. Other examples could be quoted.

In spite of the wide range of annual rainfall totals found over the British Isles, there is a strong link in southern areas between the western and eastern parts. This gradually decreases northwards as topography becomes more important in modifying rainfall totals. In Scotland, the areal extent of significant correlation coefficients becomes much smaller for this reason though significant links with Helensburgh, for example, do extend as far as Swansea and probably to parts of western Ireland with somewhat similar locations.

For smaller areas the study of rainfall regimes is more complex. Gregory (1956) distinguished four regions in the British Isles which exhibited similar trend characteristics of annual rainfall for the period 1881–50 and Barrett (1966) found nine distinct regions in the north of England alone based on the time at which annual rainfall totals reached their maximum using 10-year running-mean values. They appeared to be the result of the varying interaction between topographic effects and the constantly changing atmospheric circulation.

Synoptic origins of rainfall

Brief mention of the synoptic origins of precipitation has already been made in Chapter 2. This is a theme into which much investigation has taken place in recent years and which provides an appropriate conclusion to this section on rainfall. So far in this chapter implicit emphasis has been placed on the essential links between rainfall and its synoptic origins at all time scales. The significance of the association has been appreciated by many researchers (summarized in Barry and Perry, 1973), but the methods of classifying the atmospheric circulations have varied. One possible method is to classify the circulation over the British Isles on any specific day in terms of the main pressure and wind systems involved. This forms the basis of Lamb's (1950) classification from which much further work has developed (Lawrence, 1971b, 1972, 1973a; Perry, 1970). In this system the amount of rain which fell during each category can be summed and the synoptic origins of the precipitation can be obtained for any time period. These have been summarized in Table 2.7. There is a clear dominance of westerly and cyclonic categories as the main contributors to rainfall in England and Wales especially when the frequency of occurrence is included. This approach has also been used for specific areas (Barry, 1967), but not, as yet, for Ireland or Scotland. However, it seems likely that similar influences will operate there to make the westerly and cyclonic categories still most important, the former contributing a higher proportion in the west and the latter in the east. Exposure to open sea is usually relevant to assessing the significance of the other categories. One result of this type of work has been to show that as a result of the increase in frequency of days of cyclonic weather type, there has been an increase in the number of days with high values of areal rainfall ($\geqslant$ 20 mm) over England and Wales since about 1960 (Lawrence, 1973b).

An alternate method of relating precipitation to its synoptic origins has been taken by Thomas (1960b), Sawyer (1956a) and Jackson (1969). In this system, greater stress is placed on depressions and their tracks as causative factors of precipitation. Jackson distinguished nine categories associated with moving depressions and fifteen stationary types incorporating both anticyclones and depressions as they affected North-East England. With a large number of categories a long period of records is required to ensure that significant conclusions can be drawn about the less frequent systems.

Precipitation has also been classified in terms of the precise synoptic events initiating the rainfall as opposed to the general types of Lamb (Shaw, 1962). This developed from an analysis of the charts from a Dines Tilting-syphon Rain Recorder when it was noticed that the varying character of the rainfall trace could be related to the current weather pattern. These were sub-divided into warm, cold and occluded fronts, warm sectors, maritime polar, continental polar and Arctic airstreams, non-frontal depressions (called 'polar low' in the original classification although it covered any low pressure area filling slowly after frontal activity had ceased) and thunderstorms. As this classification is not precise, almost all rainfall could be allocated to one of the categories by use of the Daily Weather Report. From the analysis of a sequence of raingauge charts, the synoptic origins of the precipitation for that site could be obtained. This method is outlined in Chapter 2 with a west to east transect from Snowdonia to Lincolnshire. Similar work has also been published for Scotland (Smithson, 1969b) which enables a comparison of north to south records as well as east to west changes. From this it is found that the contribution from warm fronts decreases northwards, partly accounted for by their lower frequency, and the contribution of occluded fronts, Arctic and continental polar airstreams and non-frontal depressions increases northwards. It must be added that as well as this north-south contrast, great diversity exists between western and eastern parts of both areas with a greater proportion of rainfall from warm sectors and maritime polar airstreams in the former, and from occluded fronts, continental polar airstreams and non-frontal depressions in the latter, associated with the reduced rainfall from direct westerly influences and the process of occluding from west to east.

The use of synoptic classification clarifies the causes of the temporal and spatial variations of rainfall that have been outlined in earlier sections. The amount of rain which falls on an area is closely controlled by its atmospheric origins and the interaction between the ground surface and the lower atmosphere. If the changes in atmospheric circulation are known, a better understanding of rainfall patterns and regimes is obtained.

Snowfall

The bulk of this chapter has been concerned with liquid precipitation but in the introduction, explicit recognition was given to hail and snow. The former has already been given brief consideration and the latter, despite its romantic appeal and potentially important economic effects, must, for want of space, receive similar treatment. Within such an analysis of precipitation as this chapter presents, however, we hope to avoid a major climatological injustice to snowfall as one of the comparatively minor atmospheric outputs over the British Isles.

Table 6.19 *Mean number of days with snow or sleet observed to fall (UK data 1921–50 Atlas of Britain, 1963; Irish data 1961–70)*

	Oct.	Nov.	Dec.	Jan.	Feb.	March	April	May	Year
Armagh	0·1	0·8	2·1	4·1	3·9	3·2	1·0	0	15·2
Balmoral	1·2	2·9	4·9	6·8	6·2	5·3	5·0	0·8	33·1
Birmingham	0·2	1·3	4·2	5·8	5·6	4·8	2·2	0·2	24·3
Buxton	0·3	1·5	4·8	6·7	6·5	5·0	2·8	0·2	27·8
Durham	0·3	1·4	3·7	5·1	5·0	4·1	1·6	0·2	21·4
Dyce	1·1	2·9	5·1	7·4	7·6	6·7	4·4	1·1	36·3
Edinburgh	0·1	0·8	3·4	5·0	4·9	4·0	1·5	0·1	19·8
Eskdalemuir	1·6	3·7	8·0	10·9	10·2	8·6	5·8	1·1	49·9
Kew	0	0·1	2·2	4·0	4·5	2·3	0·4	0	13·5
Lerwick	1·0	3·8	5·9	9·1	8·9	8·2	5·1	1·2	43·2
Lowestoft	0·1	0·9	3·5	5·4	5·1	3·2	0·7	0·1	19·0
Oban	0·2	0·9	2·0	4·8	4·2	2·6	0·6	0	15·3
Plymouth	0	0	0·9	1·6	2·2	0·5	0·1	0	5·3
Renfrew	0·2	1·0	4·2	6·0	5·5	4·7	1·6	0·1	23·3
Rhayader	0·1	0·8	3·3	5·0	4·4	3·6	1·8	0·1	19·1
Stornoway	0·1	2·1	3·4	5·2	5·5	5·0	3·1	0·2	24·6
Belmullet	0	0·7	2·8	2·9	4·4	3·3	1·9	0·2	16·2
Birr	0·1	0·8	3·4	4·6	4·2	1·9	0·8	0	15·8
Claremorris	0·2	1·8	3·4	5·8	6·2	4·8	1·9	0·2	24·3
Clones	0·1	2·7	6·1	6·6	7·9	5·1	2·5	0·3	31·3
Dublin Airport	0·2	1·0.	4·2	6·1	7·2	4·2	1·4	0·1	24·4
Kilkenny	0	1·1	3·4	4·9	5·8	3·1	0·7	0·2	19·2
Malin Head	0·2	2·5	4·7	4·8	6·7	4·8	2·5	0·1	26·3
Mullingar	0·1	1·7	4·7	6·2	7·0	4·2	2·2	0·1	26·2
Roches Point	0	0·4	2·0	2·8	3·7	1·4	0·4	0·2	10·9
Rosslare	0	0·2	2·1	3·4	5·2	2·7	0·7	0·1	14·4
Shannon	0·1	0·5	2·0	3·2	3·4	1·0	0·5	0·1	10·8
Valentia	0	0·2	1·6	1·3	2·2	1·1	0·4	0	6·8

Snow reaches the surface whenever the freezing layer is at the ground or sufficiently close for melting of the ice agglomerations to be inhibited. Its occurrence is therefore closely linked to air temperature and not surprisingly the frequency of snow falling increases towards the north, the east and with altitude. When the winds are from the west or south-west, snow is unlikely in Britain as the temperature of the Atlantic in winter results in the freezing level being at a sufficiently high altitude for any solid precipitation to be melted before it reaches the ground. Areas exposed to northerly or easterly winds suffer most from instability snow showers especially where high ground is close to the coast, such as the North Yorkshire Moors and even north Norfolk. Table 6.19 shows the frequency of days with snow or sleet observed to fall, though the alertness of the observer must be taken into account for this type of data, as explained by Manley (1969). The data for Ireland and Great Britain are not strictly comparable as they are based on different periods during which the frequency of snowfall changed appreciably.

Table 6.20 *Mean number of days with snow lying on the ground at 09.00 hrs (UK data 1921–50 Atlas of Britain, 1963; Irish data 1961-70)*

	Oct.	Nov.	Dec.	Jan.	Feb.	March	April	Year
Armagh	0	0·4	1·4	3·8	1·5	1·4	0·1	8·6
Balmoral	1·0	4·0	9·8	14·1	12·5	10·3	3·3	55·0
Birmingham	0	0·2	2·1	5·0	3·4	2·3	0·3	13·3
Buxton	0·1	1·0	4·6	7·5	7·3	4·7	0·1	25·3
Durham	0	1·1	2·4	4·7	5·0	2·7	0·1	16·0
Dyce	0·1	2·1	5·5	8·8	8·3	5·1	0·1	30·0
Edinburgh	0	0·6	1·9	4·6	3·6	1·7	0	12·4
Eskdalemuir	0·1	0·8	5·0	7·9	5·4	4·2	0·4	23·8
Kew	0	0·1	0·9	2·3	1·9	0·8	0	6·0
Lerwick	0	1·0	2·6	4·0	4·0	2·6	0·2	14·4
Lowestoft	0	0·2	1·2	4·4	2·1	1·1	0	9·0
Oban	0	0·2	0·9	2·0	1·0	0·1	0	4·2
Plymouth	0	0	0·3	0·9	0·5	0	0	1·5
Renfrew	0	0·1	1·5	3·9	4·2	1·9	0·2	11·8
Rhayader	0	0·5	2·1	4·6	3·8	2·8	0·7	14·5
Stornoway	0	0·3	1·2	2·7	2·3	1·2	0·2	7·9
Belmullet	0	0·1	0·6	0·7	1·1	0·5	0·3	3·3
Birr	0	0	1·1	4·3	2·9	1·2	0·3	9·8
Claremorris	0	0·5	1·6	2·1	2·3	0·8	0·3	7·6
Clones	0·1	1·1	1·7	6·1	4·5	1·6	0·8	15·9
Dublin Airport	0	0	0·8	1·6	1·6	0·6	0·1	4·7
Kilkenny	0	0·2	0·6	1·6	1·4	0·6	0	4·4
Malin Head	0	0·1	1·1	1·3	1·5	0·8	0·2	5·0
Mullingar	0	0·8	2·0	5·3	2·7	1·6	0·2	12·6
Roches Point	0	0	0·3	0·1	0·9	0·1	0	1·4
Rosslare	0	0	0·3	0·6	0·3	0·5	·0	1·7
Shannon	0	0	0·4	1·0	0·7	0·2	0·1	2·4
Valentia	0	0	0·3	0·1	0·6	0	0·1	1·1

Table 6.21 *Days of substantial snowfall for selected stations, 1954–69 (after Lowndes, 1971)*

	No. of occasions with >7 cm of snow	With subsequent snow cover For 24 hrs	For 3 days	For 5 days	For 10 days
Stornoway	41	22	11	6	0
Wick	48	35	19	10	1
Benbecula	22	10	5	4	0
Tiree	4	1	1	1	0
Dyce	37	26	16	9	3
Leuchars	19	10	5	4	2
Abbotsinch	9	1	0	0	0
Prestwick	0	0	0	0	0
Eskdalemuir	52	31	20	15	8
Tynemouth	31	11	5	3	0
Aldergrove	13	7	3	1	0
Ronaldsway	9	2	0	0	0
Squires Gate	9	6	4	1	0
Kilnsea	13	4	3	0	0
Finningley	9	7	4	2	1
Ringway	11	8	3	1	0
Valley	5	3	2	0	0
Watnall	15	13	7	2	1
W. Raynham	18	12	7	3	2
Gorleston	15	7	2	1	0
Mildenhall	2	2	1	0	0
Elmdon	13	10	6	3	1
Shawbury	15	12	8	4	2
Aberporth	5	3	2	1	0
Ross-on-Wye	13	8	2	1	1
Heathrow	8	5	3	1	1
Boscombe Down	7	6	3	2	2
Hurn	5	2	1	1	0
Plymouth (Mt Batten)	5	2	0	0	0
St Mawgan	10	4	1	1	0
Scilly	2	0	0	0	0

As even the coldest month in Britain has a mean temperature above 0 °C, snow cover is very variable and intermittent. The data on snow cover depend upon an observer's assessment of whether the ground is more than half-covered by snow at 09.00 hrs. This may or may not last until the next observation time. In the milder parts of the British Isles, snow lying at 09.00 hrs will frequently have melted by sunset unless the ground is particularly cold. However, in Scotland and upland areas, the total of days with snow cover at 09.00 hrs is probably close to the number of days duration of snow. Table 6.20 shows the frequency of days with snow lying at 09.00 hrs and confirms the greater importance of snow cover in the north, the east and on high ground (see Fig. 12.8). The duration of snow cover is also a function of the original depth of snow. Data for Scotland (Dunsire, 1971) show that for the western coastal areas even when snow is lying it is usually less than 2 cm deep, whereas to the east and north much greater depths are probable. At Wick 10·1 per cent of the reports of snow lying had depths greater than 16 cm, 9 per cent at Dyce, 19·1 per cent at Balmoral, 9·2 per cent at Baltasound (Shetland Islands), whereas values on the west coast are less than 2 per cent. Local factors and snow drifting can be significant in such figures.

A survey of substantial snowfalls (at least 7 mm water equivalent) by Lowndes (1971) for the period 1954–69 confirms this pattern of snowfall and snow cover, but anomalies do exist. The lowest number of occasions was not the Isles of Scilly as might be expected, but Prestwick in Ayrshire with no substantial falls (Table 6.21). There is also a very steep gradient from Tiree with only four occasions, to Stornoway with forty-one. Even these falls rarely persisted for long, the severe winter of 1962–3 being the only exception; in this example, snow cover lasted at Elmdon for 32 days. This winter apart, a ground cover of snow will rarely last 5 days except over high ground and in north-east and parts of east Scotland.

The synoptic origins of snowfall may be divided into three groups, (i) frontal; (ii) polar lows or troughs; (iii) instability showers (Clarke, 1969). The relative importance of these categories varies throughout the country with instability showers of snow being most frequent on coasts and hill exposed to onshore winds between north-west and south-east via north. They rarely provide substantial falls. Polar lows are most significant in the north and north-west as they frequently occur in northerly airstreams and provide the main source of substantial snowfall at Stornoway and Wick. In the frontal category, it is warm fronts and warm occlusions which are the main source of snowfall. The heaviest falls of this nature occur when the front remains stationary with the area in the cold air to the north and east receiving continuous snow. If the front eventually crosses the area, the influx of warm air is usually sufficient to turn any remaining precipitation to rain and for the surface snow to start melting.

As snowfall is very dependent on a few synoptic conditions, it is not surprising that falls are very variable from year to year. It is likely to be most frequent when depressions have southerly tracks up the English Channel or become stationary off southern Ireland whilst most of the British Isles remain in a cold air mass. Such conditions can produce heavy falls over extensive areas contrasting with the polar low or trough situations where falls are often heavy but much more localized. The number of days with snow has increased since about 1940 relative to the mild period of 1900–40 and is now more in accord with values in the mid-nineteenth century. However, there has been a tendency in the last decade for snowfall to be more important in October and November and decrease slightly during the main winter months (Green, 1973). Increasing frequencies of snowfall have been forecast, but recent years (1970–76) have failed to confirm these fears.

Introduction

Atmospheric water vapour, which is present in the atmosphere in relatively small and variable amounts, is derived solely from the evaporation process (including transpiration). Globally, and in terms of the British climate, the oceans, which cover 75 per cent of the earth's surface and contain 97 per cent of its total water supply, represent by far the largest humidity source, although evaporation also takes place from the surface of fresh water bodies and from moist land surface, while transpiration occurs from plants.

The water vapour in the atmosphere represents only 0·0001 per cent of the global water supply (Nace, 1960) and if precipitated uniformly over the earth's surface would yield only 25 mm (Bernard, 1942) and yet this minute amount plays an important climatological and hydrological role. It is the source of all precipitation, although clearly large-scale horizontal and vertical air movements are needed to supply the large quantities of moisture commonly precipitated in major storms and rainy periods. Through the medium of both latent heat storage and the 'glasshouse effect' it is a major factor in the global energy budget. Finally, the amount of atmospheric moisture is one of the factors influencing the rate of evaporation, and the reciprocal tendencies between humidity and evaporation have long been the basis for attempts to estimate evaporation from standard climatological measurements.

Evaporation is thus a transfer process resulting in the addition of moisture and latent heat to the atmosphere and the removal of moisture and heat from the earth's surface. In the case of the oceans and most surface water bodies, this depletion by evaporation is made good by precipitation and streamflow; in the case of land areas, it can be made good only by precipitation. The extent to which phase inequalities between precipitation and evaporation result in periods of water deficit and surplus have important repercussions in agriculture and water resource management, as well as more generally in climatology and hydrology, and studies of this and other aspects of the water balance have now assumed considerable importance.

Evaporation

Definitions and influencing factors

The vaporization of water may take place from the surface of a water body (free-water evaporation), from the films of water held around and between soil particles (soil evaporation), within the stomata of plant leaves (transpiration) and from the wetted surfaces of vegetation (interception loss). Although the term *evaporation* is appropriate to describe all these aspects of water loss there has been a tendency in recent years to use, instead, the more cumbersome composite term *evapotranspiration*. In this chapter the latter will be preferred unless reference is being made specifically to either free-water or soil evaporation. A further distinction must be made between *potential* evaporation or evapotranspiration (*PE*) and real or *actual* evapotranspiration (*AE*). *PE* is the (often hypothetical) maximum water loss which would occur in given climatological conditions from a moist. surface, e.g., water surface or well-watered vegetation, whilst *AE* is the usually lower water loss which takes place from vegetation and soil surfaces which dry out from time to time and from which, therefore, despite favourable climatological conditions, little or no evapotranspiration actually takes place.

PE, although it is often an unreal value, is dependent on far fewer influencing factors than is *AE* and is therefore easier to estimate or measure. Of the influencing factors, solar radiation and temperature are the most important since the change in state of water from a liquid to a gas involves the expenditure of approximately 590 calories per gram of water at ordinary field temperatures. The reciprocal relationship between evaporation and humidity has already been noted and in this context wind and its associated turbulence play an important part in removing the moist layers of air in contact with the evaporating surface and replacing them with drier air. In contrast, *AE* is also dependent on a number of complex plant and soil factors which influence the resistance to water movement through the soil–plant–atmosphere continuum.

Available data

Ward (1971) reviewed the problem of measuring evapotranspiration in broad terms and suggested that, in the case of *PE*, the three main approaches were likely to involve (i) the extrapolation of measured data from water surfaces; (ii) similar extrapolation from irrigated vegetated surfaces; (iii) the use of theoretical estimates based on climatological data. In the British Isles these three approaches are most commonly represented by, respectively, the standard Meteorological Office (or Symons) evaporation pan, simple irrigated evapotranspirometers popularized by Green, and the potential evaporation formula devised by Penman.

Results from Symons evaporation pans first appeared in *British Rainfall 1870* and for almost 100 years were the only regularly published data for the British Isles. *British Rainfall* currently includes results from more than twenty standard evaporation pans of the Symons design. It has been suggested that, in the British Isles, evaporation from the open water surfaces of properly exposed sunken pans may approximate the potential water loss from moist vegetation surfaces (Holland, 1967). Since 1961 data from an increasing number of irrigated, turf-covered evapotranspirometers or lysimeters have also been published in *British Rainfall* and at present more than forty stations are represented. Unfortunately, instrumental problems have limited the usefulness of this technique and have made it difficult to derive consistent mean values for even 5-year periods for more than about twenty stations.

Of the few successful formulae for estimating *PE* that devised by Penman (1948) appears to be the most appropriate for conditions in the British Isles, and the Meteorological Office uses an amended version of the revised formula (Penman, 1962). The calculations of *PE* are for a vegetated surface having an albedo of 0·25 which is generally considered representative of grass, most green agricultural crops and deciduous woodland (Grindley, 1972). Penman calculations of *PE* for seventy-one stations have appeared in *British Rainfall* since the 1963 volume.

In the case of *AE* there are again three main direct approaches to its measurement involving (i) the use of theory; (ii) water-balance calculations for known areas, or volumes of vegetation, or vegetation-covered ground surface (e.g. in a lysimeter or river catchment), in which evaporative loss is derived as the residual item; (iii) the measurement of moisture flux above the evaporating surface. In the British Isles little systematic work on the direct measurement of *AE* has been done until comparatively recently and much of this, excepting water-balance calculation, is still at the research stage. In fact, water-balance calculations are not particularly satisfactory in the case of lysimeters and although far more satisfactory when applied to entire river catchments there are, in the British Isles, few rivers which have been continuously gauged for a sufficient period to establish reliable mean values. Most of our 'official' work is based on the indirect approach suggested by Penman (1949) using the concept of the root constant, i.e. that amount of soil moisture which can be extracted from the soil without difficulty by a given vegetation on a given soil (Grindley, 1972), to calculate soil moisture deficit. *PE* values are then converted to *AE* according to the degree of soil moisture depletion on the basis that the greater the soil moisture deficit the greater will be the shortfall of *AE* below *PE*.

Temporal variations of evapotranspiration

The over-riding influence of solar radiation and temperature on *PE* is reflected in the quasi-sinusoidal curves defined by the changing rate of *PE* both diurnally and seasonally. In clear-sky conditions the daily curve peaks in the early afternoon and falls close to zero for much of the night. Examples of the mean seasonal variation of measured *PE* at Valentia and Sutton Bonington for the period 1966–73 are shown in Fig. 7.1. The general form of the two curves is similar, with peak values being reached in June or July and minimum values in December or January. The maritime location of Valentia, in contrast to the inland location of Sutton Bonington, is reflected in the smaller seasonal range accompanied by both lower summer and higher winter values. At both stations water losses in summer considerably exceed those in winter, so that at Valentia 75 per cent and at Sutton Bonington 81 per cent of annual *PE* occurs during the April–September period. The variability of annual totals of *PE* again relates closely to variations of solar radiation and hours of bright sunshine particularly during the summer months. Interest in such annual variations has increased in recent years as a result of demonstrations that *PE* is a good weather index to crop growth when sufficient moisture is present to meet the potential demand (Penman, 1962; Smith, L. P., 1967). Highly significant correlations have been obtained between *PE* during periods of low soil moisture deficit and both hay yields and the incidence of grassland farming in the British Isles (Smith, L. P., 1967).

Fig. 7.1 *Mean annual potential evapotranspiration (*PE*) in mm, with mean monthly measured* PE *at Valentia and Sutton Bonington* inset. *(Based on data published in* British Rainfall, *and in Ministry of Agriculture, Fisheries and Food, 1967)*

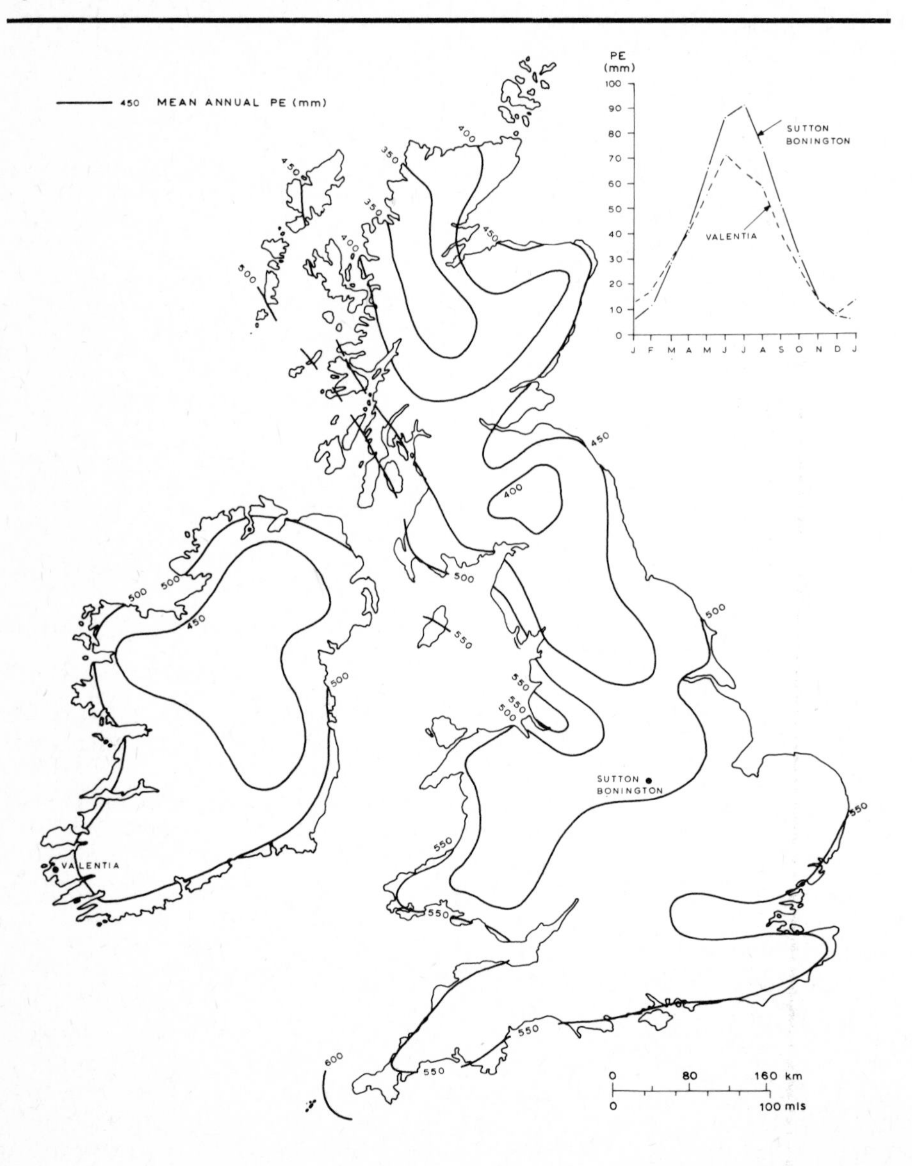

It is more difficult to generalize about temporal variations of *AE*. Clearly, however, during wet periods *AE* variations, like those of *PE*, will be closely related to energy availability whereas, during periods when the vegetation and upper soil layers dry out, moisture availability will tend to exert the greatest influence. Normally moisture availability will be directly related to the frequency of rewetting by precipitation. Penman and Schofield (1941) suggested that, in the light of such considerations, the situation in the British Isles would be that during the winter – when soil and vegetation remain moist – *AE* values would be fairly consistent, albeit low, from year to year, and that, during the summer, *AE* would be greater in wet years than in dry ones, and again greater in conditions of frequent showers than where the same amount of precipitation falls in a smaller number of prolonged rains.

Spatial variations of evapotranspiration

A number of maps of Penman *PE* for parts of the British Isles have been produced during the past 25 years, (e.g. Penman, 1950; Ministry of Agriculture, Fisheries and Food, 1964, 1967; Grindley, 1972). The earlier maps exemplified the main data problem which still exists, namely that there are insufficient long-period stations to permit the construction of a valid, detailed map of *PE* distribution. The deficiency is particularly marked in upland areas and Grindley (1972) noted that of the 150 stations for which Penman *PE* data were available 75 per cent have an altitude of less than 100 m. Grindley's map for England and Wales is probably the most detailed one in general circulation and was based on seven-year averages within the period 1954–66. The Meteorological Office intends to produce a definitive Penman *PE* map based on 20-year mean data for a larger number of stations for the period ending 1975, although this is unlikely to become generally available until some years after that date. Until then there appears to be some merit in attempting to produce a tentative general map using as wide a variety of corroborative evidence as possible.

Figure 7.1 presents such a map which is based on four main sets of data: mean free water evaporation from fifteen Symons pans in England and Wales for the period 1962–6; mean measured *PE* from nineteen widely distributed irrigated evapotranspirometers based on 5 years of record within the period 1964–70; mean Penman *PE* for seventy-one stations in Britain and Northern Ireland based on data published in *British Rainfall* for the period 1963–6; and, finally, long-period mean Penman *PE* for each of the 172 former counties and for seventy-four coastal locations throughout the British Isles as presented in Ministry of Agriculture, Fisheries and Food (1967). Allowing for the commonly accepted margin of error for *PE* estimates of 10 per cent, the data were remarkably consistent and of the 350 plotted points through which the isopleths in Fig. 7.1 are drawn only eight (two evapotranspirometers and six evaporation pans, all of which yielded low values) lie on the 'wrong' side of an isopleth. It is felt, therefore, that the isopleths drawn in Fig. 7.1 represent a reliable and internally consistent, albeit somewhat generalized, pattern of mean *PE* distribution in the British Isles.

In contrast to most earlier isopleth maps of *PE*, it reflects not only a broad latitudinal decrease of *PE* from south to north but also a marked decrease away from the coast. Both characteristics reflect the availability of net radiation (there is a strong resemblance between the *PE* pattern in Fig. 7.1 and the pattern of bright sunshine distribution for the British Isles (see Figs 4.12 and 4.13), although the higher coastal values are probably also a reflection of the additional drying power associated with higher wind speeds in coastal areas. Conversely, reduced *PE* values in the main upland areas are largely associated with low radiation availability. It will be observed that the highest values, 550–600 mm, occur more or less continuously around the southern coasts from Suffolk to Dyfed and also in the Merseyside and Cheshire lowland lying in the lee of the northern Welsh mountains. The lowest values, below 400 mm, occur over central areas of northern Scotland and the Southern Uplands and there is a tongue of low to intermediate values extending southward down the Pennines and into the Welsh mountains. Similar intermediate values are found over much of Ireland. Coastal areas of Ireland and virtually the whole of England and Wales south-east of a line from Hull to Cardiff experience annual *PE* values in excess of 500 mm.

The mapping of *AE* for the British Isles, even in a very general way, has been virtually impossible in the past because of the lack of suitable data. Results from a few simple lysimeters (percolation and drainage gauges), mostly located in south-east England, are published in *British Rainfall*. Because these are non-weighing lysimeters and because of the boundary errors associated with small isolated soil blocks in conditions of varying soil moisture, these instruments yield only annual *AE* values of unknown reliability. The most sensible approach is undoubtedly to use information on soil moisture deficit to moderate long-term *PE* data. This is the approach which the Meteorological Office is currently developing. A map of maximum soil moisture deficit based on calculated deficits at grid intersections has already been prepared and this information will be applied to long-term mean *PE* for the period ending 1975. The resulting map will be detailed, will take into account altitude, soil and vegetation type and will be of great value to a wide range of users. Unfortunately it is unlikely to be generally available much before the end of the present decade and in the meantime a less rigorous approach may be justified if it yields at least a plausible distribution of *AE*.

Such an approach is attempted here using the only readily accessible data, namely, mean precipitation and streamflow for river catchment areas published in *The Surface Water Yearbook of Great Britain*. Assuming that storage changes over annual periods are negligible, the difference between precipitation input and streamflow output should represent the annual actual evaporative losses from the catchment area. This approach suffers from several inherent weaknesses. First, few British rivers have been gauged for more than 20 years; many set up in the 1960s have been gauged for less than 10. Catchment means exist, therefore, for widely different and at first sight incompatible time periods. If, however, it is accepted that winter *AE* values are low and that summer values are much influenced by the amount of precipitation it may be reasonable to assume that if the mean precipitation for the short period of record differs only slightly from the long-period mean for the same area then *AE* for the short period may closely resemble *AE* for the long period. In the present work estimates of catchment *AE* based on precipitation minus streamflow have been considered valid where the short-period precipitation mean deviates less than 5 per cent from the long-period (1916–50)

Fig. 7.2 *Mean annual actual evapotranspiration (*AE*) in mm, with a frequency histogram of the* AE *values for the 129 catchments used in the map construction* inset. *(Based on data in Water Resources Board, 1971)*

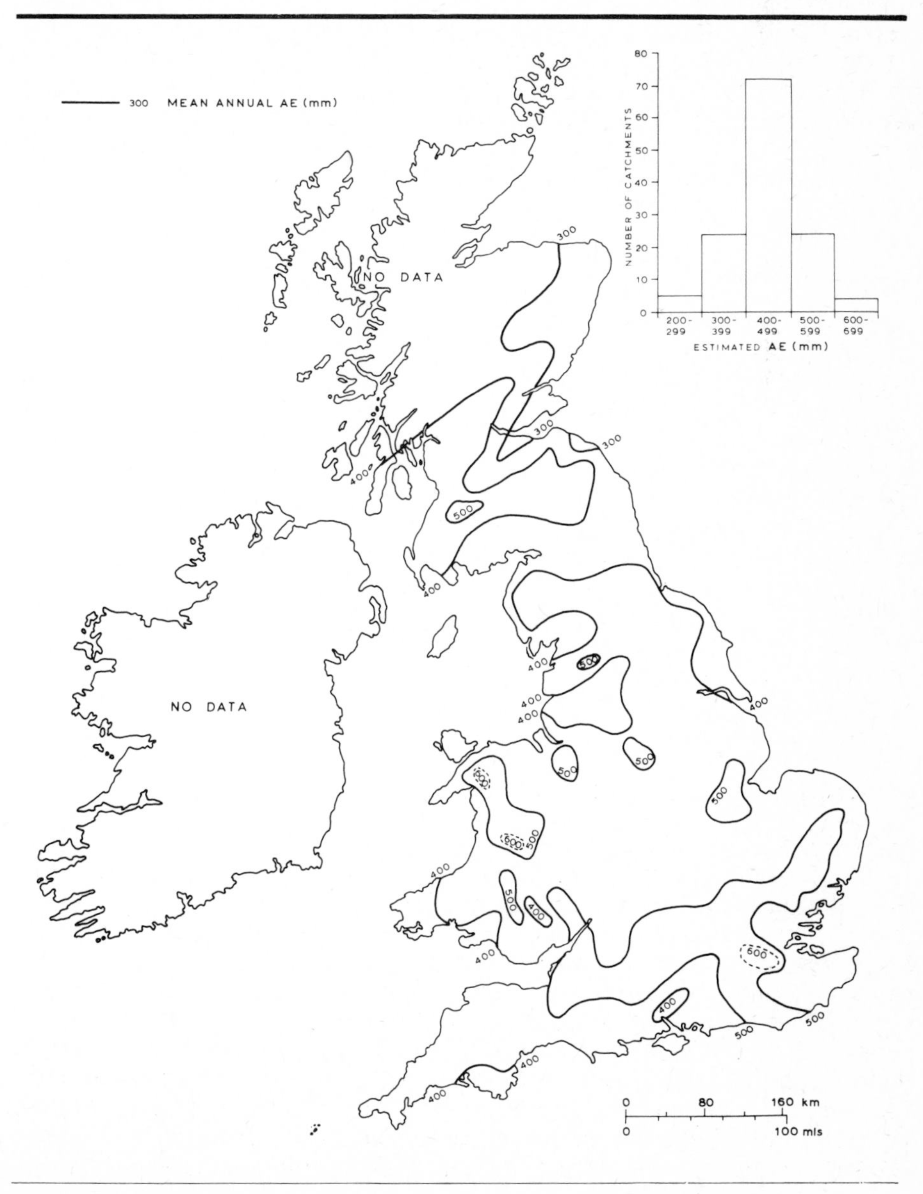

mean. A second problem is that large catchments are normally gauged at several places on the main stream and on the larger tributaries. These gaugings do not, however, provide independent results since downstream stations incorporate upstream flows. In the present work, therefore, it has been the general principle as far as possible to use the 'independent' results from tributary streams and upstream stations on the main stream, in order to define spatial variations, rather than the generalized integrated results from downstream gauges. A third weakness is that there are data gaps for north-west Scotland and for Ireland although it is likely that *AE* closely approximates *PE* in these areas, except perhaps on eastern margins.

The isopleth map of *AE* (Fig. 7.2) is based on data from 129 river catchments which, as the inset to Fig. 7.2 shows, are normally distributed with 56 per cent of the values lying between 400 and 499 mm. The map shows that most of England and Wales is represented by this particular range thereby confirming, at least in principle, the old-established engineering practice of making a fixed deduction from precipitation for annual evaporation. Broadly speaking the *AE* distribution shown is a plausible one, with high values in the south-east, particularly associated with the sedimentary aquifers of the English plain, and lower values in the north and west. The main variations from this pattern can probably be explained in the following ways. Firstly, there is a tendency for reduced values in coastal areas, where higher wind-speeds and higher radiation load result in higher soil moisture deficits, particularly during the summer months. Unfortunately the sparsity of 'independent' streamflow gaugings in coastal areas makes it difficult to substantiate this apparent trend. Secondly, high values are also associated with lowland areas where evapotranspiration may draw more heavily on shallow groundwater supplies (cf. Somerset Levels, the Fens and the Cheshire Plain). Thirdly, higher values also occur in wet areas, provided that there is sufficient energy available (cf. Welsh mountains, Pennines, Lake District and the Southern Uplands). Even allowing for a 10 per cent margin of error, however, some of the high values seem excessive (cf. those in excess of 600 mm in the Welsh mountains and in south London).

Humidity

Definitions

That part of total atmospheric pressure which results from the presence of water vapour is known as the *vapour pressure*. When air at a given temperature is brought into contact with a water surface in confined conditions evaporation will take place from the water surface until the overlying air is saturated with water vapour, i.e. the vapour pressure of the air will increase until it reaches the *saturation vapour pressure* (*SVP*) for that temperature. It follows that air is unsaturated if the *actual vapour pressure* (*VP*) is less than the saturation vapour pressure at the prevailing temperature and in this situation the humidity of the air may be expressed in various ways. The most common expressions are the *saturation deficit* (*SD*), which is the difference between the actual vapour pressure and the saturation vapour pressure, and the *relative humidity* (*RH*) which is the percentage of the actual vapour pressure to the saturation vapour pressure. The latter term has achieved great popularity among meteorologists and climatologists as an indication of how close the air is to saturation and subsequent discussions will be devoted largely to a consideration of variations of actual vapour pressure and relative humidity. Finally, the *dewpoint* is the temperature for which *VP* equals *SVP*. Because the source of atmospheric humidity is evaporation from the earth's surface there will be normally a moisture gradient in the atmosphere reflecting a rapid decrease of water vapour with height away from the earth's surface. This vertical zoning will not be discussed further in this chapter.

Temporal variations of humidity

SVP increases rapidly with temperature, ranging from approximately 2·5 mb at −12 °C, through 10 mb at 7 °C, to approximately 40 mb at 29 °C. The evaporation 'ceiling' represented by *SVP* thus tends to increase as the amount of energy available for evaporation increases, i.e. diurnally in the afternoon and seasonally in the summer. It follows that the actual moisture content of the air will show both diurnal and seasonal variations with maximum *VP* tending to coincide with high temperatures. The mean seasonal variation of VP for Kew and Valentia, representing eastern and western margins of the British Isles, are illustrated in Fig. 7.3*a*. The curves show a simple seasonal progression with minimum values occurring in mid-winter and maximum values in mid-summer. The smaller range at Valentia results from higher winter values associated with an oceanic influence, whilst summer values at the two stations are virtually identical. Mean diurnal variations in January, July and for the whole year are shown only for Kew in Fig. 7.3*a*. Although the diurnal range is small, only 1 mb even in July, the curves suggest a more complex situation than that affecting the seasonal rhythm. Thus, although the minimum values occur consistently in the early morning hours, at about the time of sunrise, maximum values occur between early afternoon in January and late evening in July. The relative flattening of the July curve from late morning through early afternoon probably results from the convective turbulence which transports much water vapour upwards away from the earth's surface during the heat of the day. Only after this turbulence dies down do vapour pressures attain maximum values.

Temporal variations of *RH* are illustrated for the same stations in Fig. 7.4*a*. In general terms *RH* variations are the inverse of temperature variations. Thus seasonally the highest values occur in winter and the lowest in summer, although the range is small, particularly at Valentia where, in association with the moist maritime location, winter temperatures are relatively high and summer temperatures relatively low. Diurnal variations, on the other hand, are much greater and normally exceed seasonal variations. So at Kew the mean diurnal range is 20 per cent, with a maximum in July and a minimum in January, whilst the mean annual range is only 13 per cent. Maximum values are

Fig. 7.3 *Vapour pressure in mb.* (a) *Above: mean 2-hourly values for January, July and the year at Kew. Below: mean monthly values at Kew and Valentia (based on data in Bilham, (1938):* (b) *Mean annual values at 15.00 hrs, 1961–70;* (c) *Mean January values at 15.00 hrs, 1961–70;* (d) *Mean July values at 15.00 hrs, 1961–70; (based on Meteorological Office data, and on Meteorological Office, 1952)*

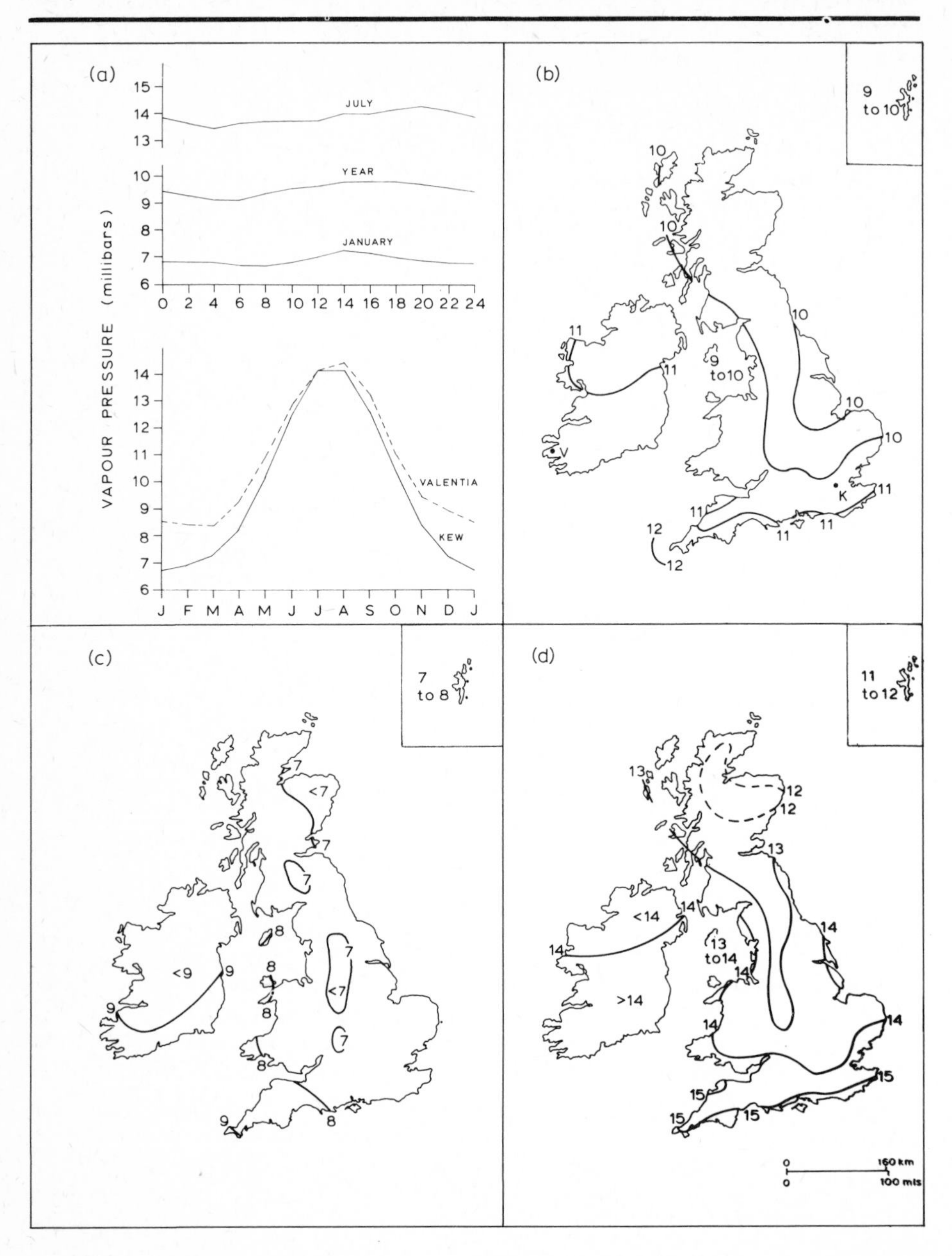

attained at about the time of sunrise and minimum values at about 14.00 hrs, coinciding with maximum daily temperatures.

The preceding discussion of diurnal and seasonal mean values should be placed in perspective by at least a brief reference to extreme

Fig. 7.4 *Relative humidity in percentage values:* (a) *Above: mean 2-hourly values for January, July and the year at Kew. Below: mean monthly values at Kew and Valentia (based on data in Bilham, 1938).* (b) *Mean annual values at 15.00 hrs, 1961–70;* (c) *Mean January values at 15.00 hrs, 1961–70;* (d) *Mean July values at 15.00 hrs, 1961–70; (based on Meteorological Office data, and on Meteorological Office, 1952)*

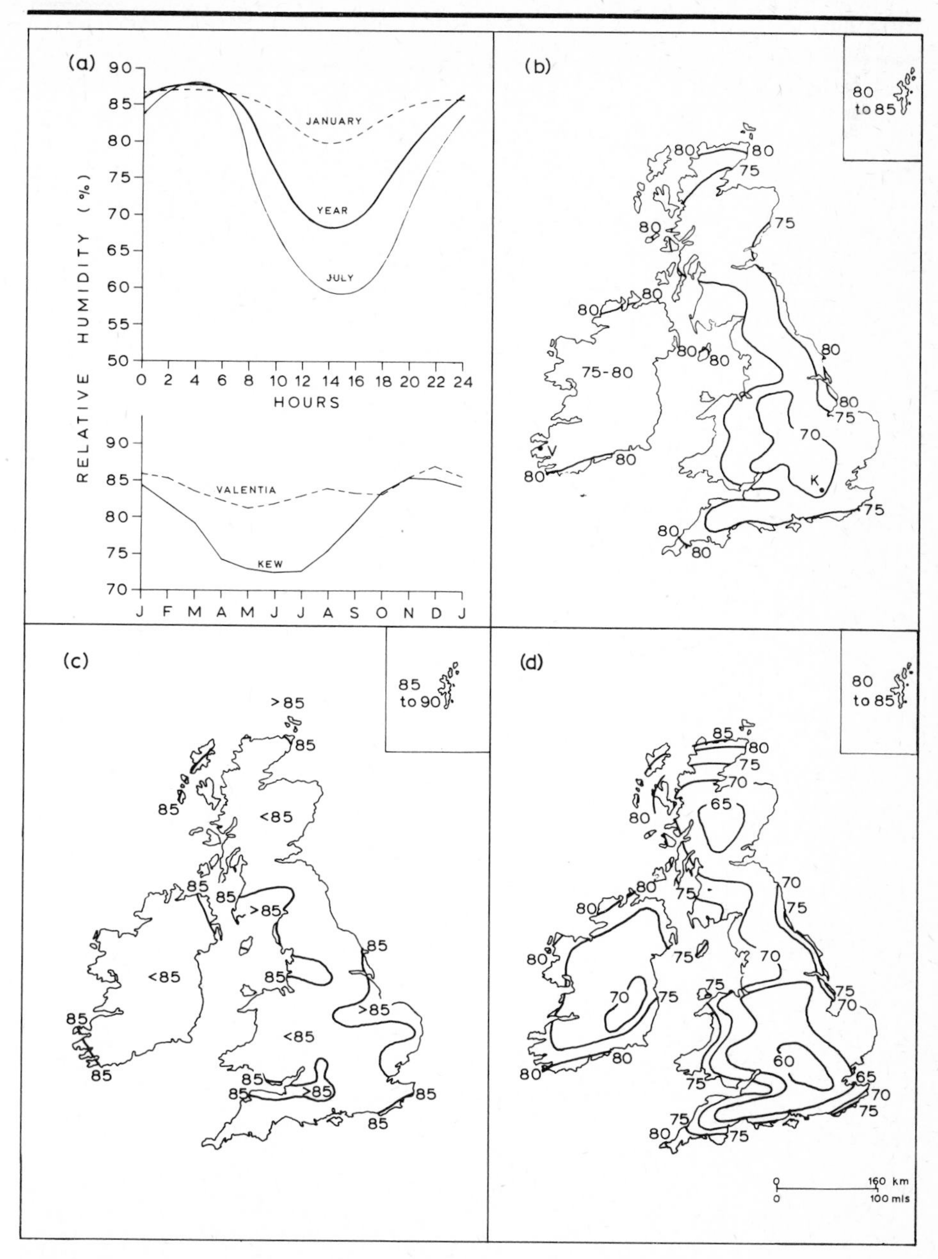

values of VP and RH. The relationship between SVP and temperature suggests that extreme VP values will be associated with extreme temperature values. Thus vapour pressures of almost 24 mb have been recorded in south-east England in July and August, whilst vapour pressures of less than 1 mb have almost certainly occurred on a number of occasions, particularly in north central Scotland, although information is difficult to obtain because of the unreliability of the wet-bulb thermometer at sub-freezing temperatures. Bilham (1938) suggested that in the London area the vapour pressure has probably not fallen below 2 mb very often. In terms of *RH*, values of 100 per cent are quite common, particularly at night and in the early morning, although in late autumn and winter fog conditions they may occur during the daytime as well. *RH* values below 50 per cent are uncommon in winter and values below 30 per cent normally occur only in spring and summer at inland stations where the diurnal temperature range is large. Occurrences of *RH* of less than 20 per cent are very rare and have been associated with either high temperatures in June or July or with easterly winds in April (Meteorological Office, 1952).

Spatial variations of humidity

The spatial distributions of mean annual *VP* at 15.00 hrs and mean monthly *VP* at 15.00 hrs for January and July are illustrated in Fig. 7.3*b*, *c*, *d*, and once again the over-riding influence of temperature can be seen clearly. For the year there is a small, generally north-eastward, decrease in *VP* from 12·0 mb in the Scillies to 9·1 mb at Lerwick. This pattern is moderated by a maritime influence resulting generally in higher coastal values, particularly in the south and west. The patterns for January and July are broadly similar although both the individual magnitudes and the range of values are larger in July, (from 15·6 in the Scillies to 11·6 at Lerwick) and smaller in January (from 9·3 in the Scillies to 6·7 at Aberdeen).

The spatial distribution of *RH*, also at 15.00 hrs, is shown for the corresponding periods in Fig. 7.4*b*, *c*, *d*. Over the year, and particularly in July, the increasing diurnal range of temperature with distance from the sea exerts a dominating influence on mid-afternoon *RH* values and results in a decrease from high values at the coast to substantially lower values inland. This contrast is less marked in Ireland where inland values are considerably higher than in either England or Scotland. During the winter months the contrast between coastal and inland temperatures largely breaks down. Other factors, such as air-mass type and prevailing wind direction, then become more important and usually result in only very small variations of *RH* from place to place. This is well illustrated by the January map in Fig. 7.4*c*. There is an increasing tendency now to map *RH* not in terms of monthly means but in terms of the length of time that certain values of *RH* are exceeded. This information has direct application in the fields of agriculture, particularly with regard to the incidence of certain pests and diseases, and in heating and ventilation engineering. Smith, L. P., (1962) presented maps of this type and it is likely that information will be presented in this form in some future Meteorological Office publications.

The water balance

Definitions

Over most of England more than 50 per cent of precipitation is 'lost' as evapotranspiration. Elsewhere in the British Isles the proportion is smaller, falling to less than 20 per cent in the Welsh mountains and the Western Highlands of Scotland. The remaining precipitation either goes into more or less temporary storage as soil moisture or groundwater, or it runs off as streamflow. So the water balance for any river catchment can be expressed as:

$$P - E - \triangle S - \triangle G - Q = 0$$

or

$$P - E = \triangle S + \triangle G + Q$$

where P is precipitation, E is evapotranspiration, $\triangle S$ and $\triangle G$ are respectively changes in soil moisture and groundwater storage, and Q is streamflow. In other words the relationship between P and E (which Thornthwaite (1948) used as the basis for his classification of climate) directly influences the quantities of water remaining in storage or in transit on and under the earth's surface ($\triangle G$, $\triangle S$, Q). Of these the last two are more dynamic, whereas groundwater changes more slowly and is of less direct relevance to the climatologist. For this reason subsequent discussion will concentrate largely on soil moisture storage and streamflow as influenced by the balance between precipitation and evapotranspiration.

When P continually exceeds PE soil moisture storage will be at a maximum and there will be a surplus of water available for streamflow and for the recharge of deeper groundwater supplies. When PE exceeds P there will be some withdrawal from soil moisture storage and if these conditions continue soil moisture storage will become exhausted and AE will fall below PE, i.e. there will be a deficit of water. Magnitudes of water surplus and water deficit are a major aid in defining 'total' climate (wetness and dryness) and in applied terms they may provide valuable indications of, say, the suitability of an area for water supply purposes on the one hand, or of the need for irrigation on the other.

Water deficit

Water deficits may occur at any time of the year in the British Isles as evaporation dries out the soil during rainless periods. These deficits are normally small during the winter half of the year and are, in any case, frequently 'smoothed out' if monthly values of P and PE are used. It is during the summer that substantial deficits may build up, often reaching maximum values in late September and early October.

Two aspects of the spatial distribution of water deficit are illustrated in Fig. 7.5. Mean annual potential soil moisture deficit (Fig. 7.5*a*) is calculated on the assumption that there is no limit to soil moisture depletion and it therefore represents a virtually unattainable extreme. In other respects, however, the map presents a useful indication of dry summer conditions and of the marked contrast between the dry south-east and the wet north-west and west of the British Isles. Significantly, it has been shown that south-east of a line from the Humber to Lyme Bay irrigation is needed in at least 7 years out of 10 in order to promote optimum crop growth (Ministry of Agriculture, Fisheries and Food 1954). It has also been suggested that there is a relationship between potential soil moisture deficit and crop distribution, with the ratio of grass to crops-and-grass decreasing in a definable way with increasing soil moisture deficit (Hogg, 1965).

Fig. 7.5 (a) *Mean annual potential soil moisture deficit in mm (based on Green, 1964);* (b) *Estimated soil moisture deficit at 09.00 hrs, 29 October 1969 (based on Grindley, 1972)*

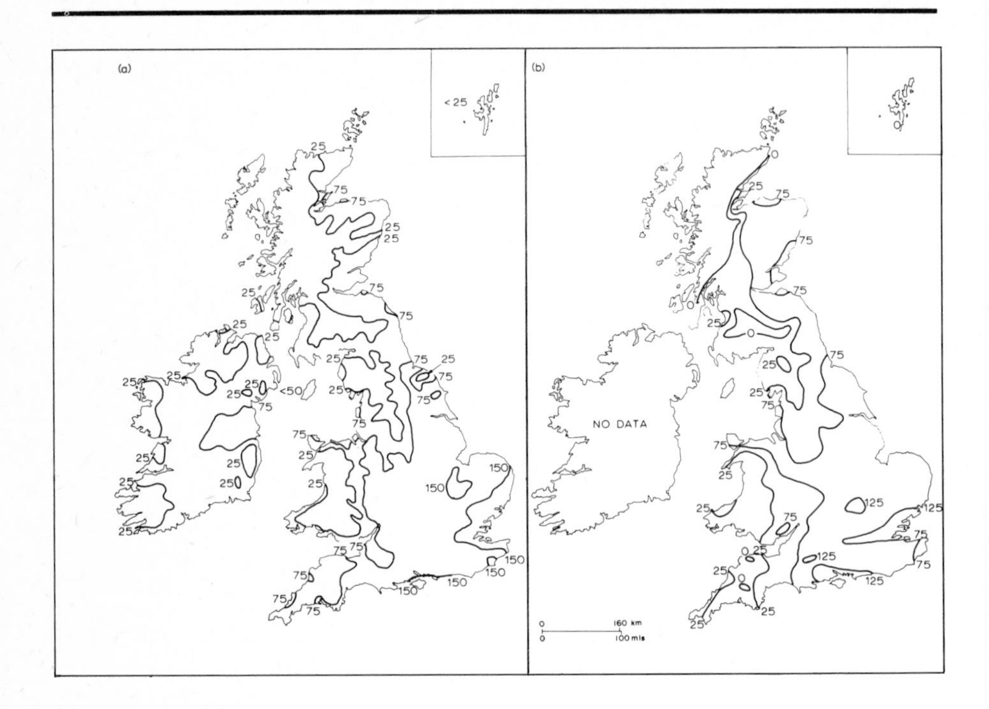

It is interesting and instructive to note the close similarity between the generalized, hypothetical distribution (Fig. 7.5*a*) and the very specific distribution of estimated actual soil moisture deficit at 09.00 hrs on 29 October 1969 (Fig. 7.5*b*). This latter map is one of those produced regularly by the Meteorological Office using the root constant approach suggested by Penman (1949). The definition of this was given earlier, but it should be noted that a further 25 mm of moisture can be extracted with increasing difficulty and extraction thereafter becomes minimal. The Meteorological Office method assumes that each station is representative of a typical river catchment where 50 per cent of the area is covered by short-rooted vegetation (root constant=75 mm), 30 per cent of the area is covered by long-rooted vegetation (root constant=200 mm) and the remainder of the area is riparian with moisture always freely available so that evapotranspiration is never restricted (Grindley, 1972).

Water surplus

Water surpluses may also occur at any time of the year in the British Isles since during a normal rainstorm P exceeds PE. Again, however, the use of monthly or weekly values may 'smooth out' the short-term surpluses, particularly during the summer months. The three main repercussions of water surplus are leaching, particularly

Fig. 7.6 *Mean annual streamflow and mean winter leaching in mm (based on Water Resources Board, 1971, and Ministry of Agriculture, Fisheries and Food, 1964)*

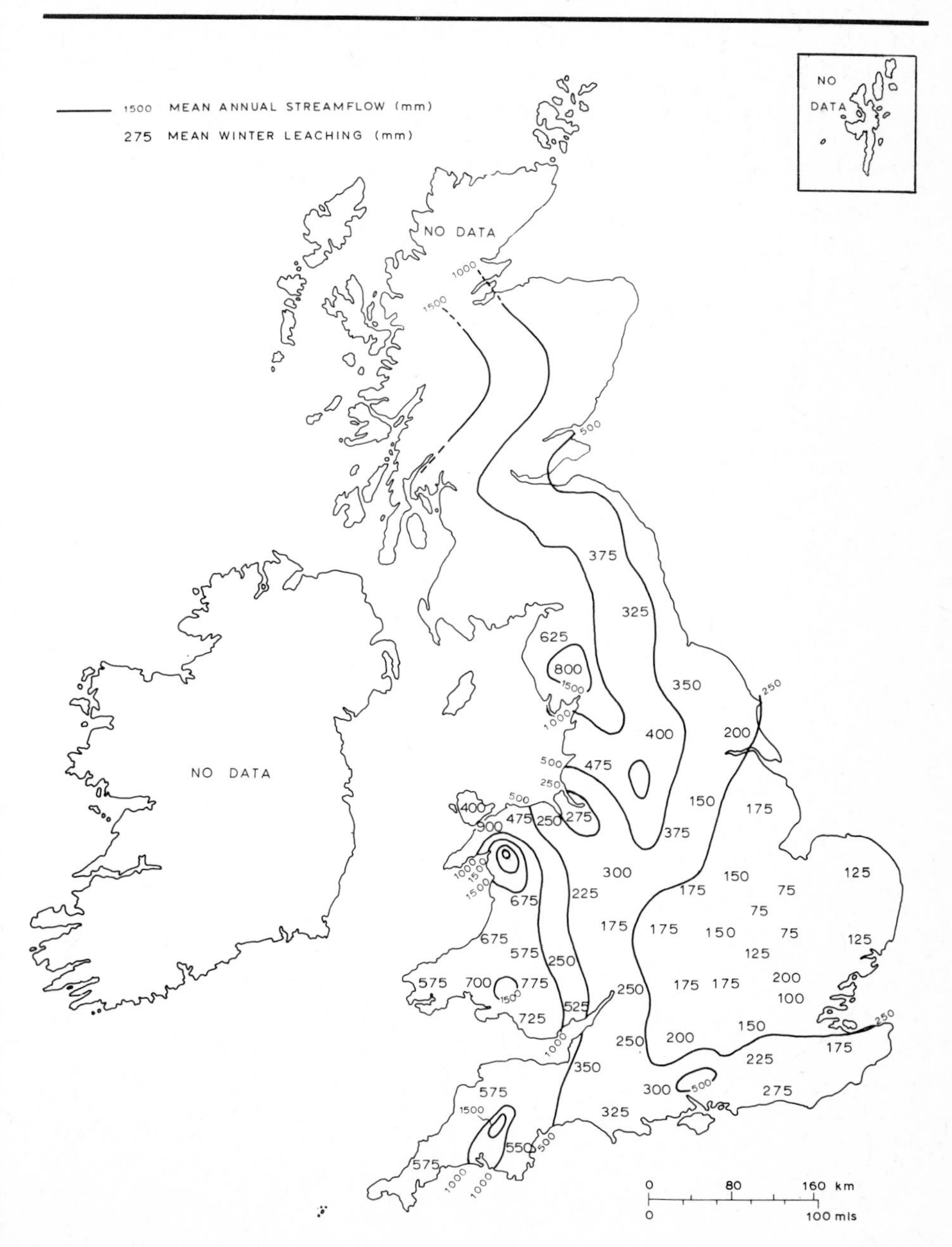

during the winter period, as excess water infiltrates through the soil profile, groundwater recharge as some part of that excess percolates deeper beneath the surface, and streamflow as, on a catchment scale, *P* exceeds *PE*.

The spatial distribution of two aspects of water surplus are mapped in Fig. 7.6. Firstly, areal averages of winter leaching in England and Wales, expressed as mm of precipitation, are shown numerically. These values range from a minimum in East Anglia and around the Thames estuary, where average excess winter rainfall may be as low as 100 mm, to maxima in the western upland areas of Devon and Cornwall, Wales and Cumbria, where average excess winter rainfall may be as high as 1 350 mm. Secondly, the map illustrates, by means of isopleths, the mean annual distribution of streamflow over England Scotland and Wales. These isopleths have been derived from the same basic data as those used to construct Fig. 7.2 namely, catchment values of runoff for those river catchments where the mean annual precipitation for the period of record deviates less than 5 per cent from the mean annual precipitation for the standard period 1916–50. The result accords well with previous attempts to map runoff in Britain (Ward, 1967, 1968) with the added advantage that the greater number of stations made available by this method permits a more detailed treatment of streamflow distribution in western districts. The resulting pattern is an extremely plausible one with low values (below 500 mm equivalent precipitation) in the south-east, in the lee of the Welsh mountains and along eastern coastal areas of northern England and southern Scotland, and high values (exceeding 1 500 mm) confined to Dartmoor, the Welsh and Cumbrian mountains and the Western Highlands of Scotland.

Introduction

Seen from a distance, the Earth has been called the Blue Planet. Unfortunately for the British Isles, the predominant blueness of the Earth has its principal origins elsewhere, and here, the suitable aphorism might be the Elusive Archipelago. Situated off the north-western coasts of the Eurasian continent, the British Isles are dominated by maritime influences. These, operating in sympathy with a never-ending succession of mid-latitude depressions, largely determine that the region is always humid and often very cloudy. So, for the man in space, the frequency of sightings of the British coasts is rather low, whilst for the man on the ground, the dominant colour in this environment is grey. Here, the intensity of solar radiation – whether direct or diffuse – is lower than in most of the world.

The simple distinction between clouds and fog is not without ambiguity: low cloud may be hill fog where the tops of hill or mountain ranges protrude into it. Here we are concerned with cloud not fog, which is dealt with in Chapter 9, but upland influences are important and will be outlined and discussed. Thunder, a weather event singularly hard to monitor objectively, is associated with a certain sub-section of the total population of clouds. Thus it will be considered subsequent to our examination of clouds and cloudiness in general.

Observations of cloud in the British region are reported in terms of oktas, or eighths of the sky covered by cloud, whether the observations relate to low, middle, or high cloud, or total cloud cover. This facilitates the transmission of the information to the Meteorological Office at Bracknell, since, for each category of cloudiness, only a single digit is required to indicate the presence and amount of cloud (1–8 oktas inclusive), the absence of cloud (0 oktas) or the state reported as 'sky obscured' (represented by 9 on the okta scale). In this case, the sky is invisible owing to manifestations such as fog and smoke, or the observer is unable to estimate the cloud amount owing to darkness (Meteorological Office, 1969). The ensuing discussion is based upon data expressed in these terms.

The dominant types of cloud which affect the British Isles may be summarized as follows:

1. *Thin layer clouds.* These are associated with stable air and are usually due to the advection of warm air over colder surfaces. Fog, stratus or stratocumulus may occur depending upon the vertical structure of the air including the strength of the wind. Daytime insolation heating may disperse or break what is otherwise a characteristically complete or nearly complete cloud cover, though sometimes, especially overland in summer, penetrative cumuliform clouds may develop embedded in an otherwise sheet-like cloudiness.
2. *Thick layer clouds.* These are associated with frontal depressions and jetstreams and, especially in the former instance,

are often multilayered. The major cause is the widespread uplift of moist air in convergent situations. This is enhanced in upland regions. However, convective cloud elements are frequently present also, especially along cold fronts, intermingled with stratiform cloud.

3. *Convection clouds*. These may be either of local or more distant origins. Local heating, especially in summer, often leads to the development of thunderclouds, especially in the warmer south and east. Other convective clouds commonly develop over land anywhere in Britain along cold fronts, in cold sector air at the rear of Atlantic depressions and in outbreaks of Arctic air. Under unstable conditions, the depth of instability – and consequently the height of the cloud tops – increases with the distance travelled by the airstreams over warmer surfaces. The incidence of convection cloud due to local heating is greatest in late summer, whilst the incidence of such cloud accompanying the advection of unstable air is highest in spring and autumn.

The geographical pattern of cloud cover

The cloud statistics set down in Table 8.1 summarize the mean annual pattern. It is evident that the amount of cloud cover is generally greater in the north (Lerwick, 6·3 oktas, and Stornoway, 6·2 oktas) and the west (Valentia, 6·0 oktas) and smaller in the south (Kew, 5·6 oktas) and east (Gorleston, 5·6 oktas). However, it is interesting to note that the maximum among the eleven stations listed is for Renfrew (fractionally higher than at Lerwick), whilst the minimum is at Aberdeen (5·4 oktas).

Although other stations not listed in Table 8.1 may be even more extreme, those selected suggest that, although latitude and position are basic influences, topography has a considerable effect upon cloud amount as upon so many more meteorological elements (see Chapter 12). At Renfrew, cloud cover may be strengthened by the influence of the Glasgow conurbation (see Chapter 14), whilst at Aberdeen the unusually low cloud average is in large part due to the effects of the Scottish Highlands. Cloudiness is generally enhanced over high ground, especially on the windward sides of mountain ranges. It is generally less over low ground and along coasts, particularly those which are sheltered from the west. At Aberdeen these factors combine to give clearer skies than are experienced on average even much further south; for example, at Gorleston on the East Anglian coast.

Dividing the year into summer and winter halves, it emerges that the northern stations (Lerwick, Stornoway and Renfrew) experience little or no seasonal variation in mean cloudiness. Coastal stations in the east and south (Aberdeen, Gorleston and Plymouth) have more cloud in summer than winter by as much as 5 per cent (70 per cent cover as compared with 65·4 per cent at Aberdeen). This is due in large part to the higher incidence of advection fogs along the southern and eastern coasts, especially in late spring and early summer, associated with gentle onshore winds. These winds are cooled below saturation level on their passage across the English Channel and North

Sea from the continent of Europe and its warmer coastal waters (see Chapters 9 and 11). This fog may be lifted during the day by a slight increase in the wind, leading to the development of an overcast of stratus cloud which may be slow to clear. Statistically, this effect is greater than the overland strengthening of convective cloud which, though a more frequent occurrence in summer, also occurs in winter when it may be very strongly marked; for example, in showery cold sector air (Meteorological Office, 1964 and 1965).

All the other stations listed, representative of most of England, Wales and Ireland, are more cloudy in winter than summer, e.g. by over 7 per cent (73·8 per cent compared with 66·6 per cent) at Kew. Mean cloudiness in winter is generally enhanced by the deeper and more frequent frontal depressions which approach from the North Atlantic. These usually travel less rapidly across the British Isles than their summer counterparts, and may become almost stationary here, due to blocking anticyclones over western Europe. Filling *in situ*, they may cause periods of particularly gloomy weather. Most of Scotland and the northern isles are dominated throughout the year by cyclonic weather and show less seasonal variation in mean cloud cover.

The cloudiest months of the year differ much from region to region, but Table 8.1 indicates that December–February is generally the cloudiest period, with secondary maxima in some areas in July–August. The clearest months generally are May and June. But considerable local differences occur in detail, due to special regional and local influences.

Mean cloudiness apart, both meteorologically and psychologically the incidences of relatively complete cloud cover and relatively clear skies are of special significance. Meteorologically, the distribution of mean cloud cover is hard to rationalize satisfactorily in a few words because of the many factors which have a bearing on it. Moreover, some factors tend to mitigate against or counterbalance others. Psychologically, mean cloud cover is probably less important than the incidence of the individual cases of which it is composed. Consequently, maps which depict the frequencies of days of much or little cloud may be both easier to explain and easier to appreciate than any of mean cloudiness. Figures 8.1*a* and 8.1*b* are two such maps. These underline the essentially cloudy nature of the British Isles, which is all the worse since everywhere the sky is heavily overcast (7–9 oktas cloud cover) more frequently than not, whilst relatively clear skies (0–2 oktas cloud cover) are enjoyed on average only about once a week in lowland and coastal areas, and less than one day in ten in the mountainous regions of Wales, Northern England and Scotland. Ireland is probably much the same in this respect as the upland areas of Britain.

The parallelism between the coasts of Britain and the isopleths of overcast conditions is particularly striking (Fig. 8.1*a*). We saw earlier how, on average, the northern and western regions of Britain are more cloudy than the south and east. This underlying trend is evident in Fig. 8.1*a* but the map reveals also that the frequency of very cloudy days is as high in parts of southern England only a few miles inland, as in the Outer Hebrides in the far north-west. Indeed, for the period 1957–70,

Table 8.1 *Average amount of cloud in oktas, at selected stations for stated hours. Period of observations, generally 10 years (after Manley 1970b; data from the Meteorological Office 1964,1965 and the W.M.O)*

Station	Time	J	F	M	A	M	J	J	A	S	O	N	D	Year (oktas)	Year (per cent)
Renfrew	07·00, 13·00	6·2	6·2	6·2	6·3	6·1	6·2	6·4	6·6	6·2	6·3	6·2	6·3	6·3	78·2
Lerwick	07·00, 13·00	6·1	6·4	6·3	6·2	6·2	6·0	6·6	6·4	6·1	6·2	6·2	6·3	6·3	78·1
Stornoway	07·00, 13·00	6·2	6·3	5·8	6·0	5·9	6·1	6·7	6·4	6·4	6·5	6·0	6·0	6·2	77·4
Valentia	07·00, 13·00	6·2	5·8	5·7	5·7	5·5	6·0	6·4	6·2	5·8	6·2	6·2	6·0	6·0	74·7
Dublin	09·00	5·9	5·8	5·7	5·6	6·0	5·5	5·8	5·6	5·4	5·4	5·7	6·1	5·7	71·4
Plymouth	07·00, 13·00	5·9	5·9	5·5	5·6	5·6	5·3	5·5	5·5	5·4	5·9	6·0	6·1	5·7	71·0
Kew	07·00, 13·00	6·0	6·2	5·7	5·8	5·4	5·2	5·0	5·2	5·4	5·6	6.2	5·7	5·6	70·2
Gorleston	07·00, 13·00	5·8	6·2	5·2	5·9	5·6	5·3	5·3	5·1	5·2	5·6	6·2	6·0	5·6	70·2
Birkenhead	07·00, 13·00	5·6	5·8	5·4	5·6	5·6	5·1	5·6	5·5	5·5	5·7	5·7	5·9	5·6	69·8
Cardiff	09·00	5·5	5·9	5·4	5·7	5·4	5·0	5·1	5·4	5·0	5·5	5·9	5·8	5·5	68·3
Aberdeen	07·00, 13·00	5·0	5·3	5·4	5·8	5·7	5·4	5·7	5·6	5·4	5·3	5·2	5·2	5·4	67·1

Gatwick (64·1 per cent) had a higher frequency of very cloudy days than Benbecula (63·3 per cent), whilst Hurn, the airport for Bournemouth on the supposedly favoured south coast of England, had a frequency of 61·1 per cent of very cloudy days, only just exceeded by Stornoway with 61·9 per cent. At the same time, the north-east coast of Scotland is favoured as much as many locations along the English south coast. The general pattern emphasizes the significance of cloud enhancement over land, caused partly by orographic uplift of moist air, and partly by stronger convective processes even when these might be expected to be at their weakest, namely at 09.00 hrs.

Fig. 8.1 (a) *Percentage frequency of days with 7–9 oktas cloud cover at 09.00 hrs, 1957–70 (Meteorological Office data)*; (b) *Percentage frequency of days with 0–2 oktas cloud cover at 09.00 hrs, 1957–70 (Meteorological Office data)*

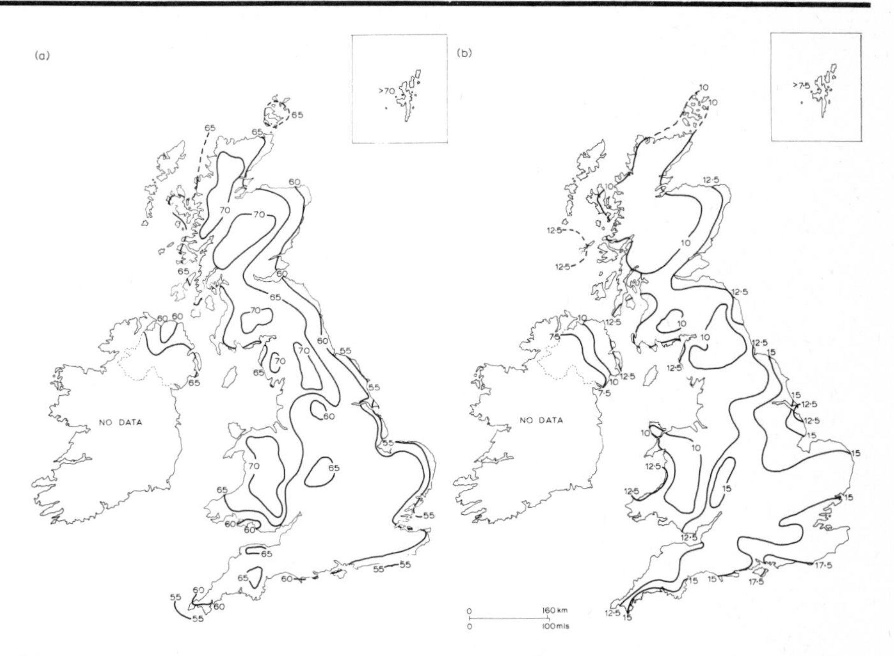

Not surprisingly, the pattern in Fig. 8.1b is broadly the inverse of that in Fig. 8.1a, with certain departures, not all of which are easy to explain. Here the south-eastern corner of England does appear in a more favourable light by comparison with the remainder of the British Isles. It is certain, therefore, that along more northerly coasts more of the days not characterized by very cloudy conditions must be more cloudy than in the south-east, pointing to higher frequencies of days with 3–6 oktas cloud over (not mapped here). Notable local embroideries on the national pattern appear in central western England and at the mouth of the Humber. In the former, subsidence of relatively dry air in the lee of the Welsh mountains may be invoked to explain the higher incidence of relatively clear skies. In the latter, a simple physical explanation for the lower incidence is more elusive, but particularly frequent fogs and low stratus under the appropriate synoptic conditions outlined earlier may be the cause.

Frequency distributions of cloud amounts

Until now, no analysis has been made of the percentage frequency distribution of cloud at a variety of locations across the British Isles. It has been widely accepted that 'cloud amount in temperate regions is generally characterized by a U-shaped distribution . . . .' (Brooks and Carruthers, 1953). Results for Kew have been shown to conform to that pattern.

As a result of computations by the Meteorological Office we are able now to consider the picture for Britain much more comprehensively (see Fig. 8.2). The eighteen stations represented in the map fall naturally into three groups, distinguished by the detailed forms of their percentage frequency distribution curves.

Fig. 8.2 *Cloud provinces determined by the forms of the percentage frequency distributions of total cloud in oktas for all daily data at selected stations, 1957–70 (Meteorological Office data). Province A: 'V' shaped distributions; Province B: 'U' shaped distributions; Province C: 'J' shaped distributions*

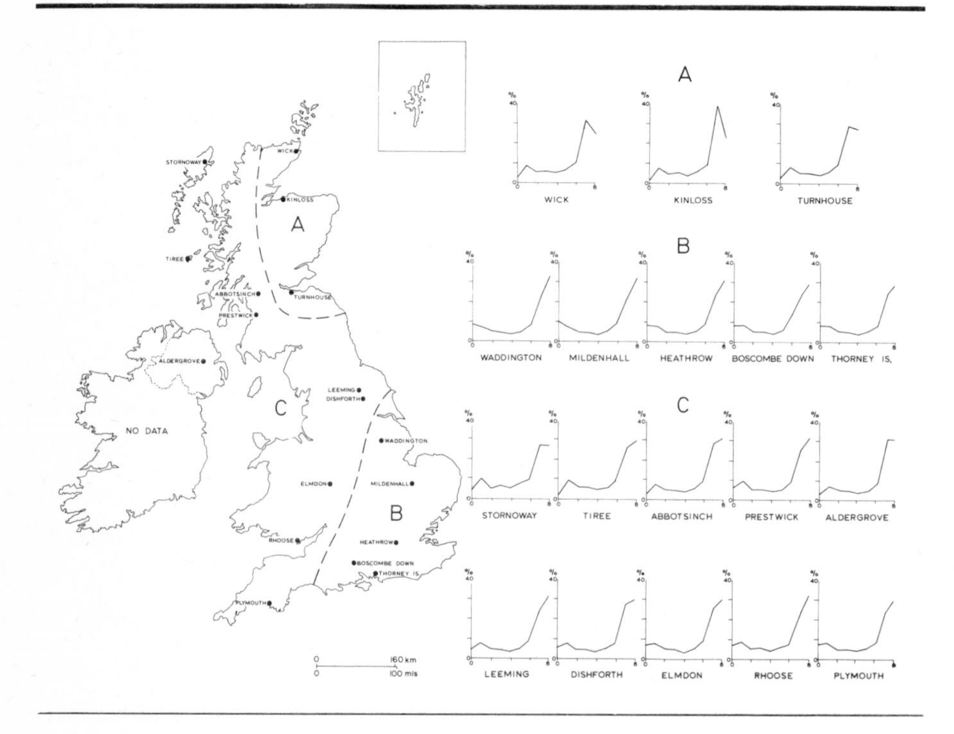

1. Those with U-shaped frequency distributions. These are all located in lowland areas of central, southern and south-east England (B).
2. Those with modified J-shaped frequency distributions. These have their principal peaks at 8 oktas, but differ from the classic J-shaped curve in possessing secondary peaks at 1 okta. These stations are all in northern and western Britain (C).
3. Those with frequency distributions which, extending the range of alphabetical analogies suggested by Brooks and Carruthers (1953),

may be described as (longhand) V-shaped curves. These curves are bimodal, but their antimodal troughs are broader than in the classic bimodal case. The stations are confined to eastern Scotland (A).

It is noteworthy that all thirteen stations in 'highland Britain' rarely report that cloud is completely absent.

It would seem that the higher frequencies of completely clear skies in the south and east reflect a higher frequency of anticyclonic conditions, plus the absence of upstanding relief features above which cloud may persist even when valleys and coasts are clear overhead. This argument is supported by Fig. 8.3. Here we see that considerable differences may occur in detail in the percentage frequency distributions of clouds at different times of the day. The examples chosen, namely Heathrow (London Airport) and Wick, belong to the extreme classes in Fig. 8.2. It is apparent that the forms of the curves for 02.00 hrs differ most, clear skies being much more frequent at Heathrow than Wick. The curves for 14.00 hrs are both of essentially the same form although that for Wick peaks much the more strongly at 7 oktas.

Fig. 8.3 *Percentage frequency curves of total cloud amount in oktas at Heathrow and Wick, at 02.00 and 14.00 hrs, 1957–70 (Meteorological Office data)*

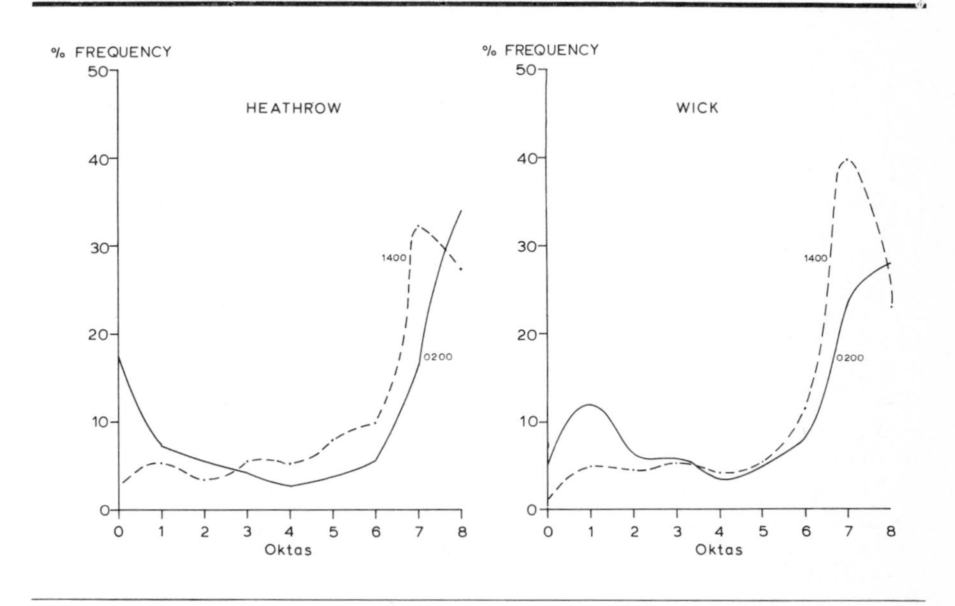

The preponderance of occurrences of 8 oktas over 7 oktas in south-east England is more difficult to explain. Probably a number of factors, including meteorological and topographic, are involved.

Diurnal cycles of cloud cover

Contrasts between south-east England and north-west Britain emerge quite strongly when the diurnal pattern of percentage frequencies of selected categories of cloud cover are plotted for selected localities (see Fig. 8.4). The basic similarities between Heathrow (London Airport) and Tiree include:

(*a*) The diurnal distribution of very cloudy conditions (6–8 oktas) peaks in the early afternoon and falls away at night. This pattern is caused by an enhancement of convective processes during the daytime.

(*b*) The diurnal distribution of relatively clear conditions (0–2 oktas) is inversely related to (*a*) so that conditions of little cloud cover are rarest in the early afternoon.

(*c*) Broken cloud (3–5 oktas) is fairly evenly distributed throughout the day.

Fig. 8.4 *The diurnal cycle of cloudiness at Heathrow and Tiree, in percentage frequencies of selected categories of cloud cover, 1957–70 (Meteorological Office data)*

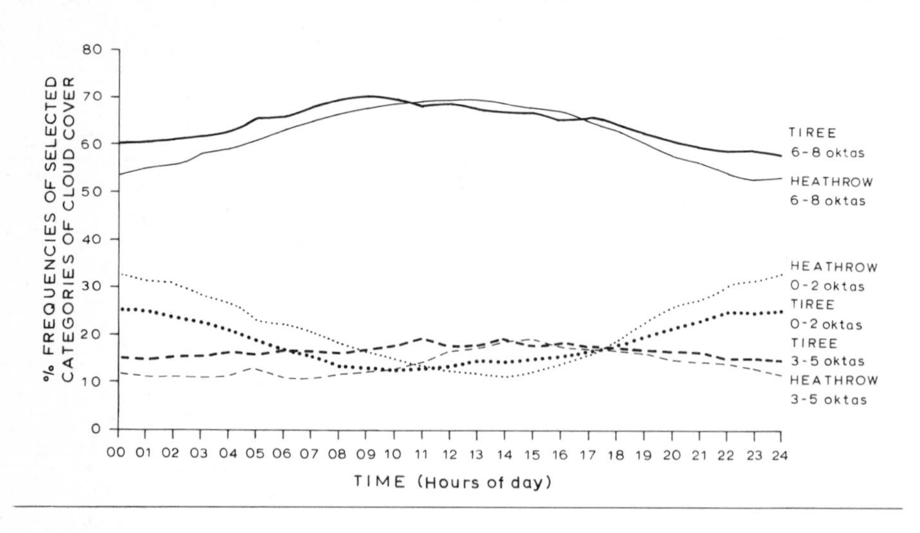

The greatest differences apparent between Heathrow (London Airport) and Tiree are: (*a*) very cloudy conditions are more frequent at Heathrow than Tiree in the early afternoon, but less frequent at night; (*b*) relatively clear skies are more frequent overall at Heathrow, peaking especially high at night; (*c*) Broken skies during the day are more frequent at Heathrow than at Tiree.

These contrasts are all commensurate with more maritime conditions and a greater dominance of cyclonic weather in northern and western Britain, and with more continental conditions and less dominant cyclonic weather in the south and east.

Thunder

If difficulties surround the observation and mapping of cloud cover because of the nature of the element itself, even greater problems are associated with thunder. Recordings of thunder are usually summarized in terms of 'thunder days', but difficulties arise in mapping these whether the network of observers is fine or coarse. Where few observers are used a misrepresentative picture may emerge since outbreaks of thundery weather are often highly localized. However, comparable dangers may accompany the employment of a large number of observers since the sound of thunder can travel many miles, and some outbreaks may be subject to multiple reporting. The problems of thunderstorm observation may be exacerbated by a network which is close in some parts and not in others. Detailed observations of thunder over Britain have been made for many years by the Electrical Research Association, and helpful information of a climatological nature has emanated from that source (e.g. Bowell *et al.*, 1966). Unfortunately the number of observers involved has not been constant (rising from about 650 in 1955 to about 1600 in 1960, with little variation thereafter), and the added effects of closeness and irregularity of the observing network cast further doubt upon the maps produced.

Fig. 8.5 *The frequency of thunder expressed in terms of the mean annual number of days when thunder was heard (Meteorological Office data):* (a) *1901–30;* (b) *1931–60*

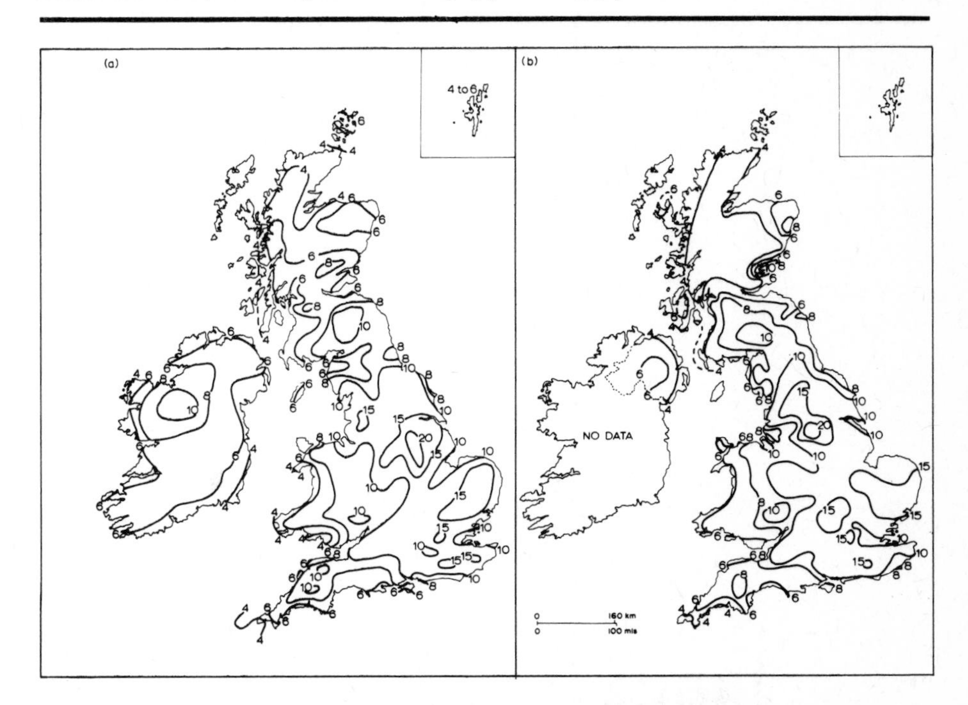

A general view of the distribution of thunderstorms is afforded by Fig. 8.5, based on Meteorological Office data for 1901–30 and 1931–60. It emerges that thunderstorms – defined as 'combinations of thunder and lightning with or without precipitation' (Meteorological Office, 1969) – are most frequent in the south and east of Britain, with higher incidences generally over land than water and over high rather than low relief. The thunderstorms in lowland Britain mostly develop locally, or originate on the continent of Europe.

Table 8.2 *Thunderstorm figures at Heathrow Airport (London), July 1946–June 1968, and at Manchester Airport (Ringway) January 1946–December 1972, (Prichard, 1973)*

	Mean number of hours per year			Mean number of days per year			Mean number of hours per thunderstorm day		
	Summer	Winter	Year	Summer	Winter	Year	Summer	Winter	Year
Heathrow (22 years)	29·05	2·86	31·92	12·50	1·86	14·36	2·32	1·54	2·22
Manchester (27 years)	29·56	4·04	33·59	12·93	2·74	15·67	2·29	1·47	2·14

Those in highland Britain are associated more with the cold fronts of North Atlantic depressions and showery west to north-west airstreams, whose instability is increased by the hill and mountain ranges.

The differences in the maps for 1901–30 and 1931–60 cannot be attributed to any single factor. The strengths and occurrences of anticyclones, depressions, cols and fronts are all involved, along with the ever-changing patterns of airstream frequencies. Even rural land-use and urbanization may have played a part through the changes they impose on the all-important thermal properties of the earth-atmosphere interface (see Chapter 14).

Seasonally, the most severe thunderstoms occur in summer (May–August) in the Midlands, eastern, south-eastern and southern England, often after a few days of fine weather (Lamb, 1964). The highest frequency of these summer storms is early- to mid-afternoon. However, the most impressive (often originating over the continent) are delayed by the time taken for them to travel north and are usually nocturnal (Bower, 1947). Further north and west, winter storms are more frequent though probably never dominant in any area. These are better distributed through the day, and are relatively brief. The indications are that the common ratio of duration of summer:winter thunderstorms over Britain as a whole is about 4:1 (Bowell *et al.*, 1966), although a lower ratio emerged from a comparative study of thunderstorms at Heathrow (London Airport) and Manchester Airport (see Table 8.2).

Detailed research in south-east England has demonstrated that an urban area such as London may increase locally the frequency of thunderstorms (Atkinson, 1968, 1969) and that the London conurbation has had a similar effect to the warm chalk uplands around it upon a thunder-prone atmosphere. Various factors, including the roles of the conurbation as a mechanical obstacle and heat source, and the more abundant supplies of water vapour and condensation nuclei over the built-up area, may all be involved (see Chapter 14).

The basic differences between 'thunder' and 'mean' atmospheres have been outlined in terms of familiar parameters for 1951–60 at Larkhill/Crawley (Atkinson, 1967). These can be summarized as follows:

(*a*) ***Temperature.*** The thunder atmosphere is warmer in two layers, from the surface to 850 mb, and above 300 mb; elsewhere it is cooler.

(*b*) ***Mixing ratio.*** At all seasons the thunder atmosphere has higher mixing ratios up to and including the 700 mb layer, and consistently lower at and above 600 mb. In winter and spring the lowest 3 000m is as much as 10 per cent more moist than the mean.

(*c*) ***Stability.*** Potential instability in the thunder atmosphere was found to occur up to 600 mb in summer and autumn, up to the 700 mb–600 mb layer in spring, whilst being non-existent (on average) in winter. The mean atmosphere is potentially stable except in the 900 mb–850 mb layer.

Analyses of the synoptic types in which thunder was observed revealed that a variety of situations were involved. Predominant were anticyclones, though the thunder outbreaks associated with them were usually small and peripheral to the high pressure systems. Next in rank were non-frontal cyclonic airstreams where large-scale upward motion makes for a more unstable atmosphere. Less than two-fifths of all thunderstorms were frontal in origin, associated with cold, occluded, warm and quasi-stationary fronts in order of incidence. Most of the storms coincided with the actual passage of the fronts, but other thunder outbreaks occurred pre- or post-frontally. A small fraction of the total (about one-eighth) were associated with troughs and slack circulations. These results give a uniquely detailed insight into thunder activity in one region of Britain. However, they may not be representative of the whole: in order to expand the picture, much further work needs to be done.

Visibility observations

Like temperature and humidity, visibility is a continuously varying quality of the atmosphere, but it has proved to be more difficult to measure objectively. Before the First World War it was customary to refer indirectly to atmospheric visibility by such terms as 'haze', 'mist', or 'fog' or to use qualitative descriptions such as 'exceptional visibility'. During that war and afterwards, the growing needs of aviation led to the development of numerical observations of visibility.

During and following the Second World War, visibility measurements were made instrumentally by the use of visibility meters such as the Gold Visibility Meter and by the transmissometer, the latter being used automatically at some large airports. The Gold Visibility Meter contains a window of varying transparency. The more sophisticated instruments for measuring visibility mostly use photo-electric cells which measure the attenuation coefficient of a long column of air or the scattering of light from a small volume of air. Instruments have proved especially useful at night, particularly at special locations such as runways, but the sampling of small volumes of air leads to errors arising from insufficiently representative samples. It is not surprising therefore that, in spite of instrument development, human optical observations of visibility remain a fundamental part of meteorological routine. However, good quality visibility data are scarce compared with data for, say, temperature and rainfall.

Causes of variation of visibility

Physical factors such as the colour and illumination of an object in relation to its background, and personal characteristics of an individual's optical vision, may be regarded as sources of error in the observer's estimate of the visibility and not as factors affecting the property of the atmosphere which the term 'visibility' is intended to indicate. Broadly speaking, visibility is related to the concentrations of water droplets and solid particles, and visibility will improve or deteriorate according to whether conditions favour a decrease or an increase in these constituents of the atmosphere.

When water droplets are present in suitable concentration and sizes, mist or droplet fog may be formed. The occurrence of fog is associated with meteorological conditions which are conducive to the cooling of air below its dew point. This cooling may be produced in various ways. A current of warm moist air passing over colder sea or land may produce sea and coastal fog, the cooling spreading upwards by eddy-motion; air may undergo dynamic cooling when forced to rise by high ground thereby causing cloud at ground level, on windward slopes or beyond; or air may be cooled by nocturnal radiation from the land surface and the lower air layers to form 'radiation fog' in low-lying areas, provided that winds and turbulence are not strong enough to bring about air-mixing through too deep a layer; finally condensation may occur as the result of mixing of two moist air masses at different temperatures.

In large towns and industrial complexes, visibility may be seriously reduced by the presence of solid particles. When visibility is thereby reduced to 'fog' limits, the fog is often referred to as urban fog or smoke fog (smog). In urban fog, the relative humidity may be well below 100 per cent but not necessarily so, because urban fogs are often the result of both water droplets and 'smoke' particles.

The latter term 'smoke fog' should not be confused with 'arctic smoke', 'frost smoke', 'sea smoke', 'steam fog', 'warm-water fog', 'water smoke', 'the barber' – various terms used for fog produced by the passage of cold air over warm water which is followed by excessive evaporation and condensation. If there is a temperature inversion just above the water surface, this type of fog may be dense. In Britain, however, such fogs are usually shallow and transient, though when conditions are suitable for more widespread fog, 'warm-water fog' may help to worsen visibilities.

Foggy conditions are characteristically associated with particular synoptic situations. For example, radiation fogs and urban fogs commonly occur in anticyclones where winds are light and also in cols. Sea fog and fog on coasts and hills are usually associated with moist south-westerly airstreams which often occur in the warm sector of a depression. Fog, especially hill fog, may occur also in the frontal areas of depressions.

Fog or poor visibility may drift in the direction of light winds, determined by the general synoptic pattern or by local topography and diurnal changes. Drift is more likely to extend over long distances in stable air masses. For example, poor visibility over the London area may affect much of the region between London and Reading, under appropriate conditions.

In the British Isles, excellent visibility is more commonly associated with unstable polar airstreams, particularly if these come directly from more northern latitudes and across sea tracks rather than urban areas. Even to the lee of industrial areas, a brisk polar airstream will disperse urban pollution so that concentrations fall off quite sharply.

Diurnal variation in visibility

As minimum temperatures occur around dawn and because surface cooling is broadly associated with the development of low-level inversions of temperature, it is not surprising that frequencies of poor visibility often tend to reach a peak around dawn. Table 9.1 shows the percentages of time when visibility at given times relative to dawn is less than 200 m for the winter and the summer half-years for 'smoky' sites, 'clean' inland sites and coastal sites. In each of these three categories of topography, there is a peak percentage around dawn or shortly afterwards. For higher visibility thresholds, for example, less than 1 km or less than 10 km, the results are similar, but in smoky areas in winter the peak is a little delayed and on average occurs some 2 to 3 hours after dawn.

Fig. 9.1 *Diurnal variation of the percentage frequency of visibility below specified limits at Heathrow (London Airport), Valley (Anglesey) and Boscombe Down (Hampshire), at different times of the year, 1962–71*

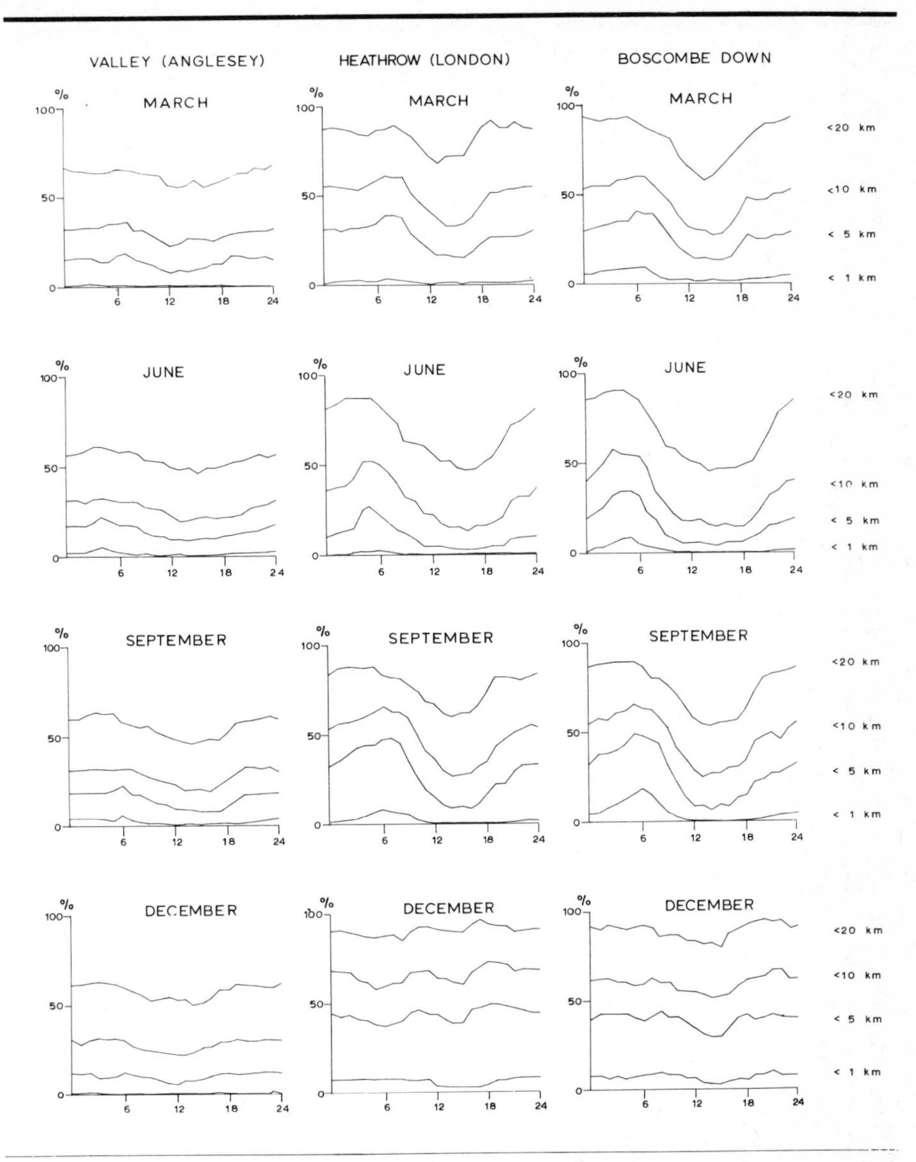

The mean diurnal variation throughout the 24 hours is shown clearly in the graphs of Fig. 9.1. The curves show that the highest values of visibility tend to occur in the afternoon while poor visibility tends to build up during the night. The basic difference between the sites is that Heathrow (London Airport) has considerably higher fre-

Table 9.1 *The percentage of time that given visibilities occur at hours around the dawn period, 1957–66 (based on Grant, K., at the Meteorological Office) M.S.*

Percentage frequency of visibility less than 200 m: thick fog

	Hours after sunrise												
	—6	**—5**	**—4**	**—3**	**—2**	**—1**	**0**	**+1**	**+2**	**+3**	**+4**	**+5**	**+6**
Winter half-year													
Smoky	3·23	3·39	3·52	3·72	3·80	4·02	4·45	4·47	3·88	2·89	2·11	1·50	1·16
Clean inland	3·72	3·92	4·05	4·16	4·20	4·31	4·90	4·68	3·82	2·79	1·81	1·27	0·95
Coastal	0·95	0·94	0·98	1·04	1·02	1·05	1·23	1·17	1·00	0·76	0·59	0·44	0·40
Summer half-year													
Smoky	0·38	0·45	0·49	0·73	0·88	1·11	1·55	1·56	1·21	0·68	0·32	0·24	0·21
Clean inland	0·60	0·84	1·09	1·35	1·63	1·89	2·49	2·52	1·81	0·85	0·35	0·11	0·04
Coastal	0·80	0·92	1·01	1·15	1·19	1·36	1·61	1·51	1·24	0·88	0·64	0·51	0·44

Percentage frequency of visibility less than 1 000 m: fog

	Hours after sunrise												
	—6	**—5**	**—4**	**—3**	**—2**	**—1**	**0**	**+1**	**+2**	**+3**	**+4**	**+5**	**+6**
Winter half-year													
Smoky	8·60	8·66	8·55	8·50	8·49	8·93	10·74	11·76	11·46	10·01	8·06	6·34	5·62
Clean inland	9·31	9·69	9·98	10·17	10·21	10·52	11·51	10·96	9·42	7·46	5·79	4·46	3·83
Coastal	2·74	2·80	2·96	3·00	2·97	2·96	3·49	3·40	2·91	2·47	2·06	1·75	1·67
Summer half-year													
Smoky	1·07	1·30	1·64	2·03	2·42	2·94	4·30	4·68	3·88	2·50	1·47	0·78	0·57
Clean inland	1·89	2·65	3·32	4·14	4·84	5·88	7·04	7·00	5·36	3·27	1·42	0·53	0·27
Coastal	2·32	2·65	2·84	3·15	3·40	3·73	4·21	4·15	3·43	2·64	2·04	1·74	1·51

Percentage frequency of visibility less than 10 km: haze

	Hours after sunrise												
	—6	**—5**	**—4**	**—3**	**—2**	**—1**	**0**	**+1**	**+2**	**+3**	**+4**	**+5**	**+6**
Winter half-year													
Smoky	64·68	62·74	61·24	60·19	59·75	61·69	68·18	74·02	76·53	76·02	72·83	69·28	66·47
Clean inland	61·37	61·43	60·93	60·36	60·14	60·14	60·45	61·10	60·60	58·77	55·45	51·79	48·79
Coastal	39·91	39·69	39·04	38·74	38·34	38·44	39·83	40·61	41·26	40·61	39·32	37·64	36·60
Summer half-year													
Smoky	50·25	49·22	48·71	48·88	49·51	52·01	57·21	59·54	59·60	57·75	50·75	43·49	37·59
Clean inland	40·22	42·95	45·97	48·43	50·57	53·36	54·92	54·19	51·92	46·78	38·71	31·76	26·72
Coastal	30·60	31·29	31·97	32·56	33·00	34·21	35·52	35·67	35·26	33·71	31·03	28·05	25·95

quencies of poor visibility and more marked diurnal variation than Valley, a coastal site in Anglesey; the 'clean' inland site of Boscombe Down has intermediate frequencies. However, Heathrow in winter shows a double peak, the visibility tending to deteriorate shortly after dawn, with the increased wind and turbulence at this time tending to bring down the air pollution accumulated beneath an inversion during the night, a process known as fumigation: the deterioration of visibility may be worsened by the disturbance of any ground fog and the influx of air pollution at this time of the day.

Diurnal variation of visibility is very dependent on local topography. Where the topography inhibits air movement, such as in sheltered valley basins, insolation may be insufficient in winter to disperse fog during the day. In this way, winter smogs may last several days, as in December 1952 and December 1962 in the London region. If the local topography is such as to permit the development of local katabatic (slope) winds and land breezes, these winds may help to maintain an improved visibility after a deterioration during the earlier part of the night.

Seasonal variation in visibility

As most land fogs, especially urban fogs, are associated with nocturnal radiation cooling, land fog is chiefly a winter half-year phenomenon. On the other hand, coastal fogs, in so far as these are caused by sea fog conditions, occur in spring and early summer, when sea temperatures around Britain are low and the excess of air temperature over sea temperature is generally high.

Summer coastal fogs are relatively infrequent as compared with winter land fogs. Thus, in the graphs of Fig. 9.2, frequencies of poor visibility are strikingly lower at Valley than at Heathrow (London Airport) and Boscombe Down, and also the magnitude of the seasonal variation is markedly lower at Valley. However, all sites, including the one near the coast, have a summer minimum and a winter maximum frequency of poor visibility, though sites with basically coastal characteristics and very limited inland effects will tend to have a flat maximum in summer. The graphs of Fig. 9.2 suggest that the peak in the winter half-year is not smooth and that several factors are operating.

Day-of-week variation in visibility

Industrial processes and other urban activities which normally vary in a definite way through the week are reflected in the variation of meteorological elements (Lawrence, 1971c). It is found, for example, that Sunday, on average, had better visibility than Friday during the winter half-year at Finningley, Nottinghamshire (Ratcliffe, 1953).

Geographical distribution of visibility

There are important variations in visibility over short distances, due partly to meteorological causes and natural variations in topography and partly to artificial variations associated with urban and industrial complexes. Thus the natural hollow at the valley site of Rickmansworth in Hertfordshire may have about fifty morning fogs (visibility less than 1 000 m) in a year, several times as many as at the more exposed site of Rothamsted, only a few kilometres away.

Fig. 9.2 *Seasonal variation of the percentage frequency of visibility below specified limits at Heathrow (London Airport), Valley (Anglesey) and Boscombe Down (Hampshire) at different times of the day, 1962–71*

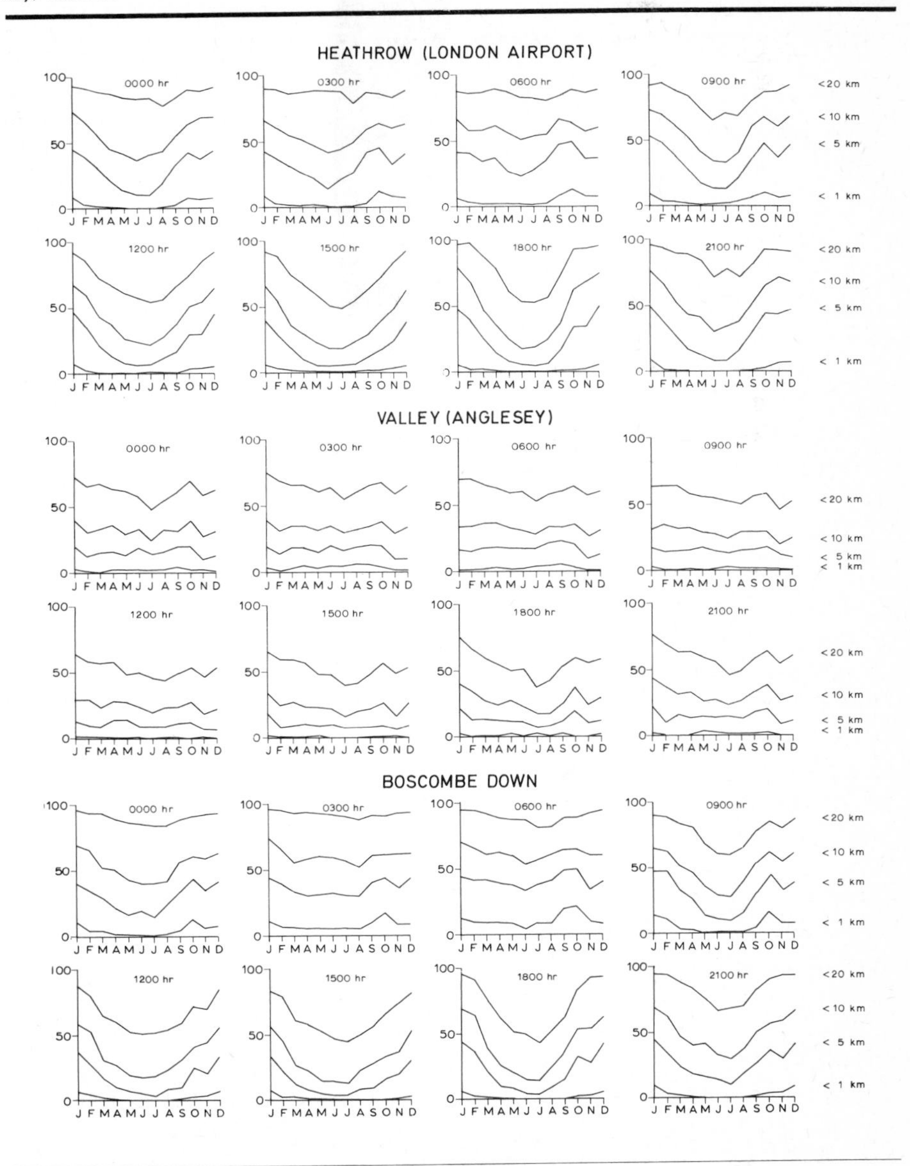

The effect of ground slope may be seen in a shallow valley which acts as a 'drain' for cold air. Where the valley bottom is almost horizontal a shallow layer of thick fog may accumulate; somewhat steeper slopes may have a deeper layer of fog while even steeper slopes may have no fog at all (Lawrence, 1954). It is believed that many sites remain free of radiation fog because of the development of katabatic (down-slope) winds and land breezes.

Fig. 9.3 *The percentage of time with visibility less than 200 m (220 yds) at various sites in the British Isles, based on observations at 03.00, 09.00, 15.00 and 21.00 hrs, 1962–71. Altitudes are given for certain specifically referred to in the text*

There are important differences between urban areas and surrounding rural areas. Rural areas are liable to water fog while urban areas are more prone to smoke fog (see Chapter 14): however, in quiet and damp conditions, general widespread 'heterogeneous' fog is likely to occur. Because of such variations over short distances, any general geographical distribution must be regarded only as a very broad outline.

For low-level areas below about 150 m, the foggiest zones, with about 40 days or more per year with fog (less than 1 000 m) at some time of the day are (1) the zone from south-east England across the Midlands to South Lancashire and West Yorkshire, to the extreme north-east of England and (2) the lowland area of Scotland from the Clyde basin to the Firth of Forth. The least foggy areas are the extreme northern areas of Scotland and western Ireland, with about 10 or fewer foggy days (visibility less than 1 000 m) per year.

With the advent of air travel and faster road travel, practical interest has shifted to frequencies of lower thresholds of visibility. A motorist is more likely to regard fog as being somewhere in the range below 200 m – referred to as thick fog. The percentage of time with less than 200 m is shown in Fig. 9.3. The values are based on observations at 03.00, 09.00, 15.00 and 21.00 hrs.

In Fig. 9.3, the effect of height is demonstrated very clearly. The site at Great Dun Fell, 848 m, in the Pennines, is 'foggy' or in cloud for about two-thirds of the time. Even sites at more moderate heights have noticeably more frequent occasions of poor visibility, for example:

Little Rissington, 223 m, 4·7 per cent;
Bovingdon, 163 m, 2·5 per cent;
Lyneham, 157 m, 2·1 per cent;
Wattisham, 87 m, 2·4 per cent;
Waddington, 70 m, 2·0 per cent;
Gatwick, 59 m, 2·3 per cent.

Apart from these higher-level sites, it can be seen that the general pattern for visibility less than 200 m is rather similar to that described for visibility less than 1 000 m.

The areas of least fog are generally the areas with the highest frequencies of good visibilities. Visibility frequencies of over 10 km at midday in winter vary from about 20 per cent in the Midlands to about 90 per cent in the west of Ireland. In summer, good midday visibility is more general and percentage frequencies of 80 to 90 per cent occur in parts of south-west England and East Anglia as well as much of Ireland and Scotland.

Changes of visibility with time

With regard to changes of visibility over long periods, the period of most interest has been that associated with the application of the Clean Air Act, 1956. Changes of air quality may affect sunshine, solar radiation, illumination and other meteorological elements but the element most readily affected by changes in smoke pollution is visibility. Relationships between air pollution and visibility are complicated by many factors, one of which arises from the fact that increased urban pollution is often associated with increased urban heating which tends to reduce the accumulation of air pollution at or near ground level.

Changes of visibility in the London area may be seen from the average annual number of hours of fog during three consecutive periods 1947–54, 1955–62 and 1963–70 (Table 9.2). The results are given for thick fog (less than 200 m) and dense fog (less than 50 m) and both sets of figures indicate a general decrease. However, fog frequencies over England, for the period 1958–67, show a decline generally, even for rural or semi-rural sites (Dinsdale, 1968), and the improved visibilities since 1957 cannot be attributed entirely to the Clean Air Act (Lawrence, 1971a).

Table 9.2 *Average annual number of hours, and the percentage decrease, of thick fog and dense fog at London Weather Centre, Kew Observatory and Heathrow (London Airport) during the periods 1947–54, 1955–62 and 1963–70 (Betts, 1972)*

	Average annual number of hours of fog			**Decrease (per cent)**	
	1947–54	**1955–62**	**1963–70**	**1947–54/1955–62**	**1955–62/1963–70**
Thick fog (less than 200 m)					
London Weather Centre	68	52	17	24	68
Kew Observatory	241	189	83	22	63
Heathrow	202	177	67	12	62
Dense fog (less than 50 m)					
London Weather Centre	27	13	3	52	77
Kew Observatory	95	73	20	23	73
Heathrow	57	54	10	5	82

The general decline in the frequency of poor visibility is consistent with the general decrease in smoke concentrations illustrated by Kew Observatory data, during the period 1962–67; this period is the descending part of a 10–11 year air pollution cycle following peaks around 1952 and 1962, etc. (Lawrence, 1966). The latter two peaks coincide with the well-known London smog episodes. The number of foggy hours during December 1952 and December 1962 (the disaster months) are compared in Table 9.3 with data for December 1972, at Heathrow (London Airport) and the London Weather Centre.

Table 9.3 *Number of hours of thick fog (less than 200 m) and dense fog (less than 50 m) at London Weather Centre and Heathrow (London Airport) in December 1952, December 1962 and December 1972*

	London Weather Centre		*Heathrow*	
	No. of hours Thick fog	**Dense fog**	**No. of hours Thick fog**	**Dense fog**
December 1952	81	69	113	61
December 1962	63	30	101	73
December 1972	1	1	39	5

Regional case studies

The most studied occasions of fog are urban radiation fogs because these have had the most disastrous results on public health and mortality rate. Such fogs are normally associated with anticyclones or ridges of high atmospheric pressure and consequently tend to persist for several days. During this time, visibility tends to decrease and to remain very poor until the pressure system begins to weaken.

A notable example of such a spell occurred in the London area from 3–7 December 1962. Visibility at Heathrow decreased from 2–4 December, remained below 100 m (and mostly below 50 m) until the afternoon of 7 December and then improved throughout the next 48 hrs. From 2–6 December, daily values of air pollution increased generally and finally decreased as the spell began to break down. The spell was particularly marked by a persistent low-level air temperature inversion and very light winds. From early on 2 December to early on 7 December, the midnight and midday inversion bases at Crawley, Sussex were below about 300 m (Lawrence, 1967). Similar but rather shorter spells of low-level inversion of air temperature occurred in December 1963 but with winds reaching 2·5–5·0 ms^{-1} daily. During these spells, fog occurred much less frequently and visibility was usually 1 000 m or more at midday.

Studies of days in summer with high values of air pollution at Kew showed that such days were associated with poor visibility (914–3 658 m between 18.00 hrs and 09.00 hrs), light easterly winds from the central London area and general air conditions conducive to the accumulation of nocturnal air pollution (Lawrence, 1969). A further case study of radiation fog, at Cardington, Beds. is described elsewhere (Roach *et al.*, 1973).

Other visibility terms

Vertical visibility

The vertical visibility may be taken as the height above ground at which a pilot balloon, if rising vertically, would disappear from view in daylight. If the balloon does not rise vertically but is at an angular elevation E at the moment when it disappears from view, the vertical visibility should be taken as the distance along the line of sight, and this is equal to h cosec E, h being the height of the balloon at the moment of disappearance.

Vertical visibility may be impaired by low-level fog or haze which tends to accumulate especially just below low-level inversions of temperature. Little overall data is available for vertical visibility but measurements made during research projects (Stewart, 1955) indicate that surface fogs in central England may extend 150–300 m or more vertically from ground level. In such fogs, visibility may be poorer near the fog top than near the ground. A typical growth pattern of the depth of such fogs is fairly rapid growth at first, followed by a more or less static period and a final brief period of upward growth before the ultimate dispersal of the fog by rising wind and temperature.

Table 9.4 *The mean ratio* (R) *of runway visual range* (V) *to the meteorological visibility* (M) *for given ranges of meteorological visibility, and the number of occasions* (N) *in each range*

	Metres	0–100		100–200		200–300		300–400		400–500		500–600		600–700		700–800		800–900		900–1000		1000–1100		1100–1200		1200–1300		1300–1400		1400–1500	
		Values of the mean ratio R (= V/M) and N																													
		R	N	R	N	R	N	R	N	R	N	R	N	R	N	R	N	R	N	R	N	R	N	R	N	R	N	R	N	R	N
Day																															
Gatwick	Oct.–March	3·04	70	1·90	154	1·69	74	1·59	44	1·54	32	1·48	26	1·40	19	1·30	24	1·15	16	1·03	10	0·94	11	0·89	7	0·85	8	0·82	0	0·31	7
	April–Sept.	4·00	76	2·77	153	2·38	87	2·02	38	1·70	23	1·44	11	1·28	9	1·18	7	1·11	8	1·04	5	0·97	3	0·91	0	0·86	4	0·83	0	0·32	3
	Year	3·70	146	2·36	307	2·03	161	1·76	82	1·60	55	146	37	1·34	28	1·24	31	1·14	24	1·04	15	0·96	14	0·90	7	0·85	12	0·83	0	0·32	10
Heathrow	Year	2·27	387	1·72	289	1·47	191	1·33	102	1·22	178	113	118	1·06	91	0·99	44	0·93	93	0·87	160	0·83	229	0·78	128	0·74	68	0·69	62	0·65	165
	SD*	2·62		1·17		0·72		0·55		0·45		0·38		0·41		0·26		0·25		0·31		0·20		0·19		0·15		0·16		0·14	
Night																															
Gatwick	Oct.–March	3·81	208	2·65	209	2·10	115	1·84	74	1·66	54	1·49	37	1·36	40	1·25	20	1·14	12	1·04	11	0·94	4	0·87	3	0·83	7	0·81	0	0·80	9
	April–Sept.	>5·00	102	3·25	69	2·60	35	2·00	27	1·68	12	1·50	8	1·37	4	1·25	3	1·14	4	1·05	0	0·97	0	0·90	0	0·85	1	0·83	2	0·31	0
	Year	4·50	310	2·91	278	2·33	150	1·87	101	1·66	66	1·50	45	1·36	34	1·25	23	1·14	16	1·04	11	0·95	4	0·88	3	0·84	8	0·82	2	0·30	9
Heathrow	Year	2·63	362	2·10	253	1·83	195	1·63	121	1·47	146	1·36	94	1·25	122	1·15	25	1·06	76	0·98	89	0·90	77	0·83	36	0·76	20	071	22	0·68	31
	SD*	1·72		1·05		0·92		0·73		0·53		0·40		0·40		0·16		0·29		0·29		0·24		0·29		0·24		0·23		0·25	

*SD=Standard deviation of the ratios in the stated range of meteorological visibility

Slant visual range

The slant visual range may be regarded as the distance of the furthest approach light on the central line of an aircraft runway, which can be seen from a position on a 3° glide path. This term is similar to the runway visual range described later.

Meteorological optical range

The meteorological optical range is the length of path in the atmosphere required to reduce the intensity of light from a defined standard source to 0·05 of its original value. This measure of the optical state of the atmosphere is comparatively new, being recommended for adoption by the World Meteorological Organization in 1957.

Runway visual range

Runway visual range (V) is defined in the Air Navigation (General) Regulation 1960 thus, 'runway visual range in relation to a runway or landing strip means the maximum distance in the direction of take-off or landing, as the case may be, at which the runway or landing strip or the markers or lights delineating it can be seen from a point 15 feet above its centre line'.

This is the most important of visibility terms apart from the horizontal visibility (M) normally measured at meteorological stations. The V depends on the intensity of the runway lights and whether it is day or night. For M values up to about 1 500 m, the *mean* ratio of V to M is about two, but mean ratios vary from around unity at 1 000 m to more than two at 50 m, as shown in Table 9.4. The ratios are higher at Gatwick (with lighting intensity 10 000 candela) than at Heathrow (London Airport) (lighting intensity 4 000 candela). Also the ratios are higher at night (at one-third lighting intensity) than by day, when lighting contrast is poorer.

In a survey (Harrower, 1963) of six fogs at Heathrow (London Airport) and Gatwick in the late 1950s and early 1960s, it was found that the rates of decrease in V for the 'country' fogs at Gatwick can be much higher than those of Heathrow (London Airport) but the rates of clearance were less at Gatwick. The maximum decreases quoted are 128 m min^{-1} at Gatwick and 52 m min^{-1} at Heathrow, while the maximum rates of increase are quoted as 27 m min^{-1} at Gatwick and 37 m min^{-1} at Heathrow. It is of extreme importance in aviation that such decreases in visibility are reported without delay. On occasions of increasing V, great care must be exercised, as in some cases, a fog appears to be clearing but subsequently thickens again.

Global changes of climate

This chapter is concerned with the climatic changes that have taken place in the British Isles during the last 100 years, and in particular since 1910. It is important to remember that climate is global, and that climatic changes occur on all time scales, so the recent changes in Britain need to be seen in the contexts of global and longer-period changes.

The largest climatic fluctuations have been the glacial cycles, in which world climates fluctuated from somewhat warmer than today (interglacial) to a state in which the North polar ice cap extended to about 50°N. (glacial). The last glacial was at its most intense around 18000 B.C.; world temperatures then increased until the *Climatic Optimum* centred around 4000 B.C. The change was not simply a uniform warming; variations in temperature gradients at the Earth's surface cause changes in atmospheric circulation, and these then react on and modify the temperature patterns that produce them. Rainfall, wind, sunshine, etc., vary in response. Lamb (1971) has discussed the most probable circulation patterns over the Northern Hemisphere during various epochs since 20000 B.C.

During the last 3 000 years world temperatures have fallen and some glaciers have reappeared or regrown; however, there have been important fluctuations and regional variations superimposed (Lamb, 1967, 1971). In Britain a marked warm epoch around A.D. 1150, and relatively minor ones around A.D. 400 and A.D. 1930, were probably associated with poleward shifts of the middle-latitude westerlies, whereas during the intervening colder periods the circulation belts probably lay in lower latitudes and blocking interludes were more frequent.

Considering now the last century, Fig. 10.1 shows that the mean temperature over the surface of the Earth rose from the 1880s to the 1940s and has been falling since. A variation in the frequency of westerly winds in Britain (and perhaps in the same latitudes throughout the hemisphere) was probably connected with this, but the maximum of westerlies preceded the maximum temperature by about 20 years. The lag can be explained on the hypothesis that strong westerlies favour warming, especially at higher latitudes. The westerlies have decreased since the 1920s. Temperature levels in several areas of the world, especially in high latitudes, have decreased in recent decades; for example, at Franz Josefs Land (80°N., 53°E.) the mean temperature for 1960–9 was −14·5°C compared with decade means during the previous 30 years in the range −10·0° to −10·8°C. Other large-scale modifications of global pressure patterns this century have probably been linked with the variations in westerlies in the latitude of Britain. Since the 1930s pressures have become, on average, higher around 60°N. with less frequent subpolar lows, and lower around

Fig. 10.1 *Changes in global mean surface temperature based on 5-year averages (after Lamb, 1937b)*

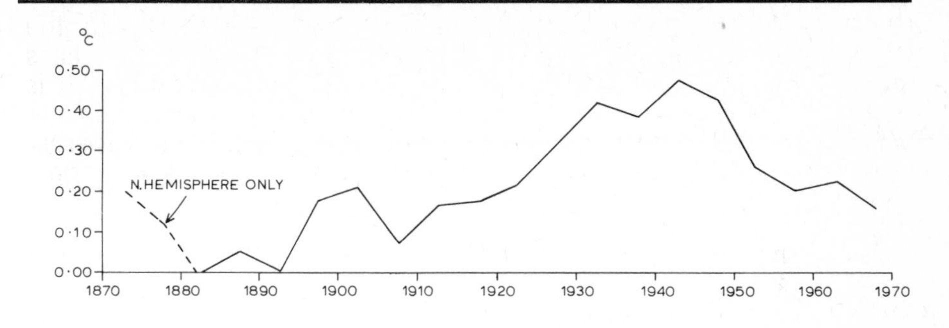

30°N. with higher rainfall in the zone of 'Mediterranean' climates. The Equatorial trough has reduced its range of annual movement, associated with increased rainfall near the Equator and a southward shift of the desert zone in Africa and parts of Asia. Roughly similar changes are believed to have occurred in the Southern Hemisphere. The effects of these and other circulation changes on the climate of the British Isles will be discussed in the following sections.

Problems of data analysis

Climatic trends are usually small in amplitude compared with year-to-year variations, and to determine them it is necessary to obtain homogeneous data series. In theory a series of observations from one reliable station should be suitable, but often the station will have suffered changes of site, instrument or observing procedure for which details may not be available. For longer periods use must be made of records from different stations, collated by means of comparison of an overlapping period; Manley (1974) described the difficulties. Standard tests for homogeneity (World Meteorological Organization, 1966) cannot distinguish between an accidental cause (e.g. change of site), a genuine but local climatic change (e.g. growth of a town) or a climatic change characteristic of the whole region. Therefore conclusions about climatic changes should be based on evidence from several stations rather than one only. One should also be suspicious of a comparison of maps analysed at different times, such as those in different editions of the 'Climatological Atlas of the British Isles', because: (i) different stations may have been used, and it may not be known which; (ii) the drawing of a map involves subjectivity, and different amounts of smoothing may have been used; (iii) gradients of meteorological elements near coasts and across hills are often strong; maps may be seriously in error in some places if the intensity of a gradient was not fully recognized.

The changes in temperature, precipitation and sunshine in the British Isles described in the following sections have been deduced from analyses of small numbers of stations. The choice was dictated by availability of data in a form suitable for processing by computer, but the records from all stations used are complete and believed to be homogeneous, and they are reasonably distributed over the British Isles. It is considered that most of the important climatic changes that occurred during the period 1911–70 are represented. Any variations are likely to be relatively local and a much larger number of stations would be necessary to reveal these or to present meaningful maps. The reader who wishes to know about past climatic changes at any particular place should analyse data for that place; if this is not possible, he should interpolate between the stations used here, making allowance for altitude, exposure, etc., in accordance with the circulation changes described in detail in this chapter. For the purpose of estimating future changes, only the broad-scale features are likely to be amenable to prediction and so a coarse network is probably adequate.

In this chapter the term 'trend' will be used to mean the net effect of climatic changes with period of about 80 years or more; shorter-period changes will be referred to as 'fluctuations'. This is a purely arbitrary distinction, because a longer period of data would show many trends to be part of long-period fluctuations. In examination of trends, it must be remembered that shorter-period fluctuations may be of large amplitude and may confuse the identification of the trends.

There is no such thing as 'the normal climate'. If the term 'normal' is used (for convenience) it should always be referred to a standard period of years such as 1931–60. The terms 'reference period' or 'datum period' are preferable.

Among various indirect methods of estimating climatic trends, one of the best is the date of flowering of wild plants, because its relation to weather is unlikely to change with time. Evidence based on fixed reference points (such as the height of a flood relative to a particular bridge) is also generally reliable. The human memory alone, however, is a poor guide; it may be regarded as an instrument which measures quite accurately over the range to which it has become accustomed through experience, but it is not sufficiently sensitive for accurate assessment of trends which are mostly small compared with year-to-year (or perhaps more to the point, day-to-day) variations.

Circulation, pressure and wind

Circulation

The variations in the atmospheric circulation in the neighbourhood of the British Isles are described, providing a background against which changes in the various weather elements may be better understood. Trends and fluctuations over the past century are examined in general terms, then the changes from 1911–40 to 1941–70 are described in more detail.

Circulation variations may be conveniently studied with the aid of indices derived from Lamb's daily circulation type classification (Chapter 2). The indices used are: W (Westerly), closely correlated with the P index; DW (Days Westerly) as used by Lamb (1972); S (Southerly), C (Cyclonic), M (Meridional), all mentioned in Chapter 2; DC (Days Cyclonic) and DA (Days Anticyclonic). The W index is a measure of 'westerly minus easterly' by contrast with DW which measures westerly only. (North-westerly types are assigned a score of zero in DW because NW was regarded by Lamb as a basic type, whereas SE, SW and NE were regarded as 'hybrids'.) Similarly C ('cyclonic minus anticyclonic') should be distinguished from DC (cyclonic only).

Fig. 10.2 *Decade mean values of various circulation indices derived from Lamb's daily circulation type classification*

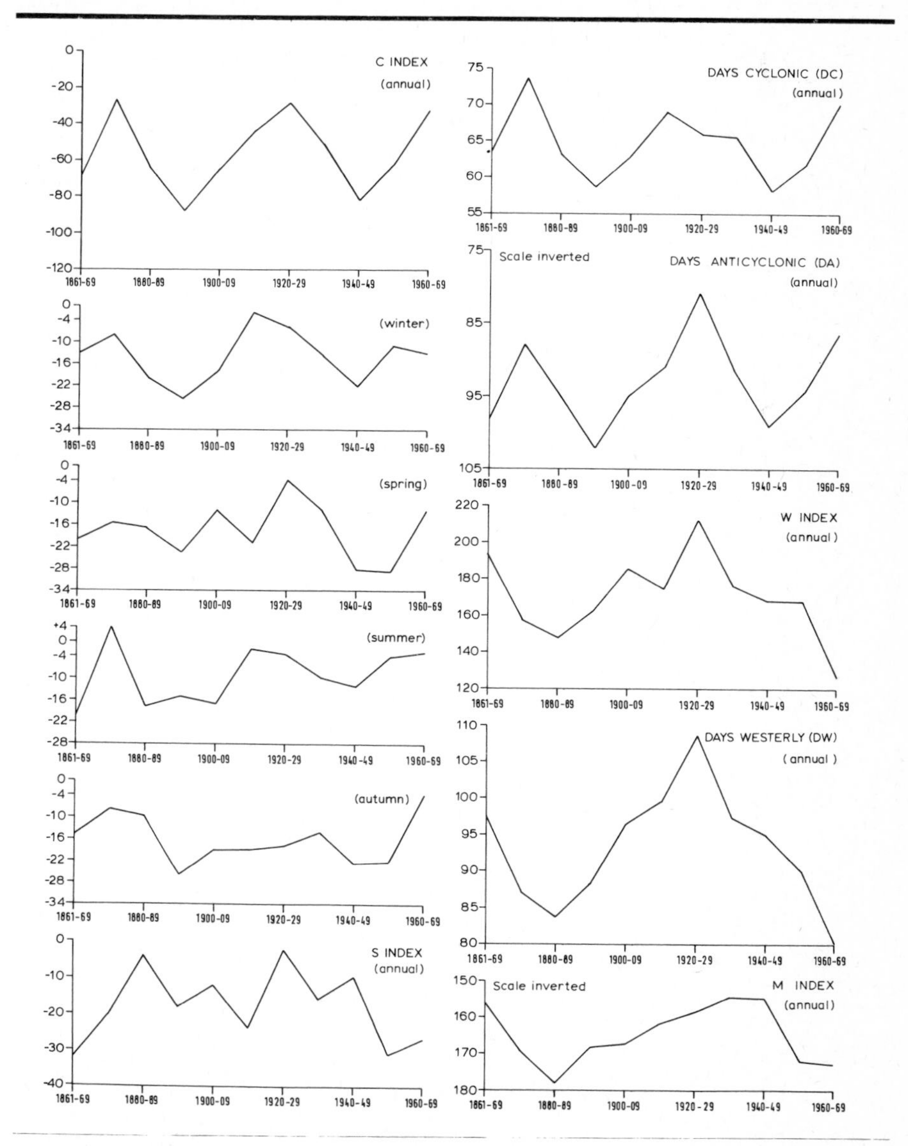

The value of an index for a given period is calculated as the sum of the scores (Table 10.1) corresponding to the Lamb types for each day of the period ; thus for example if every day of a given January were of type 'south-west', the indices for the month would be : W = 31, DW = 15·5, C = 0, etc. An analysis was made of decade mean values of these indices (monthly, seasonal and annual), supplemented by closer examination with the aid of cumulative departures from the long-period average. Selected results are presented in Figs. 10.2 and 10.3.

Fig. 10.3 *Monthly values of three circulation indices for 1911–40 and 1941–70*

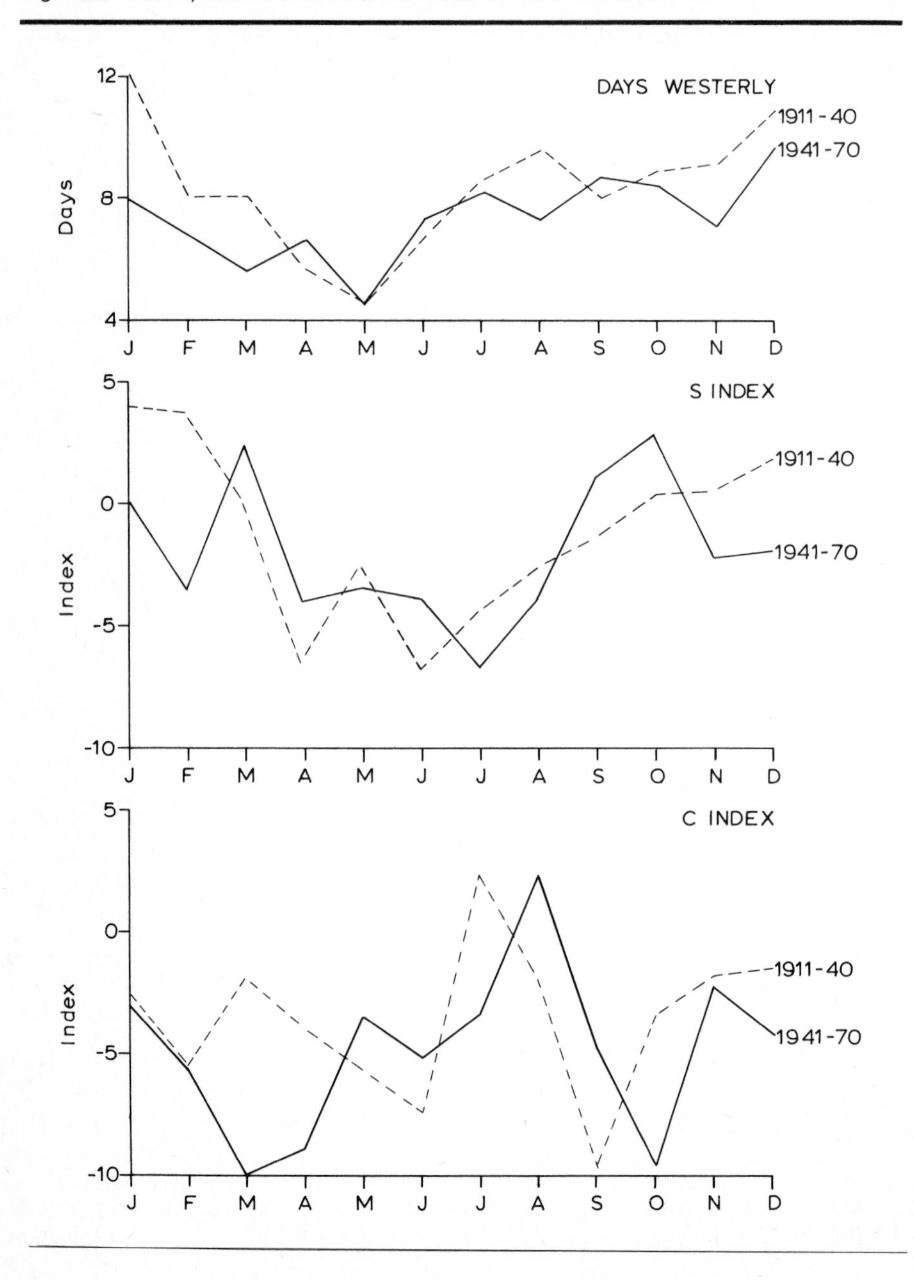

Table 10.1 *Scores associated with each 'type' of Lamb's classification for various circulation indices*

Lamb type	Index						
	W	DW	C	DC	DA	S	M
A	0	0	—2	0	1·0	0	0
ANE	—1	0	—1	0	0·5	—1	1
AE	—2	0	—1	0	0·5	0	0
ASE	—1	0	—1	0	0·5	1	1
AS	0	0	—1	0	0·5	2	2
ASW	1	0·333	—1	0	0·5	1	1
AW	2	0·5	—1	0	0·5	0	0
ANW	1	0	—1	0	0·5	—1	1
AN	0	0	—1	0	0·5	—2	2
NE	—1	0	0	0	0	—1	1
E	—2	0	0	0	0	0	0
SE	—1	0	0	0	0	1	1
S	0	0	0	0	0	2	2
SW	1	0·5	0	0	0	1	1
W	2	1·0	0	0	0	0	0
NW	1	0	0	0	0	—1	1
N	0	0	0	0	0	—2	2
C	0	0	2	1·0	0	0	0
CNE	—1	0	1	0·5	0	—1	1
CE	—2	0	1	0·5	0	0	0
CSE	—1	0	1	0·5	0	1	1
CS	0	0	1	0·5	0	2	2
CSW	1	0·333	1	0·5	0	1	1
CW	2	0·5	1	0·5	0	0	0
CNW	1	0	1	0·5	0	—1	1
CN	0	0	1	0·5	0	—2	2
U	0	0	0	0	0	0	0

Figure 10.2 presents decade values of selected indices. Summarizing these and other results:

1. W and DW: minima in 1880s, maxima in 1920s, minima in or after 1960s; variations in January and August similar to annual, those in November, December, February and March rather similar to annual.
2. S: minimum in 1860s, maximum in 1920s, minimum in 1950s; trend similar to that for W in latter half but reverse in first half of period, and much greater inter-decade variations; variations in Winter, and each month November to February, similar to annual; trends in spring, March, April, June, September and October sometimes reverse of annual.
3. M: maximum in 1880s, minimum in 1930–49, maximum in 1950–69; trend similar to reverse of W but less symmetrical; winter, December, March and August similar to annual; summer, November, January and February rather similar; June reverse trends at times.
4. C, DC and DA: an oscillation of about 50 years, with maxima of C and DC and minima of DA in the 1870s, 1920s and 1960s; trends in winter similar to annual; trends in spring, March, April, July and

October to December rather similar to annual; DC and DA good negative correlation.

The indices W, DW, S and M behaved broadly in concert, with maxima of W, DW and S and a minimum of M around the 1920s. The winter months and August showed best agreement with the annual trends. The smoothness of the variation of DW hints that this may be a good indicator of some basic feature of the global circulation, confirming Lamb's (1972) use of it for this purpose to obtain insight into the climates of past centuries. It is particularly fortunate that January alone seems to reflect this feature quite well, as January happens to be better documented than most months. The relatively erratic behaviour of W compared with DW probably indicates shorter-term variations of easterly types (more easterlies in 1910s, fewer in 1950s). The oscillation in the C index has been notably smooth on a decade basis, although sometimes several successive years went against the trend. A period of high values ended abruptly in 1970; a similar event occurred in 1882. In the 1920s the low value of DA was not compensated by a high value of DC; the balance was made up by a high frequency of westerly types. It seems that in the 1920s the circulation features tended to be on a larger scale than in other decades, with both cyclonic and anticyclonic centres generally remote from Britain which was frequently under a straight, usually westerly, flow. The reverse seems to have been the case in the early 1970s which showed high values of both DA and DC. While definite trends are apparent in the annual values of most of the indices, seasonal values at times differed widely. It would therefore be unwise to forecast an individual season, or even the average of that season over several years, by extrapolation of these trends and fluctuations.

The changes in the indices from 1911–40 to 1941–70 are summarized below, with selected results in Fig. 10.3.

Days Westerly In the later period by comparison with the earlier, each season showed a decrease but six of the months showed little change. The mean annual value fell by 28·6 days, of which January and August contributed more than half (9·2 and 6·3 days respectively). The westerliness of the peak months was reduced to about the level of most other months; only spring was consistently lower than other seasons. The W index behaved similarly.

S Index In view of the large fluctuations with period around 20 years in this index, comparison of two 30-year periods may not be very revealing in terms of the underlying mechanisms, but it is necessary for later comparison of weather elements. The substantial decrease in the annual value of the index (i.e. trend to more northerly and/or less southerly) was mainly contributed by the months November to February. The equinoctial months became more southerly. In 1911–40 the index exhibited a simple annual variation with a maximum in January and a minimum in June, whereas in 1941–70 there were two peaks of southerly, in March and October.

M Index (*not included in Fig. 10.3*) Values of this index increased in November to March and August, decreased in June and September, but otherwise showed little change. The annual mean showed an increase. This is similar to the reverse of the W index.

C Index The presence of the marked oscillation of about 50 years means that comparison of 1911–40 with 1941–70 may not show the fundamental changes to best advantage. March, April, July, October and December became more anticyclonic in the second period, while May, June, August and September became more cyclonic. The indices DC and DA showed corresponding changes, except in January when they both increased (at the expense of DW).

Pressure

Comparison of the mean pressure during 1941–70 with that during 1911–40 (Fig. 10.4) shows that, over the British Isles region, pressure rose in winter, spring and autumn but showed little change in summer. This is probably a reflection of the approximately 50-year cycle in cyclonicity. In winter there was a marked decrease in south-westerly wind component; the pattern of change implies a weakening without change of position of the characteristic winter pressure pattern. The patterns of change in the other seasons were much weaker, and the annual pattern (not shown) is similar to the winter one. Looking wider afield, a set of hemispheric charts of the mean pressure for each month during 1951–66 compared with 1900–39 (this choice of periods should reduce the effect of the 50-year cycle) has been presented by Lamb *et al.* (1973). The patterns for January (Fig. 10.5) and February are similar near the British Isles to Fig. 10.4a. March, November and December experienced a strong rise of pressure over the Davis Strait, whereas in September and October the Icelandic low increased in intensity in 1951–66 compared with 1900–39.

Wind

The changes in circulation type frequencies and mean pressure patterns imply changes in the frequency distribution of wind directions. Thus since the 1920s the main trend has been an increase of north-westerly and northerly winds, mainly at the expense of westerly and south-westerly. Wind speeds, however, depend on transient disturbances, and the mean flow patterns cannot give much indication; little analysis seems to have been done, probably due to the difficulty of obtaining a long homogeneous record since wind speed is extremely sensitive to slight changes in exposure. Changes of direction distributions have some implications for speeds in particular localities; for example, if the decreased westerlies and increased easterlies have resulted in no change of mean speed at a site in the Midlands, then places on the east coast will have experienced an increase in mean speed, and west coast stations a decrease. There are implications also over the sea; Weiss and Lamb (1970) reported that in the North Sea the frequency of waves of at least 5 m height increased during 1950–69, attributed to a greater frequency of winds with a long sea fetch (between west and north) at the expense of winds with a shorter fetch (from between west and south). Harris (1970) stated that the periods 1934–8 and 1961–5 were notably stormy in the British Isles, the numbers of widespread severe gales averaging 2·0 and 4·4 per year respectively; by contrast, during 1939–50 there were only 0·4 per year.

Fig. 10.4 *Mean pressure near the British Isles, 1941–70 minus 1911–40 (in mb).* (a) *winter,* (b) *spring,* (c) *summer,* (d) *autumn*

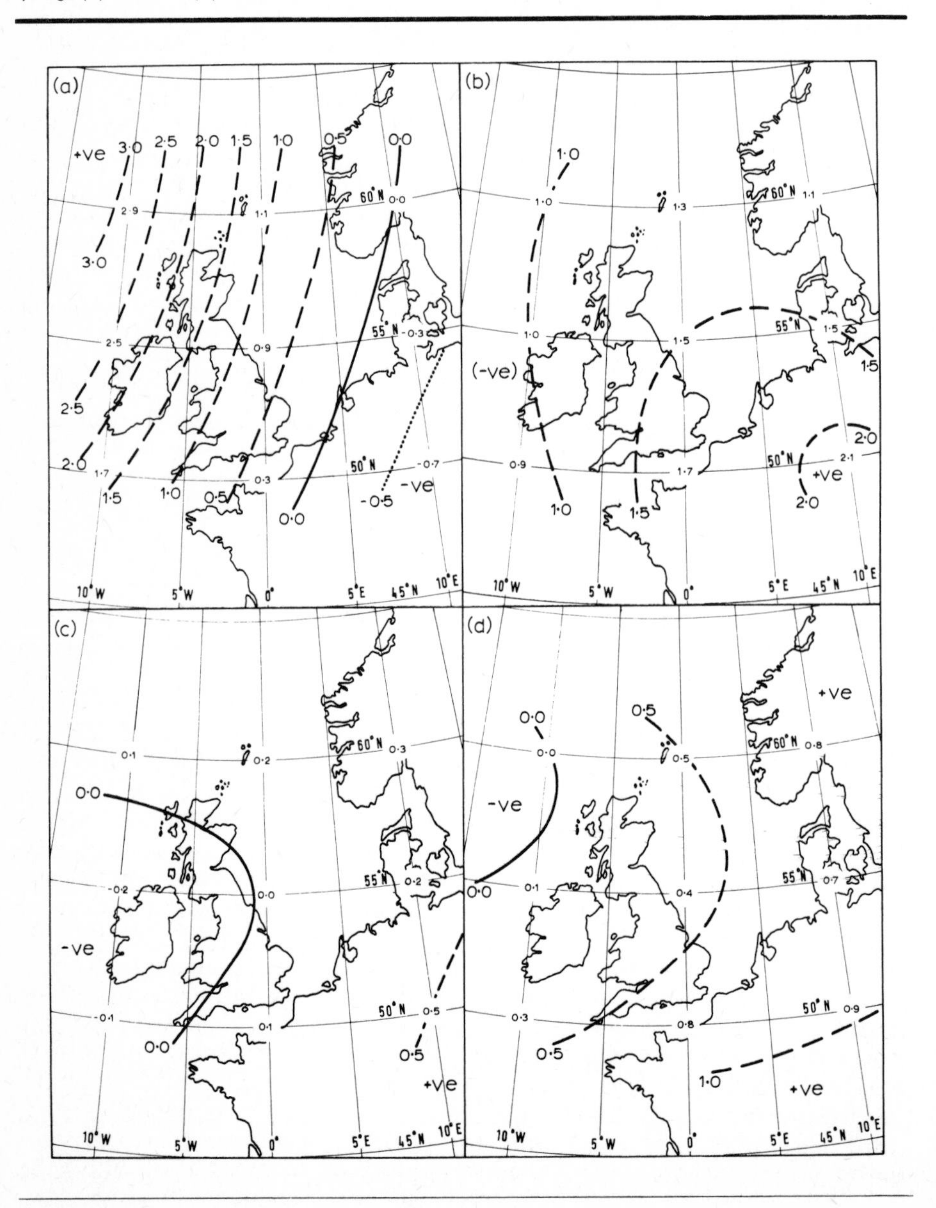

Fig. 10.5 *Average January pressure, 1951–66 minus 1900–39 (after Lamb* et al.*, 1973)*

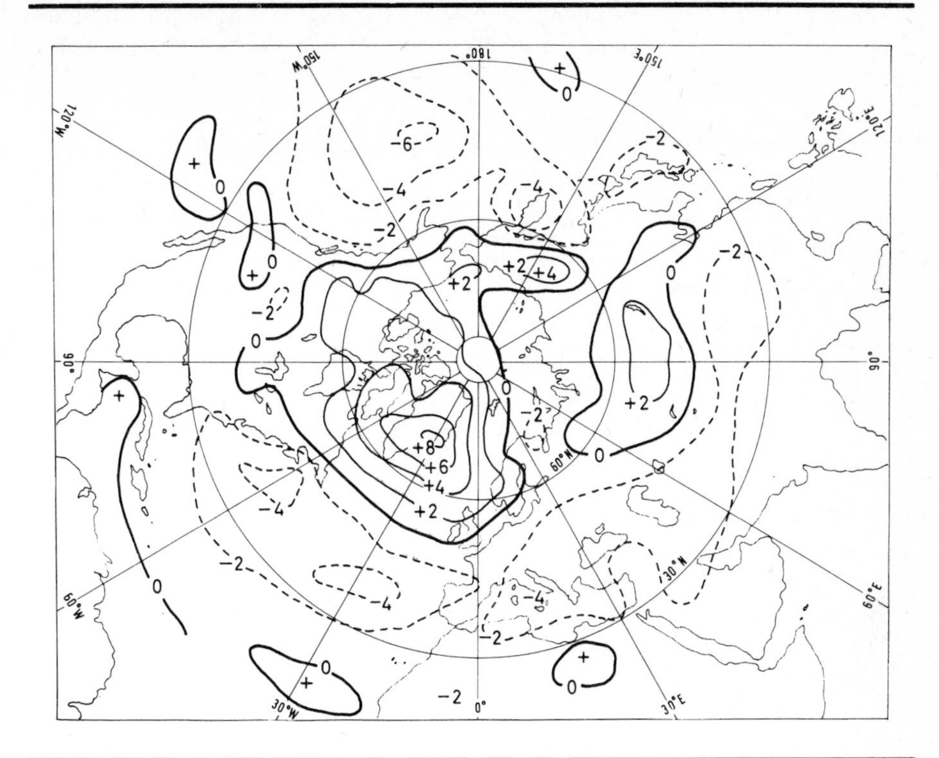

Sea Surface Temperatures (SST) around Britain

Smed (1949, and other papers) found that, over most of the areas with line shading in Fig. 10.6, there was a general rise in SST from the 1910s to the 1930s, and the higher levels continued at least until the late 1940s; over the stippled areas, SST changes were only slight. In more recent years, the mean temperature at the nine North Atlantic weather ships (indicated on Fig. 10.6) varied as in Table 10.2.

Table 10.2 *Mean sea surface temperatures at the nine North Atlantic weather ships (after Rodewald, 1973, and Zverev, 1972)*

Period	**1948–50**	**1951–5**	**1956–9**	**1960–4**	**1965–9**	**1968–72**
temp (°C)	11·3	12·0	11·8	11·7	11·7	11·5

The decrease of SST from 1951–55 to 1968–72 was greatest south of Newfoundland (Fig. 10.6) with little change near Iceland; however, Table 10.2 implies that large-amplitude short-term fluctuations can occur (compare 1948–50 with 1951–55). At Mykines (Faeroes) the temperature level rose from 1923 to 1940, fluctuated until 1958 then fell until 1965. Summarizing, there seems to have been a general increase in the SST of the seas around and west of Britain until about 1950, then a decrease. Some large changes occurred, but not simultaneously in all areas. The varying patterns of SST must have affected the temperature and moisture contents of air-streams reaching Britain, as well as influencing the large-scale circulation patterns which will in turn have had even greater implications for the weather in Britain (see Chapter 2).

Fig. 10.6 *Sea surface temperature (SST) changes in the North Atlantic, being the mean SST for 1968–72 minus the mean for 1951–5 (after Rodewald, 1973). The explanation of the shaded and stippled areas is given in the text (p. 233), and the positions of the nine weather ships are marked on the map*

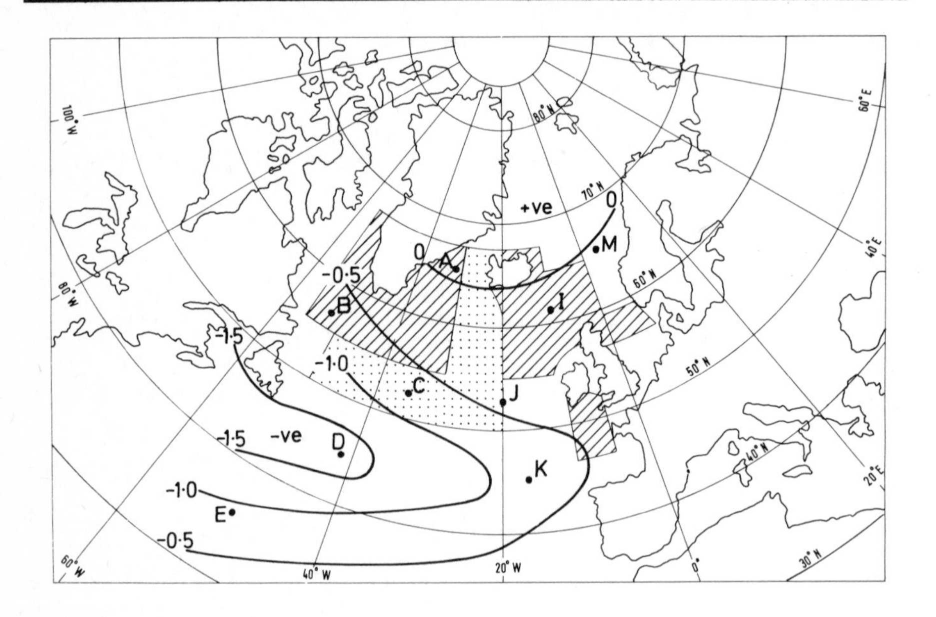

Sunshine

Figure 10.7 shows the mean monthly number of hours of bright sunshine in 1941–70 minus that for 1911–40 at Bradford and Edinburgh. The general pattern (supported by data for Southampton for 1911–60) is of a trend towards more sunshine in winter and spring and less in summer and autumn; however, there were some marked differences between adjacent months. Figure 10.8a suggests a general maximum of sunshine in the 1950s. The decade mean values for March to June at Southampton (Fig. 10.8b) illustrate the marked differences in the time variations in different months. The variations show general similarity with those of the C index, high cyclonicity being associated with low sunshine. Glasspoole and Hancock (1951) found that the sunshine over the British Isles as a whole decreased from 1909–30 to 1931–48 by 29 hours per year; but there were large regional differences, for during almost the same periods Southampton sunshine decreased by 53 hours per year while Bradford and Edinburgh increased by 8 and 2 hours respectively. There were also marked fluctuations with periods of around 5 years.

The parameter 'number of hours of bright sunshine' is very sensitive to urban effects, especially to smoke, which, even when not dense enough to affect visibility, may cause significant diminution of radiation at the ground. Comparative studies of sunshine trends in city and adjacent country, carried out for London and Manchester, provide evidence of effects of pollution (see Chapter 14).

Fig. 10.7 *Mean monthly sunshine, 1941–70 minus 1911–40*

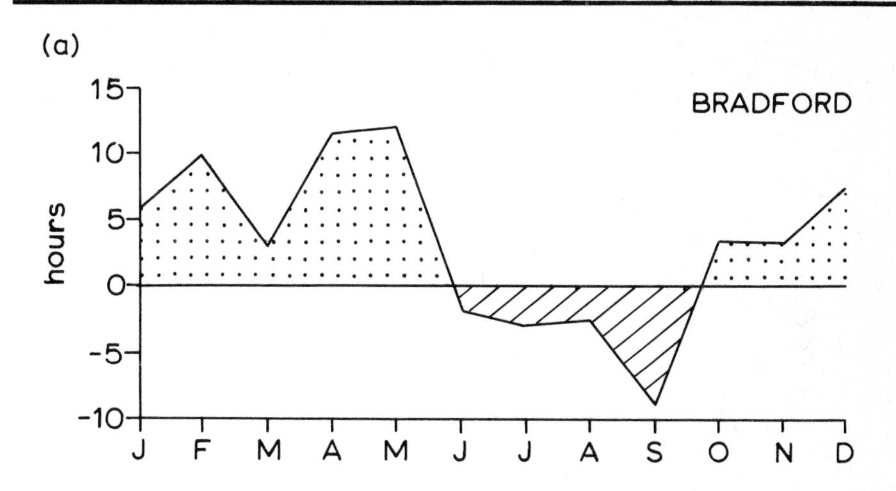

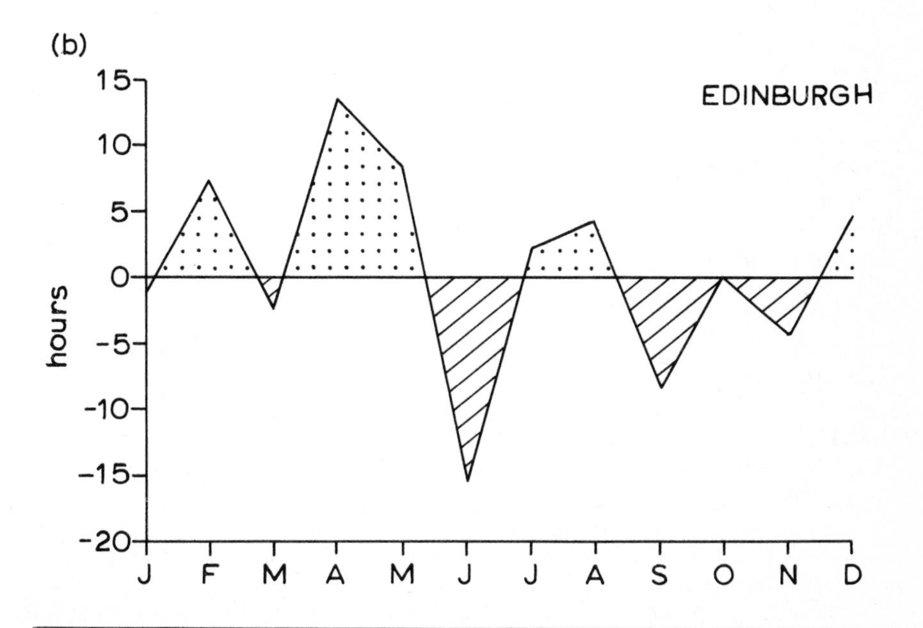

Fig. 10.8 (a) *Decade mean annual sunshine totals;* (b) *Decade mean sunshine at Southampton for March, April, May and June*

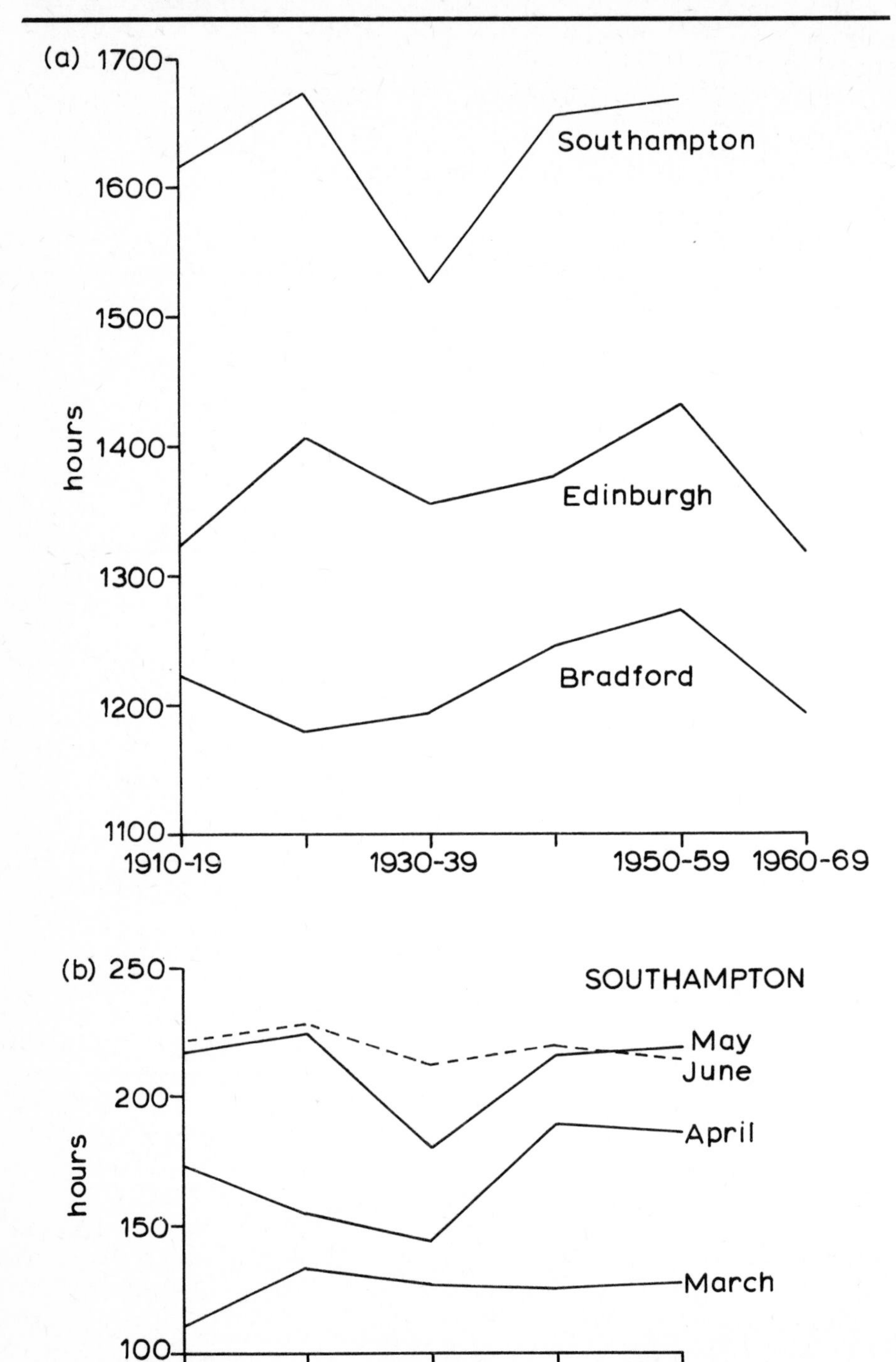

Air temperature

Relationships between monthly mean air temperature and circulation type in Britain were summarized in Chapter 2. Figure 10.9 shows for each month the change of mean temperature from 1911–40 to 1941–70 at nine British Isles stations. ('Central England' is a synthetic series based on several stations (Manley, 1974).) Differences between stations were slight but spatially systematic.

Fig. 10.9 *Mean monthly temperature, 1941–70 minus 1911–40*

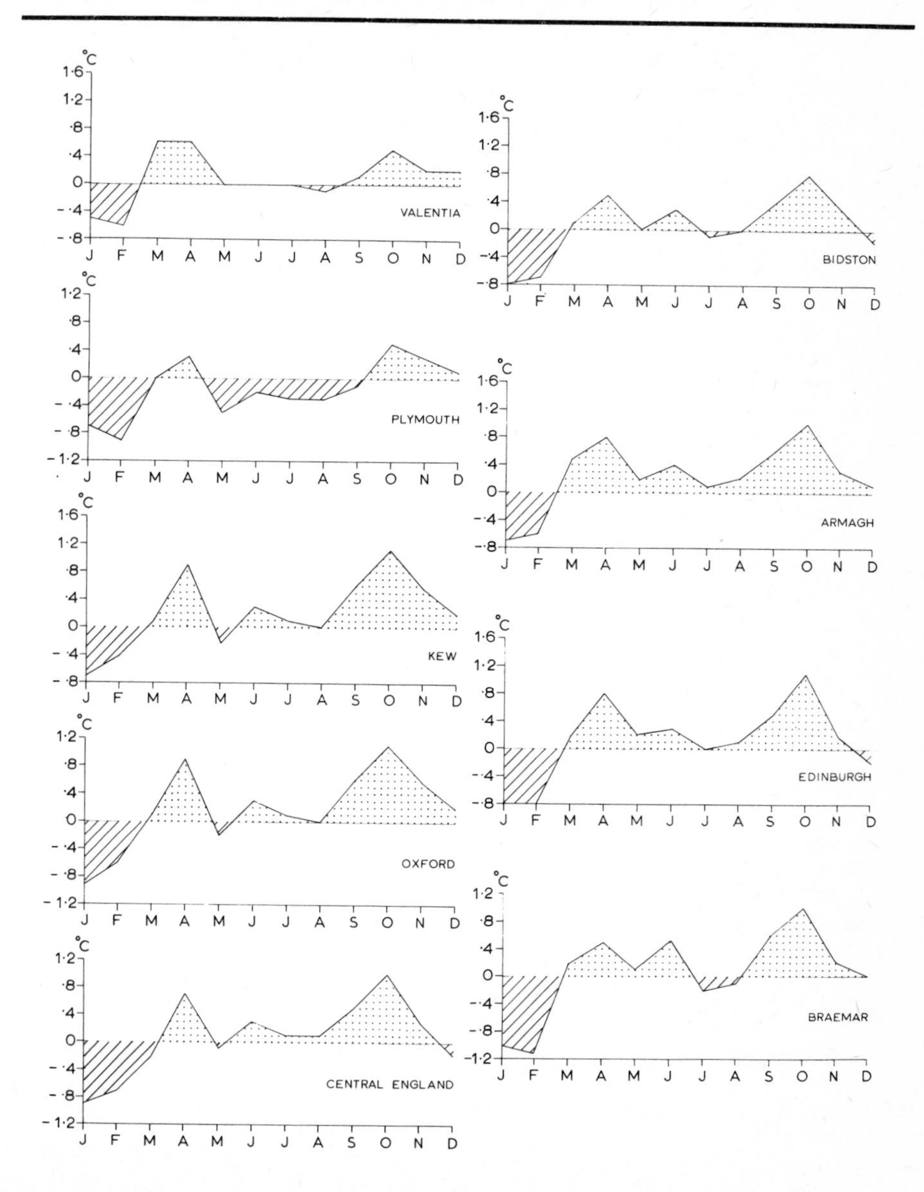

January and February became markedly colder, especially in the east, reflecting the decreased W and S indices. November and December however, although they underwent similar (though smaller) circulation changes to January, showed little change of temperature. March became warmer in the west but not in the south-east; this can be related to more south-easterly (i.e. higher S and lower W indices) and lower C index (implying on the whole more southerly winds in the west but not in the east). April and October were warmer everywhere, probably due to more anticyclonic and southerly types. June and September were warmer everywhere except the south-west; this was associated with increases in the S index, the warmth of which would be tempered in the south-west by increased cyclonicity relative to other areas. May, July and August showed only small temperature changes. Thus most of the changes of temperature between the 30-year periods can be related to circulation changes.

The decade mean temperatures for each season at six stations (Fig. 10.10) indicate the following patterns of time variation:

Winter: The warmest decade was the 1920s. Temperature decreased thereafter, most rapidly from the 1930s to the 1940s, with a temporary slight warming in the 1950s in the south.

Fig. 10.10 *Mean temperature of each decade for each season, 1910–69: V—Valentia; P—Plymouth; B—Bidston; O—Oxford; E—Edinburgh; Br—Braemar*

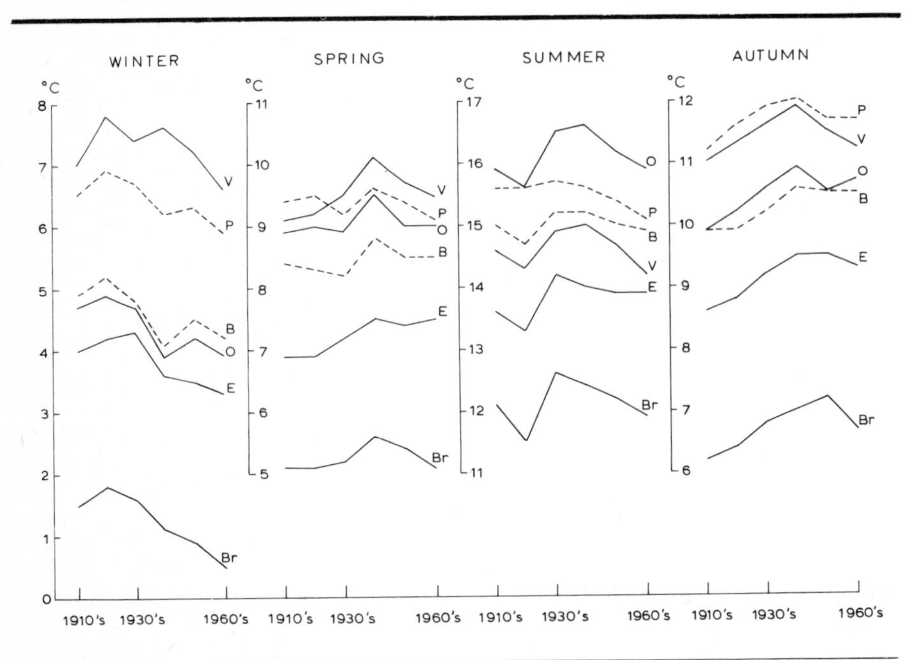

Spring: Changes varied from place to place, but everywhere the 1940s were the warmest decade.

Summer: The 1920s were colder than the adjacent decades everywhere except Plymouth; this could be attributed to the high W index in the 1920s. The 1930s and 1940s were warmest everywhere.

Autumn: Temperatures everywhere rose gradually from the 1910s to the 1940s, then on average decreased slightly. (Octobers however continued to become warmer until around 1970.)

The changes in winter agree well with the corresponding variations in the W index. In the other seasons the minimum of the C index in the 1940s (together with the 1950s in spring and autumn) could account for the main temperature variations. Alternatively, sea surface temperatures around Britain were highest in the 1940s and 1950s, so perhaps this was a factor in the observed trends. This inference is at first sight supported by Perry and Barry (1973), who concluded that temperature trends in April, July and October were due mainly to within-type changes, whereas those in January were due to changes of prevailing type frequencies. However, their study used only limited period and stations and their results were complicated; other possible factors (e.g. changes in mean wind speed of a given circulation type) could cause within-type temperature changes.

It can also be shown that Valentia and Armagh were consistently cold relative to the more easterly stations, Plymouth and Bidston, in the first two decades, but consistently warm during the 1940s, and that the changes of regime were notably abrupt. Comparison with the C index shows a very close similarity. This implies that days defined as anticyclonic on the Lamb classification are associated with relatively warmer weather in Ireland than in Great Britain, not altogether surprisingly because the western side of an anticyclone is generally the warmest. The effect was most marked in winter, moderate in spring and summer but negligible in autumn.

Estimates of the variations in temperature in Britain since about 1700 can be made from studies of various long-period records. These results should, however, be treated with caution because most such records have been affected by the growth of cities. Winter temperatures at Edinburgh (Thomson, 1964) rose steadily from a minimum in the 1810s to a maximum in the 1920s (the difference between these two extreme decades was 2·0 °C), apart from a minor reversal from the 1850s to the 1870s. London showed a similar pattern but with a marked minimum around 1890. In Central England (Davis, 1972a) the winter months during 1873–1970 were on average substantially warmer than those during 1698–1872; in the summer months, by contrast, the general level of temperatures in these two periods hardly differed, but this result disguises some marked short-period variations. The period 1740–1810 showed a consistently high level of summer temperatures, and 1933–52 gave the warmest 20-year sequence of summer and annual temperatures during the last 300 years in Europe generally (Rudloff, 1967).

One of the most important influences of man on climate is in the 'urban heat-island' (Chapter 14). It is difficult to distinguish large-scale climatic changes from those due to the growth or development of a city, because most sites in cities have suffered changes in their immediate environment and because a change in regional climate may alter the local intensity of the heat-island. Craddock (1972) was able to eliminate these effects for London by comparing two stations whose immediate surroundings (within 1 km) remained unchanged over the period 1880–1968. At Kew, compared with Rothamsted (40 km N.N.W. of Kew and hardly affected by the heat-island), the temperature rose over the period by about 1 °C in all months. In summer most of the rise occurred between 1880 and 1930, coincident with the period of change from mainly rural to built-up conditions in the neighbourhood of Kew; after a period of no real change a further rise began in about 1955, suggested as being due to increased sunshine consequent on the Clean Air Act. In winter the rise was fairly steady throughout the period, suggesting a steady increase in the heat-island effect.

Precipitation

The spatial distribution of precipitation (rainfall) in an individual month is very uneven, just as it is on an individual day. Rainfall in Britain is associated with several distinct synoptic situations; all places may get rain from most such types, but some areas get more from some types than others. It is therefore not surprising that patterns of temporal variation of rainfall are complex.

Smith, C. G., (1974), in a thorough investigation of changes in monthly and seasonal rainfalls at a single station (Oxford), found that 'At certain periods in the past, some months have become noticeably wetter or drier while other months, frequently adjacent months, have had an opposite tendency.' He also concluded: 'A first attempt to relate these fluctuations to circulation indices has proved rather disappointing and difficult, in that expected relationships have sometimes been found, while at other times the relationship has been the opposite of that expected.' Among Smith's conclusions were: (i) averaged over the period 1815–1964, summer and autumn comprised the wetter half of the year; (ii) there seemed to be a gradual trend from very wet summers and autumns in the 1830s to a more even distribution of rainfall this century, but with marked differences between decades. Reynolds (1956a) analysed the rainfall of Bidston and came to the conclusion that 'The rainfall regime of the station is a series of distinct and inhomogeneous groups of years, with quite abrupt changes of regime between them.' His analysis employed cumulative deviations from a long-term mean, a good tool for identifying abrupt changes. His results may be summarized as:

Annual: 1870–86 wet, 1887–1917 dry, 1918–31 wet, 1932–45 dry, 1946–55 average.
Winter: 1884–1914 dry, 1915–50 wet.
Spring: 1867–1901 dry, 1902–24 wet, 1925–46 dry.
Summer: No marked trends; 1872–82 and 1927–31 were wet.
Autumn: 1869–86 wet, 1898–1922 dry, 1923–33 wet, 1934–50 average.

The annual pattern shows quite a good correlation with the C index, but none of the seasons do so. In autumn the rainfall shows some correlation with the S index; this could be due to a local sensitivity to small changes of wind direction whose effect elsewhere would be negligible. Other comparable studies of rainfall trends have been made by Milner (1968) for Sheffield and for Kew by Wales-Smith (1973).

Some workers have analysed different rainfall parameters. For example, Craddock (1965b) showed that the annual number of days with more than 5 mm rain at Kew showed the following fairly simple pattern:

1871–80 wet, 1891–1900 dry, 1911–20 wet, 1941–50 dry.

This resembles the variation in the C index. Similar but less regular patterns were obtained for thresholds of from 3 to 10 mm in place of 5 mm. Because of the irregular distribution of rainfall, clearer patterns usually emerge when averages over a number of stations are used. Commonly employed are indices of regional mean rainfall; for example, the 'England and Wales' and 'Scotland' indices. However, trends in these indices must be interpreted with some caution because of their method of definition. For example, if in a given month the wettest fifty stations were 10 per cent below normal and the driest fifty were 10 per cent above normal, the value of the index would be registered as 'normal' although the total amount of water falling on the region was below normal.

The frequency of days when the 'England and Wales' rainfall was at least 20 mm increased during the period 1950–72. Widespread heavy falls were frequent in summer during 1968–73 in southern Britain; in 1973 there were eight occasions from May to September when 25 mm or more fell over a large area (Crumb, 1973). Finch (1972) discovered an increase in the frequency of occasions of more than 30 mm rainfall in two days from 1957–61 to 1967–71, especially in south-east England. This can be linked with the increase in the number of days cyclonic over the same interval. (In north-west Britain heavy falls at a station are more commonly associated with south-westerly or westerly types, which decreased during the interval.) Davis (1972a) has also studied variations in 'England and Wales' rainfall.

The change in rainfall in each month from 1911–40 to 1941–70 at six stations and in two regions is shown in Fig. 10.11. The main conclusions are:

1. The summer half of the year (except July) became generally wetter, the winter half drier.
2. The patterns of all eight curves are fairly similar, leading to confidence about their reality. Where they differ, the geographical pattern is what would be expected from the circulation changes in the month concerned.
3. Superimposed on the above patterns, each station showed a uniform trend common to all months; for example, Plymouth became wetter, East Dereham drier. However, evidence based on single stations should not be relied on.
4. The regional indices showed changes in each month which accord quite well with the C index, except in January and February when

the trend to drier conditions is more likely to have been due to decreased westerlies.

Fig. 10.11 *Mean monthly rainfall, 1941–70 minus 1911–40*

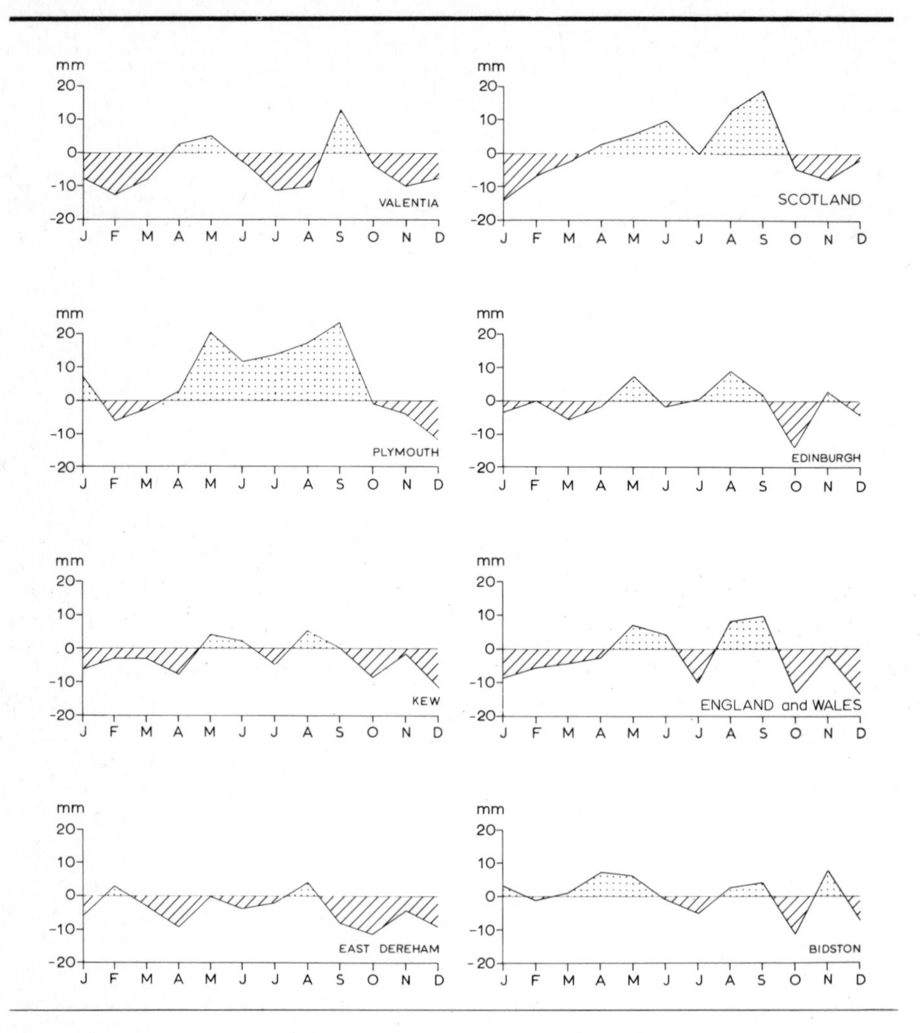

Figure 10.12 shows decade mean rainfalls for the year and for 2 half-year periods. The two regional indices both indicate generally downward trends in winter (mainly from 1910s to 1940s) and upward trends (but with fluctuations) in summer. These half-yearly values show no obvious relation to any of the circulation indices, but the annual values show a reasonable relationship with the C index in the expected sense. The individual stations exhibit substantial differences in temporal patterns, and few relationships with the circulation indices can be discerned. Patterns become clearer when several stations are averaged. Figure 10.13 shows decade mean annual rainfalls averaged over eighteen stations (twelve in 1861–9) evenly

distributed south-east of a line from Hull to Bournemouth and including Kew and East Dereham. A good relationship is indicated with the C index, and a rather better one with the number of cyclonic days.

Fig. 10.12 *Decade mean rainfall totals, 1910–69:* (a) *the year;* (b) *April to September;* (c) *October to March*

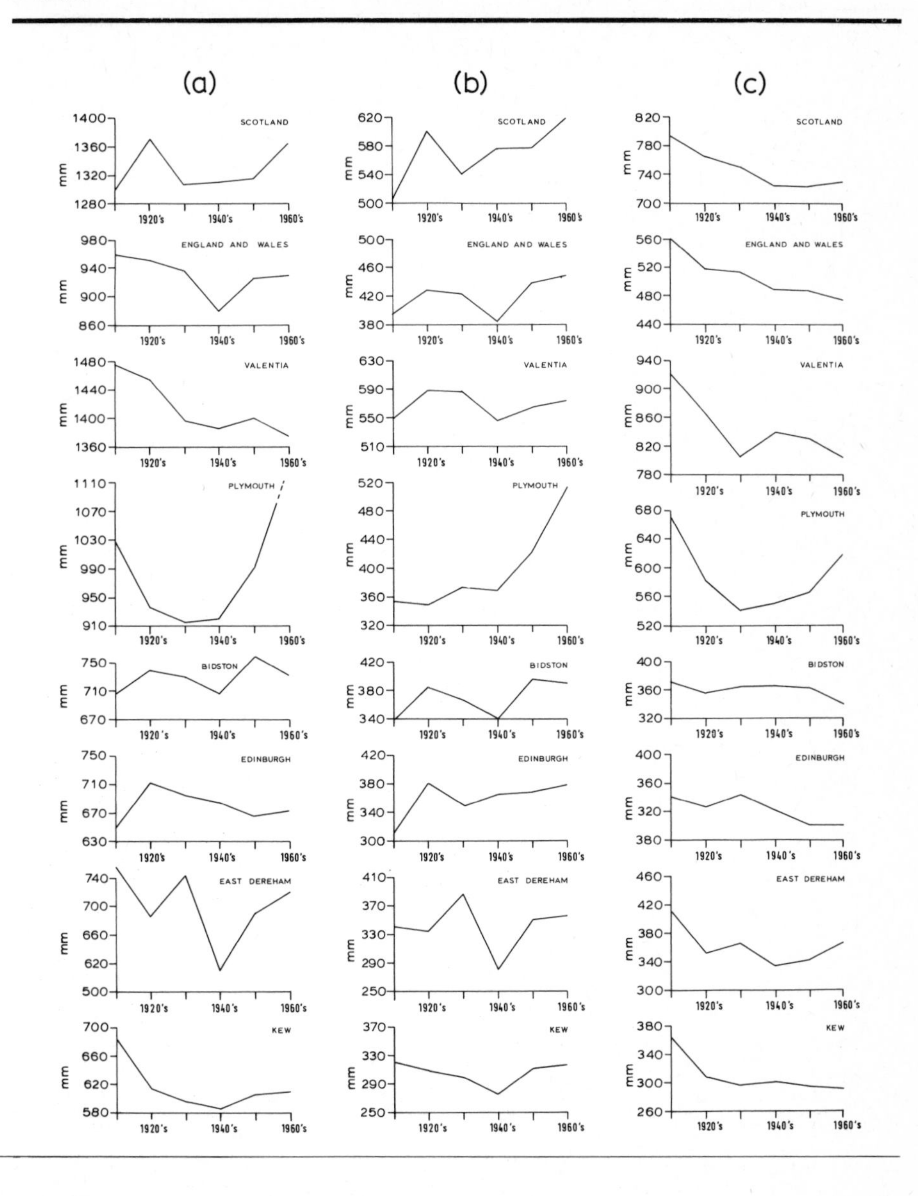

In many parts of the British Isles no particular season has been consistently the wettest. For example, at Kew, East Dereham and Bidston, in some epochs the summer half of the year was wetter than the winter half, in other epochs the reverse. In an individual year, of course, an annual cycle in rainfall is rarely observed at most places in the British Isles.

Fig. 10.13 *Decade mean values of:* (a) *annual rainfall averaged over 18 stations (12 in 1861–9) in the south-eastern half of England;* (b) *the C index;* (c) *the number of cyclonic (C) days*

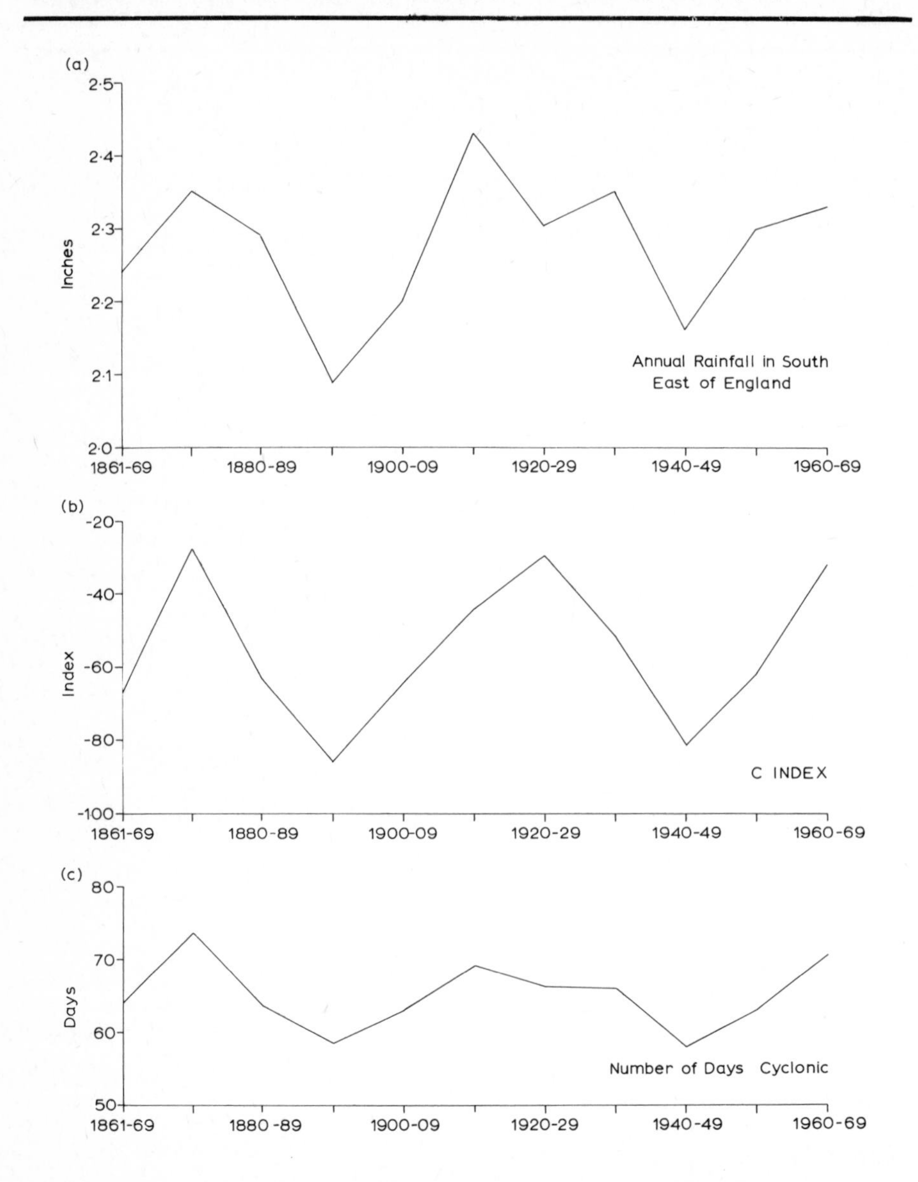

Senior (1969) found that, over the country as a whole, the year-to-year variability of annual rainfall did not change greatly from 1901–30 to 1931–60 but the distribution changed. In England and Wales highest variabilities became more concentrated in the east, whereas in Scotland the highest values shifted from the east to the west.

Thus rainfall changes are complex, especially when a single station is considered. On the whole, the main factor in rainfall trends is the variation of cyclonicity; in winter, variation in westerliness is also important. One must be exceedingly cautious about attempting to forecast future rainfall trends except in very general terms. Perhaps indeed the causes of rainfall have an inherent large 'random' element; that is to say, rainfall is very sensitive to minor details of the atmospheric circulation pattern, unpredictable more than a few hours ahead. If so, then only the grossest features will ever be predictable. However, parameters such as regional average rainfall or the number of days with at least 5 mm rain, although less useful for some practical applications, may be more amenable to prediction; they deserve further study.

Snow

General

With a surface air temperature above 4 °C nearly all precipitation falls as rain; with temperatures below 1 °C it generally falls as snow. In Britain, prevailing temperatures in winter and early spring are frequently around these critical values; therefore a slight change in mean temperature can make a great difference to the frequency of snow at a given place and to the altitude of the snow line. A decrease in the mean winter temperature will (unless accompanied by a decrease in precipitation) lead to a marked increase of snow on low or middle ground, but little increase on high ground where most of the winter precipitation is already snow. Snowfall accompanies different synoptic systems in different areas, so changes in the frequencies of particular synoptic types could have much more effect on snowfall in some areas than in others. Snow lying depends not only on the amount of snowfall but also on the persistence of cold weather. A small decrease of winter temperature could cause a marked increase of snow cover on low ground. On high ground, however, spring temperature is more significant because, as well as favouring additional snowfall, it favours the persistence of winter snow (autumn temperature is less important). A decreased persistence of lying snow in urban areas is favoured both by higher air temperatures and by artificially-warmed ground.

A valuable homogeneous descriptive series of the snowiness of each winter from 1876 to 1965 in Britain has been produced by Bonacina (1966, and previous articles in *British Rainfall* for 1927, 1936, 1948, 1955).

Snowfall

Although there are difficulties in comparing snowfall data because of the different vigilance of different classes of observers, Manley (1969) has made a careful study. For several stations on low ground in east Britain the general trend has been a decline from a maximum during 1750–90 to a minimum in 1900–40. The decade values show good correlation with the mean temperature of winter and spring. For south-east England, Manley gave the following mean dates of the snow season:

Period	Mean date first snowfall	Mean date last snowfall
1901–30	25 November	15 April
1931–60	8 December	1 April

implying a shortening of the season at both ends. In more recent years this trend has been reversed; Clarke (1969) presented the following values for the number of days of snowfall each month in south-east England:

Period (year of Jan. used)	N	D	J	F	M	A+M
1955–9	0	10	37	43	21	5
1960–4	4	23	41	38	27	5
1965–9	10	29	39	41	27	9

Green (1973) found a steady increase in late autumn precipitation at Kinlochewe (north-west Scotland) during the period 1954–72, which would imply increased autumn snowfall and snow lying on higher ground.

Snow lying

Lamb (1966) presented graphs of variations in the number of days of snow lying at seven stations in Britain (Lerwick, York, Cambridge, Kew, Ross-on-Wye, Dalwhinnie and Eskdalemuir). Most stations experienced a marked decrease about 1920, consistently low values during the 1920s and 1930s, a marked rise about 1940, and fluctuations around a relatively high level since. The 1930s had rather lower values than the 1920s. However, at Dalwhinnie (360 m) the trends tended to be the opposite. These trends agree with the changes of westerliness in January, since westerlies are associated with less snow (due to their relatively high temperature) except at Dalwhinnie which is sufficiently far north and high to receive more snow from westerlies than from other types. In the 1930s the Scottish snowbeds disappeared completely in several summers. Manley (1969) showed that from 1925–38 to 1939–67 snow lying at low altitudes increased (Craibstone (100 m) from 26 to 37 days), whereas little change occurred at higher altitudes (Huddersfield (240 m) 24 to 26 days). Green (1973) showed that throughout Scotland, the 1960s compared with the 1950s showed a marked increase of snow lying in November and April, a lesser increase in December, February and March, and a decrease in January. In the eastern Highlands the January decrease was less marked because there the early winter snow can usually persist through January mild spells.

Longer snow lying generally implies later snow-melt, and therefore an increase during April and May in run-off from mountain rivers fed by snow. Green (1973) demonstrated such an increase for two

Scottish rivers (the Avon and the Dulnan) from the late 1950s to the early 1970s.

Visibility

In London there was a steady increase in the frequency of both 'fog' and 'dense fog' during the nineteenth century (Brazell, 1968); the number of days of 'fog' roughly tripled from the 1810s to the 1880s. More recently, there was a general decrease of fog in London from 1949 to 1970 (Kelly, 1971; Freeman, 1968). Synoptic factors played some part in this, because rural stations also showed a decrease during 1958–67, and because there is a large year-to-year variability (the winter of 1952–3 had nearly five times as many fogs as 1966–7). Smoke concentration decreased from the 1920s to 1960s, with fluctuations (Lawrence, 1966). Some interesting details indicate the important part probably played by the reduction of smoke pollution in reducing the frequency of fogs in recent years. Wiggett (1964) pointed out that the percentage decrease in frequency had been greatest in the visibility range 400–1 100 m, nearly 40 per cent from 1948 to 1960; he also quoted evidence that visibilities in this range are usually associated with smoke pollution.

Climatic forecasting

The subject of climatic change has, as well as its intrinsic interest and fascination, an obvious importance for historians and workers in many other fields. But more practically, people ask about the past because they think it will help to forecast the future; and this is right, because all forecasting is based on past experience. Of course, this does not mean that we need only extrapolate the recent trend and oscillations to obtain a reliable forecast, though even this will be better than assuming climatic constancy. It is necessary to discover the patterns of the underlying processes, then to understand their dynamics, before we may forecast with confidence. Progress in understanding is steady but slow, since the atmosphere is immensely complex and many processes interact, most of them on a global scale. Much work has been done in documenting trends in specific features of the climate in various regions of the world in great detail. There is now a need for a careful collation of this material, to discover the global patterns of the fluctuations that are occurring and to determine the various time scales on which they operate.

General considerations

Like all boundary climates, resulting from a physical discontinuity at the earth's surface, the climates of coastal areas provide certain problems of identification and definition. Thus, all parts of the British Isles lie within some 160 km of the sea and oceanic characteristics of mildness and wetness influence every locality. On the other hand, it is equally well known that boundary climates are associated with edge effects which become progressively less marked with distance from the surface discontinuity. In the case of coastal climates, this leads to a general reduction in mean wind speed, a larger range of temperature and relative humidity, plus a higher frequency of thunderstorms and winter fogs with distance inland. Many of these transitional features may also be, at least partially, associated with the large-scale regional concept of continentality and may have little direct linkage with the meso- or micro-scale processes which have been generated in the lower atmosphere by the presence of particular sea or lake shores. The initial problem, therefore, is to define the landward extension of such boundary climates based, ideally, on the nature of the specific mechanisms involved.

In broad terms it is the smaller-scale processes which control the fluxes or rates of transfer of heat, moisture and momentum across the land–water interface and which are consequently largely responsible for restricting the main edge effects to a relatively narrow littoral zone. The inland penetration of the marine air forms a convenient yardstick for delimiting coastal climates. According to the available evidence, the most vigorous sea breezes may encroach up to about 65 km from the coast and this range gives a reasonable indication of the maximum expanse of genuinely coastal conditions in Britain. In many instances, the actual distances involved will be very much shorter, especially in the case of small inland water bodies such as lakes or reservoirs the horizontal influences of which may well decay within a few hundred metres of the shoreline. Nevertheless bearing in mind the length of the British coastline, the atmospheric conditions described in this chapter represent one of the most extensive of all local climates.

As already implied, the distinctive nature of this group of climates arises from the juxtaposition of land and water surfaces which, in turn, modify the overlying air layers in quite different ways. Perhaps the main difference lies in the utilization of incoming solar energy and the subsequent effects that this has on local heat budgets. Over land surfaces heat transfer into the soil is achieved almost entirely by molecular conduction, and the limited downward penetration of the heating wave ensures a marked diurnal flux of sensible heat between ground and air. In turn, this leads to diurnal fluctuations of air temperature, humidity and wind speed (see Chapter 13). In water bodies, however, mass exchange circulations transfer heat to much greater depths in addition to the very large proportion of the heat loss from the water surface that is effected by evaporation and latent heat transfer. The result is that, above the sea surface, the sensible heat flux remains small and almost constant during day and night. The same basic principles, operating over a longer time-scale, are responsible for the seasonal contrasts in thermal behaviour which are apparent between land and water surfaces.

In certain areas, coastal energy budgets may also be influenced by what appear to be anomalously high inputs of solar radiation compared to equivalent stations inland. Following a critical review of the records, Jacobs (1964) concluded that some coastal stations in the south and west of the British Isles had received relatively high values of summer radiation and this view has been largely confirmed by other workers, such as Monteith (1966) for Aberporth (Wales) and Valentia (Ireland). For example, on the basis of sunshine records, it was found that the annual mean intensity of direct radiation at Aberporth was 31 mw cm^{-2} compared with an average of 25 mw cm^{-2} at a number of inland stations, representing an excess coastal radiation receipt of 24 per cent. Comparable results emerged from a study by Rees (1968) on the south coast of England where, after eliminating differences in cloud amount, measured values of incoming radiation were approximately 15 per cent more than the amount expected from the duration of sunshine. More research is needed to determine the reasons for and the climatic significance of such excess receipts of coastal radiation. It may be that the high values are specific to a few localities only. Observations near Aberystwyth, which is less than 50 km from Aberporth and only about 3 km from the coast, have revealed an attenuation of solar radiation similar to inland stations in southern England, thus suggesting that the Aberporth receipts may be confined to a very narrow coastal strip. In terms of causal factors, Monteith (1962) has pointed out that along coasts more incoming radiation is intercepted by natural fog than by atmospheric pollution and that the feature may be primarily attributed to the dominance of relatively clean maritime air. Whilst this could well be the case along western coasts when a westerly wind is blowing, or even in other areas under sea breeze conditions, Rees has reported that on clear, cloud-free days the coastal radiation values appear normal, implying that the supplementation occurs on partially overcast days in the mid-range of sunshine duration. An alternative theory has been advanced, therefore, that limited orographic cloud, which is induced at the land–sea boundary but which does not decrease the direct radiation component by obscuring the sun, could augment diffuse radiation receipts by downward reflection.

The above energy balance contrasts account for many features of the boundary climates located at a land–water interface, but other factors must also be taken into consideration. For example, the albedo of water surfaces is normally appreciably lower than that of land surfaces, even in wet littoral areas. This may be illustrated by reference to Table 11.1, which lists measured values of albedo obtained by Bendelow (1969) during low tide situations in the area around Morecambe Bay. It can be seen, however, that highly turbid water in a shallow sandy estuary has a higher value of albedo than deep water.

Table 11.1 *Measured values of mean albedo around Morecambe Bay, July to September (after Bendelow, 1969)*

	albedo(%)
Dry sand (dry surface)	35·4
Dry sand (damp surface)	17·0
Wet sand	15·9
Salt marsh	15·7
Water-covered areas at low tide	10·8
Deep water	8·0

Differences in surface roughness between a relatively smooth ocean or lake surface and the adjacent land area often produce variations in the speed and direction of surface winds, especially when these cross the shoreline. This effect is often compounded by differences in atmospheric stability over the land and the water. From a series of simultaneous wind measurements made at Gorleston on the East Anglian coast and at a lightship some 32 km offshore, Francis (1970) reported that, for offshore winds, the ratio between wind speeds over the sea and over the land (ratio $R = Us/Ul$) increased as the initial instability over the sea increased. This increase was most apparent with light land winds and it was also noted that, when initial conditions over the sea were stable, the wind backed, and when conditions were unstable the wind veered.

In practice, however, it is the onshore advective processes which assume the greatest importance. Coastal climates are most clearly identified when marine characteristics invade the land in the onshore advective situation of a sea breeze and they effectively cease to exist when the oceanic air has been fully modified by the underlying land surface. Conversely, little attention has been given to the reciprocal process of transformation in air passing from the land to the sea and in a comparatively short account such as this it will be necessary, after a brief discussion of coastal temperatures, to concentrate mainly on the onshore advective processes.

Temperature conditions

Thermal modification is one of the most publicized aspects of coastal climates and there is ample evidence to show that coastal areas experience a more equable temperature regime than equivalent localities inland. This feature occurs on all coasts, and the effect may possibly increase with latitude due in part to stronger advection associated with higher average wind speeds in the north and in part to the influence of comparatively high sea surface temperatures. In the latter context Table 11.2, based on data presented by Blench (1967) and Irvine (1968), reveals that the average range of seasonal temperature is lower in Shetland than in Jersey and that this is largely due to the relatively high winter values in Shetland. In turn, this may be related to the warming effect exerted by the winter ocean surface temperatures off Scotland. This warming is especially significant for night minima and during winter the mean nocturnal temperature in Orkney and Shetland is similar to that along the south coast of England.

Table 11.2 *Average range of seasonal temperature (°C), in coastal locations in the north and the south of the British Isles (after Blench, 1967, and Irvine, 1968)*

Station	Mean air temperature		Range	Mean sea surface temperature		Range
	Feb.	July		Jan.	July	
St. Helier (Jersey)	6·0	17·1	11·1	8·5	15·6	7·1
Lerwick (Shetland)	3·2	12·0	8·8	7·8	11·5	3·7

The diurnal range of temperature is also suppressed near the coast, as demonstrated by Bilham (1938) with respect to selected pairs of coastal and neighbouring inland stations. Although the mean annual values are often comparable, it was generally concluded that within some 50 km of the coast the mean daily range of temperature was likely to be at least 1 °C less than farther inland. The main problem with this type of generalization is that it assumes exact comparability between the two sites. In practice, this is virtually impossible to achieve and quite small variations in topography near the coast can often obscure much of the maritime influence. This is especially true in respect of the minimum temperatures when the ground rises inland from the coast and facilitates katabatic drainage at night. Thus, Manley (1944) reported that, despite a location only 1·5 km from the sea, the Bridlington site had a frost liability more or less equal to that of an inland station because it lay near the foot of a slope leading down from the Yorkshire Wolds. Similarly, Gregory (1964) has presented information which illustrates that, even in shallow open valleys on the Wirral peninsula, nocturnal inversions of up to 5·5 °C occur not infrequently between sites only 27 m different in height, about 1·2 km apart and less than 5 km from the coast. Other work on Wirral, drawing attention to variations between Bidston and West Kirby, has been undertaken by Reynolds (1956b), whilst the highly localized effects of topography on coastal temperatures have also been stressed by Smith, L. P. (1952) for western Wales, by Spence (1936) for eastern Scotland and by Ratsey (1973) for Somerset.

Some other coastal features

Apart from temperature, there are other characteristics of coastal climates which merit discussion. For example, the general breeziness of British shores is well known and, at sea-level, the highest maximum wind speeds are experienced on exposed west coasts. Equivalent values on the east coast are noticeably lower, so that the highest mean hourly wind speed with a 2 per cent annual probability is only 27 ms^{-1} along much of the east coast compared with 36 ms^{-1} in the Outer Hebrides (Shellard, 1962). Coastal winds are often subjected to localized steering by small-scale topographic features as exemplified for south-west Wales by Oliver (1960b).

There are important differences in the frequency of poor visibility between coastal and inland stations. In general, high wind speeds and low pollution levels along coasts contribute to a relatively low incidence of fog as shown in Fig. 11.1 which depicts comparatively the percentage frequency of fog (visibility less than 1 000 m) at an east coast station (Strubby) and at an inland station (Syerston) in the Trent valley. It can be seen that fog is mainly confined to the winter half-year at both stations when it is usually two to two-and-a-half times more frequent at Syerston than at Strubby. At the coastal station the incidence of fog during most daylight hours is less than 1 per cent.

Fig. 11.1 *Percentage frequency of fog (visibility <1 000 m) at an east coast station, Strubby, and at an inland station, Syerston (after Smith, F. J., 1967)*

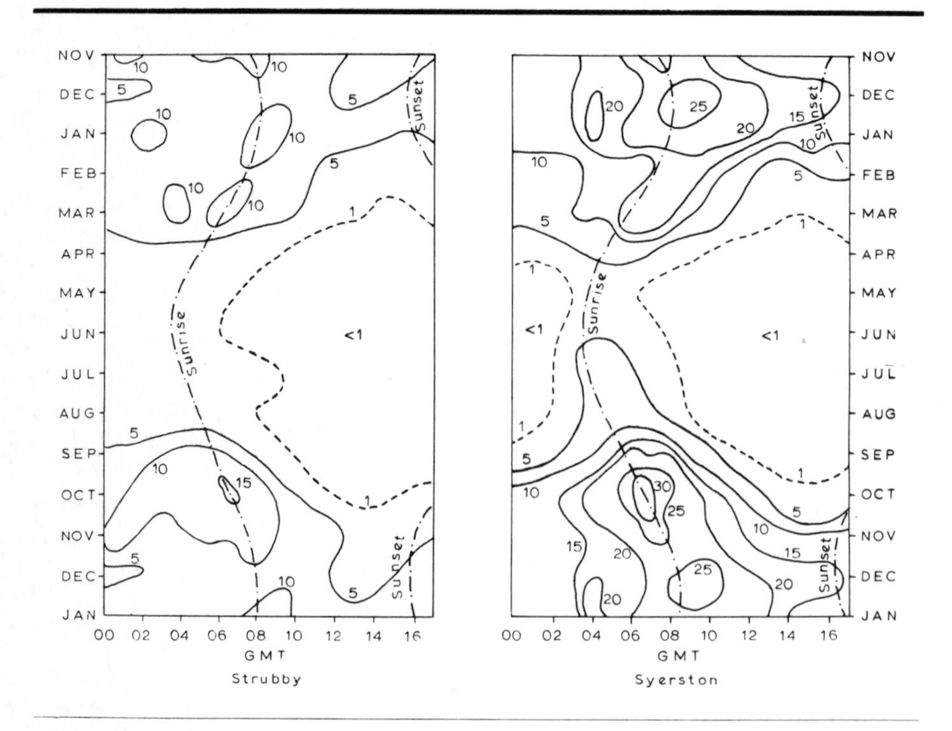

However, in contrast to inland areas, many coastal sites experience a relative increase in fog frequency during the spring and summer. Thus, in an investigation of poor afternoon visibilities throughout England and Wales, Smith, L. P. (1961) found that the south-western coastal stations from Portland Bill to Holyhead had a maximum frequency in March, the most likely time of the year for the creation of sea fog. Anticyclonic spells in summer have been associated with coastal fogs by various writers such as Ruck (1949) and Howe (1953) for Wales and Douglas (1960) for south Devon. At the Mount Batten station near Plymouth the seasonal variation in coastal fog-forming factors has been well identified by Saunders (1961). During summer most of the mist and fog occurred with moist south to south-south-west winds off the Atlantic whilst in winter poor visibility was associated with winds from between north-west and north-east which conveyed smoke aerosols from Plymouth itself.

Although smoke pollution is not normally a feature of coastal areas, the atmospheric chloride content is often high. As shown by Stevenson (1968) the greatest concentrations are measured in the winter months along north-western coasts owing to the general storminess of the sea which increases the spray content of the air and the salt content at cloud level. With strong onshore winds, appreciable quantities of marine salt can be deposited for up to 80 km inland as noted by Edlin (1957) for the south-east of England. Marine salt deposition in Wales has been reported by Edwards (1968) and Rutter and Edwards (1968).

The sea breeze

Probably the most characteristic feature of all coastal climates is the sea breeze mechanism which, in Britain, is found principally along southern and eastern coasts during spells of summer anticyclonic weather. Owing to relatively low radiation inputs and the more disturbed weather typical of the Westerlies, the British sea breeze is not usually so well developed as in lower latitudes and the objective identification of occurrences can be a difficult matter. However, in most cases a sea breeze event can be recognized when an excess of land temperature over sea temperature occurs before either a change in the direction of the surface gradient wind or an increase in the velocity of a light onshore gradient flow in the afternoon. Sometimes possible events have to be discounted owing to the existence of fairly strong pressure gradients or the passage of low pressure troughs.

Fig. 11.2 *Diurnal incidence of the sea breeze at Kinloss in relation to the times of sunrise and sunset between March and October (after Gill, 1968)*

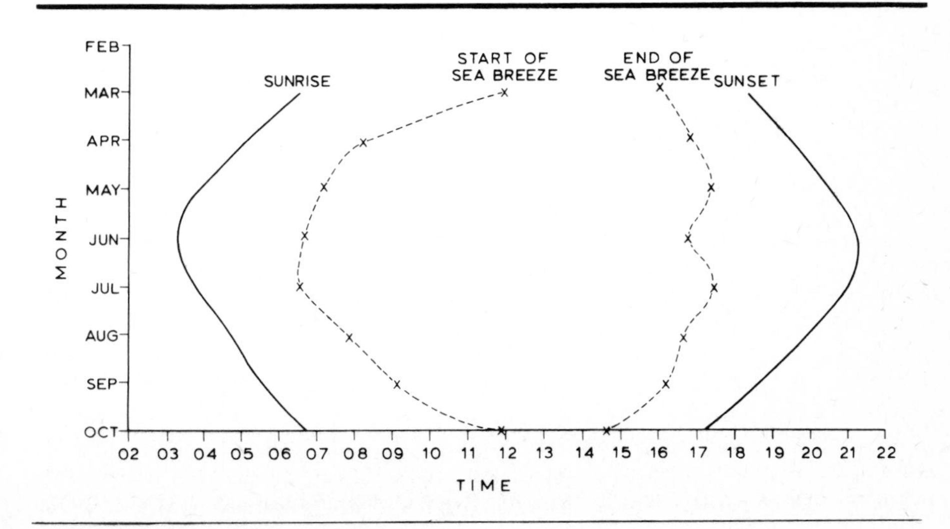

The initiation of the sea breeze is essentially dependent on differences between land and sea surface temperatures which develop on both seasonal and diurnal scales. Such contrasts emerge most strongly in high summer and, at Kinloss in north-east Scotland, Gill (1968) has shown that at the ends of the sea breeze season in March and October the local air circulations become noticeably weaker. As indicated in Fig. 11.2, on the days of occurrence, the onset of the sea breeze during the main season at Kinloss follows regularly some 3 to 4 hours after sunrise, with the lag increasing to 6 hours in March and October. The period between the end of the sea breeze and sunset was 2 to 3 hours, except in June when it increased to 4 hours.

Sea breeze events occur intermittently throughout every summer and the actual incidence varies considerably from place to place and year to year. In some localities it has been suggested that the sea breeze is sufficiently regular to promote a diurnal reversal in mean wind direction during the summer months. One of the most remarkable of such examples must be that of Manchester Airport (Ringway), over 50 km from the open sea. According to Crowe (1962), the record for this station confirms the existence of an extensive, two-way diurnal exchange of air between land and sea which may be attributed to the combined effects of land–sea breezes and hill–valley circulations from the Pennines. On the other hand, specific sea breeze investigations nearer the coast have tended to reveal a low frequency of occurrence. At Porton Down, some 40 km from the south coast of England, Elliott (1964) identified only forty-seven clear cases involving the arrival of a sea breeze front between March and October over a 6-year period. This study indicated that most of the breezes lasted from one to three hours and there is little evidence to show that an entire land–sea breeze reversal can be traced regularly throughout a 24-hour period. Thus, Stevenson (1961) reported that, in a sea breeze investigation actually on the Yorkshire coast during the exceptionally anticyclonic summer of 1959, a clear-cut diurnal wind shift was detected on only 13 days between May and September although the air temperature over the land was often 8 °–16 °C higher than that over the sea. However, many more days had a sea breeze 'tendency' marked by an increased velocity in the onshore wind in the warmer hours of the day.

Sufficient case studies have now been undertaken along the south coast to permit the evolution of sea breeze forecasting models which are normally based on the maximum expected excess land temperature plus a prediction of the speed and direction of upper level winds. At Porton Down most sea breezes occurred with high pressure to the north and east with midday winds around 900 m less than 5 ms^{-1}. With noon winds less than 5 ms^{-1}, it was found that a sea breeze could be expected when the inland screen temperature rose by 10.00 hrs to that of the mean monthly sea temperature in the vicinity of the Isle of Wight and continued to rise through the day. At Thorney Island, just east of Portsmouth, Watts (1955) concluded that a 5·5 °C temperature excess of land over sea was more than enough to initiate a sea breeze but considerable emphasis was laid on the need to obtain representative sea surface temperature data. This precaution was

Fig. 11.3 *Arrival of the sea breeze front at Porton Down on 24 July 1959 (after Elliott, 1964)*

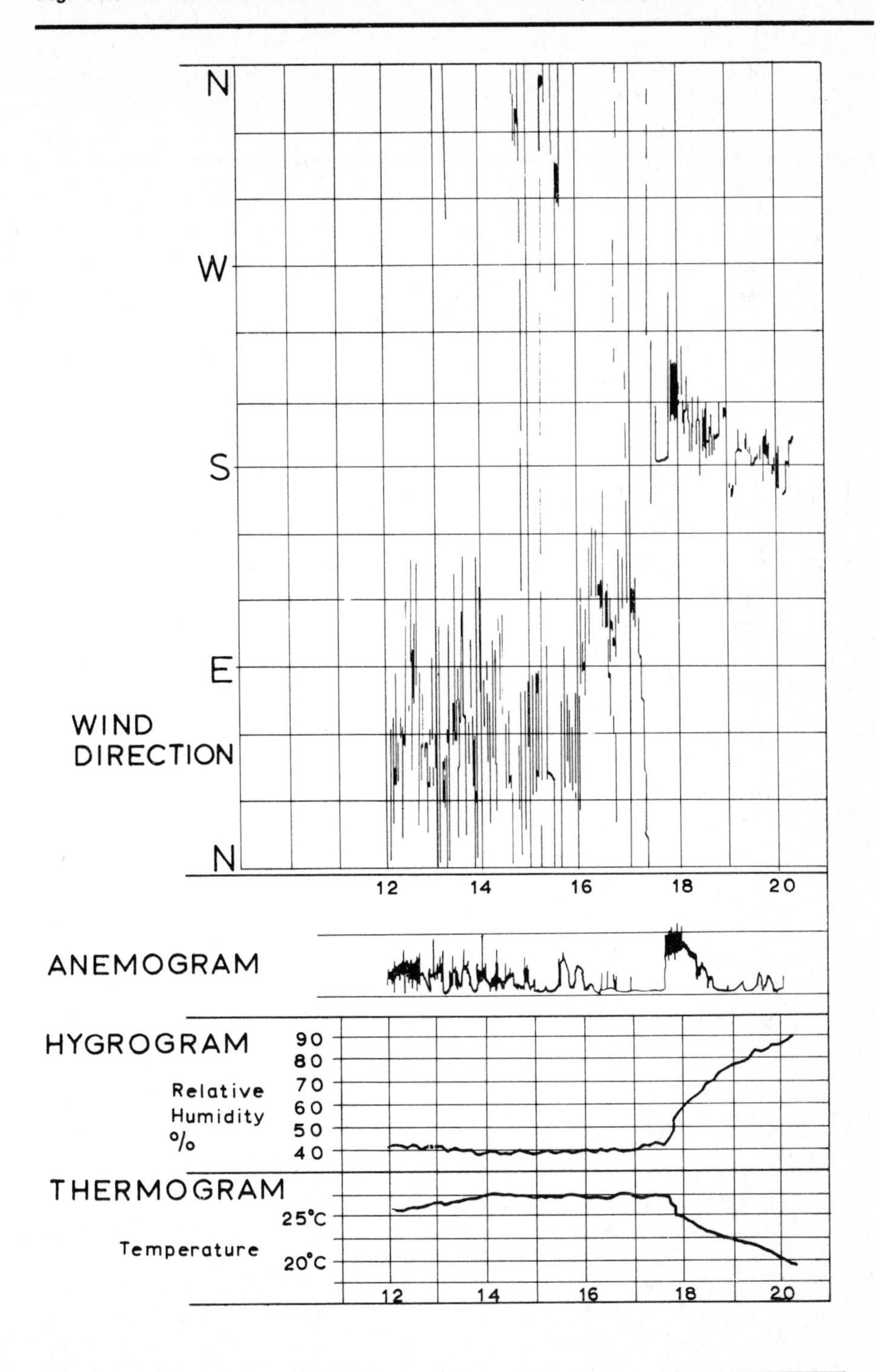

found to be particularly relevant in the surrounding coastal inlets where shallow water overlying tidal mud or sand flats could be quickly warmed or cooled leading to temperature differences of 2·8°–5·5°C between inshore temperatures and offshore published means.

Fig. 11.4 *Average isochrones for 16 sea breeze fronts which passed Lasham in 1962–3 (after Simpson, 1964). Thermo-hygrograph stations are given in CAPITALS*

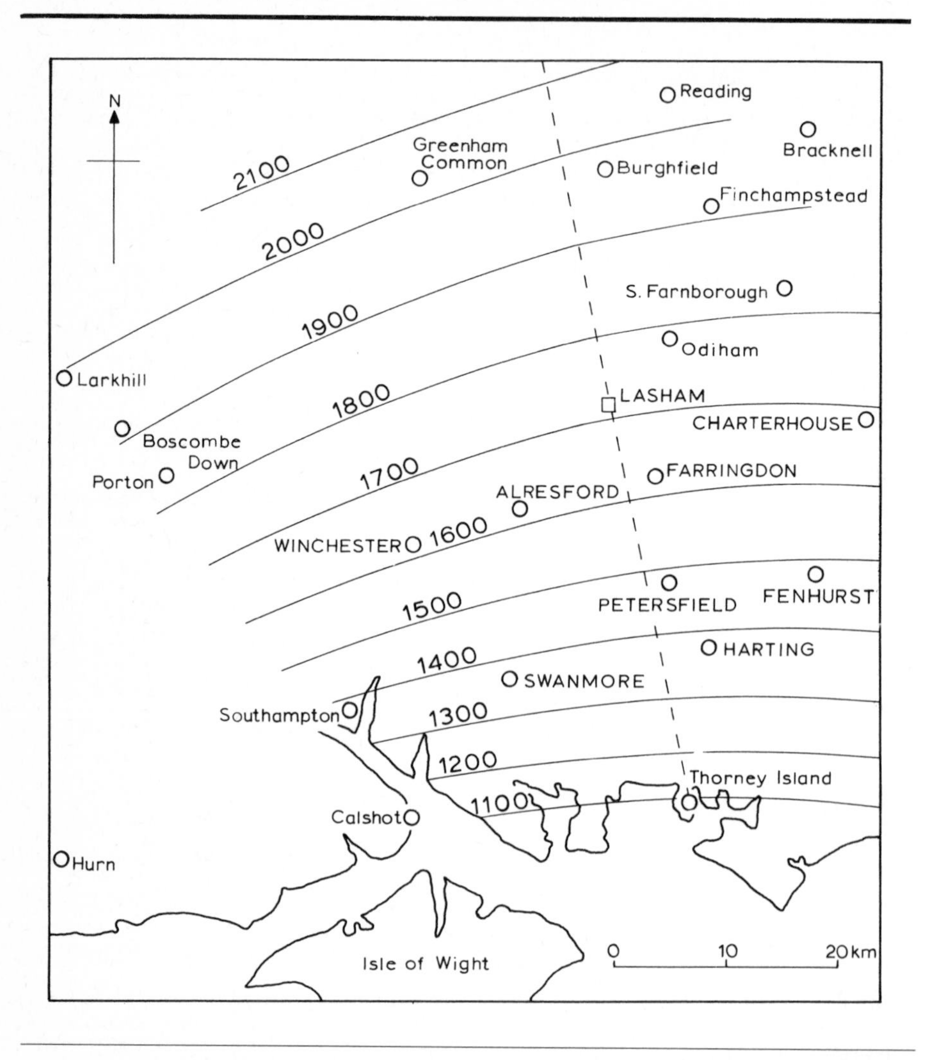

Although the sea breeze mechanism is usually discussed in isolation, it may sometimes be related to other air flows. In some coastal areas, the nocturnal land breeze component may be reinforced by katabatic flows, as implied for Ringway. At Thorney Island the presence of the South Downs provides the basis for such a combined offshore wind (Moffitt, 1956). This wind was found to be most apparent on clear nights, when over two-thirds of the occurrences arrived between 2 and 4 hours after sunset, whilst 90 per cent of all incidents were detected when the minimum temperature fell by around 8·5 °C from the daytime maximum. Conversely, Findlater (1964) has described an example of convective storms developing in south-east England in association with a landward-moving sea breeze and the intersection of a sea breeze front with an outflow of cold air from an area of thundery rain has been detailed by Rider and Simpson (1968).

Findlater (1963) has shown that British sea-breeze circulations may extend beyond 150 m above sea-level and spread well out to sea, whilst the inland penetration of the marine air produces effects similar to the passage of a cold front. This can be seen in Fig. 11.3 which illustrates that the arrival of the front near the south coast causes a marked change in wind direction, accompanied by an increase in wind speed, a rapid fall in temperature and a rise in relative humidity. Apart from the use of standard meteorological observations, the inland travel of sea breeze fronts can also be followed by means of gliding flights, radar and satellite sensing, as reported respectively by Wallington (1959), Simpson (1967) and Rowles (1972). A good deal is now known about the rate and extent of inland progress from the south coast and Fig. 11.4 depicts average GMT isochrones for sixteen sea breeze fronts which passed Lasham in 1962–3 (Simpson, 1964). It can be seen that the sea breeze front, which typically extended from at least Larkhill in the west to Gatwick in the east, was at Thorney Island on the coast at 11.00 hrs and was north of Reading by 21.00 hrs. Most of the fronts showed a gradual increase in their rate of inland progress from 1·5 ms^{-1} or less near the coast to about 4·0 ms^{-1} in the later stages. Although 65 km was found to be the normal limit of inland penetration from the south coast, Smith, M. F. (1974) has documented a sea breeze front which was tracked for almost 160 km across East Anglia against a 5 ms^{-1} westerly component to a height of over 1 500 m.

The haar (sea fret or roke)

During late spring and early summer, the east coast of Britain experiences variable spells of onshore advection when a light easterly wind with considerable surface stability is driven across the North Sea by the development of high pressure over Scandinavia. This air is cooled by the sea to a temperature below its dew point and large scale condensation occurs bringing sea fog and low stratus, often accompanied by drizzle, across the coast. The net result is a distinctive local climate along the coastal strip northwards from the Wash, with the main effects found north of the Humber estuary.

Despite an overriding preoccupation with Scottish conditions, the most comprehensive account of the haar is still that presented by Lamb (1943), who recognized six forms of occurrence, including some associated with the sea breeze. The importance of haar along the north-east coast is largely due to the presence immediately offshore of the coldest patch of water for its latitude in the North Sea. In this narrow tongue of water isothermal conditions prevail which have been attributed to either a cold tidestream flowing from the north or the upwelling of water along the coast. The low sea surface temperatures, plus the very long length of fetch across the North Sea at this point, create efficient chilling at the base of the airmass, especially since it is during May and June that the sea reaches its lowest temperature relative to the land. Indeed, temperatures are so low during summer that, with an unstable north-north-westerly airstream along the east coast of England, normal convection cloud development is inhibited according to Hindley (1972).

Any easterly airflow has a high moisture content arising from its ocean track and Dixon (1939) made the point that, unlike other fogs, haar can exist in winds up to 9 ms^{-1}. Consequently, on the east coast, easterly winds are frequently accompanied by low stratus as demonstrated in Table 11.3 for Leuchars, in Scotland. It can be seen that stratus was most frequent in an easterly situation between March and August, and most of the cloud was associated with persistent easterly flows rather than more transient weather types.

Table 11.3 *Monthly and annual percentage of easterly winds with and without stratus at Leuchars based on hourly wind data for the 14-year period 1950–63 (after Alexander, 1965)*

	J	F	M	A	M	J	J	A	S	O	N	D	Year
Without	17	19	27	25	31	25	21	20	18	16	17	14	21
With	3	6	13	8	14	14	13	12	9	8	7	3	9

Monthly or annual number of hourly occurrences averaged over 1950–63 and expressed as a per cent of total number of hourly observations in each month or year
'Easterly' means with 'E' component, i.e. 010° to 170°
Stratus' means $\frac{1}{8}$ cloud or more, lower than 300 m

The climatic implications of the haar are chiefly confined to a fairly narrow coastal strip since the cloud base tends to lift inland during the day as a result of insolation warming the land. However, along the coast, both maximum temperature and sunshine duration are reduced and the haar is largely responsible for the fact that, between the Humber and the Tweed, lies the only stretch of the east coast which enjoys less sunshine than comparable latitudes on the west coast. Pratt (1968) examined haar outbreaks in north-east England, such as on 31 May and 1 June 1967 when on both days all coastal stations from Berwick to Bridlington, together with Cockle Park and Durham, recorded no sunshine. On the west coast, on the other hand, the sunshine duration totals for the 2 days were appreciably higher than average – Morecambe 20·2 hours, Blackpool 28 hours and Douglas (I.O.M.) 30·3 hours.

Table 11.4 details the typical conditions existing on 15 June 1967 when, from the Durham and Cockle Park values, it can be seen that a partial decay of the haar took place as a result of surface heating between 10 km and 20 km inland.

Table 11.4 *Haar conditions on 15 June 1967 (after Pratt, 1968)*

Station	Sunshine Duration (hr)		Maximum Temperature (°C)	
	15 June	Monthly mean	15 June	Monthly mean
Tynemouth	0·2	6·1	14·4	15·6
Hartlepool	2·6	5·9	14·4	17·2
Cockle Park	2·9	6·0	17·2	17·2
Durham	6·1	6·0	17·2	17·8
Dumfries	11·9	6·1	22·2	17·8
Morecambe	12·4	6·9	22·2	17·8

The advance or decay of the haar is very dependent on local conditions. For example, valleys aligned at right angles to the coast facilitate deep inland advances and, following the Central Valley, the Scottish haar may penetrate as far west as Glasgow. More locally, Alexander (1964) has drawn attention to the effects of inshore temperatures around the Eden estuary in eastern Scotland. Here the haar may recede with the cooler waters as the ebbing tide exposes estuarine sands and muds which then warm up fairly rapidly. The poor visibility similarly re-advances with the next tide. In East Anglia, Sparks (1962) has emphasized the importance of falling temperatures inland in controlling the spread of the haar. For a landward advance to occur the coastal temperature must fall below the clearance temperature which, under the existing conditions of onshore airflow, is the surface value which is just sufficient to evaporate the leading edge of the stratus sheet as it is continuously formed offshore. If temperatures inland should fall below this clearance threshold, as often happens during the evening or the night, then the haar is advected inland at the speed of the gradient wind at its own level. Freeman (1962) has described such an event when, after an afternoon temperature difference of over 8 °C between coastal and inland stations, the progressive fall of temperatures over East Anglia in the evening permitted an advance of the haar from the Wash coast at 18.00 hrs to 00.01 hrs inland. The generally low level of atmospheric pollution near coasts means that virtually all of the condensation and poor visibility associated with the haar occurs quite naturally. However, evidence presented for the heavily-industrialized area of Teesside by Garland (1969) and Eggleton (1969) suggests that man-made aerosols may well be a contributory factor in certain localities.

Inland water bodies

Very little work has so far been undertaken in this country on the local atmospheric modifications resulting from the presence of inland water bodies such as lakes, reservoirs and rivers. Most of the active participants in this broad field of study have, in fact, been fresh-water biologists for whom the thermal condition of a water body is often the primary physical factor controlling the distribution and well-being of aquatic communities. Consequently, whilst there are numerous studies of temperature variations within inland water bodies, there have been extremely few investigations which have considered either the climatic implications of such variations or even the role of the atmosphere in influencing the thermal variations themselves. This account must, necessarily, reflect the balance of published work but it should be stressed that, although the direct climatic impact of inland water bodies tends to be both small and intermittent, the theme offers considerable scope for the local climatologist.

In general it may be stated that, because of the larger surface area and the greater capacity for heat storage arising from the volume of impounded water, the local climatic modification potential of lakes and reservoirs is greater than that of most rivers. Certainly, as the size of the water body increases, it is less likely to conform to the broad regional characteristics of the surrounding area. In some instances the thermal lag will operate on a seasonal scale and, when Gorham (1958) analysed the temperature observations obtained during the bathymetric survey of the Scottish lochs, he found that, although a maximum temperature was reached in June in the smaller lochs, the maxima did not occur until September in the large ones. In addition to a thermal lag, large water bodies also exhibit a low amplitude of temperature variation. Even so, Jenkin (1942) has reported that the maximum temperature recorded in Lake Windermere over a 6-year period reached 23·1 °C, although the record was from Wray Castle harbour and therefore probably slightly higher than the maximum near the centre of the lake.

With the approach of winter the surface water cools progressively, becoming denser than the other layers, and then begins to sink from the surface to be replaced by warmer water from below. By the autumn, isothermal conditions exist due to this vertical exchange of water but, during the winter, water cooled to 4 °C (the temperature of maximum density) begins to accumulate at the bottom of the lake or reservoir. This produces an inverse thermal stratification with colder but less dense water near the surface but, even during the coldest winters, the temperature difference between surface and bottom water cannot exceed the 4 °C represented by freezing conditions near the surface and maximum density conditions at the bottom.

During spring, another overturn of water occurs as the surface temperatures warm up to the 4 °C threshold and the resulting convective sinking re-establishes isothermal conditions at this season. As the surface temperatures continue to rise a well-defined and highly stable summer stratification usually develops which is resistant to mechanical disruption by wind or other agents, such as the flow of water through a reservoir system. The warm, low-density water is now concentrated in a shallow surface layer known as the *epilimnion* and Thompson (1954) has asserted that, if the surface temperature reaches 20 °C, the stratification is likely to be extremely stable under British conditions. The base of the epilimnion is defined by an increasingly distinct *thermocline* as the temperature discrepancy between the top and bottom waters increases through the summer. Below the thermocline most of the water exists in a largely stagnant state in the *hypolimnion*, which is cut off from wind action and sunlight by the layers above. Here temperatures probably remain at around 7 °C and, at more than 30 m below top water level, temperatures will hardly ever rise above some 13 °C.

Fig. 11.5 *The effect of a direct-supply reservoir on water temperatures in the Lune valley during a dry anticyclonic spell in mid-June 1969 (after Lavis and Smith, 1972). (A) temperature of inlet water; (B) near-surface temperature of reservoir; (C) temperature of outlet water*

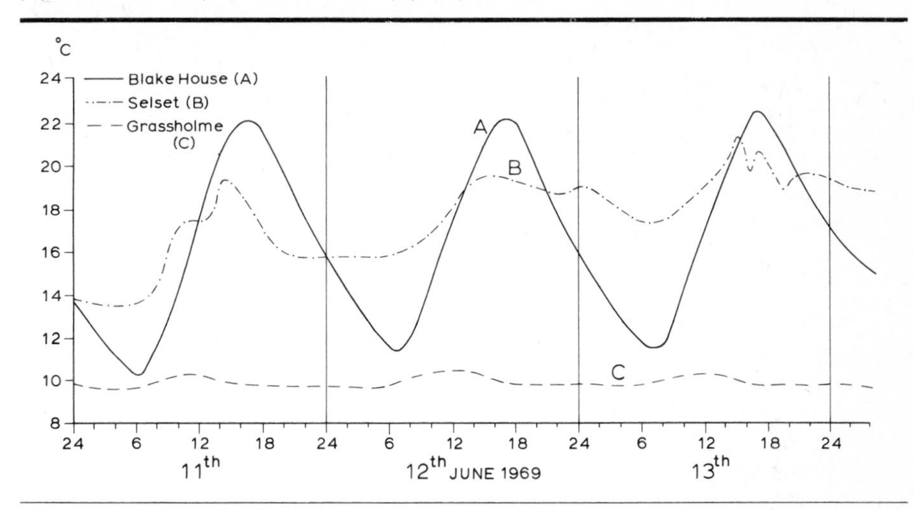

It has been noted by Lavis and Smith (1972) that summer stratification can produce large changes in downstream river temperatures below water supply reservoirs when compensation water is drawn off at depth from within the hypolimnion. Figure 11.5 shows water temperatures recorded over a dry, anticyclonic period in mid-June around the Selset and Grassholme reservoirs on the River Lune in northern England. A stable stratification existed with surface reservoir temperatures (Selset) rising to about 20 °C, whilst the temperature of the compensation water (Grassholme), drawn off at 28·3 m below top water level, remained around 10 °C. In contrast, the temperature of the River Lune immediately above the reservoir complex (Blake House) showed a large regular diurnal cycle with maximum values reaching 22 °C and a daily range of 10°–12 °C. The principal effect of the reservoir system was to suppress the natural range of variability of stream water temperature and ten day means for inlet and outlet river temperatures differed by about 4 °C during the warmest and coldest parts of the year. During summer daily maxima downstream of the reservoir were depressed by up to 12 °C.

As far as local atmospheric modifications arising from lakes or reservoirs are concerned, Smith, I. R. (1973) has examined the wind structure at three sites around Loch Leven in central Scotland, whilst Sheppard *et al.* (1972) and Garratt (1972) have initiated a sophisticated micro-scale experiment in order to produce vertical profiles of wind speed, potential temperature and humidity up to a height of 16 m above Lough Neagh in Northern Ireland. In England, Gregory and

Fig. 11.6 *Hourly variations in air temperature, relative humidity and vapour pressure between the northern and southern edges of Selset reservoir for a period of persistent northerly winds from 13.00 hrs, 31 August to 09.00 hrs, 2 September 1965 (after Gregory and Smith, 1967)*

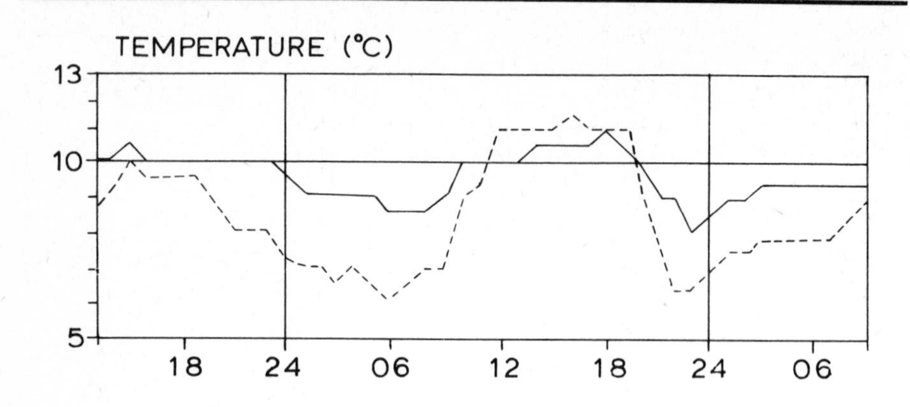

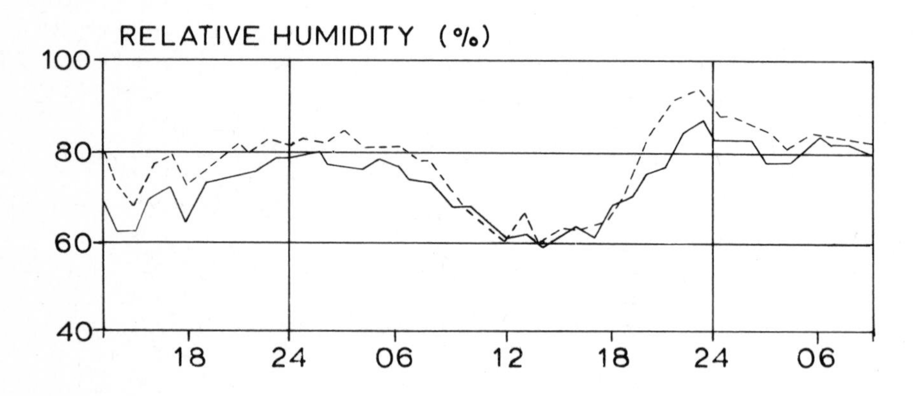

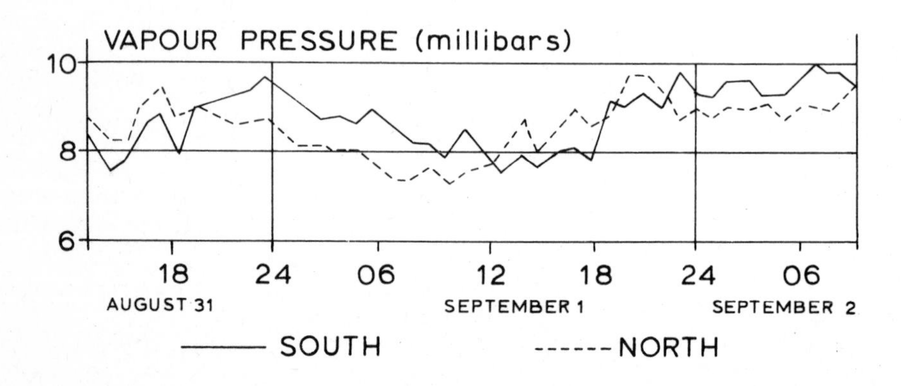

Smith (1967) reported on the advective influences resulting from winds blowing across Selset reservoir during a two-month period in late summer. Owing to difficulties involved in finding fully comparable observing sites around the reservoir edge, the analysis had to be confined to periods with either northerly or southerly winds, and for over 43 per cent of the time it was found that the temperature difference between windward and leeward stations was less than 0·5 °C. On no occasion was the leeward station more than 3 °C warmer or cooler than the site to windward. Nevertheless, a diurnal cycle of reservoir influence, showing daytime cooling and night-time warming downwind, was observed on some occasions. Figure 11.6 illustrates the variations in temperature, relative humidity and vapour pressure during a period of persistent northerly winds between 13.00 hrs on 31 August and 09.00 hrs on 2 September. It can be seen that the diurnal range of temperature for the leeward (northerly) station is only 3·3 °C (7·8 ° to 11·1 °) as opposed to the windward range of 5·8 °C (5·6 ° to 11·4 °), thus indicating a recognizable moderation of temperatures across the water surface. The vapour pressure values show a clearer diurnal variation than those for relative humidity. From 19.00 hrs on 31 August to 10.00 hrs on 1 September, and from 22.00 hrs on 1 September to 09.00 hrs on 2 September, moisture is taken up as air crosses the reservoir. From mid-morning until late evening, however, on both 31 August and 1 September, the air over the leeward (southern) station was drier than that over the one to windward.

Smith, K. (1972) has reviewed the progress of research into river water temperatures and has shown that, in Britain, there has been a remarkable concentration of effort on thermal variations in small upland streams in northern England as evidenced by the work of Macan (1958), Edington (1966) and Crisp and Le Cren (1970). In all these studies the influence of the atmospheric environment has been implicitly rather than explicitly stated, although more recent work by Smith and Lavis (1975) has included measurements designed to relate stream water temperatures to both climatic and hydrologic variables. As in the case of lakes and reservoirs, water volume is very important in determining the nature and timing of the river's thermal response to changing atmospheric conditions (Smith, K., 1968). In the lower reaches of some industrial rivers the temperatures may be artificially raised by waste heat rejection from power stations and other sources as shown by, amongst others, Gameson *et al.* (1957, 1959) and Langford and Aston (1972).

12 Upland Climates

J. A. Taylor

Inadequacy of Primary Data

The horizontal dimensions of the climate of the British Isles have been regularly monitored and are reasonably well understood. The altitudinal component, in contrast, has never been comprehensively studied and is less well understood, despite the sharper control that changes in relief exert upon gradients of climate. Descriptions of the British upland climate must perforce be probablistic at best since the meteorological data available are so localized and fragmentary. In Britain the location and distribution of operational meteorological stations (January 1974) shows concentrations at low elevations, and near the coasts to the particular neglect of the uplands (Fig. 12.1). This is continuing testimony to the extreme paucity of primary information on our upland climates which necessitates the substitution of crude and unavoidable upslope extrapolation from lowland stations. The pioneer records of the Ben Nevis observatory at 1 343 m for 1883–1904 (Buchan and Omond, 1905–10; Paton, 1954) still constitute the longest continuous source of primary data. Manley (1936, 1938, 1942, 1943 and 1952) assembled data for Moor House, Westmorland (561 m) from 1931 (see *Monthly Weather Report* from 1953), and also compiled a record for Dun Fell, Westmorland (857 m) for the period 1937–41. Dybeck and Green (1955) provided additional short-term data for the Cairngorms, whilst Tagg (1957) has collated wind data for a number of upland locations in Britain. For upland Wales, Smith, L. P. (1950, 1952), Oliver (1960a, 1961, 1964), Harrison (1973, 1974) and Job (1974) have assembled and analysed climatic data for a number of altitudinal transects. From 1967 the Institute of Hydrology (Rodda, 1971) has established the most intensive network of hydrological and meteorological stations in the world in the upper Severn and Wye catchments just eastward of Plynlimon. The investigation is to continue for some years and will clearly provide a major data source of the future. The highest station (Eisteddfa Gurig) lies just above 610 m. on the slopes of Plynlimon.

The Ben Nevis observatory was one of a number of mountain top sites adopted for meteorological purposes in the later nineteenth century, e.g. Mt Blanc and Sonnblick in the European Alps (Talman, 1934), Mt Washington and Pikes Peak (4 301 m) in the United States of America (Stone, 1934), and Mt Wellington and Kosciusko in Australia (Paton, 1954). They were mostly disestablished in the early twentieth century but by the 1930s the requirements of aviation and the introduction of radiosonde techniques promoted the collection of 'upper air' data, thus replacing the mountain observatories for some purposes. But 'free air' climates are not strictly comparable with adjacent mountain-top or slope-climates of identical altitude for which purposes high-altitude surface observations are still needed.

Fig. 12.1 *Frequency of meteorological stations in Highland and Lowland Britain operative in January 1974, in relation to altitude and distance from the sea*

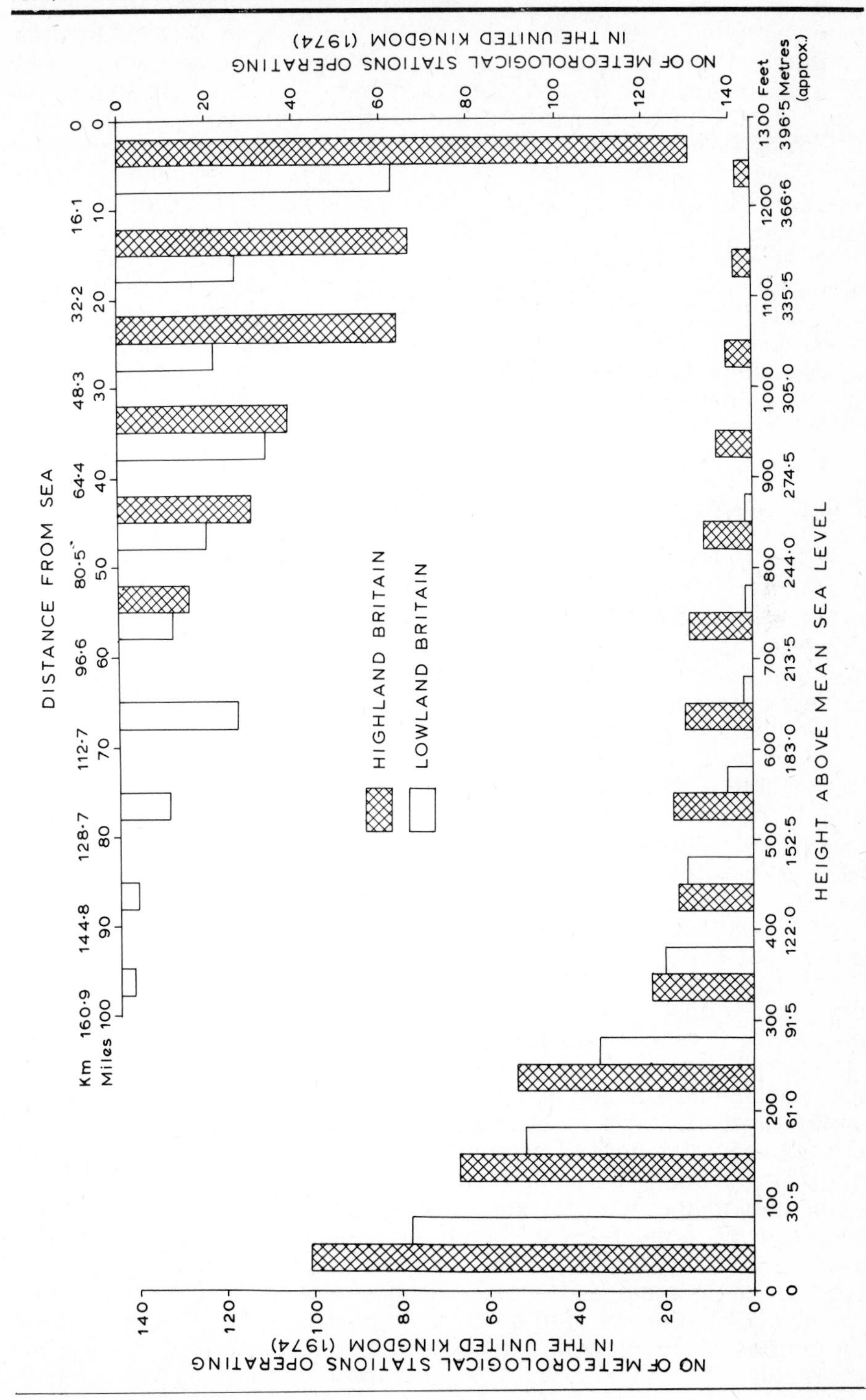

Samson (1965) compared slope and free-air observations at Pikes Peak, concluding that the former gave a good indication of the latter. Coulter (1967), reviewing New Zealand mountain climates, confirmed that free-air data could provide a general basis for the study of adjacent surface climates, but advocated the *in situ* measurement of particular mountain climates because of the great diversity imposed by the complexity of relief. Ingham's (1967) work, based on stations in north-west England, including Dun Fell (857 m), implied a similar variation of scale when free air and hill station data are compared. The contradictory evidence emerging from parallel investigations in Norway (Eide, 1942), the Soviet Union (Bordovskaja, 1963) and Mexico (Hastenrath, 1963) is also due to differences in climatic type and data reference. However, to attempt an analysis of the long-term deviations of radio-sonde data from comparable records for high altitude stations in Britain would produce a generalized model of the British upland climate as related simply to altitude or less simply to terrain geometry.

Basic problems of interpretation

Ekhart (1948) and Geiger (1965) have suggested that mountains not only modify the weather and climate prevailing around them but may also generate their own discrete air layers (Fig. 12.2). The British uplands, however, would appear to be too low, narrow and discontinuous, even by European standards, to modify weather systems in depth or generate their own particular climate with any regularity. Again, the type of synoptic situation most conducive to the development of separate topoclimatic zones, i.e. the persistent anticyclone with stable air and light winds, is relatively infrequent. At the same time it should be emphasized that in the British Isles the temperature lapse-rates with altitude are among the sharpest in the world (Manley, 1970b). Consequently, the British upland climates may be clearly, if variably, differentiated from their lowland counterparts at least on the regular basis of temperature and the immediate effect it has on the growing season.

Furthermore, conditions below Stevenson screen level are often in marked contrast to the prevailing macroclimatic regime, and may dictate the mesoclimatic zonations which commonly develop in upland terrain (Harrison, 1973, 1974). The radiation, heat and moisture balances are all affected by slope, aspect, vegetation, soil, peat cover, snow cover etc., and by topographically-induced local winds. Thus the representativeness of screen data for high-altitude sites must be constrained, more particularly for bioclimatological reference, notwithstanding the need for standardisation of instrumental exposure. The use of conventional meteorological instruments also raises problems at these sites which may suffer from great extremes of weather and climate, as the pioneer observers first discovered on Ben Nevis, and as the Institute of Hydrology is currently discovering on the slopes of Plynlimon with sophisticated as well as simple instruments.

Fig. 12.2 (a) *Model of a 'mountain atmosphere' (after Ekhart, 1948)*; (b) *Diagrammatic representation of slope temperatures (after Geiger, 1965)*

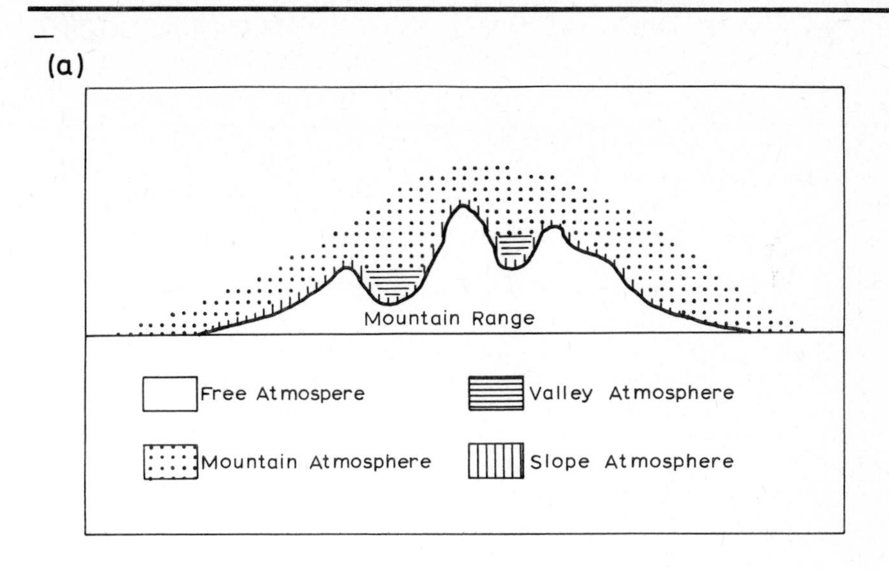

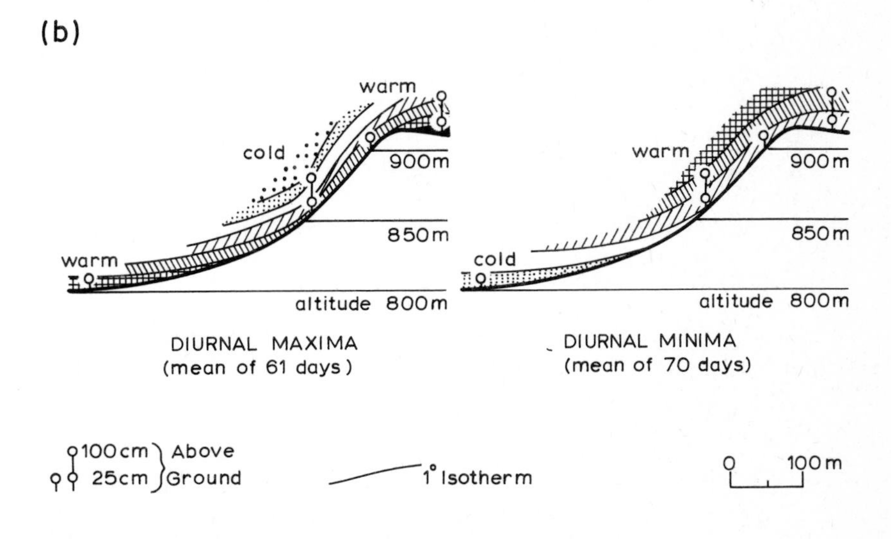

The extent to which data for a given high-altitude observatory are representative of its precise elevation will vary with topography and exposure, which induce 'station errors'. Even when symmetrical relief transects are available, which is rare, it is dangerous to assume that derived climatic gradients will be linear in any sense. Lapse rates are liable to variation and reversal with location, aspect, slope, proximity of the sea (eastern or western), air mass or weather type, time of day or night, stage of season or year, etc. They are therefore subject to extreme variability over short distances and brief time periods. This is not to deny, however, the general utility of any average or standard lapse rates of temperature, rainfall or the other elements, provided they are not abused.

Fundamental to the climate of the British uplands is their general emplacement in the north and west of the British Isles, with the major divides virtually overlooking the western seas across which the majority of our weather systems come. The relatively high sea temperatures off the west and south-west coasts as compared with the eastern coasts (particularly in winter and spring) and the high frequency of polar-maritime air masses (with inherently large temperature lapse rates) are factors which promote a rapid fall of temperature with altitude on our relatively steep, exposed, maritime westerly slopes. Again, much heat, especially from the characteristically warm tropical-maritime air masses, is advected atmospherically and oceanically to our western littorals, bestowing a mild winter and early spring at low altitudes. However, temperatures fall as the air rises over the uplands and temperature lapse rates are generally substantial. Large cloud amounts, high humidities and heavy precipitation result in regular water surpluses. Downslope movements into the rain-shadowed eastern lowlands involve warming, increased evaporation and periodic moisture deficiencies in summer.

Had the British Isles been entirely lowland, the maritime factor, i.e. the relative distance from western and eastern seaboards, would have exerted a broad longitudinal control on the climate. The patterns of average daily maximum and minimum temperatures, reduced to sea-level (Figs 5.3, 5.4, 5.6 and 5.7), afford some indication of the hypothetical contrasts that would have existed between the sharp thermal gradients to the west and the gentler ones to the east, more particularly in winter.

Harrison (1973) examined mean temperature data (1931–60) for seventy stations for a west–east transect of varied relief across England and Wales between 50° and 55° north. He concluded that altitude was the vastly more significant factor in temperature variation than either distance from the west coast or latitude, both of which played very minor roles. This contrast was also shown in a study of temperature data for the slope transect between the coast of Cardigan Bay and the Plynlimon plateaux where altitude was responsible for more than 98 per cent of the temperature gradient. In the east of the country, however, only locally does altitude enhance the effect of distance from the North Sea on temperature gradients, e.g. in the North York Moors (Dimbleby, 1953) and the Lammermuir Hills (Parry, 1975), etc.

The effect of the interposition of the British uplands in the immediate path of eastward-tracking weather systems is most pronounced: (*a*) where there is a rapid and large change in elevation; (*b*) where such a change lies at right angles to the predominant winds; and (*c*) where the upland blocks are compact and extensive, and attain their highest elevations.

Altitudinal modification of air flows

General modifications

In the lower atmosphere, wind speeds are reduced by the friction of the earth's surface in a layer which may extend up to 500 m–2 000 m under British conditions and varies greatly with the synoptic situation. However, the greater the surface roughness and the higher the altitude, the deeper and more distorted the friction layer becomes, and the greater the wind speed.

The effect of a mountain barrier on airflow is lateral deflection round its extremities and elevation over it. Wallington (1961a) has shown, however, that air may sometimes descend over mountains. The passage over the summit concentrates the airflow and increases wind speed. To the lee of the upland, eddies may form and possibly also a small orographic depression. With wind speeds of about 20 kt, stationary eddies develop especially to leeward but also to windward of the uplands. At 25 kt and above, turbulence is greatly increased and mobile frictional eddies form extensively on the lee slopes where the wind direction may be the reverse of that at higher levels. The upward displacement of air flow as it negotiates an upland block is occasionally followed downstream by a series of 'standing waves' depending upon temperature and wind profiles. These waves are frequently marked by lenticular clouds which may very occasionally occur at great altitudes beyond the tropopause (Scorer, 1949, 1951, 1953, 1954, 1955; Corby, 1954, 1957; Nicholls, 1973). These features are typical of those parts of the British uplands lying immediately east of the north–south watersheds – or with easterly winds, of those zones lying just to the west (see Manley's classic study (1945b) of the 'helm wind' of the northern Pennines).

Synoptic variations

The uplift of frontal structures over mountains causes increased cloud and rain. The lower sections of cold fronts may be deformed especially when the front parallels the long axis of the mountain barrier. This may happen on a grand scale over the Alps and Pyrenees (Godske *et al.*, 1957) but locally it has been observed to happen in north-west Wales where cold fronts may decelerate and be slow to clear. Occluding depressions as they pass over the British uplands undergo marked structural changes and additional mixing of air which leads to some of the heaviest and most intense rainfalls, especially when the depression is slow-moving.

When a potentially unstable air mass rises upslope, instability is triggered, often resulting in deep clouds and heavy precipitation. The inherently sharp temperature lapse rates in polar maritime air and the steep western slopes of upland Britain accentuate this process. Lee slopes are less showery as the air descends and tends to become stable. Tropical maritime air, characterized by stability and low-level stratiform cloud, undergoes a lowering of cloud base and often gives light precipitation as it ascends the British uplands. Lee areas, in contrast, have substantially higher cloud bases and less rainfall. Polar

Fig. 12.3 *The relationships between slope and valley circulations over a period of 24 hours (after Defant, 1949)*

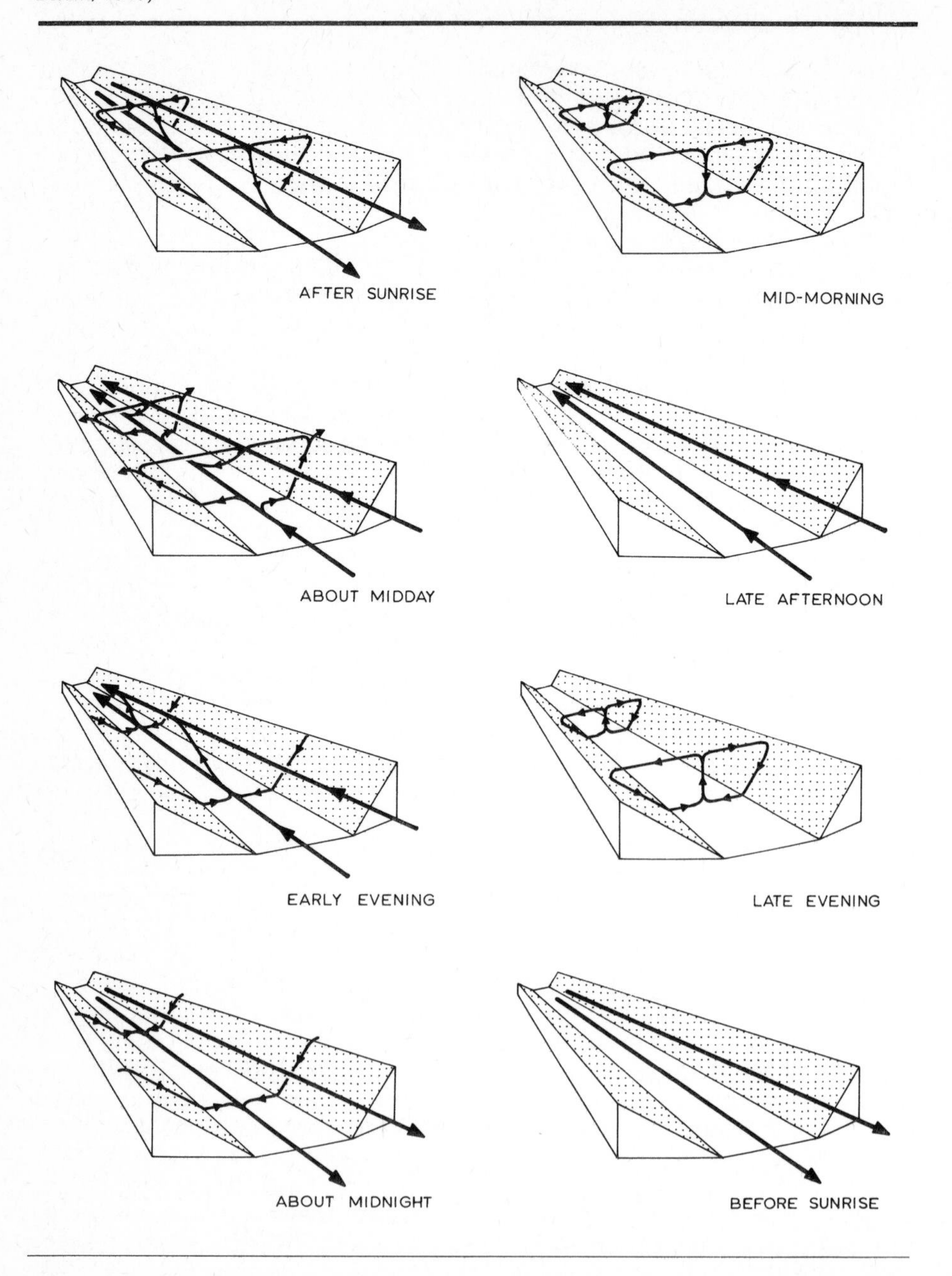

continental air arriving across the North Sea often brings low cloud, mist and rain to eastern uplands, e.g. the North York Moors, the eastern Pennines, the Lammermuir Hills and the eastern Grampians, in particular the Cairngorms; in winter, heavy snowfalls may occur. Downslope winds in western Britain may cause exceptional desiccation because of their temperatures and low humidities. Tropical continental air from the south-east is normally warm and dry, and especially so following its western descent from the slopes of the major British uplands. Minor cloud development may mark its ascent of the uplands.

In anticyclonic weather, wind speeds are generally reduced and air flow over mountains is less disturbed. Anticyclonic inversions, the high frequency of which in general was noted by Oliver (1960a, 1964), resulting in high radiation, sunshine and temperature values at high altitudes, while lower levels suffer from lower minimum temperatures and more frequent ground frosts and radiation fog, perhaps smog, especially in winter.

Thermal processes

Differential heating and cooling of contrasted elevations, slopes and aspects may modify local airflows especially in deeply cut valleys or where there is shading from direct insolation for long periods of the day. Where upland valleys open out towards the sea (e.g. in western Scotland, western Wales and western Ireland), thermally generated land and sea breezes may enhance or counter topographically-induced winds developed in the valleys.

Large-scale convection is more characteristic, however, of the lowland climate, as is convectional rainfall. In summer, however, exceptional convectional storms of some severity may develop in upland areas. Eddies set up by convection may be difficult to distinguish from those due to friction. The latter may initiate or accentuate turbulence in the uplands but, micro-climatically, convectional flows may be induced by contrasting ground conditions, especially variations in the albedo and thermal diffusivity of different types of vegetation, land-use, soil or rock exposure. Interfaces which, *ceteris paribus*, tend to be relatively warm by day, thus generating rising air, include bare rock; dry, coarse soil; sand, gravel, scree or rock debris; arable or cleared land; patches of land without snow cover; and dry ground or dry vegetation. In contrast, surfaces which tend to be relatively cool by day and not to generate rising air include vegetated surfaces; wet, heavy, peaty soils; grassland and woodland; patches of land with snow cover; and wet ground or wet vegetation. The relative disposition of these features will induce marked temperature variations in the upland environment both diurnally and seasonally.

Thus a variety of upslope (anabatic) and downslope (katabatic) winds of considerable velocity may be generated by orographic, frictional, gravitational and thermal factors (Fig. 12.3), often in combination (see Chapter 13). They are a recurrent local feature of the British upland climate. Clearly, vertical temperature gradients are very variable in both space and time in upland areas, especially near the ground.

Altitudinal climate gradients

It is understandable that the British upland environment has been likened to the 'tundra' by more than one European scientist (Manley,

1952). The contrasting amplitude of maritime and continental-type temperature curves and their seasonal variations with altitude have been discussed elsewhere (Manley, 1945a; Gloyne, 1958; Taylor, 1961). Figure 12.4 illustrates the sharp abbreviation of the length and the weakening of the intensity of the 'growing season', and the complementary prolongation and strengthening of winter, which occur with increasing elevation at the maritime site. This applies most forcefully in the highest British uplands. Changes with altitude are more gradual at the continental site where intensity compensates for shortness of season at all altitudes.

Fig. 12.4 *Variation in the amplitudeof the annual curves for mean monthly temperature with increasing altitude, at a maritime and a continental site: a—lowland stations; b—upland stations; shaded areas indicate appropriate growth potential; x and y indicate length of growing season*

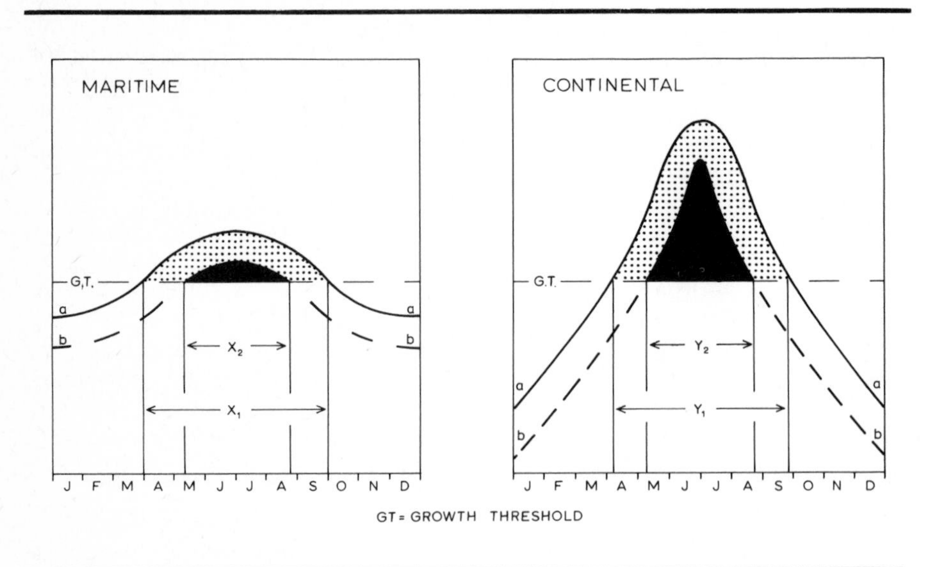

Two additional elements accentuate the rate of climatic deterioration in the British uplands: first, windiness which increases general exposure and second, wetness including heavy and prolonged precipitation, high cloud amounts and humidities, and wetness of soil, especially when peaty, and of vegetation especially when compact. Because of fairly persistent cloud, radiation and sunshine inputs at the ground are reduced. Upland climates are thus rendered marginal since diurnal weekly and monthly temperature curves commonly oscillate (*a*) across the 6·0 °C growth threshold especially in spring and autumn, thus causing intermittency in growth and insecurity of harvest, and (*b*) across the 0 °C freezing threshold in winter (and at night throughout the year), e.g. at a basin site at Tanllwyth (350 m) on Plynlimon during the period from June 1973 to June 1974 when a grass minimum frost temperature was recorded for each and every month of the year (Smart, 1974).

Radiation

Radiation data are available for only short-term periods for British stations. The best available records for an upland station are for Eskdalemuir in southern Scotland, which is located at 242 m (Table 12.1). Day (1961) has constructed very generalized maps, based on short-period data, for the British Isles area and Taylor and Smith (1961) have calculated estimates of radiation and illumination for eighty-six stations, based on sunshine and visibility, but the vast majority of these stations are of low elevation and are located outside the British upland areas.

As Hogg (1971) observes, net radiation values decrease northwards across Britain, notwithstanding the longer summer days in the north. They also decrease inland from the coast and especially so with increasing altitude because of the excessive cloud amounts and reduced visibilities of the uplands which otherwise would receive (and in clear, fine weather spells do receive) more net radiation than the lowlands. These horizontal spatial variations apply to direct radiation (insolation) in particular although the mean annual values for diffuse radiation (that received under cloudy or similar conditions) exceed those for direct radiation at all stations.

Table 12.1 *Radiation data for selected stations – monthly and annual mean values of total radiation on a horizontal surface in cal cm^{-2} day^{-1}, and the percentage ratio of direct/total radiation*

		J	F	M	A	M	J	J	A	S	O	N	D	Year
Lerwick	82 m													
Total		24	70	140	268	347	397	352	277	184	86	32	15	183
% Direct		25	31	36	41	38	38	34	34	38	32	27	18	36
Eskdalemuir	242 m													
Total		47	94	159	266	361	384	340	277	205	116	56	34	195
% Direct		35	34	36	39	42	41	35	35	39	38	37	33	38
Aberporth	133 m													
Total		63	114	227	334	453	496	429	377	272	153	73	51	253
% Direct		34	36	42	45	48	50	43	46	47	41	33	36	45

Note: The Aberporth data are incomplete for 1957 and 1958.

Both Borisov (1959) and Barry and Chorley (1968) quote a standard radiation increase with height of 5–15 per cent (mean 10 per cent) per 1 000 m. Theoretically, this means that the potential radiation inputs at 610 m (cloudiness, etc., apart) could be 6·0 per cent greater, and on Ben Nevis (1 343 m) 13·2 per cent higher, than at sea-level. Preliminary analysis of radiation data for high-altitude stations at the head of the Wye catchment on Plynlimon supports the view that radiation inputs at, say, 400 m and above, are at least as high in fine summer spells as at stations in lowland England. Like the 'tundra' regime, the intensity of summer-season radiation receipts compensates for its brevity, a feature which would be evident in accumulated temperature data for upland stations, were they more abundant (Gregory, 1954; Shellard, 1959). This advantage could clearly apply during clear, fine spells especially near the summer solstice and during the longer daylight periods operating particularly in northern Britain. Unfortunately, this illumination maximum occurs too early for vegetation growth in the uplands where ground conditions delay warming until late summer, more especially in wet, peaty areas (Taylor, 1965), by which time the days are shortening especially in northern Britain.

Barometric pressure

Table 12.2 gives details of pressure and density for the first 1 524 m of Standard Atmosphere. At 1 524 m on the international and theoretical average, pressures are 17 per cent lower and densities 13·8 per cent lower than at sea-level. The vertical gradients of pressure and density are of course subject to variation depending on the alternation of cyclonic and anticyclonic systems and air mass types.

Table 12.2 *Data for the 0 to 1 524 m of the International Standard Atmosphere (source: Meteorological Office, 1971)*

Height (m)	Pressure (mb)	Relative density (per cent)
1 524	843·1	86·2
610	942·1	94·3
0	1 013·25	100·0

Wind

Gloyne (1967) has demonstrated how windforce becomes greater and strong winds more persistent with altitude. Pearsall (1950) and Pears (1967) quote individual data for Ben Nevis and the Cairngorms respectively, and Shellard (1962) and Hardman *et al.* (1973) have examined the pattern of extreme wind frequencies and gusting for recent periods.

The general increase in windiness and turbulence with altitude accentuates the biological impact of temperature lapse rates. Gloyne (1967) indicated that for altitudes up to 305 m wind speeds are approximately comparable to those on the coast but at 610 m, wind speeds are similar to those over the open sea. Table 12.3 gives available data for some high-altitude stations in Britain, with some indication of the scale of increasing wind force with altitude. Pearsall (1950) noted that data for Ben Nevis over a period of 13 years showed an annual average of 261 gales of force greater than 50 mph (43 kt). This contrasts with a figure of forty gales at sea-level. Wind stress alone on these terms is very severe at high altitudes and is a major constraint on vegetation development, especially tree-growth, and also on the efficiency and productivity of animal husbandry (Gloyne, 1957; Caborn, 1957; Roberts, 1972), particularly when strong winds are combined with driving rain, drifting snow, cold, raw air or warm, dry air (which may rapidly desiccate upland grazings).

Fig. 12.5 *Various temperature lapse rates.* (a) *for continental and maritime Europe;* (b) *seasonal variation in lapse rate (Ben Nevis)*

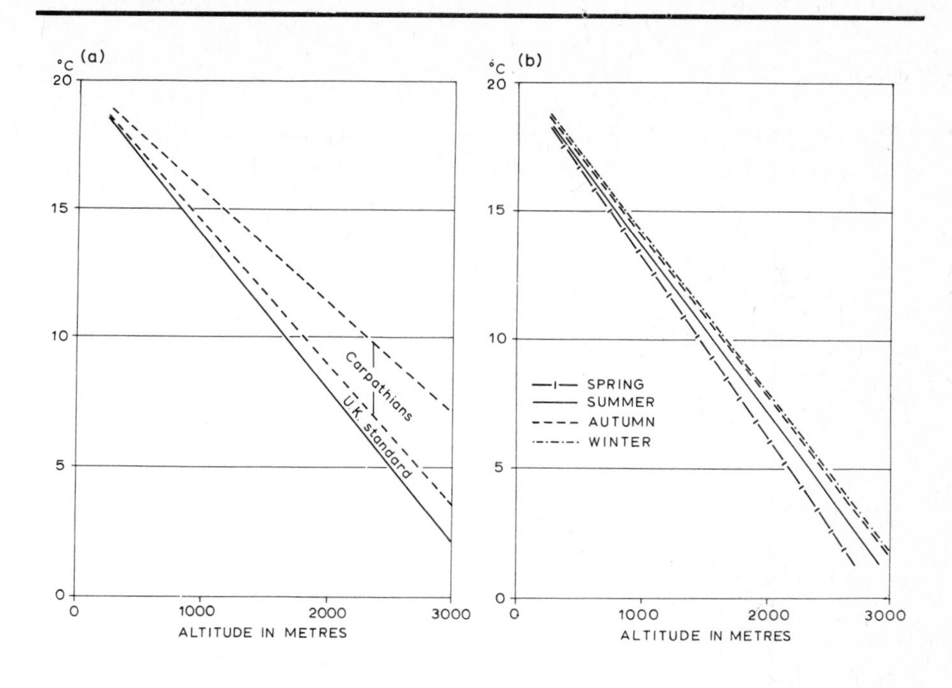

Temperature

Figure 12.5 shows a selection of seasonal temperature lapse rates for Ben Nevis, as well as a comparison of UK conditions with those for the Carpathians (Hess, 1968). The diagrams, often based upon the records of only two terminal stations, assume that temperature lapse rates are linear with height, which is clearly open to question. Lapse rates vary with latitude, longitude, altitude, aspect (Birse, 1971; Taylor, 1974a), type of air mass, time of day, season and year, etc. Much depends too on the level at which temperatures are taken. The use of Stevenson screen data alone must be criticized. The representativeness of a screen observational height of 1·25 m cannot be regarded as constant with either altitude or latitude. Thermometer mounts, exposed at for example 20 cm above ground, or soil temperatures, are often more appropriate measures of the vegetation or grazing environment of the uplands (Alcock *et al.*, 1974; Harrison 1973, 1974). A further and quite regular source of error is the close control exercised by the local topography at and near the stations used. Harrison (1973, 1974) has demonstrated siting errors and discussed the problem of the extent to which stations are truly representative of the specific altitudes of their sites (Fig. 12.6).

The Meteorological Office has adopted a standard lapse rate of 6·0 °C per 1 000 m rise in elevation for mean temperatures. For maximum temperatures, however, 7·0 °C is used and for minimum temperatures, 5·0 °C per 1 000 m. These rates are slightly greater than

Table 12.3 *Percentage frequency of winds at selected stations (source: Meteorological Office)*

Mean wind speed				**Great Dun Fell (857 m)**	**Eskdalemuir (242 m)**	**Rannoch (284 m)**
Force	*Knots*	*mph*	*ms⁻¹*	%	%	%
0	0	0	0	2·2	13·9	18·09
1	1–3	1–3	0·3–1·5	1·4	17·4	
2	4–6	4–7	1·6–3·3	4·3	18·9	27·16
3	7–10	8–12	3·4–5·4	13·9	20·7	
4	11–16	13–18	5·5–7·9	25·0	18·7	33·73
5	17–21	19–24	8·0–10·7	19·1	6·6	
6	22–27	25–31	10·8–13·8	17·3	3·0	5·83
7	28–33	32–38	13·9–17·1	9·4	0·6	
8	34–40	39–46	17·2–20·7	5·2	0·1	0·29
9	41–47	47–54	20·8–24·4	1·7	0·0+	
10	48–55	55–63	24·5–28·4	0·362	0·0+	
11	56–63	64–72	28·5–32·6	0·069	—	Defective 14·90
12	>63	>72	>32·6	0·034	—	
				Total (*per cent*) 100·00	100·00	100·00

Fig. 12.6 *Accumulated deviations of observed weekly values of:* (a) *maximum air temperature (°C);* (b) *minimum air temperature (°C);* (c) *precipitation (mm); from predicted values derived from linear regressions of each element respectively on altitude*

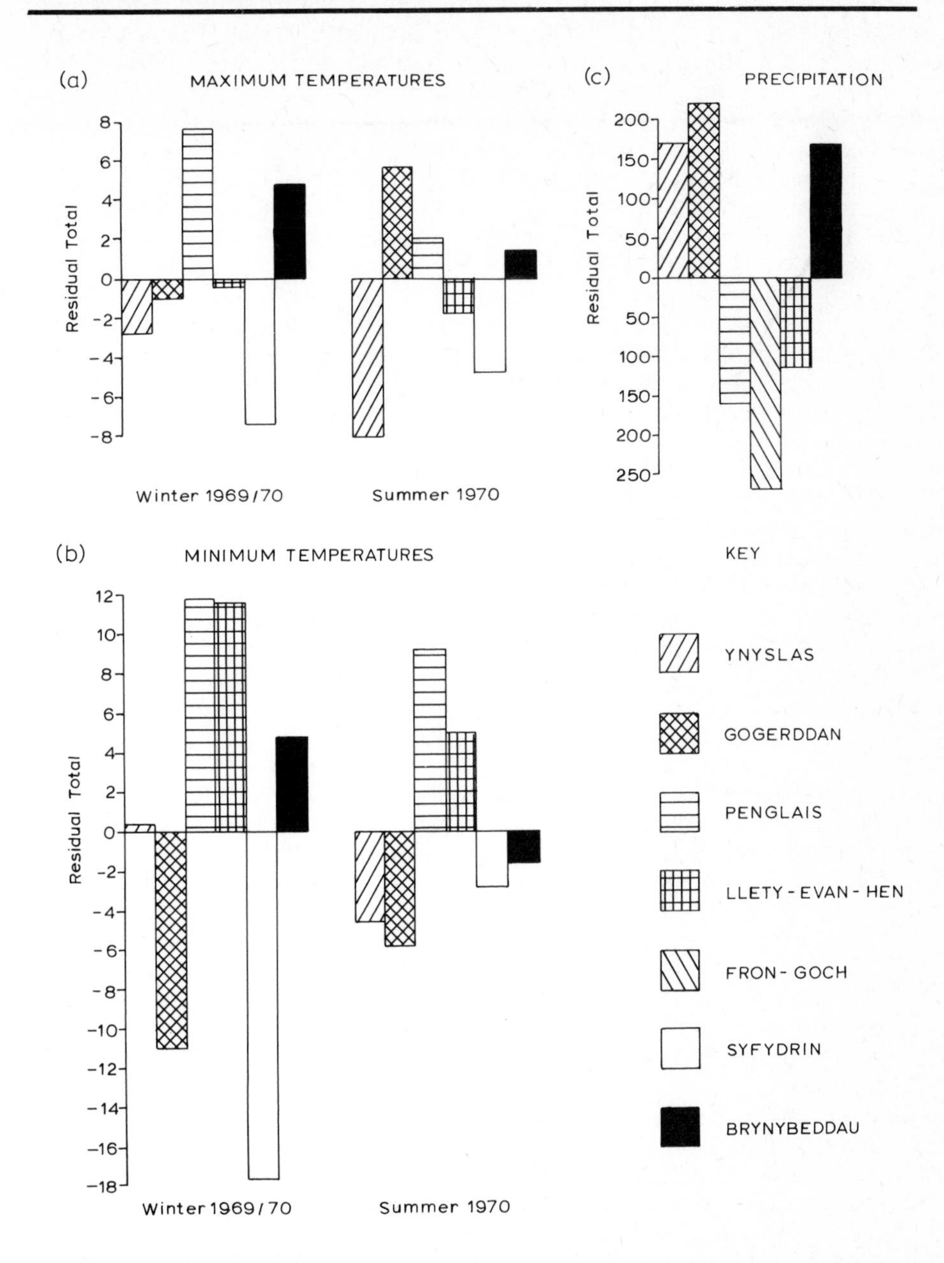

those for free saturated air : the saturated adiabatic lapse rate (S.A.L.R.) ranges from 4·0 °C to 7·0 °C with median occurrences approximating to 5·0 °C per 1 000 m. The dry adiabatic lapse rate for dry, unsaturated air (which is relatively less frequent over Britain, on the whole, than saturated air) is much greater, being 10 °C per 1 000 m. Regional or local lapse rates are subject to variation. Manley (1943) estimated a rate of 6·9 °C for the northern Pennines whilst Smith, L. P. (1950) derived a rate of 6·7 °C for the Aberystwyth hinterland and Oliver (1960a, 1964) a rate of 7·3 °C for part of South Wales. Previously the data for the Ben Nevis observatory had yielded a rate of 6·4 °C per 1 000 m. It is impossible to detect areal or regional pattern in these rates which are too few in number and based on linear extrapolation between two or, at most, three stations.

The evidence for seasonal fluctuations in temperature lapse rates is varied and conflicting. Manley (1943), Pearsall (1950) and Oliver (1960a, 1964) diagnosed that the greatest lapse rates occurred in spring; Harrison (1973, 1974), however, for the Aberystwyth–Plynlimon slope transect, discovered that lapse rates were least in spring and at a maximum in autumn and winter. Shaw (1955), admittedly using captive balloons to obtain data for free air above Cardington, Bedfordshire, discovered that the sharpest lapse rates up to 609 m occurred in summer. These apparent contradictions suggest great variability over time and space in temperature lapse rates which only future large-scale research can clarify and resolve.

In calm, clear weather, night-time temperature lapse rates near the ground tend to be negative especially in areas of concave relief. Cold air drains from the sloping ground above which is warmer and relatively frostfree. At the highest altitudes, e.g. above 600 m, minimum temperatures fall below freezing on most winter nights. The average annual frequency of days with a minimum screen temperature of 0 °C or less (1913–40) ranges from about 50 days on the edges of the British uplands to 100 and more over the highest ground (Meteorological Office 1952). Manley (1970b) reports a figure of 131 days (1956–65) for Moor House at 561 m in the Pennines. The length of the frost-free season decreases generally with altitude but the gradient is more sharply expressed in terms of surface and near surface temperatures as illustrated in Table 12.4. At screen level the reduction in 1969 amounted to only 2 weeks and in 1970, 4 weeks. The grass minima, however, reveal a loss of 9 weeks and 13 weeks, respectively, which is equivalent to a loss of 53 per cent and 63 per cent. Equally well shown is the reduction in the length of frost-free season at the ground as compared with the screen. Reductions at Ynyslas are of the order of a half (1969) and a third (1970); at Brynbeddau the corresponding figures are three-quarters (1969) and two-thirds (1970). Also noteworthy is the increase with altitude in the occurrence of glazed frost.

Available evidence suggests the following general properties about temperature lapse rates in Britain. Firstly, they are among the most rapid in the world. Secondly, the rate for maximum temperatures is the greatest, implying cool afternoons and summers in upland areas thus limiting growth potential. It follows that diurnal temperature ranges at screen level are reduced with altitude but the progressive confinement of the heat exchange to shallow zones in the upper layers of the soil preserves relatively low minima and high maxima at the ground interface more particularly in dry, sunny weather. Thirdly, although the minimum temperature lapse rate is smaller, partly due to the effects of inversions, the lowest minima of all must occur in deep, peaty hollows at the highest elevations. Fourthly, as on all 'late' land, a heat storage effect could maintain some reserve of soil warmth in July, August and September especially in the drier, warmer slopes (Taylor, 1965, 1967; Job, 1974). Fifthly, as Manley (1951) has discussed, the variability of temperature in upland Britain is greater than in lowland Britain, even more so with increasing altitude and diversity of relief. This can be advantageous in making 'early' and 'late' grazings available over short distances for the upland stock farmer but in spring and autumn it spreads the climatic marginality of many cold or exposed slopes and widens the risk and impact of weather extremes. Sixthly, the sensible temperatures, so far as plants and animals are concerned, are frequently lower than the air temperatures because of the general increase in windforce and gustiness with increasing altitude. Strong winds in combination with driving rain, snow, low temperatures or very high or very low humidities, constitute a less imprecise definition of the vague notion of excessive 'climatic exposure' in the British uplands.

Table 12.4 *Length of frost-free season at Ynyslas and Brynbeddau (after Harrison, 1973).*

Station and Year	Instrumentation	Last frost (temp. less than 0°C.) Date of week beginning	First frost (temp. less than 0°C.) Date of week beginning	Length of frost-free season in weeks	Vertical difference in weeks between Ss and Gm
Ynyslas (3 m)					
1969	Ss	24 March	24 Nov.	35	18
	Ctm	19 May	6 Oct.	20	
	Gm	2 June	29 Sept.	17	
1970	Ss	6 April	14 Dec.	36	12
	Ctm	27 April	16 Nov.	29	
	Gm	25 May	9 Nov.	24	
Brynbeddau (450 m)					
1969	Ss	31 March	18 Nov.	33	25
	Ctm	2 June	29 Sept.	17	
	Gm	30 June	25 Aug	8	
1970	Ss	27 April	16 Nov.	29	20
	Ctm	27 April	26 Oct.	26	
	Gm	25 May	24 Aug.	9	

Ss = Stevensons screen. Ctm = cocoa-tin mounted 20 cm above ground. Gm = grass minimum.

Seventhly, the greater the frequency of polar-maritime air, the greater the average lapse rate, *ceteris paribus*: but this factor operates periodically and locally rather than seasonally and regionally (Manley, 1942; Harrison, 1973, 1974). Lastly, temperature lapse rates are probably exponential rather than linear, and liable to reversals because of their sensitivity to variations in the local topography of slopes.

Humidity and cloud

Increases in humidity and cloud occur with increasing elevation, contributing to the general wetness of the upland environment. Radiation inputs are much reduced as a consequence and diurnal and seasonal warming are seriously delayed, often well after the diurnal and seasonal maxima of illumination have occurred, all detracting from growth potential. Thus, ground wetness, as aggravated by high atmospheric humidities and abundant peaty top-soils, must reduce surface and air temperature ranges in the British uplands. Vapour pressure increases according to Smith, L. P. (1952) at the average rate of 0·467 mb per 100 m of elevation away from the immediate vicinity of the coast at Aberystwyth, and at least up to 290 m. The proportion of the sky covered by low cloud (base below 2 000 m) increases gradually from south-east to north-west, on the average, across the British Isles. In the vicinity of the uplands the increase is sharpened, however, especially on upper west-facing slopes.

Rainfall

Since Salter's (1921) pioneer study of orographic rainfall gradients, a number of specific local studies have indicated, with one exception (Harrison, 1973), that British average rainfall gradients are approximately linear with altitude. (Table 12.5).

Table 12.5 *Rainfall gradients with altitude, in millimetres of rainfall per 100 m rise of altitude*

Source		Area	General gradient	West slope gradient	East slope gradient
Pearsall	(1950)	Central Pennines	—	188	98
Gloyne	(1958)	Scottish Highlands	—	253	83
Rodda	(1962)	Ystwyth catchment N. Ceredigion	—	167	—
Unwin	(1969)	Snowdonia	458	—	—
Harrison	(1973)	Slope–Plynlimon to Cardigan Bay	—	228	—

Preliminary analysis of rainfall data to the east of Plynlimon suggests a linear relationship between rainfall and altitude (Fig. 12.7*a*), while in contrast, Harrison (1973) indicates that rainfall gradients, at least for one maritime slope analysed over a 2-year period, may be curvilinear with rainfall increasing exponentially with altitude (Fig. 12.7*b*).

Occasional convectional rainfall may help to create deviations between isohyets and contours (Smithson, 1969a), but it is such a minor component of upland rainfall regimes that mean gradients with altitude are scarcely affected at all. Gloyne (1968) has calculated that whilst the eastern coastlands of Scotland have an average annual duration of rainfall of 500 hours, the north-west Highlands have as

Fig. 12.7 (a) *Rainfall and altitude relationships in the Upper Wye and Severn catchments, 1970–4 (private communication from M. D. Newson, Institute of Hydrology);* (b) *Relationship between precipitation total for 1970 and altitude for coastal west Wales (after Harrison, 1973)*

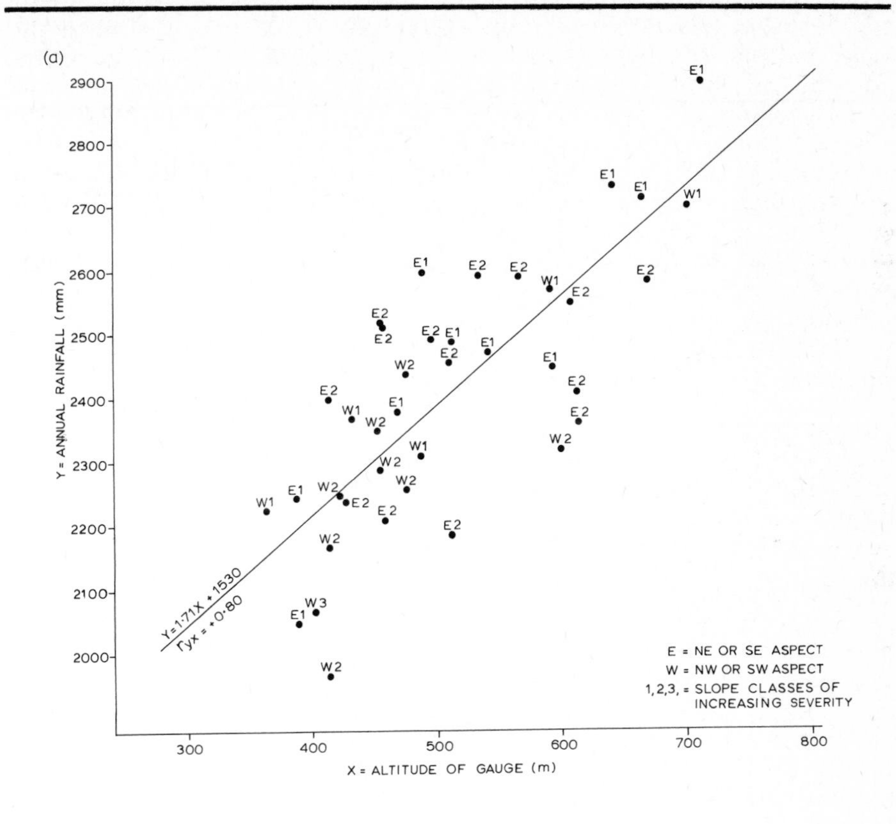

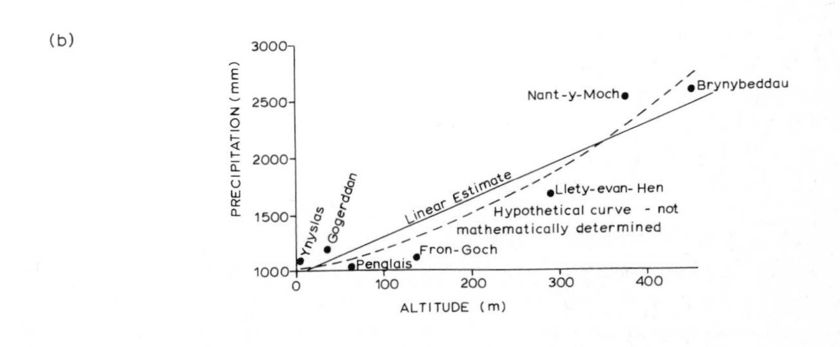

much as 2 000 hours or more. The total average annual rainfalls for the two areas differ by a factor of nine, suggesting that intensity of rainfall increases with altitude. Whilst it is true that rainfall gradients on west-facing slopes are generally greater than on east-facing slopes in Britain (Salter 1921; Pearsall 1950; Gloyne 1958), it should be remembered that west slopes are generally steeper and shorter themselves than the gentler and longer easterly ones. This helps to explain the substantially higher rainfalls normally found on west slopes than at comparable altitudes on east slopes, as many isohyetal maps demonstrate. Maximum rainfalls commonly occur just to the east of major watersheds but without doubt the wettest parts of the British uplands are those where deep valleys are open to the convergence of wet, westerly air streams and penetrate well inland towards the highest peaks, e.g. the Dovey estuary in Wales.

Rodda (1967) has examined the pattern of intense rainfalls in the United Kingdom and shows that the largest daily rainfalls occur in the most elevated areas of Britain. There is some local evidence that intense rainfalls have increased, at least in parts of upland Wales, since the 1950s and 1960s, and this has caused severe landslips and soil erosion from time to time.

Significantly, the uplands are areas of major water surplus and very rarely suffer from water stress during their short growing season except in nationwide droughts (as in 1955 and 1959) or on slopes (Taylor, 1967) where intermittent soil drought may occur. Gregory (1955) has also shown how the rainfall of upland Britain is subject to relatively less annual variability compared with that of lowland Britain.

Snowcover

Snow frequency and intensity increase with elevation in Britain as does the risk of blizzards and prolonged snow lie and prolonged frost. The generally small crystal size and strong winds aid drifting in upland areas. Figure 12.8 reveals the sensitive relationship between snow cover and altitude. A number of estimates Gloyne, 1968; (Manley, 1971) of the rate of increase of snow cover with elevation range between 5 days and 15 days, with an average of 8 days, per 100 m rise. The lower rate is applicable to the Devonshire moors in the south, and the higher rate to the Central Highlands of Scotland. Uplands exposed to east and north-east winds emerge as the snowiest parts of their respective upland regions, e.g. Cairngorms, the eastern side of the northern Pennines, the Peak District and the uplands of north-east Wales. Dartmoor too is liable to heavy snowfalls when for example a depression over the Bay of Biscay or western France causes strong easterly winds to blow along the English Channel.

Microclimatically and mesoclimatically, the effect of persistent snowcover (and to a lesser extent hail or hoar frost cover) is to change the albedo in upland areas. Much more incoming radiation is reflected back and air temperatures are therefore lowered, with extremely low minima occurring just above the snow surface. Soils and peat cover, although initially relatively warmer under snow-cover, may be deeply penetrated by frost and the ultimate effects of frost-heave and the disintegration at thaw, even on gentle slopes, on the stability of soil can be locally severe and have probably been underestimated to date.

Evaporation

Although available data are limited, Hogg (1967a), adapting

Penman's (1948) original formula for the estimation of evaporation, has produced a series of maps showing the distribution of theoretical irrigation-needs on the basis of selected specific soil moisture deficits. Although the maps refer to England and Wales only, the Pennines and Lake District, and the Welsh uplands and finally Dartmoor and Exmoor emerge, in that approximate order, as areas with consistently low or

Fig. 12.8 *Average annual number of mornings with snow cover, 1931–60 (after Manley, 1970b)*

zero irrigation-needs. In reality, the British uplands are normally areas of water surplus except during dry sunny spells or on severe, south-facing slopes liable to soil drought.

Sunshine

The distribution of 'bright sunshine' is, in general terms, the converse of that for cloud and is better documented, although isohels in upland areas often have to be crudely extrapolated (Meteorological Office, 1952). On the fringes of the British uplands, about 30 per cent of the possible, annual bright sunshine is received (1901–30). In the uplands proper this is reduced to less than 25 per cent in the southern Pennines and to less than 20 per cent over parts of the western Scottish Highlands. In the latter region the average annual mean daily duration of bright sunshine (1901–30) is less than 3 hours but in parts of the uplands of Wales and south-west England, a range from 3·5 to 4·5 hours is available.

For a short transect just inland from Aberystwyth, Smith, L. P. (1950) concluded that sunshine hours were curtailed with altitude at a rate that is virtually linear. Some selected data for representative stations at various altitudes are given in Table 12.6.

Table 12.6 *Average daily duration of bright sunshine in hours, 1931–60, for selected stations, together with the altitude of the recorder. (derived from Manley, 1970b)*

Station		J	F	M	A	M	J	J	A	S	O	N	D	Year
Braemar (1933–60)	339 m	0·8	2·0	3·0	4·3	5·4	5·5	4·6	4·0	3·5	2·1	1·0	0·6	3·1
Eskdalemuir	242 m	1·4	2·3	3·1	4·3	5·6	5·6	4·5	4·2	3·3	2·5	1·7	1·1	3·3
Edinburgh (Blackford Hill)	134 m	1·7	2·7	3·6	4·9	5·8	6·3	5·2	4·6	4·2	3·1	1·9	1·4	3·8
Stonyhurst	115 m	1·4	2·1	3·3	4·8	6·2	6·3	5·1	4·9	3·8	2·8	1·6	1·1	3·6
Lerwick	82 m	0·8	1·8	2·9	4·4	5·3	5·3	4·0	3·8	3·5	2·2	1·1	0·5	3·0
Aldergrove	68 m	1·5	2·3	3·3	5·0	6·3	6·0	4·4	4·4	3·6	2·6	1·8	1·1	3·5
Cardiff	62 m	1·8	2·7	3·9	5·7	6·3	7·0	6·1	6·0	4·7	3·4	2·0	1·5	4·3
Dublin	48 m	1·8	2·6	3·7	5·3	5·8	6·0	5·5	4·9	4·3	2·3	2·3	1·5	3·9
Plymouth (Mt Batten)	27 m	1·9	2·9	4·3	6·1	7·1	7·4	6·4	6·4	5·1	3·7	2·2	1·7	4·6
Inverness	4 m	1·4	2·3	3·5	4·5	5·4	5·5	4·4	4·2	3·8	2·8	1·6	1·0	3·4
Aberystwyth	4 m	1·8	2·8	4·1	5·4	6·5	6·6	5·1	5·3	4·5	3·4	1·9	1·5	4·1
Stornoway	3 m	1·1	2·2	3·5	4·7	6·3	5·8	4·1	4·3	3·7	2·5	1·5	0·9	3·4

It is difficult to integrate the individual component gradients of climate with altitude because of the inadequacy and in consistency of records and the variation in instrumental exposures and errors. It is possible, however, to estimate the total impact of the climatic deterioration with altitude as reflected in biological responses and land-use potential and productivity, especially: (i) the length and intensity of the growing season; and (ii) growth rates and yields in agriculture and forestry, provided the management variables can be controlled.

Variations in growth potential and in biological responses and land use

Estimates of the loss of 'growing days' with altitude vary. The Meteorological Office has indicated that, on the average, 2 days are lost for every elevation of 15·2 m, or 13·1 days per 100 m (Meteorological Office, Agricultural Branch, 1955). This gradient is no more than a rough generalization and should be treated as such. It clearly underestimates the rates of reduction of growing season on west and north slopes and inevitably exaggerates those on east and south slopes. Other estimates, including cartographic portrayals of variations in the growing season, are those of Gloyne (1958), Gregory (1954, 1964), Shellard (1959) and Hurst and Smith (1967). Indices adopted include 'accumulated temperatures' and 'grass-growing days', the latter being defined as the number of days between April and September, inclusive, when the soil moisture deficit does not exceed 50·8 mm. Figure 12.9 reveals the advantages of the British uplands in having long periods in summer without moisture deficits although they suffer from an excessive frequency of periods with heat deficits in winter, spring and autumn (Gregory, 1954, 1964; Shellard, 1959).

Fig. 12.9 (a) *Number of grass-growing days in England and Wales, 1962 (after Hurst and Smith, 1967);* (b) *Average annual floral isophenes for the British Isles, 1891–1925 (after Clark* et al, *1935)*

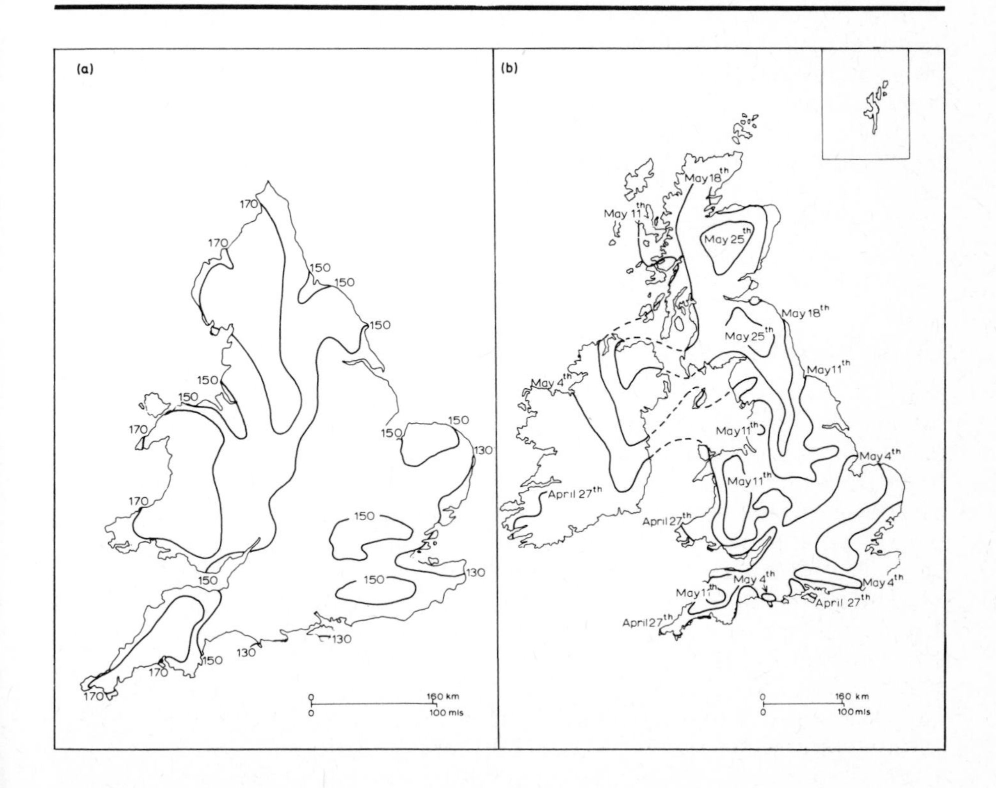

The distribution of isophenes indicates, if only crudely, the bioclimatic lateness of the British uplands (Fig. 12.9b). Pearsall (1950) analysed data for: (i) length of flower stalk; (ii) number of flowers produced; and (iii) number of mature capsules, for the ubiquitous heath-rush (*Juncus squarrosus*) in the central Pennines. All three criteria varied with altitude along gradients which, intriguingly, were not quite linear. He attributed these gradients to climatic deterioration, mainly decreasing temperatures. More quantitatively, Jones (1972) and Harrison (1973) have calculated variations in timber yield with altitude in spruce plantations in west-central Wales. Reservations must be made, however, as soil also changes upslope and because there are many variables, not least genetic ones, within the trees themselves, and to do with their management and calibration.

Alcock *et al.* (1974) have contrasted pasture production at high and low elevations in northwest Wales. Hunter and Grant (1971) have expressed variations in grass growth in eastern Scotland in relation to climate and altitude. They conclude that the floral development of grasses is delayed, on the average, by 4·3 days per 100 m. Yield over the year as a whole fell by 2 per cent for each 30·48 m rise. There were seasonal variations. In spring, yields were depressed by 5 per cent for every 30·48 m rise, and in autumn by 1·8 per cent. This contrast between severely delayed spring and only moderately delayed autumn is also revealed in the work of Cooper (1964) and Peacock and Sheehy (1974). In summer the trends could be either non-significant or even reversed, and, on occasions, the highest yields were recorded at the highest altitudes, i.e. 686 m, thus demonstrating very convincingly the periodic potential of uplands to maximize growth rates in summer even to the extent of temporarily outpacing the lowlands. The work directed by Munro (1974) in central Wales suggests that the agricultural productivity of the British uplands, under scientifically-improved but conventional management, could be increased by a factor of at least four, which is a further indication of the true capacity of the notoriously short, late but occasionally intensive, upland growing season (Taylor, 1974b).

Conclusion

The British uplands are so sharply differentiated in meteorological terms from the lowlands that they may justifiably be regarded as possessing their own particular meso-climate. The differences between the climates, for example, of Ben Nevis, Scafell and Snowdon on the one hand and their adjacent lowlands on the other are at least comparable in range with the more familiar contrasts between the climates of western Ireland and East Anglia or Cornwall and Caithness.

The British upland climate in its ultimate form is primarily expressed in a combination of low temperatures, severe wind exposure, excessive precipitation, cloud and humidity, persistent winter frost and snow cover, deficiency of sunshine, poor visibility, continual ground wetness and low evaporation. These are characteristic of the highest summits and plateaux above say 600 m. Downslope to the westward, a more maritime regime prevails with accentuated climatic gradients with altitude, eventually merging with the contrasting 'early' climates along our western littorals. Downslope to the east, a less maritime, and less humid regime occurs with gentler climatic gradients, which merges eventually with the harder, more continental climate of eastern Britain.

However, within this general framework, the variety of relief induces sharp changes in local mesoclimates in both space and time. Furthermore, general topography and ground conditions – especially wetness, peatiness (Taylor, 1974a) and snow cover – create additional microclimatic deviations.

However, despite the shortage of primary meteorological, climatological, and parallel bioclimatic data, our subjective evaluation of the British upland climate as one of low energy availability and markedly reduced growth potential, is currently being modified in the light of local research findings. Extremes of weather and climate are typical of the uplands. The summer, especially the late summer, is however liable to fine weather spells and intermittent spurts of growth to the direct benefit of the grazings, woodlands and forests. It must be emphasized that just as the tundra climate and ecosystem is unstable, vulnerable to pollution and therefore delicately poised (Hare, 1970), so the climatic environment of the British moorlands and mountains is very sensitive to contemporary climate change, short-term weather stress (e.g. soil and peat erosion, floods, etc.) and long-term climatic disadvantages (e.g. cold, shaded slopes ; chronic frost hollows ; wind-blasted summits and lee-slopes).

It behoves us to introduce more direct and comprehensive measurements of the British upland climates which must be envisaged as more than aggregations of wet, cloud, wind, cold and snow and more than regional reservoirs of water surpluses (however nationally and economically important that fact may be), young farm stock and introduced soft timbers. The emancipation of its wet peaty soils by draining and fertilizing would bring a micro-climatic amelioration and ultimately an opportunity to attain the full potentialities of its deceptive summers, as was last realized in medieval and earlier times (Taylor, 1973, 1974a).

Introduction

Many of the exchanges of energy and moisture in the atmosphere take place at its lower boundary. The relative magnitude of these exchanges is governed by the characteristics of the surface, its shape and covering. The direct and diffuse solar radiation reaching the surface is partly reflected, according to the albedo of the surface, and partly absorbed. The absorbed portion raises the temperature of the surface until the rate of energy losses from it balances the rate of energy gained. Energy is transferred away from the surface by conduction to the soil below by an amount dependent on the characteristics of the soil. Energy is emitted from the surface by long-wave radiation, and energy is transferred, by convection, to the turbulent atmosphere above. At a moist surface, the latent heat of vaporization is extracted, the surface is cooled and moisture is transferred as water vapour to the atmosphere by convection. Comparatively small energy gains, typical of night-time conditions, occur at the surface with the convection of vapour from the atmosphere, and the release of the latent heat of condensation, and with the convection of heat from the atmosphere, when the surface is cooler than the air above. Counter radiation is an important night-time energy source; a further energy transfer is the conduction of heat from the sub-surface to a cooler surface at night. At any time, advective transfers across the surface can occur and modify these energy exchanges. Locally, advection is very important where the surface of the ground is not level, and where there are obstructions, such as trees, buildings and hills, to the general air flow across the surface. A covering over the surface – for example grass, crops or trees – modifies the physical conditions at the ground; the major exchanges of energy often occur between the atmosphere and the upper surface of the cover, with additional transfers of energy within the covering and between the covering and the ground surface below.

Soils

The capacity of any soil to retain heat and moisture, and thus influence the temperature and humidity of the air above it, depends upon the composition, arrangement and cohesion of the soil particles. The texture of a soil varies according to the proportion of sand, silt and clay in a dry sample; this proportion affects the thermal response of the soil to incident radiation. A sandy soil has at least 85 per cent sand, a loamy soil has material containing 7–27 per cent clay, 28–50 per cent silt with less than 52 per cent sand; and a clay soil has more than 40 per cent clay together with less than 45 per cent sand and less than 40 per cent silt (Townsend, 1973). Soil types are derived from the sand and silt content of the parent material, together with the clay fraction associated with weathering.

Sandy soils are derived from granite, for example, and very white bleached sandy soils from Lower Greensand. Loamy soils occur on Silurian schists, and brown loamy soils from Old Red Sandstone. Clay soils are found on Triassic Marls, and on Carboniferous Limestone. Keuper Marls, Oxford Clays and Weald Clay produce heavy and greasy clays. Various soils develop on glacial deposits: in central England, they grade from light sandy soils to heavy red clays.

The individual particles of a soil are arranged with intervening pore spaces of air and water. Extreme temperature regimes develop at the surface of a dry soil which possesses a large proportion of air; subsequently, extreme temperatures are propagated to the air above. However, a soil which holds a large proportion of water attains only moderate surface temperatures, and with additional evaporative cooling by day, leads to only moderate temperatures in the air above. At Rothamsted, in clear summer weather, the average daily temperature range at a dry soil surface was measured at 35 °C, but only 20 °C when the soil surface was wet. (Russell, 1961). The cohesion of a soil depends on its colloidal structure and the surface tension of water films. This affects the capillary potential of the soil and thus the ability of the soil to retain and store water. Fifty per cent of a clay soil is composed of many small capillary pores. This type of soil has a high surface area per unit mass (called specific area) and retains water well. A sandy soil has a 25–30 per cent pore space, formed by a few large non-capillary pores. This soil has a relatively small specific area, one-seventh of that for a clay soil, and has a limited water-holding capacity. More moisture may be retained in a soil by increasing the specific area, i.e. by ploughing. Most soils in the British Isles have a moisture content between field capacity and the permanent wilting point. Field capacity is defined as a moisture tension exerted by the soil particles on the available water, in units of atmospheres, or pF (pF refers to the logarithm of the equivalent hydrostatic column; a 1 000 cm column is approximately one atmosphere, and 3 pF). The field capacity for medium-textured soils, like loamy soils, is one-third of an atmosphere (2·53 pF), and for sandy soils, one-tenth of an atmosphere (2 pF). The field capacity of sandy soils occurs with a soil moisture content of 20–25 per cent, that of clay soils occurs with moisture contents from 26–35 per cent; the field capacity of organic soils is associated with moisture content of over 50 per cent (Townsend, 1973). Permanent wilting point, when the soil is so dry that plant life cannot be maintained, is reached at about 15 atmospheres (4·18 pF). Measurements of clay soils and sandy soils at Rothamsted, Hertfordshire, indicate that the moisture contents at wilting point are 9 per cent and 5 per cent respectively, (Townsend, 1973). In London clay, under meadow grass at Uxbridge, wilting point is reached when the moisture content reaches 14 per cent (Coleman and Farrar, 1966). The amount of water held by a soil between wilting point and field capacity is usable water; clay soils hold seven times as much usable water as sandy soils, and twice as much usable water as loamy soils.

The composition and structure of a soil affects its retention and storage of heat and moisture derived from the exchange of energy and moisture with the atmosphere. These properties, in turn, determine the climate of the soil surface and the local climate immediately above the soil.

The proportion of radiant energy reflected by the soil surface, and consequently the portion remaining for heating the surface, is indicated by the albedo of the surface. The darker the soil, the greater is the proportion of radiant energy absorbed. A moist soil has a lower albedo than a dry soil (Table 13.1). Red soils, such as red clay of the Keuper Marl, absorb more radiant energy than yellow or white soils, such as the white sands of the Lower Greensand.

Table 13.1 *Albedos of soil (wavelength less than 4·0 μm) (after Sellers, 1965)*

Surface	Albedo (per cent)
Dark soil	5–15
Moist grey soil	10–20
Dry clay or grey soil	20–35
Dry light sandy soil	35–45

The capacity of a surface to heat up and to transfer heat to lower levels depends on the thermal characteristics of the soil. Values of various physical parameters for air, water, and for three types of soil are given in Table 13.2.

The thermal capacity of a soil is a convenient measure of the ability of a soil aggregate to absorb heat, for it is derived from the volume fractions of solid matter and air and water in the soil (de Vries, 1963). The greater the proportion of moisture retained in the soil, the higher is the resultant thermal capacity. Consequently, a moist surface responds less readily to radiant energy exchanges. Moist clay soils with a higher thermal capacity than dry clays require more heat to raise their temperature by a particular amount. Clay soils are called 'cold' or 'late' soils, for they warm only slowly in spring and cool relatively slowly in autumn. Sandy soils are less retentive of water; they require comparatively less heat to raise their temperature in spring – hence the descriptions 'warm' and 'early' of sandy soils. Loams and sandy soils were compared in an experiment on Salisbury Plain (Johnson and Davies, 1927). Observations were made daily in June 1925. The maximum temperature recorded at 1 cm below the sandy surface was 31·1 °C, and was 24·6 °C in the loam; the air temperature in the screen at 1·2 m was 21·8 °C. However, the minimum temperature near the loam surface was 13·1 °C, whereas 9·1 °C was recorded at 1 cm below the sandy surface. Monteith and Sceicz (1962) estimated the apparent radiative surface temperature of clay soil at Rothamsted on three summer days. In 1961 the maximum and minimum soil surface temperatures were 44·1 °C and 8·7 °C. The extreme temperatures in the Stevenson screen at 1 m were 25·4 °C and 12·0 °C. On clear days, dry bare surfaces attain maximum temperatures just after midday whereas air temperatures reach their lower maxima somewhat later. On 27 July 1941, between 14.00 and 15.00 hrs, Penman (1943) recorded a maximum air temperature of 27 °C at 1·2 m

Table 13.2 *Properties of soils, air and water (after Geiger, 1965, and Hanwell and Newson, 1973)*

Substance	Density (g cm^{-3})	Specific heat (cal g^{-1} c^{-1})	Thermal conductivity (cal cm^{-1} s^{-1} c^{-1}) $\times 10^{-3}$	Thermal capacity (cal cm^{-3} c^{-1})	Thermal diffusivity (cm^2 s^{-1}) $\times 10^{-3}$
Still air	0·0012	0·24	0·055	0·00028	200
Water	1·0	1·0	1·3–1·5	1·0	1·3–1·5
Dry sand	1·4–1·7	*0·20	0·4–0·7	0·1–0·4	2·0–5·0
Wet sand	*2·6	*0·20	2·0–6·0	0·2–0·6	4·0–10
Dry loam	2·0	0·16	0·7	0·3	2·3
Dry clay	*2·3–2·7	*0·17–0·20	0·2–1·5	0·1–0·4	0·5–2·0
Wet clay	1·7–2·2	*0·17–0·20	2·0–5·0	0·3–0·4	6·0–16

* These values are for solid soil particles.

over dry clay soil at Rothamsted; the maximum soil surface temperature, recorded at about 13.00 hrs, was 47 °C.

There is normally a lapse of temperature with height away from the surface. Higher surface temperatures occur on dry sandy soils, with consequently higher air temperatures above, than on other dry soils because of the small thermal capacity and low conductivity of sandy soils. On clear nights, dry bare soils lose heat slowly until sunrise. The air above does not cool as much and so an inversion of temperature develops after sunrise between the ground and the air. On 7 July 1941 at Rothamsted, for instance, the minimum soil temperature was 11·5 °C at 04.00 hrs, while the air temperature at screen level reached 12·0 °C between 04.50 and 05.00 hrs. However, minimum temperatures have sometimes been recorded above the soil surface when heat losses from the air through radiative exchanges are strong. Lake (1956) for instance recorded a surface temperature on bare soil of −6·1 °C while at 19 cm the air temperature was −8·7 °C. Moist soils generally lead to less extreme surface temperatures diurnally, for these soils have high conductivities and relatively high thermal capacities. Consequently, the temperature regime of the air above is more moderate. On clear days, evaporation from moist soils begins after sunrise, and so vapour pressure decreases away from the ground. At night, with falling temperatures, the humidity inversion re-develops and there may be a transfer of moisture to the ground in condensation.

The amount of heat which flows through a soil depends on the thermal conductivity of the soil. Soils with a high thermal conductivity transfer heat rapidly; conversely, a soil with a low thermal conductivity, which possesses much air in its pore spaces, is a good insulator. Clay soil has a relatively higher thermal conductivity than sandy soil; loose sandy soil has a much lower thermal conductivity. A clay soil attains only moderate surface temperatures, but heat is transfered deeply into the soil. It is a good day-time heat store; in comparison, only a thin skin of dry sand is heated by day. The thermal conductivity of a soil increases rapidly as soils become moist, for air pores become replaced by water, a much better conductor.

The energy flux at any level, and at the soil surtace, is given by the product of the temperature gradient at that level and the thermal conductivity of the soil. This expression of the heat flux is applicable to steady conditions. To account for the variation of soil temperature with time, the net energy flux through a soil layer can be determined from the product of the rate of temperature change and the average thermal capacity of the soil through the layer. Thermal capacities are more regular and predictable than thermal conductivities, so the latter expression for the net energy flux through the soil to the surface is normally employed, if heat-flow transducers, which record flows very near the soil surface, are not available. In sandy clay loam under a short grass surface at Cardington (Bedfordshire), Rider (1954) estimated that the soil heat flux at 11.00 hrs in summer 1952 was 0·046 ly min^{-1}. At this time, the net radiation flux was 0·400 ly min^{-1}, and the atmospheric heat flux and the vapour flux were 0·150 and 0·204 ly min^{-1} respectively (Rider, 1954). In this fairly typical energy balance, 11·5 per cent of the net radiative energy was stored in the soil, while half of the total energy source was used in evaporation. At 21.00 hrs the soil heat flux and the atmospheric heat flux were two energy sources at the surface, 0·055 and 0·126 ly min^{-1} respectively, whereas energy was being lost by radiation (0·122 ly min^{-1}) and the vapour flux (0·059 ly min^{-1}). Here, 30 per cent of the total energy source is derived from the soil. Unusually, evaporation is still taking place. With identical thermal conditions, in time and depth, a dry sandy soil would have a smaller heat flux than Rider measured, and a wet clay soil would have a larger heat flux, because of the relative thermal capacities.

Thermal diffusivity (or temperature conductivity) determines the heating or cooling rate for a given temperature gradient in the soil. A soil with a high diffusivity transmits temperature rapidly. Soil with a high air content, such as dry, loose sandy soil, has a very low density and thus a relatively high diffusivity. This soil has a low thermal conductivity also. It is an efficient temperature transporter, but a poor conductor of heat energy. The thermal diffusivity of a soil varies with its moisture content. For most soils, thermal diffusivities are a maximum at a moisture content of 8–20 per cent by volume (Sellers, 1965). At these moisture levels thermal conductivity is high, so heat is conducted efficiently and changes temperature readily. At low moisture levels thermal diffusivities are lower, for thermal conductivity is low, while at high moisture levels thermal capacity is high, so thermal diffusivities are low again. The diurnal and annual variations of temperature at the soil surface are transmitted, modified and delayed with depth, according to the thermal diffusivity of the soil. Assuming that thermal diffusivity is constant with depth, the depth of penetration of the temperature wave is given in Table 13.3; the depth of penetration is defined as that depth where the daily or annual fluctuation is reduced to 0·01 of the surface temperature. The greater the thermal diffusivity, the deeper is the penetration.

Table 13.3 *Penetration depth of temperature fluctuations (after Geiger, 1965)*

	Thermal diffusivity (m^{-2} s^{-1}) × 10^{-6}		
	2·0	**1·0**	**0·1**
	rock	**wet sand**	**dry sand**
Daily fluctuations (cm)	108	76	24
Annual fluctuations (m)	20·6	14·5	4·6

The time-lag of the arrival of a maximum or minimum of temperature from the surface is shorter for soils with large thermal diffusivities. In the chalky loam soil at Wye (Kent), the maximum temperature, derived from 09.00 hrs readings in 1896, 0·15 m in the soil was 17·2 °C in June, the minimum was 1·1 °C in December. At 0·9 m, the maximum temperature of 16·7 °C occurred in July, and the minimum, 6·1 °C, in February. At 1·85 m the maximum temperature was 14·4 °C in September, and the minimum value was 4·4 °C in April (Hall, 1945). For the period 1921 to 1950 in Great Britain, the soil temperature at

0·31 m varies from 3 °C in the north-east to 6 °C in the south-west in February, and ranges from 14 °C in the north-east to over 18 °C in the south and south-west in August. At 1·23 m, soil temperatures range from 4 °C in the north-east to 8 °C in the south-west in February, and from 13 °C in the north-east to over 17 °C along the south coast in August (Mochlinski, 1970).

Slopes

The local climate of sloping ground depends upon the exposure of the slope, that is, its aspect, together with the position of the slope relative to neighbouring topographical influences. In particular, radiant energy exchanges are modified on sloping ground (see Chapter 4) and local air circulations develop, especially when regional pressure gradients are weak.

Fig. 13.1 *Direct radiation (cal cm^{-2} h^{-1}) on cloudless days on north, east and south slopes at 50 °N. (after Geiger, 1965)*

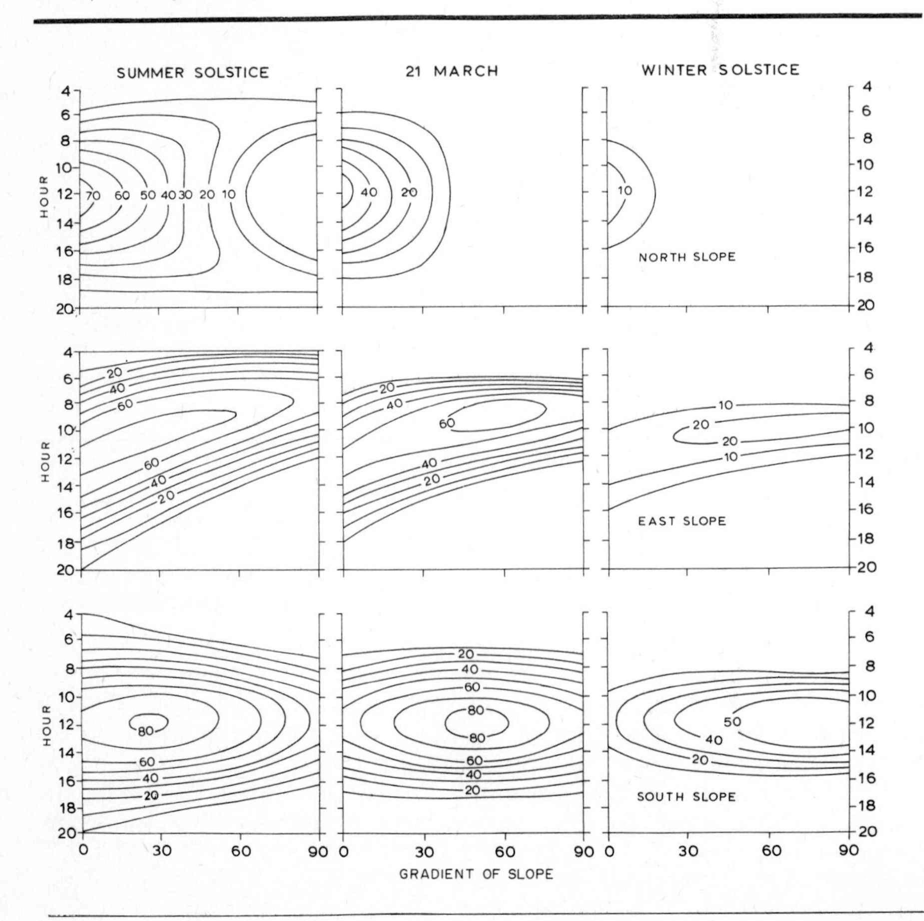

The solar radiation received directly at the ground is greatest at a surface normal to the sun's rays. The energy available for absorption is greater on a south-facing slope (adret) than on a north-facing slope (ubac) in mid-latitudes (see Fig. 4.9). The amount and duration of direct surface heating is determined by the declination and the altitude of the sun, and is modified by the shadow of intervening relief. The advantages of a south-facing slope at 50 °N. are significant (see Fig. 13.1). The greatest intensity of radiation occurs at noon on slopes of 26·5° in mid-summer, 50° at the equinoxes and 72° in mid-winter. Corresponding figures for the maximum receipt of insolation at 55 °N. are 31·5° in mid-summer, 55° at the equinoxes and 78·5° in mid-winter. Sunrise occurs on all slopes at the same time in winter and at the equinoxes, but in mid-summer the sun rises farther to the north-east, so sunrise is later on steep, south-facing slopes. Similarly, sunset is earlier on steep, south-facing slopes in mid-summer. North-facing slopes do not have radiation intensities as high as those of a level surface. Very steep, north-facing slopes have no direct radiant energy at noon in summer, although low intensities are received in the early morning and late afternoon. In mid-winter, no direct solar radiation is received on steep north-facing slopes. At 55 °N, only slopes less than 11·5° have direct insolation. On east-facing slopes, the maximum intensity of radiation occurs before noon, on flatter slopes in summer and on steeper slopes in winter. Sunrise takes place at the same time on all slopes in each season, but sunset is earlier on the steeper slopes. Conversely maximum insolation occurs after noon on west-facing slopes, and sunrise is later on the steeper slopes.

At Kinlochleven (56° 50′N.), the potential insolation income on the adret slope (about 25°) at 457 m is 7·3 g cal cm^{-2} min^{-1} at noon in summer, whereas at 396 m on the opposite ubac slope (over 30°), the potential receipt is 2·6 g cal cm^{-2} min^{-1} at noon (Garnett, 1939). Day length is about an hour longer on the ubac slope in this period. In the valley at Kinlochleven below 76 m, the potential insolation at noon in summer is 5·9 g cal cm^{-2} min^{-1}. In winter, only the adret slope is unshadowed; it receives then less than 0·1 g cal cm^{-2} min^{-1} between 09.00 hrs and 15.00 hrs. On average, in mid-latitudes, a south-facing slope of 20° is equivalent to a southerly shift of 8° to 9° of latitude; a similar north-facing slope is equivalent to a northerly shift of 12° to 15° of latitude (Crowe, 1971).

Eighty per cent of the longwave radiation from a surface is directed towards the central 30° dome of the sky. Infra-red radiation from strongly heated surfaces warms the surface and the air above steep south-facing slopes, and this warmth is retained in steep sided valleys. The high monthly maximum temperature of 32 °C in July and August 1935 in the dry valley with sandy gravel soils at Rickmansworth is associated with this reflected radiation and limited local air movements (Bilham, 1938). Counter-radiation is least from the zenith of the sky, so net radiation at night is greatest towards the vertical. Sloping surfaces lose less net longwave radiation than open, level surfaces, but this reduction is only 10 per cent for a 30° slope, and is often over-compensated by cold air drainage downslope (Geiger, 1965).

Under clear skies and weak regional pressure gradients, more direct insolation is received on all slopes, except those in shadow, than on a neighbouring plain or valley bottom. Surface temperatures increase preferentially and the air above the sunlit slope becomes warmer and less dense than air at the corresponding level above the plain or bottom of the valley. On one occasion, temperatures through 0·92 m of the stiff red clay soil on an 18° south-facing slope in southern England were 1·6 °C higher than on level ground (Hall, 1945). Warm regions develop on microscale features also. South-facing potato ridges, 26 cm high and 75 cm wide, were 3 °C warmer than the intervening furrows in clay silt at about midday on 21 July 1971 near Warboys, Cambridgeshire (Evans, 1974). With strong insolation, evaporation is greater over south-facing slopes, so these slopes tend to be drier than shaded north-facing slopes. Similarly, the humidity of the air over sunlit slopes is lower than over shaded slopes.

With light, warm air near the slopes, isobaric surfaces are raised. Away from the slopes, the isobaric surfaces are relatively low. Hence a pressure gradient is established away from the surface and a circulation develops linking upslope air flows, called anabatic winds, which are light and variable with movement away from the valley sides and above the valley floor where it subsides. The strength of the slope flow depends on the rate at which radiant energy reaches the slope, and the amount absorbed at the surface. Anabatic flows are greater on steep slopes than gentle grassy slopes (Pedgley, 1974). Slope flows are weakened in crossing rough surfaces, and in mixing with the cooler undisturbed air above. The warm rising air cools adiabatically, and the subsiding air, away from the slope, warms so that the initial density contrasts diminish. In the same may, winds develop up a valley, when density differences occur between the upper and lower sections of an open valley during daylight. There is a compensating return flow down the valley at high levels. The stronger valley winds develop after the air flows begin upslope, and persist for some time after the slope flows weaken and cease in the late afternoon (Fig. 12.3). In Snowdonia, Pedgley estimated up-slope air flows of 4·8 km h^{-1} on an August day in 1973 (Pedgley, 1974). In potentially unstable conditions, up-slope flows encourage the development of cumulus clouds and local showers.

At night, outgoing radiation is the dominant process over uneven terrain in settled weather. Air over slopes cools more than air at a corresponding height over a level surface or the bottom of a valley. A weak reverse circulation is established, with cold air sliding intermittently downslope, as a katabatic wind, with a weak return flow above. Cold, dense air sinks to the bottom of a slope, unless it is dammed up into cold pools behind obstructions such as hedges, walls and embankments. Cold air reaching the bottom of a valley may be impounded there, if air drainage from the valley is limited. This is the case of the Rickmansworth valley where the lowest minimum monthly temperature of 1935, −8 °C, was recorded in December. Between 1930 and 1942, ground frost was recorded here in every month except two (Hawke, 1944). With stable atmospheric conditions common at night, the mixing of cold air with the warmer air of the atmosphere is limited, but the downslope winds are warmed adiabatically. The cold impounded air raises the free air above, cooling adiabatically, and so very slowly reduces the initial density contrasts (Crowe, 1971). At night, differences in density between the upper and lower sections of an open valley promote the development of stronger air flows down the valley. The winds down the valley (sometimes called mountain winds) occur after air flows downslope, but continue till after sunrise, when up-slope air flows begin to form. (Fig. 12.3).

Several studies have been made of the night-time temperature regime of valleys, but few measurements of air flows have been made in the British Isles. An absolute minimum temperature of 3·3 °C was recorded in a shallow valley in the New Forest in summer observations taken by Morris and Barry in 1961. Minimum temperatures on a ridge were 2·3 °C greater than the temperature in the bottom of the valley, 15·38 m below. (Morris and Barry, 1963). In another investigation in mid-Wales, air temperatures were recorded at intervals down the side of the Upper Irfon Valley, 19 km south west of Rhayader (George, 1963). Temperatures fell from −4·5 °C at the top of the valley at 457 m, to −11 °C at the bottom (246 m). By 07·30 hrs fog filled the valley between 305 and 427 m; the temperatures in the valley below the fog were −8 °C, and −9·5 °C on the slope above the fog. Manley has compared the minimum temperatures at three sites in the middle Wear basin (Manley, 1944). The sites were Durham Observatory (103 m) on the crest of a minor ridge, Ushaw College (183 m), 5 km to the west, on the exposed crest of the main ridge, and Houghall agricultural station (49 m), 1·6 km south-east of the Observatory, on the valley floor which is 1·6 km wide. On a February night in 1941, temperatures dropped to −10 °C at Durham Observatory, −7 °C at Ushaw College and −16 °C at Houghall. The average extreme monthly minimum temperatures in January, through the period 1925 to 1940, were −6 °C, −5 °C and −8 °C respectively. The minimum temperatures at the Observatory lie intermediate between those of the exposed ridge at Ushaw College and the severe frost hollow at Houghall. In Snowdonia, a temperature difference of 2·6 °C was recorded between 385 m and 215 m in the upper Lledr valley, in August 1973. Fog was several metres thick, and down-slope drifts of 1·6 to 4·8 km h^{-1} were estimated (Pedgley, 1974). Two sets of temperature readings and the estimated slope flows were taken in a dry chalk valley near Hull, through 2 nights in 1956 (Hanwell and Newson, 1973). With a 2·5 °C difference in temperature between the valley bottom and the slope, there were katabatic flows of 3·6 km h^{-1}. Downslope flows were about twice as fast with a 3·4 °C temperature difference. Hawke (1944) found katabatic air flows reached 3·6 m s^{-1} off the Chilterns into the Rickmansworth valley. A much stronger flow, up to 8 m s^{-1}, was observed by Newham, at Benson, Oxfordshire, 8 km from the Chiltern ridge (Newham, 1918). Lighter katabatic flows, less than 1·6 m s^{-1}, have been recorded in a valley off the Cotswolds, at Leafield, Oxfordshire (Heywood, 1933).

The local climates of slopes develop most strongly on clear calm days and nights. These climates are modified by local water sources, which may have a cooling effect by day and a warming influence at night (see Chapter 11). An increase in cloud cover, or a strengthening of the overall pressure gradient, diminishes or obliterates density differences in irregular terrain. Then the strength and character of the air currents up and down slopes are related to the angle of the slope and its exposure to the regional wind. Air flows smoothly up windward slopes which are not steeper than 20° to 30°, and smoothly down leeward slopes less than 10° to the horizontal (Gloyne, 1964). Over steeper slopes, the air flow separates from the ground, leaving a small protected region at the base of the windward slope, and a larger protected region on the leeward slope. Winds are accelerated up windward slopes; on slopes between 3° and 10°, winds may be increased by 50–65 per cent (Geiger, 1965). Conversely, air currents are retarded down leeward slopes. The reduction in speed along a lee slope between 2° and 10° is between 10 and 30 per cent, and at the base of a lee slope, 7° to 12°, the reduction in wind speed is 40 per cent (Gloyne, 1964). Slope exposure to the prevailing wind influences local precipitation receipts. Geiger (1965) recorded 5–10 per cent more precipitation on the lee side of an isolated hill than on the windward side. Conversely, within a valley, Reid (1973) found on the north-facing (leeward) slope of Caydell near Helmsley, North Yorkshire, the precipitation receipt was 92 per cent of that recorded on the south-facing (windward) slope for the 50 weeks from October 1969. With strong geostrophic winds aloft, the strengthening by funnelling, or weakening by disorganized cross flows, of the circulation of the air above irregular terrain is dependent on the dimensions and orientation of the major features of the topography.

Vegetation

The British Isles are predominantly rural; in 1960, 82 per cent of the total area of Great Britain was given over to arable, permanent grass and rough grazing land (Gilchrist Shirlaw, 1966), while forest and woodlands cover 1·68 million hectares (i.e. 7·4 per cent) of Great Britain. The natural and planted vegetation of the British Isles, as elsewhere, modifies local climates, according to the type and colour of the vegetative cover, and the depth, density and area of the cover. The resultant energy and moisture exchanges change through a season, and through the years with ageing plants.

A vegetation cover, insulating the ground below, is a transition zone of living leaves and stems or trunks. Within the cover, there are internal energy exchanges, although the major exchange of energy and moisture takes place at the outer 'effective' surface of the cover with the atmosphere. With a complete cover of vegetation, incident short wave radiation is partly reflected, part of the radiation passes through the leaves (penetrability) as diffuse light below, and the rest is absorbed and raises the temperature of the vegetation or is used in evapotranspiration. The less dense the surface cover, the more shortwave radiation reaches the ground. One-fifth of the insolation at the surface of meadow grass, 1 m high, reaches the ground below (Geiger, 1965). Within a forest, 5–15 per cent of the shortwave radiation

Fig. 13.2 *Summer albedo (as a percentage) over England and Wales (after Barry and Chambers, 1966)*

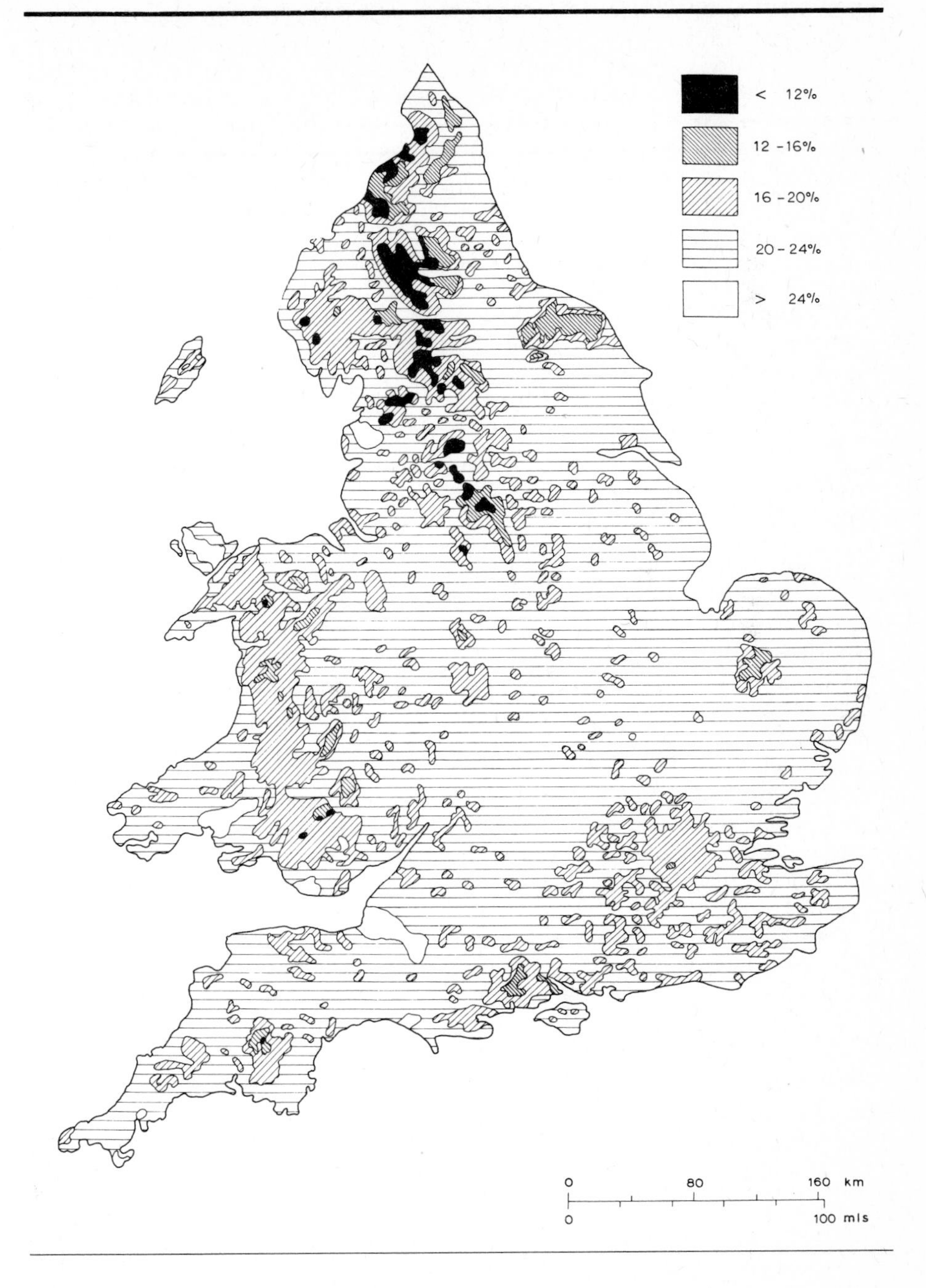

affecting the top of the trees reaches the trunk space but less reaches the ground. At 1 m above the ground under oak forest 9 per cent of the shortwave radiation at the tree-top levels was recorded under clear skies (Munn, 1966). With no foliage, the corresponding figure is 27 per cent. Under overcast skies, the percentage of shortwave radiation received at 1 m was 11 per cent with oak foliage, and 56 per cent with no foliage.

Table 13.4 *Average albedo values for England and Wales (after Barry and Chambers, 1966)*

Cover type	Albedo (%)
Festuca down	24
Agricultural land	
(1) *more than 84 per cent grassland*	24
(2) *40–80 per cent grassland*	23
(3) *less than 40 per cent grassland*	22
Molinia and acid grassland	20
Deciduous woodland	18
Towns	17
Coniferous plantation	16
Heather moor	15
Peat and moss	12

Typical reflectivity coefficients which were recorded in the summer of 1965 over parts of England are given in Table 13.4, and are employed in Fig. 13.2. Most of the country has an albedo of 20–24 per cent, corresponding with a variety of agricultural usages. The lowest albedos occur over peat and moss in the agriculturally unproductive uplands. The average crop albedos represent the mid-summer period only, lower values being more typical earlier and later in the season. Wet vegetation reflects more solar radiation, so these summer albedos may be exceeded on individual days. The albedo of wet grass in the sun is between 33 and 37 per cent, whereas the albedo for dry grass lies between 15 and 25 per cent (Sellers, 1965). Albedos are higher at the effective surface of sparse vegetation at low zenith angle of the sun. Albedos of 20 per cent were recorded late in the day at Thetford forest, whereas midday averages were only 9 per cent (Oliver *et al.*, 1974). Fourteen per cent of the incident short wave radiation reaches the undergrowth in the thinned section of this forest (870 trees per hectare) through multiple trapping. In winter, the albedo of deciduous forests is reduced. On the basis of the various albedo values, there is more energy available for absorption and penetration at the upper surfaces of coniferous forests than for deciduous forests or grasslands. This value at the outer surface declines with the decrease in density of the vegetation cover, which permits stronger energy exchanges at the exposed ground surface.

Plants absorb counter-radiation from the atmosphere and terrestrial radiation from the ground below. Radiation losses at night are least at the surface below a dense vegetation cover, and greatest at the outer surface of the plants. The interior of the vegetation cover and the ground below are shielded from extremes of heating by day and cooling by night. By day, net radiation is greatest near the top of the vegetation and least at ground surface. By night, this relationship is reversed. A net radiation profile is shown for a young spruce forest near Munich, in Table 13.5 (comparable data are not available in the British Isles).

Table 13.5 *Radiation balance in a young spruce forest in July 1952 (after Geiger, 1965)*

Height of measurement (m)	Average balance (cal cm^{-2} min^{-1}) Day	Night
10·0 above forest	0·370	—0·054
5·0 tree-top	0·370	—0·046
4·1 crown	0·164	—0·014
3·3 trunk	0·013	—0·011
0·2 floor	0·020	—0·005

Vegetation acts as a brake to the general air flow, and the rougher the surface the deeper is the surface boundary layer above the vegetation. Within the plant cover, air flow is markedly reduced; it may be less than 1 m s^{-1} and difficult to measure without special instrumentation. Reduced air movement in the plant cover limits the exchange of heat, moisture and carbon dioxide within the cover. The major exchanges occur at the outer surface of dense vegetation, or at the ground with sparse vegetation.

Many studies of the characteristics of the temperature, humidity and flow of air have been made over and within vegetation, in comparison with rarer studies over bare soil or sloping ground. The majority of these investigations, and subsequent energy balance studies, have taken place on extensive tracts of level, well-drained soil, covered uniformly with short grass, that is at ideal meteorological sites. By day, the surface of the grass cover is warmer than the air above it, so there is a strong lapse of temperature with height. Monteith and Sceicz (1962) found, at Rothamsted, on 29 June 1961, that the maximum temperature of the air in a Stevenson Screen at 1 m was 25·4 °C, when the apparent surface temperature of short grass was 35·9 °C. At Porton, Hampshire, temperature gradients were recorded from 1 August 1931 to 31 July 1933 (Best, 1935). The largest average lapse rate, over a 2-hour period, was −77 °C 100 m^{-1} between 30 cm and 120 cm, and −682 °C 100 m^{-1} between 2·5 cm and 30 cm. The greatest average 2-hour lapse rate at Rye, between 1945 and 1948, was −6·8 °C 100 cm^{-1} through the heights 1·1 m–15·2 m (Best *et al*., 1952). At night, the surface of grass is cooler than the air above; at Rothamsted, the minimum air temperature at 1 m was 12·0 °C, when the apparent surface temperature of the grass was 6·6 °C on the night of 29–30 June 1961 (Monteith and Sceicz, 1962). Similarly at Porton, the greatest average 2-hour inversion was +58 °C 100 m^{-1} between 30 cm and 120 cm, and +258 °C 100 m^{-1} between 2·5 cm and 30 cm; the greatest average inversion at Rye was +8·4 °C 100 m^{-1}. By day, the grass cover is more moist than the air above, and there is a humidity lapse away from the grass. Conversely, at night, there is a reverse, though often weak, inversion in the humidity gradient above grass. Table 13.6 indicates temperature, humidity and wind speed profiles over short grass at Cardington, for day and night conditions in 1952 (Rider, 1954). At 13.15 hrs on 10 October, there were slight lapse conditions and at 19.15 hrs on 8 October conditions were stable. At 14.45 hrs on 8 October, with relatively strong winds, conditions were isothermal in the lowest 2 m air layer.

Table 13.6 *Temperatures, humidities and wind speeds at Cardington (after Ride, 1954)*

Height	13.15 hrs on 10 Oct 1952			19.15 hrs on 8 Oct 1952			14.45 hrs on 8 Oct 1952		
(cm)	V	T	H	V	T	H	V	T	H
200	368	10·72	5·44	238	7·93	6·91	437	12·22	7·34
150	355	10·76	5·47	219	7·77	6·89	409	12·23	7·36
100	337	10·64	5·49	194	7·58	6·84	379	12·23	7·38
75	327			182			362		
50	302	11·02	5·65	164	7·32	6·82	336	12·24	7·51
37·5	286	11·07	5·75	154	7·23	6·82	320	12·24	7·60
25	262	11·21	5·94	140	7·09	6·80	270	12·26	7·73
15	230			121			250		

(*V* is wind speed in cm s^{-1}, *T* is air temperature in °C and *H* is absolute humidity in g m^{-3})

Wind speeds increase from the ground to the height of the gradient wind (300 m–600 m), through the friction layer. Below 50 m, in the surface boundary layer, the turbulent flow is affected most strongly by the roughness and temperature of the surface. The rougher the surface, then the lower is the mean wind speed near the surface, and the deeper is the friction layer to the height of the gradient wind. Below 1 to 2 m, the gradient of wind speed is greater by day than by night, under clear skies, and, correspondingly, the surface shearing stress and the rate of vertical exchange of momentum are greater also. Above this level, over warm surfaces, vertical gradients of wind speed are less steep, for in day-time lapse conditions the vertical mixing of air near the surface is increased. Over cool surfaces, vertical gradients are relatively steeper since there is little vertical exchange in stable night-time conditions. In the surface boundary layer, the mean wind speed at the lower levels is greater by day than by night. Conversely, above 10 m to 50 m, this diurnal variation of wind speed is reversed, particularly in light winds and in winter. With strong winds, in cloudy conditions and when temperature gradients are isothermal, then wind speeds increase logarithmically with height from the surface to about 8 m. Table 13.6 also indicates three wind profiles recorded by Rider over short grass (roughness length 0·32 cm) at Cardington in 1952 (Rider, 1954). Table 13.7 illustrates the variation in temperature, humidity and wind speed in a Scottish meadow of grass 50 cm high. Within the plant cover, turbulent flow does not exist.

The recordings were taken on a sunny June afternoon following precipitation. The active surface is at 30 cm; here temperatures are greatest during the day. Below this level, there is a temperature inversion by day and a lapse at night. Above and beyond the top of the grass normal daytime lapse and night-time inversion temperature gradients are recorded. Moisture is available from the soil and from the plants, and this is removed above the active surface. With evaporative cooling, the upper parts of the grass are cooled slightly. At night, dew deposition warms the upper section of the plants. So temperature extremes tend to occur just below the upper surface of the grass. The sparser the grass cover, the climate and energy exchanges become more like those over bare soil. The shorter the grass, the less protection does the soil receive from the extremes of local climates. Cutting a grass meadow from about 30 cm–45 cm to between 2 cm and 3 cm doubles the number of nights with frost (Norman *et al.*, 1957).

There are few measurements of the fluxes within or above meadow grass, but several energy balances have been computed over short grass (i.e. Rider's study at Cardington). An investigation over short grass at Kew in June 1949 indicated that the turbulent exchanges with the atmosphere were almost equal (the sensible heat flux was 0·22 ly min^{-1}, and the evaporative flux 0·24 ly min^{-1} and the energy exchange with the soil (0·08 ly min^{-1}) was small in comparison with the net radiative flux 0·54 ly min^{-1} (Rider and Robinson, 1951).

Table 13.7 *Measurements of temperature, humidity and wind speed in a Scottish meadow (after Waterhouse, 1955)*

Height (cm)	Temperature (°C)	Vapour pressure (mm Hg)	Relative humidity (per cent)	Wind speed (m s^{-1})
2	16·9	13·6	94	0
30	21·2	12·2	65	0·6
60	20·8	11·0	61	3·7

The interior climate of crops varies according to the type of crop, the depth and density of the crop, and it is also modified by irrigation. Dense wheat crops are warmer and more moist by day than thinned crops one-quarter of the density (Penman and Long, 1960). At night the thinner crop is warmer and drier. The thinner crop has a lower mean daily relative humidity and a shorter period near saturation. In a dense wheat crop, temperatures are greatest near the top of the plant. There are weak temperature gradients above, with strong vapour pressure gradients, associated with a high rate of evaporation and rapid evaporative cooling. At night, there is a transfer of sensible heat from the warmer air to the cooler plant surface, and the release of latent heat of condensation. Water vapour is transfered from the air to the crop and from the soil to the upper layers of the crop with a resultant deposition of moisture under conditions of strong radiative cooling. Burrage (1972) found that the condensate in wheat, 90 cm high, in July was dewfall above 60 cm and distillation from the soil below 60 cm. Over wheat, Penman and Long (1960) estimated that the sensible heat flux and the evaporative heat flux were 14 cal cm^{-2} and 106 cal cm^{-2} respectively between 12.00 hrs and 16.00 hrs averaged over the period 12–19 June 1957. At night, between 00.00 hrs and 04.00 hrs, these fluxes had fallen to −15 cal cm^{-2} and −0·2 cal cm^{-2} respectively.

Wheat fields have a relatively uniform horizontal structure, but many crops such as potatoes and beans have a very varied structure.

The climate within these crops is modified by row separation and by the height and density of the crops. Early in the growing season, the temperature and humidity characteristics are similar to those of bare soil, possibly varied with ridges and furrows, but these regimes become more like the exchanges of heat and moisture within wheat crops as the vegetation thickens and the surface becomes covered.

Forests have a complex vertical and horizontal structure. In dense forests, temperature profiles are like those within low vegetation, that is, an inversion of temperature by day and a lapse at night. Temperatures decrease from a maximum in the crown area upwards to the atmosphere, and downwards to the ground, by day. At night, the coolest part of the forest is just below the canopy surface, in the crown space. Within the forest, thermal conditions are moderate; on hot sunny days, temperatures within the forest can be 2°–8°C lower than in the open countryside (Barry and Chorley, 1968). Within the forest, moisture sources are the soil and undergrowth, at low levels, and the crown area above. The upper source is active by day. In the late afternoon, there is a minimum of vapour pressure near the top of the canopy, associated with diminishing transpiration. At night, conditions become more uniform, with vapour pressure increasing slowly from the ground upwards into the crown layer.

Wind speeds have been measured above and within the Scots and Corsican pine forest of Thetford (Oliver, 1971). The average height of the forest is 15·5 m, and the tree density is one per 10 m^2. The effective roughness (roughness length) of the forest is 1 m. Data from 66 summer days in 1971 suggest that the wind shear between the top of the canopy and the ground below is 29° to 35° (Oliver *et al.*, 1974). Table 13.8 shows the variation of wind speed with height in Thetford forest, derived from forty sets of recordings in 1970. Weak air movements combined with the large leaf surface area in forests lead to high relative humidities by day, and correspondingly limited rates of evaporation.

Table 13.8 *Normalized winds in Thetford Forest, Norfolk (after Oliver, 1971)*

Height (m)	Wind speed ($m\ s^{-1}$)	Location
20	1·46	
16	1·00	
12	0·38	mean canopy top 15·5 m
10	0·24	
8	0·16	mean canopy base
4	0·24	

The sources of evaporation in a forest are moisture from the soil (about 10 per cent of the total), transpiration and intercepted precipitation (Rutter, 1972). The rate of evaporation of intercepted precipitation is five times greater than that of transpiration. Oliver *et al.* (1974) found that on some days in Thetford Forest energy from the sensible heat flux of the atmosphere was required for up to eight hours in addition to the available radiation energy to enable evaporation to continue. The amount of intercepted precipitation varies with the type of tree and the amount of precipitation. The percentage of intercepted precipitation over oak trees is 32 per cent for 5 mm rainfall per day and 22 per cent for 25 mm of rain per day. The corresponding figures for pine are 62 per cent and 22 per cent (Ovington, 1954). There is 10–20 per cent more evaporation from forests than from grassland in south-east England, and this appears to depend on the amount of water intercepted by the forest (Rutter, 1972). It is a vital factor in the annual water balance of forested regions.

In a young pine forest, three-quarters of net radiation is typically transferred into the evaporative flux by day (Munn, 1966). At night, the heat stored in the soil and in the biomass of the forest are as important as the energy gains from the atmospheric heat and moisture fluxes. The energy fluxes at the forest floor are very small in comparison with those exchanges determined above the canopy.

The horizontal structure of a forest is often variable; clearings and cuttings, and the forest edges, have individual local climates which depend on the relative size and orientation. Small clearings have similar characteristics to the dense forest, but the wider the clearing the more extreme the thermal climate. If insolation reaches the floor of the clearing, then relatively high temperatures result near the ground by day. Radiative cooling also affects the ground. So, thermally, the centre of the clearing has a climate similar to that of open ground; humidity levels are higher than those over open ground because of the neighbouring moist forest. Clearings more than twice as wide as the tree height are ventilated by air brought down from the freer air above the canopy, so temperatures and humidity levels become less extreme. The climate of the edges of the forest is ameliorated with exposure to direct solar radiation, and is transitional between the conditions over open country and those within the forest (Tuller, 1973). With a combination of small clearings or thinned forest and low-lying land, extremely low temperatures can occur. In Thetford Chase no month is free of frost, and consequently newly-planted pine saplings do not survive here (Hurst, 1967).

Groups of trees and hedges, as well as fences, are employed as shelter belts to reduce exposure to strong winds. Air reaching the barrier is diverted above the leading edge, returning to the ground some distance downwind. At the leading edge, air flow separates from the streamlines, breaking into vortices below the surface of separation from the unaffected flow above. Immediately above the barrier, wind speeds become accelerated. The sheltered zone behind the barrier is characterized by disturbed flow patterns. With a dense barrier of height h, a small vortex develops in a narrow protected zone $2\ h$–$5\ h$ upwind of the barrier. Downwind, a large standing eddy develops to a distance of $30\ h$ (Gloyne, 1964). The disturbed flow may extend $50\ h$–$100\ h$ downwind. It reaches a maximum depth of $3\ h$–$4\ h$, at $3\ h$–$5\ h$ downwind of the barrier. Wind speeds are reduced by at least 10 per cent between $3\ h$ upwind and $20\ h$ downwind. The protected area behind dense barriers is relatively shallow. With moderately dense barriers, the standing eddy does not develop; there are horizontal drifts of air, in relatively smooth flow, which extend farther downward. Winds may be reduced to 70–80 per cent at 4 *h* downwind (Gloyne, 1964). The wind is accelerated at each end of the shelter belt, so groups of barriers are recommended to provide more uniform shelter (Caborn, 1957). The shelter belt affects the local energy balance. Shortwave radiation is modified by shadows near the barrier, and outgoing longwave radiation is reduced because of the limited horizon. Near the barrier, the microclimate is similar to that of the forest edge. Farther away, in the sheltered zone, soil and air temperatures are more extreme and the rate of evaporation is reduced.

Conclusion

The complex characteristics of the earth's surface interact with the atmosphere to create particular local climates. Differences in soils, slopes and vegetative covers modify heat and moisture budgets temporally and spatially. Subsequently, these differences affect the temperature, humidity and movement of the air near the surface, particularly when regional pressure gradients are weak. Local climates are of prime importance in rural land-use; any changes in land-use or land management (e.g. ploughing or irrigation) will modify natural atmospheric processes and local climates.

All surface changes must alter the thermal, hydrological and dynamic properties of the overlying air. In draining the marsh, clearing the wood, cultivating the fields and flooding the valleys, man has inadvertently changed the thermal, hydrological and dynamic properties of the earth's surface and the chemistry of the overlying air. There are therefore very few 'natural' boundary layer climates remaining, not only in the settled parts of our planet but in the world as a whole; for, even in areas utterly remote from human habitations, conditions are changing, to some degree, because of activities elsewhere.

Modifications made by man in essentially rural areas are often of limited atmospheric importance though many, such as the hydrological effects of forest planting and clearance, need further investigation. Here, as in many other cases, it is difficult to quantify the changes and there is clearly somewhat confused evidence and even theory for the precise manner and degree to which rural atmospheres have been accidentally modified by man. There is much less uncertainty about urban atmospheres, for when buildings are congregated in villages, towns and conurbations, the result is a degree of modification of local boundary layer conditions which amounts to the creation of a distinctive form of mesoclimate, the climate of towns.

In the United Kingdom, four out of every five persons live in towns and about 12 per cent of the surface area of England and Wales is built upon (Best, 1968). It is therefore of extreme importance to appreciate the role of urban areas in modifying regional and local climates. Several pioneering studies of town climates have been made in this country (World Meteorological Organization, 1970) but, except for the extensive monitoring of atmospheric pollution under the auspices of the Warren Spring Laboratory, Department of Industry, National Survey of Air Pollution, observations of other elements have been taken in only a small number of cities, based upon the records of either the few climatological stations that exist in urban areas or upon short period intensive observations taken for experimental purposes. And so, apart from very detailed information on smoke and sulphur dioxide concentrations that exist for a large number of towns in the United Kingdom, much of it summarized in the regional volumes of the National Survey of Air Pollution, 1961–71 (Warren Spring Laboratory, 1972a and b, 1973), the amount of information on urban climates in Britain is very limited, and the most one can do is to extrapolate from records taken in a small number of centres, notably London.

Another point of some importance is that urban climates are made up of a kaleidoscope of site microclimates, with the result that climatological elements and more particularly pollution, airflow and temperature, vary quite rapidly through the built-up area, often showing an immediate response to the details of very local changes in such controls as smoke emissions, building geometries and building densities. In consequence, climatological records from a single site cannot be regarded as representative of much more than the area immediately adjacent to the point of observation. Gardens, courtyards, shopping precinct, high-rise buildings, narrow nineteenth-century streets and modern, six-lane urban motorways will all exert their particular influence upon energy exchanges in the overlying air. Such detailed meteorological processes obviously necessitate rather specialized monitoring techniques such as temperature and humidity traverses using sensors and automatic recorders mounted on vehicles (Chandler, 1960).

In built-up areas, the effects of the surface's complex geometry, the particular thermal and hydrological properties of the urban fabric, the heat from combustion and metabolism and the discharge of pollutants to the air combine to create the distinctive features of urban climates. All meteorological elements are changed. Strong winds are decelerated and light winds are often accelerated as they move into towns; air turbulence is increased; relative humidities are reduced though absolute humidities are little changed and may even be higher inside towns during the night; the chemical composition of the air is changed; receipts and losses of radiation are both reduced; temperatures are substantially raised; fogs are made thicker, more frequent and more persistent; and rainfall is sometimes increased. Each of these changes will now be considered in more detail.

Airflow in urban areas

The flow of air over an urban area is affected by the rougher surface here than in rural areas and by the frequently higher temperatures of the city fabric. Buildings, particularly those in cities with a highly differentiated skyline, exert a powerful frictional drag on air moving over and around them and winds in cities are consequently more turbulent than those outside, with characteristically rapid spatial and temporal changes in speed and direction. Because of this, the mean horizontal wind profile above urban areas is changed with a more gentle gradient through a deeper boundary layer than is generally found above topographically uncomplicated rural areas and over the oceans (Davenport, 1965). Height for height, mean horizontal wind speeds in cities are less than above neighbouring flat, rural areas, but the reduction in speed is very dependent upon prevailing meteorological and topographical conditions (Jones *et al.*, 1971).

Wind speed

Studies of wind speed in this country and abroad have shown that average strengths are lower in built-up areas than over rural areas (Chandler, 1965; Graham, 1968; Landsberg, 1956; Munn, 1970), but Chandler (1965), working on the London records, showed that the difference in wind speed between town and country is a function of the regional near-surface wind speed and the wind profile. Similar results were later demonstrated in New York by Bornstein (1969) and

Bornstein *et al.* (1972) and may well prove to be typical of at least major urban areas. The foremost of Chandler's findings was that, when winds are light, speeds are greater in the centre of London than outside, whereas the reverse relationship exists when winds are strong. He also demonstrated (Table 14.1) that, on average, night-time wind speeds at Kingsway in central London were 14 per cent stronger than at Heathrow (London Airport) on the western fringe of the city, while daytime winds were 24 per cent weaker.

Table 14.1 *Average wind speeds at Heathrow (London Airport) and the excess over those at Kingsway, 1961–2*

	01.00 hrs G.M.T. m s^{-1} Mean	Excess	13.00 hrs G.M.T. m s^{-1} Mean	Excess
Dec–Feb.	2·5	—0·4	3·1	0·4
March–May	2·2	—0·1	3·1	1·2
June–Aug	2·0	—0·6	2·7	0·7
Sept–Nov.	2·1	—0·2	2·6	0·6
Year	2·2	—0·3	2·9	0·7

Chandler attributes these contrasts in the urban influence upon wind speeds to differences in the effect of increased surface roughness upon airflow having dissimilar velocity profiles and hence upon the relative sizes of the friction and momentum transfer terms (Chandler 1965). Bornstein (1972) showed that the pressure gradients, generated by higher temperatures in cities, are also important in explaining occasions of stronger winds in cities than in the country.

Such meteorological considerations are obviously relevant to the day-to-day and seasonal differences in wind speed between cities and rural areas. In autumn, winter and spring for instance, when regional winds tend to be strong, speeds in central London are reduced by about 8, 6 and 8 per cent respectively, whilst in summer, with lighter winds, there is little or no difference in mean urban and rural speeds. The overall, annual reduction in wind speed in central London for all winds is 6 per cent, but for winds of more than 1·5 m s^{-1} the reduction is 13 per cent.

Similar results to these have been found by Lee (1975) in analyses of the Birmingham and Manchester wind records, but in smaller cities the effects are likely to be less marked, since the wind takes some time to adjust its velocity profile to a change in surface roughness (Davenport, 1965). The critical fetch required for this adjustment downwind of the urban boundary is presently unknown, but is likely to be several kilometres so that small towns may show little or no effect of the type described.

In light of the above, it is not surprising that there are fewer calms in the air immediately above the buildings of a city than above the surrounding country, although between the buildings, calms might be much more frequent. Equally, however, there will be times when winds will be channelled and accelerated along streets oriented in the same direction as the wind and eddies will form across streets running at right angles to the wind. The latter will closely affect the distribution of pollution in streets with carbon monoxide and other vehicle emissions carried upwards in the rising air on the upwind side of congested city thoroughfares.

Wind direction

The direction of the wind in cities, as already indicated, is very closely controlled by the form of its buildings and the layout of its streets, by open spaces and, of course, by the topographical setting of the city.

Above the buildings of a city, winds blow at an appreciable angle to the isobars because of the intense surface friction. Angles vary but can be expected to lie in the range from 15° to 40° according to the prevailing stability. The more stable the lower atmosphere, the greater the difference. Such changes of direction will obviously affect the form of smoke plumes as they rise above the city.

Turbulence over cities

In urban areas, wind speeds and directions vary much more rapidly over space and time than they do over rural surfaces, although the details vary according to the geometrical form of individual cities. But in fact few long period records of gust speeds exist for central urban areas in this country, London being one of the few exceptions (Helliwell, 1970; Shellard, 1968a). Here, it has been shown that the ratio of maximum speeds, averaged over a few seconds, to the one minute mean value were higher, height for height, than above rural areas and fell off very sharply with elevation above general roof level. This broad pattern is likely to be repeated in other large cities. The gustiness factor, defined as the percentage ratio of the difference between the maximum and minimum horizontal wind speeds to the mean wind speed, is always higher within and immediately above towns. Over most lowland rural areas in this country, the gustiness factor varies between 25 and 100 per cent, but in towns it may reach 200 per cent. The effect is more the consequence of weaker lulls than of stronger gusts.

Local airflows

Studies in Leicester (Chandler, 1961a) and London (Carpenter, 1903; Chandler, 1965) have shown that on calm, clear nights when urban areas become considerably warmer than their rural surrounds, there is a surface inflow of air towards the areas of highest temperature, generally in the central districts of greatest building densities. These inblowing, cool winds, known as country breezes, are presumably linked to rising air over the city and a return, centrifugal flow from city to country at a higher level, though there have been no definite observations of this. The centripetal winds near the surface are very light, normally less than 4 $m\ s^{-1}$, and for this reason they become quickly decelerated by intense surface friction in the suburban areas of the largest cities. Here they occur as pulsating flows across the margins of the warmer city air or heat island.

Airflow around buildings

Environmental problems stemming from the nature of airflow above and around buildings have been intensified in recent years by the proliferation of high rise buildings in most large cities. There are five crucial components of the pattern of airflow around these tall buildings: a very intense eddy on the lower windward face with wind-speeds near the bottom of the eddy two to three times the regional wind at this level; secondly, a rather weaker eddy high up on the leeward face of the building; thirdly, an eddy above the roof of the building which often raises problems for the efficient dispersal of fumes from the boiler chimney; fourthly, accelerated streamline flow a little higher above the roof, and fifthly, concentration and thereby acceleration of flow around either side of the building.

The pattern of airflow not only around the tall slab and tower blocks but also around low, pitched-roof buildings has sometimes caused a number of structural and functional problems within the buildings themselves and, in addition, has often created intolerable conditions for pedestrians in nearby streets and shopping precincts. More recently, collaboration between architects, planners and building scientists has helped to anticipate and reduce these difficulties, but cities will always be characterized by sharp regional contrasts in airflow.

Air pollution

Atmospheric pollution incorporates a very large number, perhaps several hundred, of particulate, gaseous and liquid substances. Nearly all of these occur naturally in the atmosphere. The problem is therefore one of abnormally high local concentrations, rather than of substances alien to the natural atmosphere. The major problems concern pollutants emitted in large quantities, more especially as the result of the burning of fossil fuels. The chief of these are smoke and other particulates, oxides of sulphur, carbon dioxide, carbon monoxide, oxides of nitrogen and hydrocarbons. Stationary sources contribute most of the particulates as well as of the oxides of sulphur and carbon dioxide and a substantial proportion of the oxides of nitrogen, while motor vehicles are the main source of carbon monoxide and of hydrocarbons.

Although much more research is needed into the physics, chemistry and meteorology of atmospheric pollution, including diffusion processes, it is abundantly clear that overall, there is a fairly steep gradient of concentration away from the point of emission. The close dependence of local concentrations upon the height of emission and the associated log-normal relationship that seems to exist for many air pollution concentration gradients, produces a sharp fall-off in levels downwind of their source. Very roughly, air pollution emissions can be classified into high-, medium- and low-level sources and in proportion to the quantity of fuel, these give relative concentrations in the ratio of roughly 1 : 10 : 100 respectively. And so it is to be expected that the distribution of air pollution in many cities will on average, and in the absence of other factors such as emission controls, closely mirror the distribution of population and in detail the distribution of domestic sources. In urban areas, the modal height of emission is generally low and the patterns of emission and mean concentrations are similar. In rural areas, by contrast, even major single sources of air pollution such as power stations will generally contribute little to regional levels of sulphur dioxide.

But under suitable meteorological conditions air pollutants emitted in large industrial conurbations can travel scores of kilometres and can contribute substantially to mean concentrations in country areas. Twenty-four hourly observations of smoke and sulphur dioxide concentration at twenty-one 'National Survey' sites in country areas of England and Wales free from local anthropogenic emissions, used to calculate winter and summer mean concentrations for each wind direction over a 4-year period, show that sources such as Greater London, the West Riding of Yorkshire and industrial Lancashire are important 200 km away. None of the sites is free from exotic smoke and sulphur dioxide in winter, but a small number of monitors in Wales and western England seem to be so in summer.

Excluding country sites close (within 25–30 km) to industrial conurbations, mean winter smoke concentrations range up to 50 $\mu g\ m^{-3}$ calculated for north winds at Little Wenlock, Shropshire, and 57 $\mu g\ m^{-3}$ in north-west winds at a site in north Lincolnshire. These levels are typical of winter mean concentrations in the centre of a small provincial town such as Hereford (41–77 $\mu g\ m^{-3}$). Summer mean smoke concentrations at these more remote sites show a small (5–10 $\mu g\ m^{-3}$) exotic contribution in about half the directions (Barnes, 1975).

Country areas within 25–30 km of a major source can expect very high mean smoke concentrations in directions where they lie downwind of the source, summer and winter alike. The north Cheshire site of Delamere has a winter mean concentration of 106 $\mu g\ m^{-3}$ in north-east winds (Manchester 30 km), falling to 32 $\mu g\ m^{-3}$ in summer. The former value is within the range experienced in an industrial area of Walsall and the latter is greater than any value for central Hereford.

Many country areas of England have high mean sulphur dioxide concentrations in summer and winter. Sites in south-east England show typical winter maxima of 70 $\mu g\ m^{-3}$ and over in winds blowing from areas such as Greater London, Lower Thameside and the West Midlands. Such values are within the range (59–102 $\mu g\ m^{-3}$) experienced in central Hereford. In some cases the sources appear to be over 40 km distant. The situation in northern England, with fewer non-urban monitors, is less clear, but it would seem that some country areas experience even greater mean winter sulphur dioxide concentrations. Delamere illustrates the situation. With Widnes and Runcorn lying 15 km away to the north, a mean of 100 $\mu g\ m^{-3}$ is obtained while for east winds, with possible sources 10–30 km distant, the mean is 120 $\mu g\ m^{-3}$ (Barnes, 1975).

The winter to summer fall in mean sulphur dioxide concentrations is less than for smoke. In south-east England mean summer sulphur dioxide levels at country sites lie close to 40 $\mu g\ m^{-3}$ in the direction of important sources; in the case of Delamere the values for north and east winds are, respectively, 55 and 50 $\mu g\ m^{-3}$.

Since the air pollutants sampled at country sites have a distant origin, mean concentrations in calm conditions are low. Maxima are usually obtained in very light winds and decline progressively as wind speed increases. However, in some cases mean concentrations increase with wind speed and after reaching a maximum decline again. The maximum can occur at wind speeds greater than 8·3 $m\ s^{-1}$. The most important variable determining the wind speed at which the maximum mean concentration occurs in a given wind direction appears to be the average mixing height in relation to the distance of the major source from the monitor.

Because of the effect of intervening emissions, it is difficult to obtain a comprehensive picture of the rate at which mean concentrations decay from an area source. However, values of the mean decay length (the distance taken for the concentration to fall to 1/exp.) for smoke and sulphur dioxide in northerly winds from industrial Lancashire are 190 km and 267 km respectively. In the case of sulphur dioxide emitted in Greater London the average decay length for north-east winds is 295 km, while for south-westerlies it is 178 km (Barnes, 1975).

It must, of course, be remembered that sulphur dioxide, like several other substances, is reactive in the atmosphere and cannot by itself form an adequate measure of the diffusion of a whole complex of sulphur compounds. Recent studies have shown that, more especially under certain meteorological conditions, airborne sulphur compounds, particularly sulphates, can be carried substantial distances downwind, often being precipitated as acid rain. Such long distance transports convert local, essentially urban, problems into regional problems and regional problems are made of international importance. The United Kingdom is currently taking part in a study of rainfall acidity as part of an international investigation organized by the Norwegian Centre for Air Research.

Particulates

Though enormous quantities of fine particles are released into the global atmosphere from forest and bush fires, volcanic eruptions, surface weathering and soil erosion, in this country the main particulate problems arise from local domestic and industrial emissions of smoke (the condensed vapours of coal burning), grit, dust and ash. Figure 14.1*a* shows the sharp decrease in smoke emissions from United Kingdom towns since 1950 following shifts in the patterns of energy consumption, in part owing to the provisions of the Clean Air Acts of 1956 and 1968, and in particular the establishment of smoke control areas in towns. The 1956 Clean Air Act defined so-called 'Black Areas' of the country where smoke pollution was particularly severe and required the 325 local authorities in these areas to submit plans for smoke control using a phased programme of clean air zones. The majority of authorities in the Black Areas are now well advanced in their clean air schedules, though the more recent fuel problems have regretfully slowed down most programmes. Many towns lying outside the Black Areas, such as Exeter, Reading and Southampton, have also set up programmes of smoke control and several cities, inside and outside the Black Areas of the 1956 Act (a distinction no longer made by Government) are now completely smoke controlled. These include Salford and Sheffield (former authority area). More than 90 per cent of the area and premises in Greater London are covered by smoke control orders.

Fig. 14.1 *Trends in* (a) *smoke and* (b) *sulphur dioxide in urban areas of the United Kingdom (after Reay, 1974)*

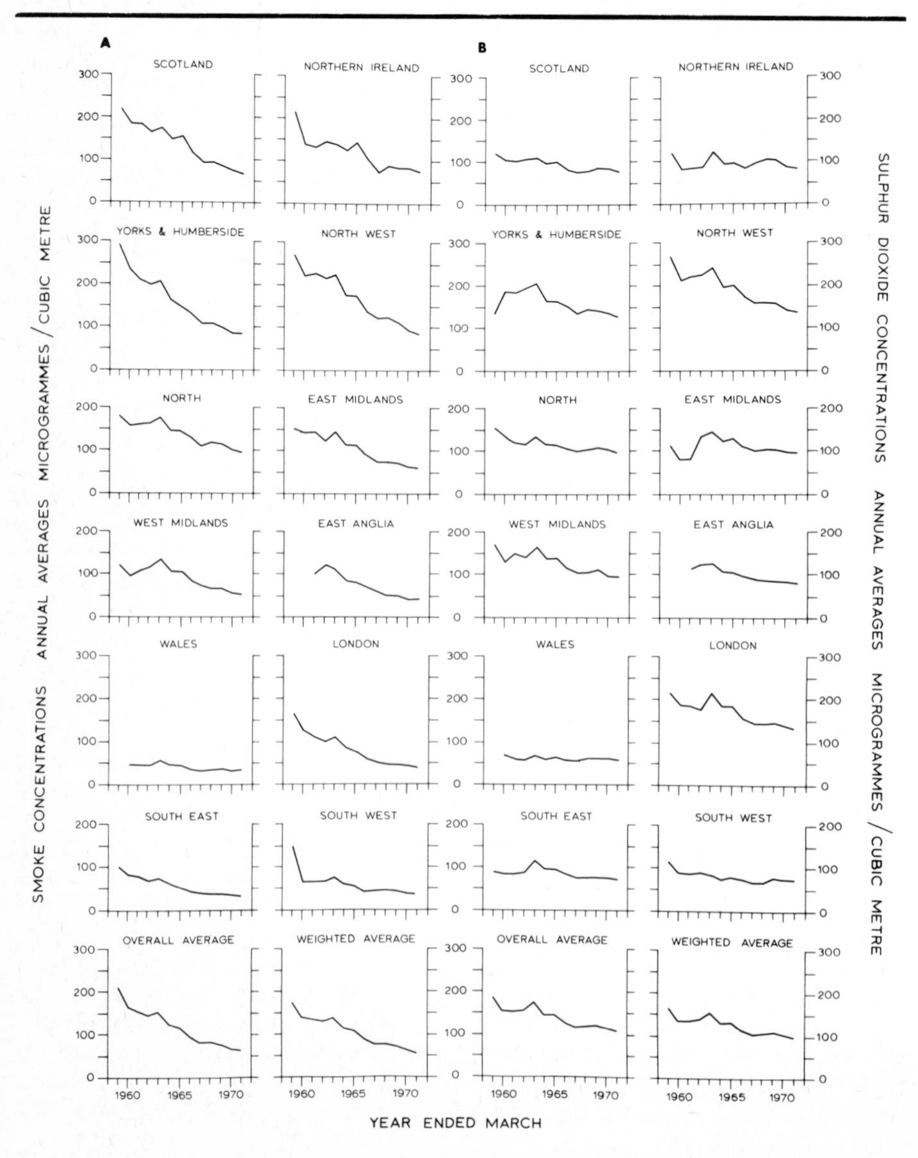

As a result of these controls and the general movement away from solid fuel as the United Kingdom's primary energy base (in 1950 coal provided 89·7 per cent of the total inland consumption of primary energy, but by 1971 this had fallen to 42·9 per cent), smoke emissions and concentrations in urban areas have fallen almost in parallel to levels less than one-third of those in 1958 (Fig. 14.1*a*). In London, emissions of smoke had, by 1970, been reduced to one-tenth of what they were in 1956, and average ground level concentrations of smoke have fallen from more than 250 $\mu g\ m^{-3}$ in 1954 to 40 $\mu g\ m^{-3}$ in 1971. This is a larger decrease than for any other region (Warren Spring Laboratory, 1972a and b).

The decrease has not, of course, been a steady one, being subject to the vagaries of the weather which exercise such a close control upon instantaneous concentrations, more particularly through the effects of temperature (via emissions), stability and wind speed (Chandler and Elsom, 1974a, b; Marsh and Foster, 1967; Garnett 1963, 1967, 1971).

Within our cities, mean smoke concentrations are closely controlled by nearby emissions, more particularly low-level emissions, those within half a mile or so of the pollution monitor contributing most to local concentrations (Williams, 1960). This means that the most polluted areas of our cities are nearly always those with the greatest domestic emissions (Chandler, 1965; Wood *et al*., 1974).

Above a city, the individual plumes of smoke merge into a broad city plume which, except in very unstable conditions, diffuses more rapidly laterally than vertically. Any high-level inversion will closely define the top of the mixing layer and valley walls will limit sideways diffusion so that valley sites are particularly conducive to high concentrations more particularly at times of stable atmospheres. This was well illustrated in London and the middle Thames valley at the time of the December 1952 smog. Many settlements in the coalfields of northern England have sites which are prone to inversions and air stagnation and are therefore singularly unfortunate as far as pollution is concerned (Garnett, 1967). Others, such as coastal sites, are generally well ventilated and abnormally high concentrations are less frequent here than might otherwise be the case, even where emissions are large, as at Liverpool.

But in general, the city plume is shallow immediately above the buildings and there is a sharp fall-off in mean concentrations at only modest elevations above the badly polluted, low lying areas. This is well illustrated in cities such as Sheffield and in London where Hampstead (137 m) and the commons of south London (about 30–45 m) are much less polluted by smoke than are the densely settled areas of north-east London (Chandler, 1965).

The asymmetric pattern of smoke distribution in many cities, often, as in London (Chandler, 1965) with the highest concentrations occurring in the north-east inner suburban areas, is normally only in part the consequence of the lateral drift of smoke into these areas by the prevailing south-westerly winds. Meetham (1945) first commented on the limited downwind shift in the position of peak concentrations in Leicester in response to changing wind directions. Even at times of stability and light easterly winds averaging 4·6 m s^{-1}, the pattern of smoke concentration in London remains essentially unchanged from the long period average (Chandler, 1965).

But although average smoke concentrations respond mainly to nearby emissions, because of the intense turbulence and wide (*c*. 30°) vertical angle of individual plumes in urban areas, under certain conditions, more distant sources will be important in determining local smoke levels. Such is the case where pollution from a single or from multiple sources some distance upwind of a city are brought down by the increased turbulence of the urban area. Such mechanical turbulence helps to explain the abnormally high smoke concentrations formerly experienced in the eastern suburbs of Reading at times of easterly (frequently stable) winds. These carry smoke at a high level from London before it is brought down by the increased mechanical turbulence of suburban Reading. For the same reasons, particulate and gaseous pollutants from industrial plants located in rural areas sometimes constitute a problem when the prevailing wind carries their plumes across settlements, perhaps many kilometres downwind.

But the generally close correspondence between emissions and concentrations of smoke is fundamental to the success of clean air zones policy within cities, for many zones were small and initially, at least, surrounded by uncontrolled areas. And yet even in these circumstances there is a marked improvement within the zones.

Clean air policies have enormously improved the air quality in many of our cities (Fig. 14.2). In Greater London, for instance, 93 per cent of the premises and 90 per cent of the area was smoke controlled by 1973. In consequence of this and the parallel changes in fuel usage (domestic emissions decreased by 92 per cent and industrial emissions by 87 per cent between 1952 and 1970), smoke concentrations have fallen by almost 80 per cent in the period since 1956, the greatest proportional decrease of any region since the passing of the Clean Air Act. In Sheffield there has been a similar quite spectacular improvement.

Table 14.2 *Regional distribution of smoke concentrations and of domestic coal consumption in the United Kingdom*

Region	Domestic coal consumption per head, tonnes, 1970	Average smoke concentration, μg m^{-3} 1969–70	1970–71
North	0·56	95	88
North West	0·52	90	81
Yorkshire and Humberside	0·50	83	80
Northern Ireland	0·54	80	72
Scotland	0·38	79	69
East Midlands	0·44	64	60
West Midlands	0·34	54	50
East Anglia	0·29	46	46
London	0·04	46	42
South East, excl. London	0·14	34	31
Wales	0·54	32	33
South West	0·18	31	29

Fig. 14.2 *Trends in regional smoke and sulphur dioxide concentrations in the United Kingdom (after Warren Spring Laboratory, 1972a)*

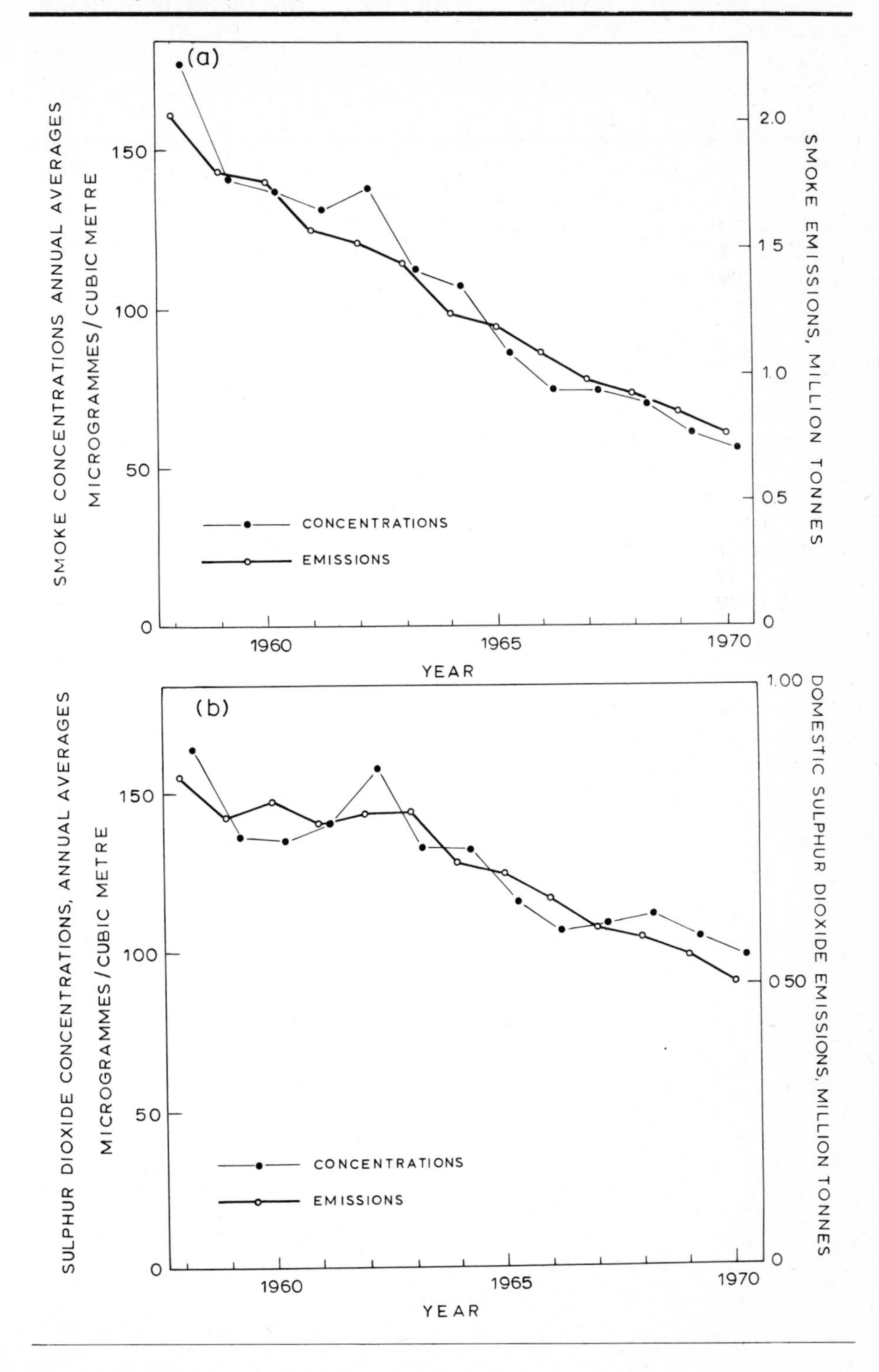

Bearing in mind the enormous variety of spatial and temporal controls upon both emissions and diffusion efficiencies, it is hardly surprising that there are marked differences in levels of (urban) air pollution in different regions of the United Kingdom (Table 14.2).

Turning to temporal changes in smoke concentrations, the highest daily readings of smoke generally occur around 10.00 and 20.00 hrs in winter (Chandler, 1965). The precise time is affected by prevailing temperature, airflow and stability conditions, so that on calm, stable winters' days, the average 24-hour concentration can rise to levels several times the average winter concentration. The ratio varies from place to place and from year to year, but in London in 1966–7 it ranged up to 12·2 and in 1962–3, the last winter occurrence of severe smog, the highest ratio was 25·8 (Warren Spring Laboratory, 1972a). In the North West Region the ratios for urban sites between 1962 and 1970 were generally between 4 and 6 with little sign of any consistent trend during the period. The number of days on which smoke concentrations rose above 500 $\mu g\ m^{-3}$ has, on the other hand, fallen sharply in the North-West Region, more particularly in smoke-controlled areas. Such instances are now very rare in cities such as Manchester and Liverpool where they were all too common a winter phenomenon no more than 20 years ago. Much the same story can be told for most cities having a policy of smoke control. In Leicester, for instance, the number of days with average smoke concentrations above 500 $\mu g\ m^{-3}$ fell from 11 in 1962–3 to nil in both 1968–9 and 1969–70 (Warren Spring Laboratory, 1973).

In London, the ratio of the highest daily reading of sulphur dioxide to the average winter concentration ranged between 1·9 and 6·3 in the winter 1966–7 in comparison with ratios up to 16·0 in the winter of 1962–3. In Manchester, the comparable ratios were 3·6 to 4·6 in 1966–7 and 5·3 in 1962–3. There has also been a sharp decrease in recent years in the number of days with a combination of smoke and sulphur dioxide levels higher than the minimum at which effects on health have been detected (250 $\mu g\ m^{-3}$ for smoke and 500 $\mu g\ m^{-3}$ for sulphur dioxide).

Sulphur dioxide

The Clean Air Acts of 1956 and 1968 proscribed the emission of smoke but not of sulphur dioxide, which was not considered so important a problem and was in any case recognized as much more difficult to control. Certain amenity and health hazards do, however, appear to be synergistic as far as smoke and sulphur dioxide are concerned; SO_2 alone can, for instance, damage certain types of vegetation (Bradshaw, 1974; Linzon, 1972; World Meteorological Organization, 1969), soils (Brandt and Heck, 1968) and natural materials. And so the trends in the levels of SO_2 are of vital interest, more especially in view of the enormous increase in recent years in the use of sulphur-rich oils. In 1971, oil provided 45·5 per cent of the total inland consumption of primary energy in the United Kingdom, having grown from only 6·2 per cent in 1950.

Figure 14.1*b* shows that there was an increase in total SO_2

emissions from 4·4 to 5·7 million tonnes between 1950 and 1969. But the increase came almost entirely from power stations and other plant having tall chimneys contributing very little to general surface concentrations. Domestic sulphur dioxide emissions have in fact fallen over the last decade or so as a consequence of modern housing development using efficient combustion systems, and the conversion from coal to less sulphur-rich fuels for space heating. In view of what has been said about the close spatial and temporal correspondence between emissions and concentrations, it is not therefore surprising to find a fall in urban levels of SO_2 which closely parallels that of emissions (Fig. 14.1*b*).

Once again there are marked local and regional differences in sulphur dioxide concentrations in the United Kingdom. Table 14.3, largely based upon urban site records, shows marked areal differences with the comparable figures for smoke.

Table 14.3 *Regional distribution of sulphur dioxide concentrations in the United Kingdom (Reay 1974)*

Region	Average concentration, $\mu g\ m^{-3}$ 1969–70	1970–1	Percentage decrease since 1963–4
London	143	132	29
Yorkshire and Humberside	140	121	23
North West	130	125	32
West Midlands	113	92	33
East Midlands	99	96	20
Northern Ireland	86	78	16
East Anglia	86	88	25
North	85	92	18
Scotland	85	85	23
South East, excl. London	79	74	23
South West	66	63	3
Wales	56	48	—3

In particular, central London, though having one of the lowest average smoke concentrations, emerges as the region most affected by sulphur dioxide because of the heavy dependence here upon fuel oil for heating and power. The City of London has recently introduced controls upon sulphur content of oil used in heating installations, but it is too early to judge the success in reducing local SO_2 concentrations. Elsewhere, the conversion from town to natural gas is contributing to a reduction in SO_2 emissions (Garnett and Read, 1972).

Sulphur dioxide is much more subject to lateral drift than smoke, but the characteristically widespread diffusion of SO_2 is, in part at least, owing to the generally more elevated emissions. Certainly there seems to be much more spatial uniformity in the distribution of sulphur dioxide than of smoke in urban areas (Chandler, 1965). But, as with smoke, instantaneous levels can depart quite markedly from the long period average.

Other air pollutants

The list of pollutants to be found in most urban atmospheres is exceedingly long, although few, other than smoke (covering a variety of particulates) and sulphur dioxide, are regularly monitored in the United Kingdom. Recently, however, a detailed study has been initiated of a number of pollutants in the busiest streets of five towns (Birmingham, Cambridge, Cardiff, Glasgow and London). In these places, levels of carbon monoxide, total hydrocarbons, lead and smoke, along with basic meteorological data, are to be measured at all sites. In addition, at two sites in London, nitric oxide, nitrogen dioxide, gaseous sulphur compounds and ozone are continuously measured.

Most previous studies have concentrated on extreme conditions. A study for over a year in six cities in the United Kingdom (Glasgow, Birmingham, Manchester, Enfield (London), Portsmouth and Cardiff), for instance, showed that even in the busiest streets the level of carbon monoxide rarely rose, and then only for a few minutes, above the 50 p.p.m. threshold for 8-hour continuous exposure (Reed and Trott, 1971). It is worthy of note that if someone is smoking inside a car with the windows closed, then CO levels are likely to be much higher inside than outside the vehicle.

Oxides of nitrogen in city streets are formed by the oxidation of atmospheric and organic nitrogen during the combustion of fossil fuels in stationary and vehicular processes. Measurements of nitrate in rain appear to indicate a substantial increase in aerial concentrations but many more measurements will be needed before we can discern the true position. Another very controversial pollutant is lead, but in this case vehicle exhausts are but one of several sources of lead in our environment, the eteology of which is exceedingly complex and far from well known. In Greater Manchester, an enquiry showed marked changes in the lead content of the air from about 130 p.p.m. ($\pm$20) in the centre of the city to 90 p.p.m. ($\pm$15) in the outer, southern suburbs and to about 70 p.p.m. ($\pm$15) in the rural areas outside the built up area (Wood *et al.*, 1974).

The Five Towns Survey of the Warren Spring Laboratory will help to improve our knowledge of the levels of atmospheric lead in city streets. Already, however, we know that compared with the quantities of other pollutants emitted by vehicles, aerosol lead compounds are minute in amounts, but it is highly persistent and toxic. The dust in a reasonably busy urban street will frequently contain 0·1 per cent to 0·3 per cent concentrations of lead (Holgate and Reed, 1974). Immediately around some industrial premises, concentrations can be ten times larger, although in both cases there are sharp gradients of concentration away from the source. The Government has recently (1974) introduced legislation to reduce progressively the amount of lead in petrol.

As with oxides of nitrogen and lead, hydrocarbons are released into the atmosphere from a variety of natural and anthropogenic sources. It is estimated that forests release 175×10^6 tons yr^{-1} of reactive hydrocarbons into the air compared with man's 27×10^6 tons (Holgate and Reed, 1974). The significance of the anthropogenic emissions is that they are generally much more concentrated and thereby damaging to plants.

Fortunately, the generally turbulent conditions and low ultra-violet radiation receipts of most United Kingdom cities prevent oxidant levels building up to the damaging levels they reach in such cities as Los Angeles, although short periods of less serious ozone levels have recently been reported in the United Kingdom (Atkins *et al.* 1972; Derwent and Stewart, 1973). The level at which throat and eye irritations become commonplace was, for instance, exceeded in London on 30 days in 1973.

Grit and dust

Dust particles of one sort or another are a very real problem in most cities. Without being a prerogative of urban areas, the intensity of the fall-out tends to be higher in towns, especially in the vicinity of heavy industrial plants, although comparable or even higher rates of deposition can sometimes occur locally in otherwise rural areas, as in the proximity of cement works and other mineral workings.

Grit is composed of large particles having a fall-out speed of a few feet per second, so that when they come from a concentrated source such as a chimney, they reach the ground within a few metres. Dust is similar in origin to grit but the size of the particles is smaller. Many are less than 20 μ in diameter so that their fall speed is less than the commonly experienced speeds of updraught in the air. For this reason they can be carried many hundreds of kilometres before impacting with earth's surface. Most grit and dust particles are composed of unburnt or partly burnt fuel or mineral dust from a variety of industrial and commercial activities.

Because most grit and much dust falls out of the air close to the source, deposit gauge readings cannot be extrapolated to give representative values for more than the area immediately around the gauge. Nevertheless, in the vicinity of iron and steel works, cement works, quarries and the like, the deposit can constitute a serious nuisance and disamenity, although the Clean Air Act of 1956 did require that grit and dust from existing solid fuel fired industrial furnaces be minimized and that new furnaces be fitted with arresters.

Conclusion

It is clear that many particulate and some gaseous pollutants are highly localized near to their source, so that the distribution of pollution in urban areas frequently parallels the pattern of emissions, with the highest smoke concentrations in the areas of older domestic, commercial and industrial development. Some pollutants such as SO_2 can disperse very widely and could constitute a problem well beyond the city boundary.

Radiation and Sunshine

Because of the blanket of pollution which shrouds many urban areas, radiation receipts at the ground are often severely reduced, more especially at times of low solar elevation, that is in the early morning and late evening in winter. Before the reductions in air turbidity following the Clean Air Acts of 1956 and 1968, many polluted British cities received between 25 and 55 per cent less winter radiation than nearby rural areas. In Manchester there has been a similar story. Here, changes in emissions reduced average smoke concentrations by about 50 per cent between 1961 and 1971, and partly in consequence of the cleaner air, December sunshine amounts rose by about 50 per cent between the 1950s and 1960s (Wood *et al.*, 1974). In central London, the loss of sunshine amounted to about 270 hours per year before the more recent improvements in air quality, there being a reduction of more than 50 per cent in December (Chandler, 1965). In central Manchester, the December deficit (1961–70) was 29 per cent (Tout, 1973). With the more recent improvements in air clarity, winter sunshine amounts in central London have increased by 50 per cent, being as much as 70 per cent in January (Jenkins, 1969). There have also been increases in radiation per sunshine hour (Monteith, 1966). The most serious reduction of radiation is in the short, ultra-violet part of the spectrum.

Investigations in London and in cities in other countries have shown an increase in radiation receipts at weekends when industrial emissions of pollution are generally reduced (Chandler, 1965).

Visibility

Because of the high aerosol concentrations in urban areas, visibilities are lower than in the country, and because of the particulates and tiny droplets, fogs form more readily and evaporate more slowly than in rural areas. The densest fogs are often found in the suburbs rather than in city centres which are warmer and have lower average vapour pressures. Central London, for instance, had, before the Clean Air Act of 1956, twice as many hours of fogs (visibilities of less than 1 000 m) per year than rural areas around the city, but only about the same number of hours of dense fog (visibilities of less than 40 m). The London suburbs, on the other hand, had only between 1·14 and 1·28 times as many hours of fog as in rural areas beyond the city, but up to four times as many dense fogs (Chandler, 1965).

The warmer air and gentler winds of city centres also delay the formation of evening fogs and their dispersal the following morning so that in the evening the fog lies like an annulus around a gradually disappearing clear centre and in the morning the fog clears in the stronger winds and more rapidly rising temperatures of rural and suburban areas to leave fog only in the centre of the city.

Since the Clean Air Acts (1956 and 1968) there have been dramatic improvements in visibilities in London and other cities where the emission of smoke has been reduced (Brazell, 1964; Freeman, 1968; Atkins, 1968; Jenkins, 1971; Wood, 1973), though some of the improvements might result partly from regional changes in climate such as stronger winds.

Temperature

The mass of warm air which more frequently than not and particularly by night covers built-up areas, is known as a 'heat-island'. On calm, clear nights the air in cities has frequently been measured as 5 °C warmer than above nearby rural areas and occasionally 10 °C differences have been recorded. Most towns with high central building densities will average between 1 ° and 2 °C warmer.

Fig. 14.3 *Temperatures in Leicester on a night with clear skies, light winds (less than 2·6 ms^{-1}) and an intense surface inversion (Chandler, 1967b)*

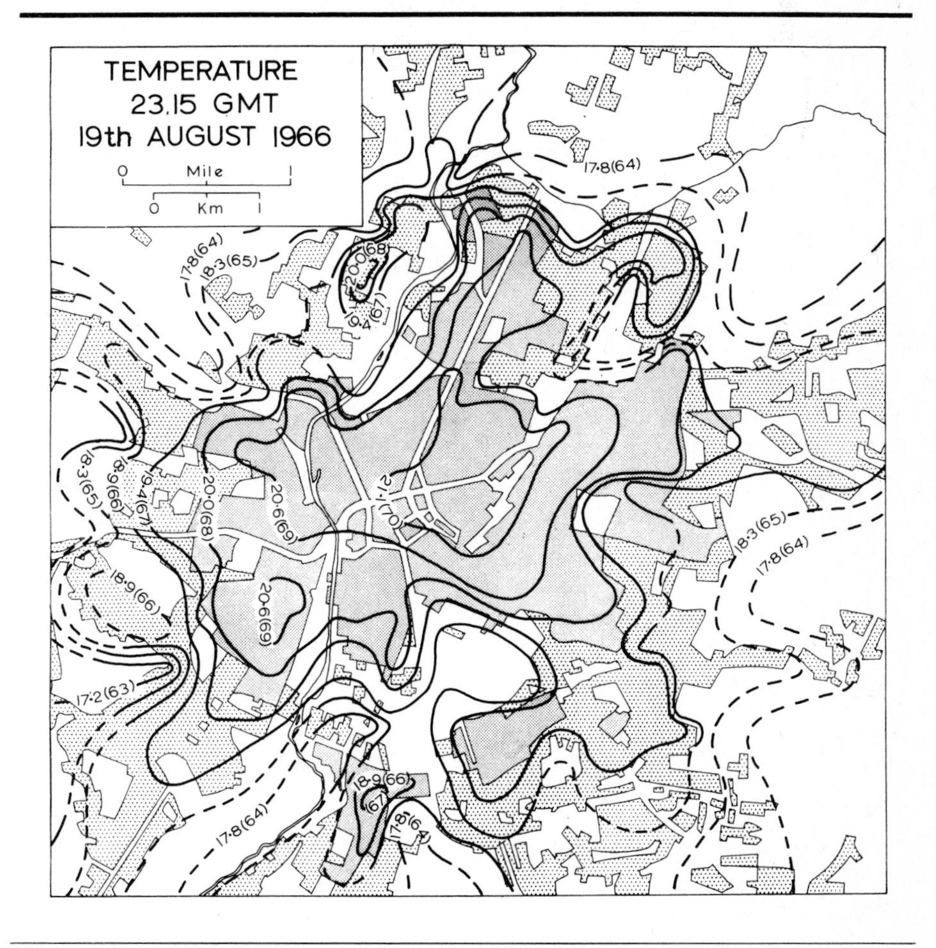

Shortly before dawn, temperatures over cities are normally higher than above the surrounding country, but after sunrise the higher heat capacity, active surface area and heat conductivity of the city fabric prevents the city warming as quickly as the vegetation-covered soils of the surrounding country. The haze hood above the city, morning fog and the rougher surface and stronger turbulence of the built-up area will also limit the rise of temperature. Thus in spite of a higher albedo and more evaporation cooling in rural areas, these soon warm to temperatures almost equalling and sometimes exceeding those of the city and in the generally unstable conditions and gentle breezes of daytime, the heat stored by the buildings and roads of the city plus that released by combustion is relatively easily dispersed. By night, several factors will prevent urban temperatures falling as fast as in the country. The heat from combustion processes of one sort or another and from metabolism can represent from one-sixth to one-third of the net all-wave radiation at the ground (Garnett and Bach, 1965). But most authors agree that it is relatively unimportant in the urban heat budget, since the heat is generally efficiently dispersed (Craddock, 1965a; Chandler, 1965). Far more telling is the release of stored heat from the urban fabric, back radiation into streets from the walls of tall buildings and the reduced turbulent diffusion of warm air trapped in the bottom of streets and courtyards (Chandler, 1967b). But although mixing between the buildings is reduced, that above the rooftops is increased by the serrated surface, often causing a downward transport of heat from a higher inversion.

The size of the city seems to be less important than building densities in controlling the maximum intensity of the heat island, large temperature differences being recorded on calm, clear nights in quite small, though densely developed towns. With moderate winds, the size of the city is clearly more important through its control upon air-flow, but even then there is a fairly close correspondence between the pattern of temperatures and local building densities (Fig. 14.3). This produces steep heat-island margins which parallel the edge of the city and a series of 'steps' in the temperature gradients marking the junction of city regions having distinctive and contrasted building developments (Fig. 14.3) (Chandler, 1967b). Any correlation between population size and heat-island intensities such as that proposed by Oke (1973) is likely to be indirect through a correlation with urban building densities and urban pollution.

Temporal variations in heat island intensity

Because of the many meteorological and topographical controls upon heat-island intensities these vary a good deal spatially and temporally. A diurnal variation with maximum intensities by night and minimum intensities or even a 'cold island' (lower temperatures in the city than outside) by day are almost universal. The air in central London parks is colder than around the city on about one day in three on average and one day in two from February to April, whilst cold-islands occur on only about one night in five with a clear winter peak. They often form when regional temperatures are rising, frequently occuring in groups of several days. Their origin is still not completely understood, but their most important cause would seem to be the more rapid increase of daytime temperatures in the open country compared with the city owing to differences in thermal conductivity, thermal capacity and the area of active surfaces.

Seasonal variations in the intensity of heat islands are more marked in some cities than others, but in the United Kingdom the strongest heat islands occur most frequently in summer and early autumn, representing a two to three months lag behind the solar regime (Chandler, 1965; Balchin and Pye, 1947; Parry, 1956). The pattern is of course a product of the heat physics of the urban fabric as affected by meteorological controls.

Studies in this country and elsewhere have highlighted the role of windspeed, cloud amount, evaporation rates, and the stability of the lower atmosphere among the several meteorological controls upon heat-island intensities. The strongest heat islands are nearly always associated with calm air, clear skies and an inversion of temperature in the lower 100 m to 300 m of the atmosphere (Lee, 1975). As wind speeds and turbulence increase, so the urban–rural temperature differences are diminished and because the size of the city is critical to changes in airflow, this might be an important fact in explaining why, on average, larger cities tend to have stronger average heat islands. The generally greater development densities found in the larger cities are also relevant here. At times of calm, city size is much less critical and even small settlements of just a few thousand people can develop strong night-time heat islands.

Oke and Hannell (1970) analysed the critical wind speeds for the elimination of the heat-island effect in several cities, including London and Reading, and derived the following relationship:

$$U = 11{\cdot}6 + 3{\cdot}4 \log P$$

where U is the critical wind speed in m s^{-1} and P the city population (used as an index of the spatial extent of the city). For London, the critical speed for the elimination of the heat island was 12 m s^{-1}; for Reading it was 4·7 m s^{-1}.

Spatial variations in heat island intensity

The pattern of temperature in a town is a function of the urban morphology as it affects the terms of the local heat balance, and the prevailing weather as it controls diffusion (Chandler, 1961b). Because cool air will be advected into windward suburbs, the highest temperatures will usually be found displaced a short distance downwind of the peak of highest building densities. But on calm nights the correlation between heat-island intensities and building densities is very close (Fig. 14.4) (Chandler, 1967a). Daytime street temperatures in the central core of cities are sometimes less than in the surrounding outer downtown area because of shading by very tall buildings, and suburban areas are cooler still, so that temperatures rise then fall in a traverse from the margins to the centre of the city.

Far fewer studies have been made of the vertical than of the horizontal structure of heat-islands though preliminary analysis of readings on the London Post Office Tower suggest that the warmer air is frequently less than 180 m deep. By analogy with enquiries in other countries, more especially the United States, above the heat island, temperatures are sometimes lower than over the country. Much more research is needed into this reversal of the urban–rural temperature gradient or 'crossover' of the urban and rural temperature profiles. At times of clear skies and light winds, an 'urban heat plume' will often rise above the city and then spread several miles downwind at intermediate heights beneath a regional inversion.

Fig. 14.4 *Mean temperatures along the line of a temperature traverse across London. The intensity of building development within 500 m of the line of traverse is also shown. This is calculated as the percentage of the area covered by buildings, multiplied by their mean height*

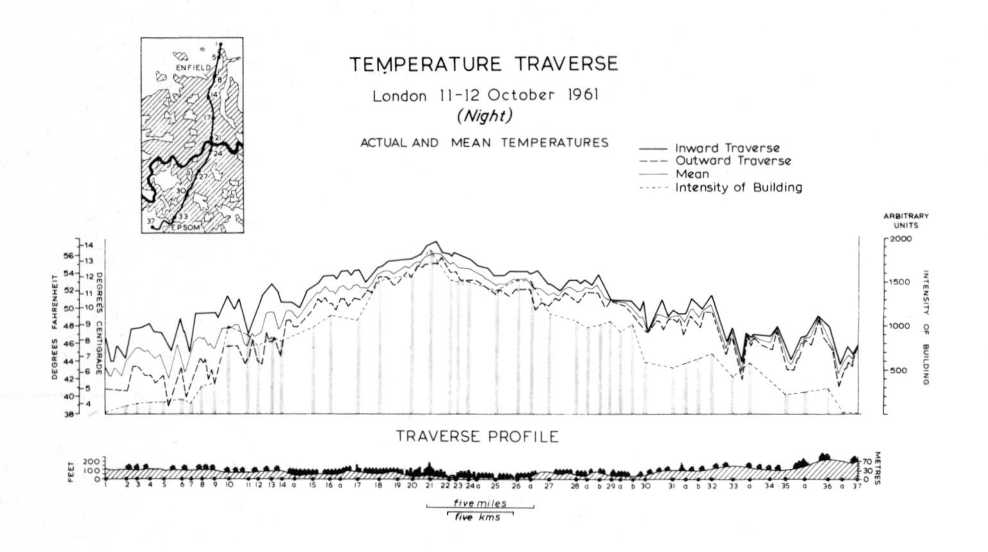

Humidity

Much less attention has been paid to the comparative moisture content of urban and rural atmospheres than to any other meteorological parameter and in consequence, there has been a far too ready acceptance of lower humidities in towns. This is by no means conclusively demonstrated, at least as a universal principle, by the limited observational evidence. Much of the urban surface is, of course, sealed by asphalt, concrete and other impervious and semi-impervious materials, but the frequently large remaining areas of open ground have often been overlooked. A great deal of rainwater is certainly led quickly underground but the quantity of water absorbed by bricks, tiles and the like is frequently underestimated. Nevertheless, there is little doubt that evaporation from cities is less than from vegetation covered soils, but this is partly compensated by the greater deposition of dew in rural areas and the large amounts of moisture added to urban atmospheres by combustion within the city.

These differences in the water balance of urban and rural areas are likely to manifest themselves in differences in absolute humidity only when the atmosphere is calm. For this reason, average annual or monthly vapour pressure contrasts between urban and rural areas are likely to be very small. Chandler (1965), for instance, found that the mean annual vapour pressure in London was only 0·2 mb lower than at a nearby rural climatological station. On calm nights, however, the vapour pressures of the air above London and above Leicester (Chandler, 1967a) were 1·5 to 2·0 mb higher in the centre than above the surrounding country (Fig. 14.5). The explanation was probably the low rates of diffusion of water vapour in air 'trapped' between tall buildings so that it retained its characteristically high daytime humidity.

Fig. 14.5 *Vapour pressure in Leicester on a night with 1 octa stratocumulus, little or no wind and a strong surface inversion (Chandler, 1967b)*

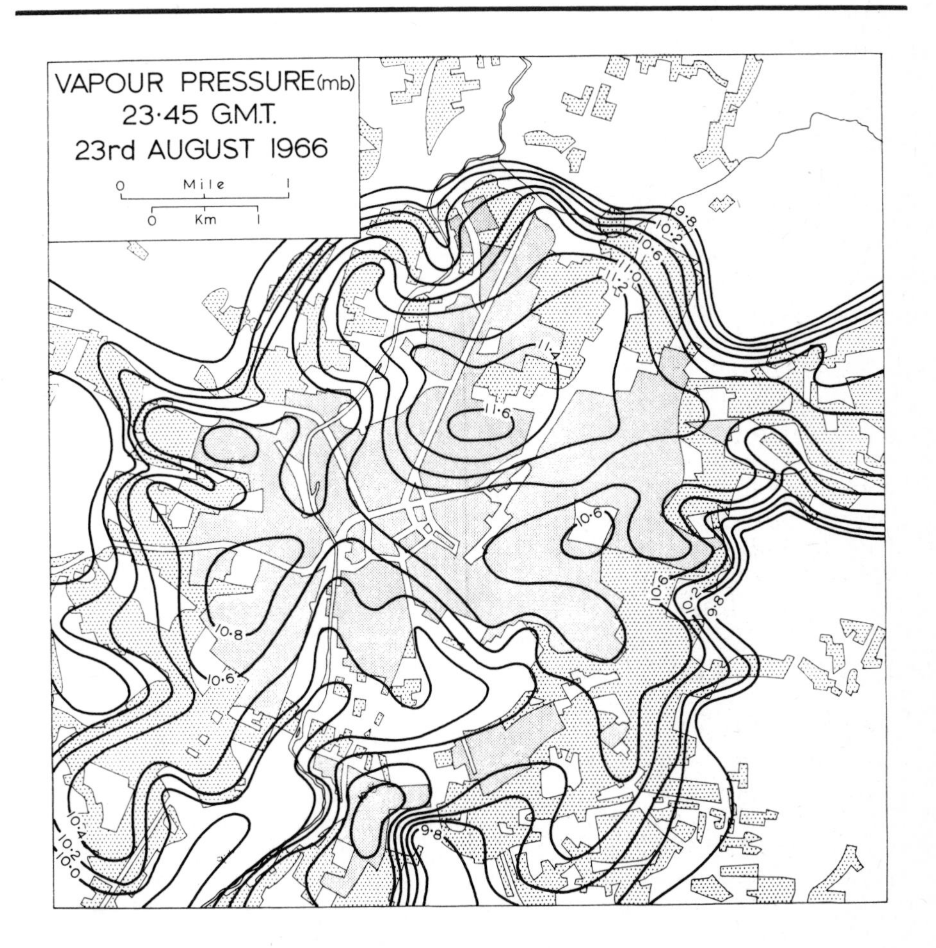

Urban relative humidities, being in part a function of prevailing temperatures, are inversely proportional to the intensity of the prevailing heat-island and average annual urban–rural differences of about 5 per cent have been widely reported. On individual nights, the pattern of relative humidites closely parallels the form of the heat-island, and this in its turn mirrors the distribution of urban building densities (Chandler, 1967a). Relative humidity differences between town and country then amount to between 20 and 30 per cent (Chandler, 1965, 1967a).

Rainfall

Atkinson (1970) has emphasized the difficulties and uncertainties which surround studies of the effects of urban areas upon precipitation. At first sight it seems plausible to postulate that because of increased pollution with active condensation and freezing nuclei, but more particularly because of more intense thermal and mechanical turbulence, there will not only be more cloud but also increased precipitation above or possibly to the lee of cities. Again, in this country, only a few cities, notably London (Atkinson 1968, 1969, 1970) and Liverpool (Palutikof, 1973) have been studied in detail.

Simple comparisons of gauge records inside and outside urban areas prove very little because of the well-known fickleness of the element, being subject to so many meteorological controls, of which the effect of the buildings is but one.

Atkinson (1968, 1969, 1970, 1971) has demonstrated the increase of heavy precipitation triggered off by the thermodynamic effects of London. Smaller towns are likely to have a more limited but as yet largely unproven effect upon rain-forming processes in air moving across them, and these effects are likely to release precipitation downwind of the built-up area.

Cities are unlikely materially to affect the frequency or amount of snowfall, but in the absence of orographic controls, fallen snow will melt more quickly in central urban parks than in suburban gardens, where it will often disappear several days before that covering the farmlands around a city.

In London, for instance, the number of days with snow lying averages about 5 to 6 days per year in the centre and about 8 days in the surrounding green belt. The more elevated parts of cities will often rise above the warmer air of the heat island and in consequence they will be subject to much more persistent snow cover.

Conclusion

Nipped by frost, scorched by the sun, buffeted by the winds and soaked by rain, man soon learnt to shelter himself against the more unpleasant forms of weather by the construction of buildings. His social and economic organizations demanded that his buildings be grouped in settlements and eventually required enclosed or semi-enclosed means of transport between them. And so man lived more and more in the controlled climates of his home, his place of work and of vehicles.

But in exercising these purposeful environmental controls he has accidentally changed atmospheric conditions outside as well as inside his buildings so that in many parts of the world, including Great Britain, the amount of time spent in 'natural' climates is really very small indeed. And so it is very important to appreciate the impact of cities upon climate and to apply this knowledge not only to prevent any further unconscious deterioration of the climate of our towns, but, whenever possible, to design more attractive atmospheric environments within and around our cities.

In the previous chapters of this book, two contrasted approaches have been adopted to the climatic characteristics of the British Isles. On the one hand there are a series of studies of the spatial distribution over the British Isles as a whole of a wide variety of individual climatic elements. These have been discussed and analysed in terms of averages, variability, extremes, intensities and frequencies; they have been related to changing synoptic situations, and examined for fluctuations during the instrumental period. On the other hand, climatic conditions peculiar to specific selected sites or environments have also been reviewed, and the inter-relationships between different climatic elements at such sites have been assessed. The latter approach accentuates the critical need to appreciate that although it is often convenient to examine climatic elements individually, the total range of such elements nevertheless co-exist in both time and space. Moreover, such characteristics as diurnal and seasonal periodicities, spatial gradients of change, and fluctuations with time, are often markedly different although they may be physically interrelated. It is this set of complex interactions that produces what we understand and appreciate as the overall climate of any particular area, and to which as individuals we respond in our daily lives. These climates, for all their internal complexities, also display a spatial pattern which can be generalized into a number of reasonably distinctive regional units. The purpose of this final chapter is to review some of the previous attempts at regional climatic description and definition, and to draw together some of the disparate but related strands of the earlier chapters into a comprehensible pattern of regional climates.

Arbitrary regions

There are, of course, innumerable climatic accounts of parts of Britain which partially integrate many of the climatic elements, and which present coherent and valuable summaries of overall climatic conditions. The areal units used, however, are likely to be arbitrary in terms of their climatic relevance, being defined by the prime purpose of the publication, e.g. a planning report; a botanical, agricultural or forestry enquiry; or a regional description aimed at the tourist market. There are, however, two sets of climatic studies along these lines that merit fuller consideration, both because of their general level and because they each succeed in covering most of the country.

The first of these sets comprises the climatic sections in the various county volumes based on the land utilization survey of Britain (Stamp, 1937–46). The county basis has no real logic in terms of climatology, but at least something is presented on all counties. The contributions are highly variable in length, content and quality, however, as they are also in relation to maps, diagrams and tabulated material. Much depends on the particular interest and approach of the author, and although it is perhaps invidious to draw distinctions between different contributors, Table 15.1 comprises the lengthier and more comprehensive accounts.

Table 15.1 *Selective list of more comprehensive climatic contributions to the county volumes of* The Land of Britain *(Stamp, 1937–46)*

Volume	County	Author	Publication Year	Pages
II	Lanarkshire, Renfrewshire and Dunbartonshire	Stamp, L. D.	1946	295–302
	Peeblesshire and Selkirkshire	Linton, D. L. and Snodgrass, C. P.	1946	387–99
	Angus	Dobson, E. B.	1946	498–510
III	North Wales (Caernarvonshire, Denbighshire and Flintshire)	Howell, E. J.	1946	637–43
IV	Yorkshire (West Riding)	Beaver, S. H.	1941	100–7
	Durham	Manley, G.	1941	193–200
	Cumberland and Westmorland	Manley, G.	1943	274–80
	Northumberland	Manley, G.	1945	429–39
V	Oxfordshire	Marshall, M.	1943	206–11
	Nottinghamshire	Edwards, K. C.	1944	439–49
	Staffordshire	Myers, J.	1945	578–84
	Warwickshire	McPherson, A. W.	1946	674–83
VI	Gloucestershire	Vince, S. W. E.	1942	327–33
	Worcestershire	Buchanan, K. M.	1944	427–33
VII	Lincolnshire	Stamp, L. D.	1942	463–8
VIII	Surrey	Stamp, L. D. and Willatts E. C.	1941	357–63
	Essex	Scarfe, N. U.	1942	412–18
	Sussex	Briault, E. W. H.	1942	481–7
	Kent	Stamp, L. D.	1943	565–71
IX	Somerset	Stuart-Menteath, T.	1938	31–41
	Hampshire	Green, F. H. W.	1940	313–22

The second of these sets is not represented over the whole country but it does cover very large areas, as can be seen in Fig. 15.1. These contributions are contained in the regional volumes for the annual conferences since 1949 of the British Association for the Advancement of Science. The location of these areas is thus arbitrary in climatological terms, their sizes vary considerably, and their boundaries are largely determined by the areal limits of interest scheduled for the volume (although the climatic boundaries are not invariably coincident with those for the rest of the volume). Moreover, the approach adopted is often very individualistic .Relatively lengthy studies of the

Fig. 15.1 *Limits of the regional climatic accounts in the volumes prepared for the annual conferences of the British Association for the Advancement of Science, for the 25-year period 1949–72*

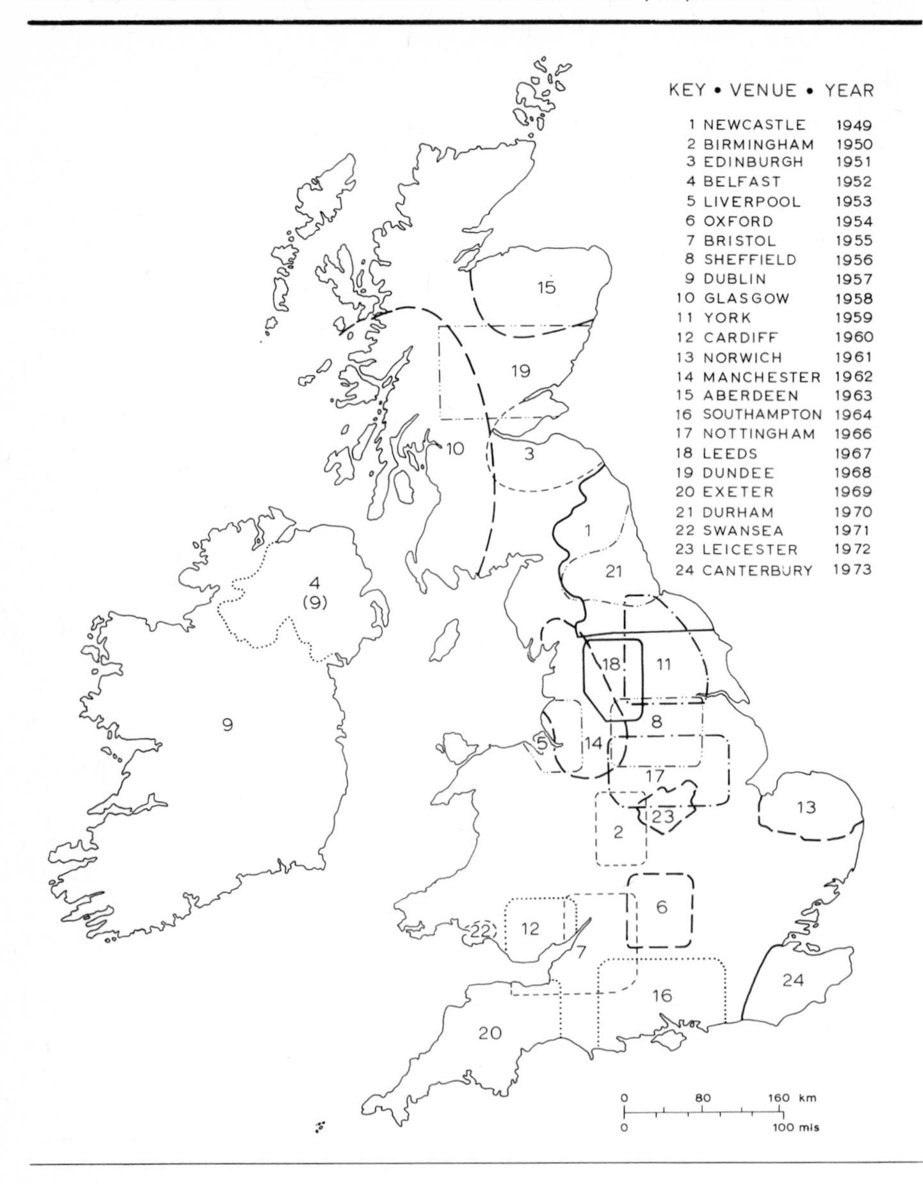

regional climate began with that for Merseyside (Gregory, 1953), and even more comprehensive reviews were presented for Bristol (Hannell, 1955), Sheffield (Garnett, 1956), Manchester (Crowe, 1962), Southampton (Barry, 1964), Nottingham (Barnes, 1966), Dundee (Berry, 1968) and Leicester (Pye, 1972). Collectively this means that the best covered parts of the country are Northern England and the Midlands with ten contributions, and south-west England and Wales with six contributions. In the cases of both Scotland and Ireland, there is a mixture of broad, very general, accounts and a few more detailed studies.

Global classifications

As a contrast to these specific studies, regional climatic contrasts in Britain can also be viewed against the background of classifications devised at a global or continental scale, for these should establish regional differences of major significance. Usually most, or even all, of the country falls into one major climatic type, but differences that are sometimes defined include areas of milder winters in the far south-

Fig. 15.2 *Regional divisions of the British Isles after:* (a) *Köppen (Miller, 1931);* (b) *Miller (1931);* (c) *Thornthwaite (1933);* (d) *Thran and Broekhuizen (1965), see Table 15.2*

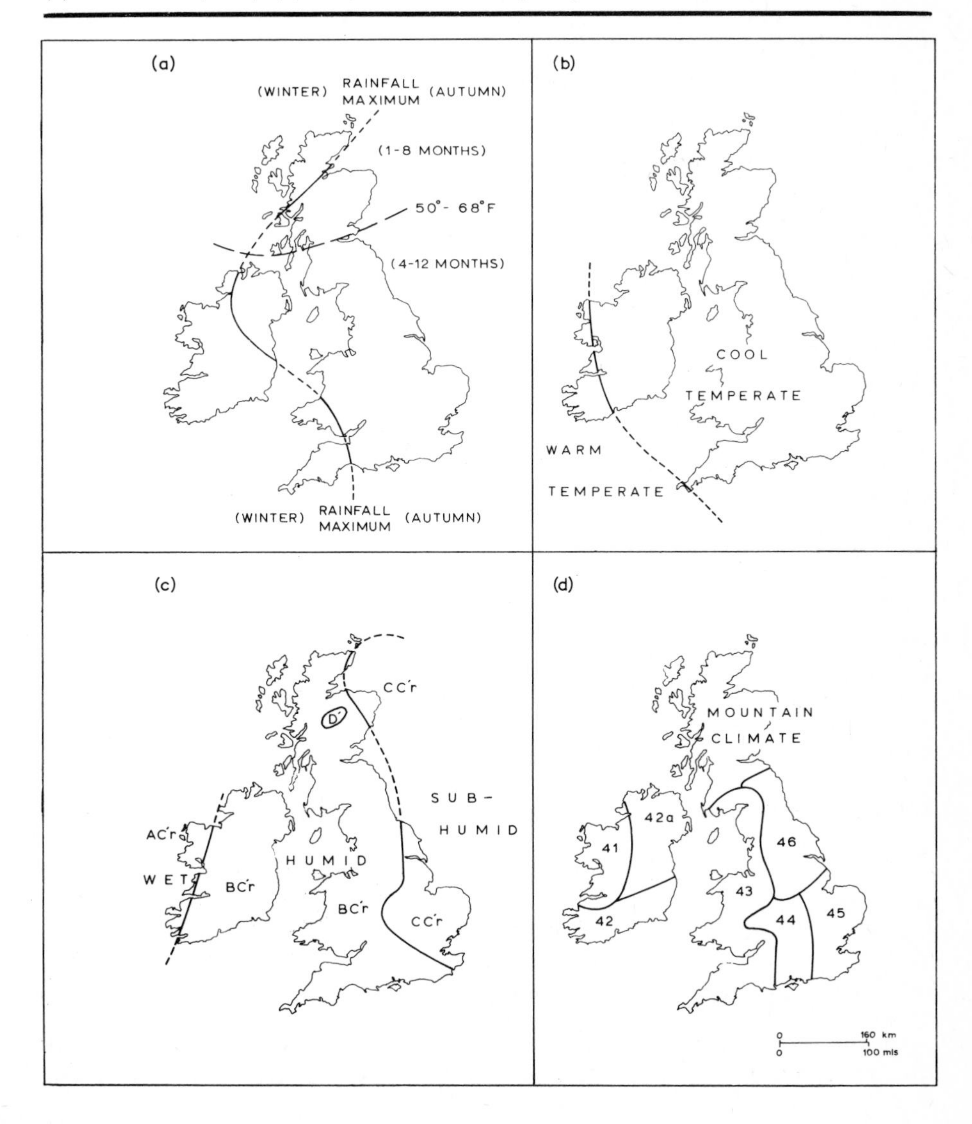

west, areas of summer rainfall maximum in limited parts of the east, areas of moisture deficit in the east and of moisture surplus in the west, and areas classed as mountain or upland climates especially but not only in Scotland (and usually where there are no observations). A reliance on lowland sites possessing long-term records is typical of all such classifications.

Thus in the early map based on Supan (Bartholomew and Herbertson, 1899) Britain was part of the West European Province, falling in the northern temperate zone between the mean annual isotherm of 20 °C and the mean warmest month isotherm of 10 °C. Köppen's classification has, of course, seen a whole host of modifications but in Köppen (1936) the country was classed as Cfb – warm temperate, rainy, marine, with warm summers and moist at all seasons. The main sub-division was between areas in the west with a mid-winter rainfall maximum, and others with an autumn maximum. In Miller's version of Köppen (Miller, 1931) this rainfall contrast was shown, while most of Scotland was distinguished as a cold zone, having only 1–8 months with mean monthly temperatures between 50° and 68 °F, the remainder of the country having 4–12 such months (Fig. 15.2a). Miller's own classification (Miller, 1931) only separated the far south-west, in terms of the mean temperature of the coldest month exceeding 43 °F, from the rest of the British Isles (Fig. 15.2b). Even less regional distinction was drawn by such other classifications as those by Trewartha (1954) where simply a few areas of summer rainfall maximum were designated; by Wissmann (Blüthgen, 1964) who separated only a few mountain climates in Scotland, Wales and Ireland: and by Troll and Paffen (1965) for whom only a few small Scottish mountain areas and the far south-western peninsulas of Ireland were distinctive.

In contrast, Thornthwaite (1933) defined three areas (Fig. 15.2c) from his P–E index, namely wet (A), humid (B) and sub-humid (C), although as regards temperature and seasonal rainfall, all areas were classed as microthermal (C^1) with abundant rain at all seasons (r). A small, ill-defined mountain (taiga) climate was also scheduled within the Scottish highlands. These were mainly repeated in Blumenstock and Thornthwaite (1941), but a somewhat different pattern was suggested by Thornthwaite's later classification (Thornthwaite, 1948) as applied to England and Wales by Howe (1956). This focused essentially on moisture conditions, with categories varying from per-humid over the mountains to dry sub-humid south of the Wash.

More regional differentiation was indicated by Thran and Broekhuizen (1965) with eight units being defined. Although the whole of Scotland was written off as a mountain climate, the other distinctions were made in some detail (Table 15.2 and Fig. 15.2d).

This type of approach was taken even further by Walter and Lieth (1967), who scheduled 16 climatic types in the British Isles, with mountain categories within some of these. There were, however, only two major categories that equated with global scale conditions, namely Region V (warm temperate, humid, with occasional frost) and Region VI (humid temperate with a cold season). The former was found only around the south-western fringes of England, Wales and

Ireland. Other differences were related to details of rainfall and temperature in terms of both magnitude and time of occurrence, but as the groupings in the original text are based on graphical presentation without clearly specified criteria, further interpretation is rather difficult in summary form. A generalization of this work is presented in Fig. 15.3, where the pattern has been amalgamated into only six units.

Fig. 15.3 *Simplified regional climates after Walter and Lieth (1967). A – warm temperate, humid with only occasional frost; B – transitional between A and C; C – humid temperate with a cold season; D – transitional from C to cooler summers and colder winters; E – transitional between C and F; F – mountainous climates*

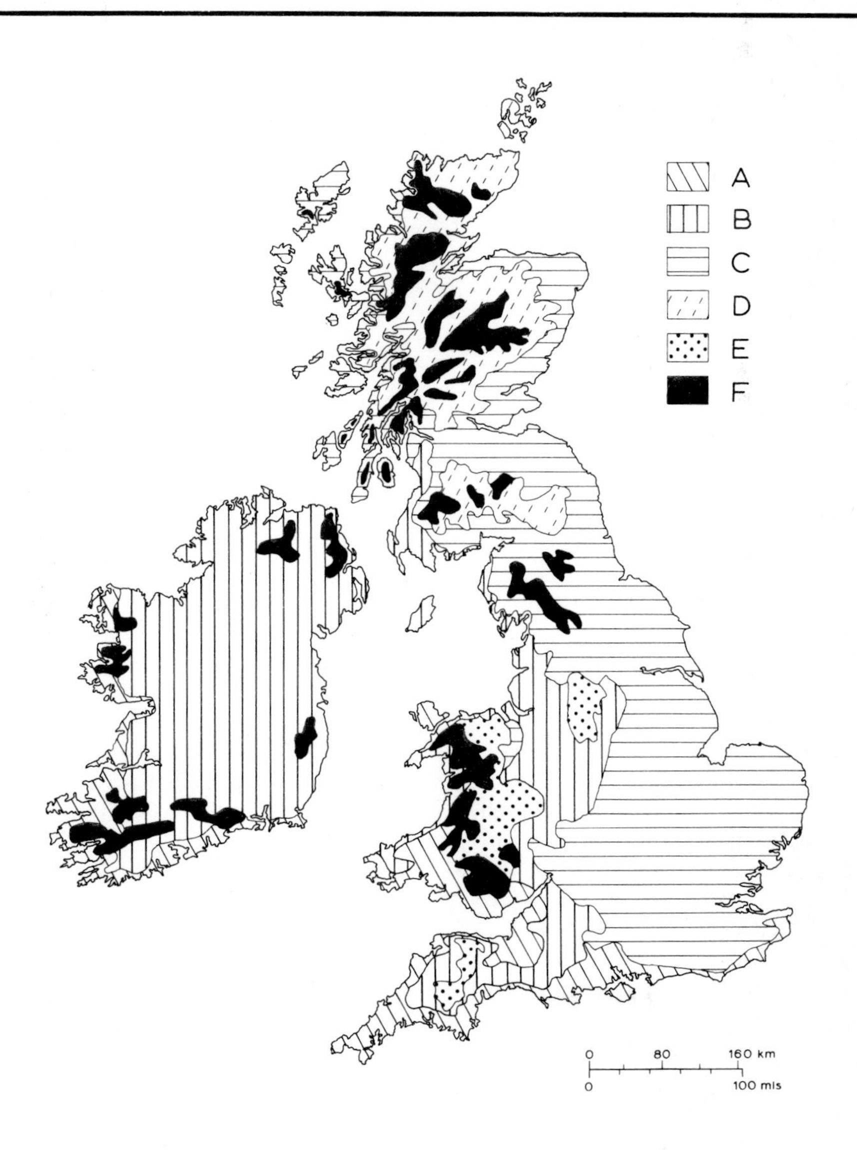

Table 15.2 *Characteristics of agro-climatic regions according to Thran and Broekhuizen (1965) – see Fig. 15.2*d

Climatic characteristics	Agro-climatic regions						
	41	**42**	**42a**	**43**	**44**	**45**	**46**
(i) Mean annual precipitation (mm)	1 100	750	700	850	700–800	600–650	650
(ii) Wettest month(s) and mean precipitation (mm)	Dec.–Jan. / 110–120	Aug. / 90	Dec. / 100	Oct–Dec.. / 100	Oct.–Dec. / 90	Oct. / 65	Oct. / 80
(iii) Driest month(s) and mean precipitation (mm)	April / 70	April / 50	April / 60	May / 45	April / 40	Feb. / 35	Feb. / 35
(iv) Warmest month mean temperature (°C)	14	15	14·5–15	15·5	16	16·5	14
(v) Coldest month mean temperature (°C)	5	7	5	7	5	4·5	3·5
(vi) Mean annual temperature (°C)	9	9	9·5	10·5	10	9·5	7·5
(vii) Number of months above 10°C	5	6	5	6	6	6	4
(viii) Number of months above 15°C	—	1	—	2	2	2	—

National scale studies

In the two latter classifications, the global scale categories were modified and sub-divided in terms of criteria specific to the British Isles, and there are many studies that focus entirely upon such 'domestic' criteria. Moreover, for a part of the world within which major horizontal contrasts are not great, whilst vertical rates of change are intense (see Chapter 12), such an approach has much to commend it. Some of the more widely used versions, however, are exceedingly generalized. Thus Mackinder (1902) divided Britain into four climatic provinces using the mean July isotherm of 60 °F and the mean January isotherm of 40 °F. The same criteria were used by Stamp and Beaver (1971), although the modifications that resulted within these four quadrants due to relief were also discussed. Rather more distinctive, perhaps, was the contribution by Brooks (1954) in which the English climate was classified in terms of the part it was assumed to play in maintaining the general level of health and the capacity for sustained and fruitful work. This was computed from monthly mean temperatures, the variability of temperature, the frequency of thunder, the number of wet days and the amount of solar radiation. The resulting map shows a general decrease in 'efficacy' from south-south-east to north-north-west.

One of the more useful and perceptive of earlier accounts was that by Tansley (1939). His major initial distinction was between coastal and inland climates, each of these then being sub-divided. The extreme Atlantic coasts were characterized by low summer and high winter temperatures for their latitude, with moderate rather than heavy rainfall, and with the rain being well distributed during the year with a December maximum. Moreover, sunshine tended to be below the average for the latitude, and windy conditions were common. The west coasts of England and Wales were seen as still mild but slightly less oceanic; the south coast of England as having above average sunshine but with oceanity decreasing from west to east; and the east coasts having lower winter temperatures and more snow, less rainfall especially in the south, and rather more sunshine than in the west. Of the inland areas, the northern parts of the country were described as receiving moderate rainfall, low sunshine values, with more extreme temperatures in the east and more foggy conditions in the west. Within this area, the mountainous parts of Scotland had markedly lower temperatures, very heavy precipitation, a large number of days with snow and with snow lying for all except mid-summer months, numerous gales, and high humidity, cloud and fog. In contrast, the English Midlands were seen as having the nearest approach to a continental-type climate, especially in the east, while in the west the rain-shadow effect of the Welsh massif was seen as producing very similar results. Finally, the inland areas of Ireland were classed as mainly maritime in character, especially in the south-west. As can be appreciated from this summary description, the regionalization was derived from an awareness of the quantitative aspects of the climate, but the final grouping and description relied heavily upon a qualitative evaluation and definition. Moreover, the assessment was strongly conditioned by the potential implications of the climate for vegetative growth.

A rather different approach was adopted by Sharaf (1954), who computed a number of derived indices for a range of stations, and then scheduled 150 potential climates, although a large proportion of these did not materialize. The criteria used were firstly, total annual accumulated temperatures above 42°F (five categories); secondly, the length of the growing season in days (five categories); and thirdly, a precipitation effectiveness index of a modified Thornthwaite form (six categories). This type of approach presents many possibilities but the particular resultant pattern is far too complicated to be of real value at a national level unless the results are published at a large enough scale. This has been done for Scotland at the scale of 1:625,000 based on accumulated temperatures and potential water deficit (Birse and Dry, 1970), and on exposure and accumulated frost (Birse and Robertson, 1970), as well as for bio-climatic regions (Birse, 1971).

One potential future approach would be through a factor analytical study, coupled with one of the many taxonomic classificatory methods that exist. This has been effected in a preliminary way for western Europe as a whole by McBoyle (1972), but to obtain a result with enough detail for differentiation within the British Isles would necessitate continuous records for the same time period for all climatic elements and for a large number of stations. These conditions are likely to obtain for only a limited number of sites, almost invariably in lowland areas, and it would prove extremely difficult to extrapolate effectively these classification patterns into areas for which data are unavailable.

Fig. 15.4 *Regional climates based on the length of the growing season, annual rainfall conditions and the seasonal incidence of rainfall (see Table 15.3)*

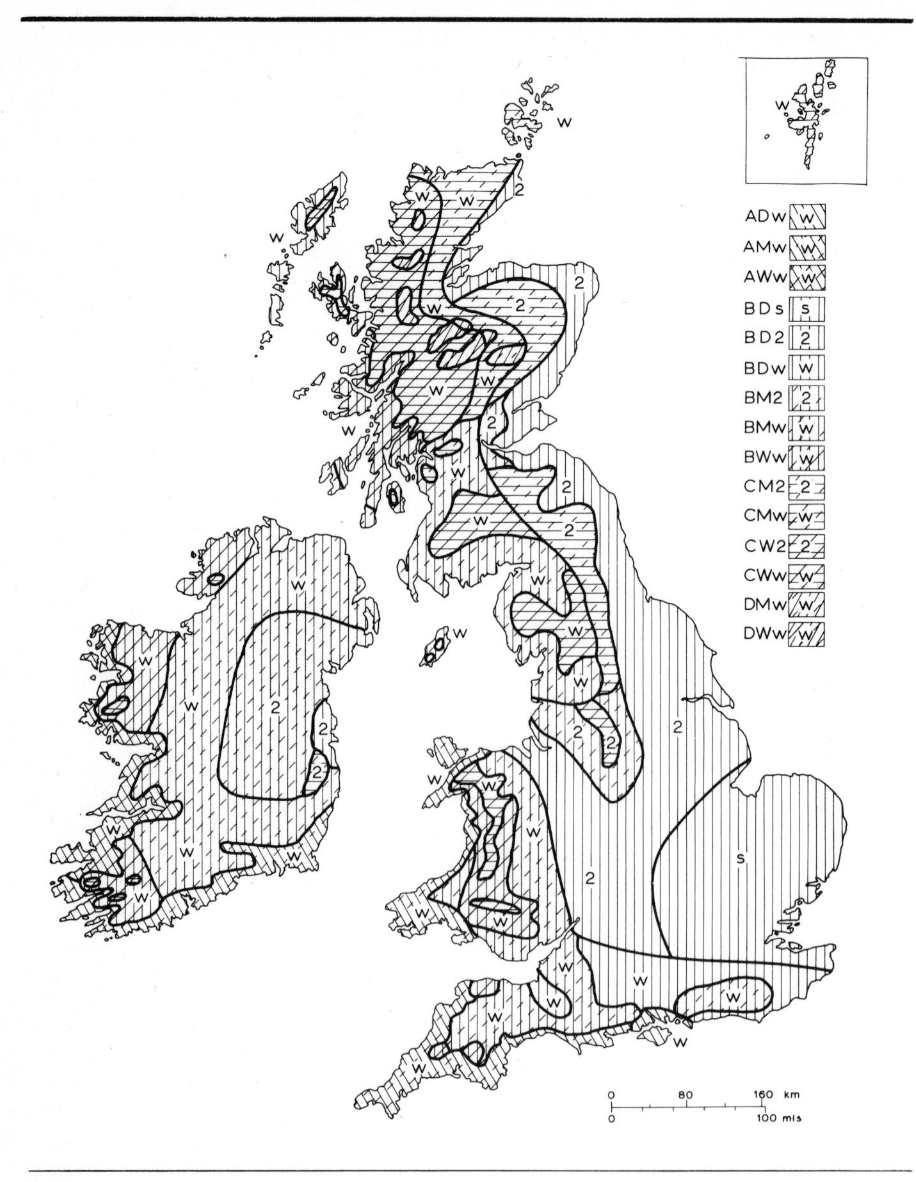

The method adopted, therefore, for the production of a pattern of regional climates in Fig. 15.4 has been essentially traditional in its approach. Three basic climatic characteristics have been selected, and their distributions superimposed. These characteristics were chosen both because of their general climatic relevance, and because they can each be mapped with reasonable accuracy for the whole country, both uplands and lowlands. Upland climatic areas can thus be defined in terms of their actual conditions, rather than by inference from altitude alone, or because of the virtual absence of data. Despite the subjectivity involved in this approach, the simplicity of the categories used, and the unavoidable generalizations that result, the pattern in Fig. 15.4 does integrate many of the detailed characteristics outlined in greater detail in earlier chapters.

The climatic distributions that are involved are as follows:

1. The length of the growing season: This is from work by Gregory (1964), based on his earlier maps of monthly accumulated temperatures (Gregory, 1954), and it depicts the number of months with accumulated temperatures above 6 °C. Actual intensity of the growing season is not included, but this is more likely to change in detail with the period of years studied than is the overall duration of such conditions, especially at the present scale of representation. Four categories are defined, and are indicated by the capital letter shown.

A A growing season of 9 or more months, this affecting the south-western coastal areas of the British Isles.

B A growing season of 7 or 8 months, which occurs over the whole of lowland Britain and Ireland, as well as over the lower hill areas. Contrasts in the total amount of accumulated warmth, and in seasonal concentration, necessarily occur between the more southerly and more northerly parts of this category, whilst differences in the intensity of temperature conditions during the non-growing (winter) season also occur between east and west.

C A growing season of 5 or 6 months, which affects the majority of the truly upland parts of the British Isles.

D A growing season of 4 or fewer months, that is really only experienced over the highest parts of the Grampians and the Western Highlands in Scotland.

2. A rainfall magnitude factor: This is based on Gregory (1959), and reflects certain aspects of the interaction of mean annual rainfall and the variability or reliability of this value. The three categories used are as follows:

W Those areas that receive at least 1 250 mm of rain a year with a probability of at least 0·7. These include the western peninsulas of Ireland and the major upland areas of the British Isles, with the exception of their eastern slopes.

D Those areas that receive 750 mm or less rain a year with a probability of at least 0'3. These comprise the bulk of lowland England and the eastern lowlands of Scotland, as well as the Somerset Levels and a small area around Dublin.

M Areas with rainfall between these two preceding sets of conditions, i.e. where high rainfall does not occur frequently yet low rainfall is itself infrequent or absent. These areas of moderate rainfall cover the lower hill areas of the country as well as most of Ireland.

3. Rainfall seasonality: This is derived from Crowe (1940), from which three general categories are obtained.

w Areas with a rainfall maximum in the winter half of the year. This includes most of Ireland, the western parts of Great Britain apart from the Dee–Mersey embayment, and southern England south of the Thames–Kennet line.

2 Areas in which the maximum is during the second half of the year, this comprising the bulk of the rest of the country.

s The area between the Thames and the Wash where there is a weakly developed tendency for a summer rainfall maximum.

The interaction of these categories implies a potential of thirty-six regional climates over the country, but in reality only fifteen of them occur. The listing in Table 15.3 should be read in conjunction with Fig. 15.4, whilst numerous further items of detail could be added by the reader from the earlier systematic chapters by using the Station Index and the associated map. Moreover, the four immediately preceding chapters are of critical relevance at various areal scales. The characteristics of upland climates (those in categories C and D on the growing season scale, and often category W on the rainfall scale) have been discussed in both greater depth and fuller detail in Chapter 12. Also, in virtually all of the regional types specified, there are coastal, as distinct from non-coastal, sections that possess the general characteristics that have been outlined in Chapter 11.

Table 15.3 *Regional climatic conditions relevant to Fig. 15.4*

Code	Climatic characteristics	Areas
ADw	Growing season ⩾9 months. Probability of ⩾0·3 of ⩽750 mm rain in a year. Winter maximum.	Southern Hampshire and the Isle of Wight.
AMw	Growing season ⩾9 months. Annual rainfall neither reliably high nor frequently low. Winter maximum.	South-western coasts of England. Western coast of Wales. Southern coast of Ireland.
AWw	Growing season ⩾9 months. Probability of ⩾0·7 of ⩾1 250 mm rain in a year. Winter maximum.	Western coast of Ireland.
BDs	Growing season 7–8 months. Probability of ⩾0·3 of ⩽750 mm rain in a year. Summer maximum tendency.	From the Wash to the Lower Thames valley inclusive, from the Vale of Oxford to the coasts of East Anglia and Essex.
BD2	Growing season 7–8 months. Probability of ⩾0·3 of ⩽750 mm rain in a year. Maximum in latter half of year.	Dee and Mersey estuaries, and Cheshire. Shropshire and the Severn valley lowlands; Cotswolds. West and East Midlands, and Lincolnshire. Vale of York, eastern Yorkshire and Humberside. Eastern coastal areas from north-eastern England to northern Scotland. Around Dublin.
BDw	Growing season 7–8 months. Probability of ⩾0·3 of ⩽750 mm rain in a year. Winter maximum.	Somerset Levels. Chalk country of southern England.

Table 15.3 continued

Code	Climatic characteristics	Areas
BM2	Growing season 7–8 months. Annual rainfall neither reliably high nor frequently low. Maximum in latter half of year.	To west, south and east of Southern Pennines. Eastern half of Irish lowlands. Eastern Scottish lowlands inland from the coast.
BMw	Growing season 7–8 months. Annual rainfall neither reliably high nor frequently low. Winter maximum.	Wales—hill country along east, and coastal plateaux of south and south-west. Inland West Country (except Levels and Moors). Western Weald. Lowlands of north-western England, south-western Scotland, and western Central Lowlands of Scotland. Low-lying Outer Hebrides. Isle of Man. Much of Ireland from north to south.
BWw	Growing season 7–8 months. Probability of ⩾0·7 of ⩾1 250 mm rain in a year. Winter maximum.	Upland moorlands in south-western England. Welsh hills around main uplands. Western peninsulas of Scottish Highlands, and low-lying Inner Hebrides. North-western and south-western Ireland inland from milder coasts.
CM2	Growing season 5–6 months. Annual rainfall neither reliably high nor frequently low. Maximum in latter half of year.	North Yorkshire moors. Eastern slopes of central and northern Pennines and Southern Uplands. North-eastern Grampians. Wicklow Mountains.
CMw	Growing season 5–6 months. Annual rainfall neither reliably high nor frequently low. Winter maximum.	Eastern slopes of Scottish Highlands and some parts central Grampians. Orkneys and Shetlands.
CW2	Growing season 5–6 months. Probability of ⩾0·7 of ⩾1 250 mm rain in a year. Maximum in latter half of year.	Main uplands of southern Pennines.
CWw	Growing season 5–6 months. Probability of ⩾0·7 of ⩾1 250 mm rain in a year. Winter maximum.	Main uplands of Wales from Snowdonia to Brecon Beacons. Lake District, Bowland Fells and western parts of northern Pennines. Western Southern Uplands, and hills in western Scottish Lowlands. Main mass of western Grampians and Scottish Highlands, and mountainous areas of Hebrides. Mountains within western peninsulas of Ireland.
DMw	Growing season ⩽4 months. Annual rainfall neither reliably high nor frequently low. Winter maximum.	Highest areas of eastern Grampians.
DW)	Growing season ⩽4 months. Probability of ⩾0·7 of ⩾1 250 mm rain in a year. Winter maximum.	Highest areas of central and western Grampians, and western Scottish Highlands

Furthermore, the larger proportion of the population lives in another sub-set of these regional conditions, where the general characteristics are modified by the particular properties associated with urban growth

(Chapter 14). Finally, the detailed aspects of the climate of any specific place are influenced by the processes related to local slopes, soils and vegetative cover, the relationships between which have been summarized in Chapter 13.

The regional grouping of climatic conditions thus presented in Fig. 15.4 and Table 15.3 provides a broadly based generalization, and although many differences of detail may occur within any one such region, differences between them are no less distinctive. These units, or others reasonably similar to them, form the basic spatial elements of the overall climatic conditions of the British Isles. The current state of knowledge and understanding of these conditions has been outlined in various sections of this book, to provide a comprehensive and up-to-date survey of the climate of these islands.

References

Abercromby, R. (1883) 'On certain types of British weather', *Q. Jl R. met. Soc.*, **9**, 1–25
Air Ministry (1959) *Handbook of Preventive Medicine*, London: H.M.S.O. (Ch. 8,' Environmental control', pp. 159–211)
Alcock, M. B. *et al.* (1974) 'Measurement, evaluation and management of climatic resources for grassland production in hill and lowland areas', *Climatic Resources and Economic Activity: A Symposium*, ed. J. A. Taylor, Newton Abbot: David and Charles, pp. 65–85 (Aberystwyth Memoranda in Agricultural Meteorology, no. 15)
Alexander, L. L. (1964) 'Tidal effects on the dissipation of haar', *Met. Mag.*, **93**, 379–80
Alexander, L. L. (1965) 'Easterly winds and low stratus at Leuchars', *Met. Mag.*, **94**, 292–8
Angström, A. (1924) 'Report to the International Commission for Solar Research on actinometric investigations of solar and atmospheric radiation', *Q. Jl R. met. Soc.*, **50**, 121–5
Atkins, D. H. F. *et al.* (1972) 'Photochemical ozone and sulphuric acid formation in the atmosphere over southern England', *Nature*, **235**, 372–6
Atkins, J. E. (1968) 'Changes in visibility characteristics at Manchester/Ringway Airport', *Met. Mag.*, **97**, 172–4
Atkinson, B. W. (1966) 'Some synoptic aspects of thunder outbreaks over south-east England, 1951–60', *Weather*, **21**, 203–9
Atkinson, B. W. (1967) 'Structure of the thunder atmosphere, south-east England, 1951–60', *Weather*, **22**, 335–45
Atkinson, B. W. (1968) 'A preliminary examination of the possible effect of London's urban area on the distribution of thunder rainfall, 1951–60', *Trans. Inst. Br. Geogr.*, **44**, 97–118
Atkinson, B. W. (1969) 'A further examination of the urban maximum of thunder rainfall in London, 1951–60', *Trans. Inst. Br. Geogr.*, **48**, 97–119
Atkinson, B. W. (1970) 'The reality of the urban effect on precipitation: a case study approach', *Tech. Notes Wld met. Org.*, **108**, 342–60
Atkinson, B. W. (1971) 'The effect of an urban area on the precipitation from a moving thunderstorm', *J. appl. Met.*, **10**, 47–55
Atkinson, B. W. and Smithson, P. A. (1972) 'An investigation into meso-scale precipitation distributions in a warm sector depression', *Q. Jl R. met. Soc.*, **98**, 353–68
Atkinson, B. W. and Smithson, P. A. (1974) 'Meso-scale circulations and rainfall patterns in an occluding depression', *Q. Jl R. met. Soc.*, **100**, 3–22
Atlas of Britain (1963) *Atlas of Britain and Northern Ireland, Oxford: Clarendon Press*
Balchin, W. G. V. and Pye, N. (1947) 'A microclimatological investigation of Bath and the surrounding district', *Q. Jl R. met.*

Soc., **73**, 297–323

Barnes, F. A. (1966) 'Weather and climate', *Nottingham and its Region*, ed. K. C. Edwards, Nottingham: British Association, pp. 60–102

Barnes, F. A. *et al.* (1956) 'The high summers of 1954 and 1955 and the long waves in the westerlies', *E. Midld Geogr.*, **6**, 9–26

Barnes, R. A. (1975) *The long distance transportation of atmospheric smoke and sulphur dioxide in and around the United Kingdom,* unpublished Ph.D. dissertation, University College London, Department of Geography

Barrett, E. C. (1966) 'Regional variations of rainfall trends in northern England, 1900–1959', *Trans. Inst. Br. Geogr.*, **38**, 41–58

Barry, R. G. (1963) 'Aspects of the synoptic climatology of central south England', *Met. Mag.*, **92**, 300–8

Barry, R. G. (1964) 'Weather and climate', *A Survey of Southampton and its Region,* ed. F. J. Monkhouse, Southampton: Southampton University Press, pp. 73–92

Barry, R. G. (1967) 'The prospect for synoptic climatology: a case study', *Liverpool Essays in Geography*, eds. R. W. Steel and R. Lawton, London: Longman, pp. 85–106

Barry, R. G. and Chambers, R. E. (1966) 'A preliminary map of summer albedos over England and Wales', *Q. Jl R. met. Soc.*, **92**, 543–8

Barry, R. G. and Chorley, R. J. (1968) *Atmosphere, Weather and Climate,* London: Metheun

Barry, R. G. and Perry, A. H. (1973) *Synoptic Climatology,* London: Metheun

Bartholomew, J. G. and Herbertson, A. J. (1899) *Atlas of Meteorology,* London: Constable, for Edinburgh Geographical Institute (*Bartholomew's Physical Atlas,* vol. 3)

Bayliss, P. L. *et al.* (1965) *Frequencies of Monthly Rainfall 1727–1964*, Bracknell: Meteorological Office (Agric. Memor., no. 129)

Beaver, S. H. and Shaw, E. M. (1970) *The Climate of Keele.* Keele: University library (Occasional Publ. no. 7)

Beckley, R. (1858) 'Description of the self-recording anemometer', *Rep. Br. Ass. Advmt Sci.*, **28**, 306–7

Belasco, J. E. (1948) 'The incidence of anticyclonic days and spells over the British Isles', *Weather,* **3**, 233–42

Belasco, J. E. (1952) *Characteristics of Air Masses over the British Isles,* London: H.M.S.O. (Meteorological Office, Geophys. Mem., no. 87)

Bendelow, V. C. (1969) 'A determination of the albedo of Morecambe Bay, north of a line from Aldingham to Bare', *Met. Mag.*, **98**, 305–9

Bernard, M. M. (1942) 'Precipitation', *Hydrology,* ed. O. E. Meinzer, New York: McGraw-Hill, ch. 2

Berry, W. G. (1968) 'Climate', *Dundee and district*, ed. Jones, S. J., Dundee: British Association, pp. 39–61

Best, A. C. (1935) *Transfer of Heat and Momentum in the Lowest Layer of the Atmosphere,* London: H.M.S.O. (Meteorological Office, Geophys. Mem., no. 65)

Best, A. C. (1951) *Some Statistics of Rate of Rainfall in Southern England,* London: Air Ministry, Meteorological Research Committee (M.R.P. no. 645)

Best, A. C. *et al.* (1952) *Temperature and Humidity Gradients in the First 100 m over S. E. England,* London: H.M.S.O. (Meteorological Office, Geophys. Mem., no. 89)

Best, R. H. (1968) 'Extent of urban growth and agriculture displacement in post-war Britain', *Urban Stud.,* **5**, 1–23

Betts, N. L. (1972) *The decrease in the frequency of London fog: a study of the relative contributions of smoke abatement and meteorological elements as causes of the decline in fog,* unpublished Ph.D. dissertation, Queen's University of Belfast, Department of Geography

Bilham, E. G. (1934) 'Notes on sequences of dry and wet months in England and Wales', *Q. Jl R. met. Soc.,* **60**, 514–16

Bilham, E. G. (1938) *The Climate of the British Isles,* London: Macmillan

Bilham, E. G. and Lloyd, A. C. (1932) 'The frequency distribution of daily rainfall', *Br. Rainf.,* **72**, 268–77

Birse, E. L. (1971) *Assessment of Climatic Conditions in Scotland, 3: The Bioclimatic Sub-regions,* Aberdeen: Soil Survey of Scotland, Macaulay Institute for Soil Research

Birse, E. L. and Dry, F. T. (1970) *Assessment of Climatic Conditions in Scotland, 1: Based on Accumulated Temperatures and Potential Water Deficit,* Aberdeen: Soil Survey of Scotland, Macaulay Institute for Soil Research

Birse, E. L. and Robertson, L. (1970) *Assessment of Climatic Conditions in Scotland, 2: Based on Exposure and Accumulated Frost,* Aberdeen: Soil Survey of Scotland, Macaulay Institute for Soil Research

Bjerknes, J. and Solberg, H. (1921) 'Meteorological conditions for the formation of rain', *Geofys. Publr,* 2(3)

Bleasdale, A. (1963) 'The distribution of exceptionally heavy daily falls of rain in the United Kingdom, 1863 to 1960', *J. Instn Wat. Engrs,* **17**, 45–55

Bleasdale, A. (1970) 'The rainfall of 14th and 15th September 1968 in comparison with previous exceptional rainfalls in the United Kingdom', *J. Instn Wat. Engrs,* **24**, 181–9

Bleasdale, A. (1971) 'The presentation of monthly rainfall', *British Rainfall, Supplement 1961–65,* London: H.M.S.O., 1, 227–59 (Met. O. 833)

Bleasdale, A. (1973) *The rainfall and flooding in Dorset on 18 July 1955 and the relationship of this fall to comparable heavy falls recorded in Britain since 1865,* unpublished (typescript available in Meteorological Office Library, Bracknell)

Bleasdale, A. and Chan, Y. K. (1972) 'Orographic influences on the distribution of precipitation', *Distribution of Precipitation in Mountainous Areas*; Geilo Symposium, Norway, 31 July–5 Aug. 1972, Geneva: World Meteorological Organization, vol. 2, pp. 322–33

Blench, B. J. R. (1967) 'An outline of the climate of Jersey',

Weather, **22**, 134–9
Blüthgen, J. (1964) *Allgemeine Klimageographie*, Berlin: de Gruyter, (*Lehrbuch der Allgemeine Geographie*, vol. 2)
Blumenstock, D. I. and Thornthwaite, C. W. (1941) 'Climate and the world pattern', *1941 Yearbook of Agriculture: Climate and Man*, Washington: US Department of Agriculture, pp. 98–112
Bonacina, L. C. W. (1927) 'Snowfall in the British Isles during the half century 1876–1925', *Br. Rainf.*, **67**, 260–87
Bonacina, L. C. W. (1936) 'Snowfall in the British Isles during the decade 1926–1935', *Br. Rainf.*, **76**, 272–92
Bonacina, L. C. W. (1966) 'Chief events of snowfall in the British Isles during the decade 1956–65', *Weather*, **21**, 42–6
Bonacina, L. C. W. (1945) 'Orographic rainfall and its place in the hydrology of the globe', *Q. Jl R. met. Soc.*, **71**, 41–9
Bonacina, L. C. W. (1948) 'Snowfall in the British Isles during the decade 1936 to 1945', *Br. Rainf.*, **88**, 209–18
Bonacina, L. C. W. (1955) 'Snowfall in Great Britain during the decade 1946–1955', *Br. Rainf.*, **95**, 219–30
Booth, R. E. (1968) 'The severe winter of 1963 compared with other cold winters, particularly that of 1947', *Weather*, **23**, 477–9
Bordovskaja, L. I. (1963) 'On the influence of the main Caucasus mountain ridge on the temperature regime of the free air', *Trudý tsent. aerol. Obs.*, **49**, 84–94 (in Russian)
Borisov, A. A. (1959) *Climates of the U.S.S.R.*, London: Oliver & Boyd (Transl. from the Russian by R. A. Medward)
Bornstein, R. D. (1969) 'Observed urban–rural wind speed differences in New York city', unpublished paper presented to American Geographical Union meeting, San Francisco
Bornstein, R. D. (1972) 'Two dimensional and non-steady numerical simulations of night time flow of a stable planetary boundary layer over a rough warm city', *Conference on Urban Environment and Second Conference on Biometeorology*; Philadelphia, Pa., 31 Oct–2 Nov., 1972, Preprints, Boston: American Meteorological Society, 89–94
Bornstein, R. D. *et al.* (1972) 'Recent observations of urban effects on winds and temperatures in and around New York city', *Conference on Urban Environment and Second Conference on Biometeorology*, Philadelphia, Pa., 31 Oct–2 Nov., 1972, Preprints, Boston: American Meteorological Society, pp. 28–33
Bowell, V. E. M. *et al.* (1966) *Thunderstorm Activity in Great Britain, 1955–64*, Leatherhead: Electrical and Allied Industries Research Association (Report no. 5168)
Bower, S. M. (1947) 'Diurnal variation of thunderstorms', *Met. Mag.*, **76**, 255–8
Bradshaw, A. D. (1974) 'The ecological effects of pollutants', *Fuel and the Environment: Conference Proceedings, vol 1, Papers*, London: Institute of Fuel, pp. 129–34 (Eastbourne conference, 1973)
Brandt, C. S. and Heck W. W. (1968) 'Effects of air pollutants on vegetation', *Air Pollution* (2nd edn) ed. A. C. Stern, New York: Academic Press, vol. 1, pp. 401–43
Brazell, J. H. (1964) 'Frequency of dense and thick fog in central

London as compared with frequency in outer London', *Met. Mag.*, **93**, 129–35

Brazell, J. H. (1968) *London Weather*, London: H.M.S.O. (Met. O. 783)

Brezowsky, H. *et al.* (1951) 'Some remarks on the climatology of blocking action', *Tellus*, **3**, 191–4

British Rainfall (1959–60) London: H.M.S.O. (published 1963) **99/100**

British Standards Institution (1972) *Code of Basic Data for the Design of Buildings; CP 3, Ch. V, Loading; pt. 2, Wind loads*, London: B.S.I.

Brooks, C. E. P. (1946) 'Annual recurrences of weather: singularities', *Weather*, **1**, 107–12 and 130–4

Brooks, C. E. P. (1954) *The English Climate*, London: English Universities Press

Brooks, C. E. P. and Carruthers, N. (1953) *Handbook of Statistical Methods in Climatology*, London: H.M.S.O. (M.O.538)

Brooks, C. E. P. and Glasspoole, J. (1922) 'The drought of 1921', *Q. Jl R. met. Soc.*, **48**, 139–68

Browning, K. A. and Harrold, T. W. (1969) 'Air motion and precipitation growth in a wave depression', *Q. Jl R. met. Soc.*, **95**, 288–309

Browning, K. A. and Ludlam, F. H. (1962) 'Airflow in convective storms', *Q. Jl R. met. Soc.*, **88**, 117–35

Browning, K. A. *et al.* (1968) 'Horizontal and vertical air motion and precipitation growth within a shower', *Q. Jl R. met. Soc.*, **94**, 498–509

Browning, K. A. *et al.* (1973) 'The structure of rainbands within a mid-latitude depression', *Q. Jl R. met. soc.*, **99**, 215–31

Browning, K. A. *et al.* (1974) 'Structure and mechanisms of precipitation and the effect of orography in a wintertime warm sector', *Q. Jl R. met. Soc.*, **100**, 309–30

Brunt, D. (1945) Contribution to discussion of Bonacina (1945) *op cit.*

Buchan, A. and Omond, R. T., eds (1905–10) 'The meteorology of the Ben Nevis observatories', *Trans. R. Soc. Edinb.*, vols 42–44

Burrage, S. W. (1972) 'Dew on wheat', *Agric. Met.*, **10**, 3–12

Butler, P. and Farley, B. C. (1973) *Surface Wind over Ireland (1961–70)*, Dublin: Eire Meteorological Service (Climatol. Note no. 2)

Caborn, J. N. (1957) *Shelter belts and microclimate*, Edinburgh: H.M.S.O. (Bull. For. Commn. no. 29)

Carpenter, A. (1903) *London Fog Enquiry, 1901–02: Report to the Meteorological Council*, London: H.M.S.O.

Carruthers, N. (1943) 'Variations in wind velocity near the ground', *Q. Jl R. met. Soc.*, **69**, 289–301

Chandler, T. J. (1960) 'Wind as a factor of urban temperatures: a survey in north-east London', *Weather*, **15**, 204–13

Chandler, T. J. (1961a) 'Surface breeze effects of Leicester's heat island', *E. Midld Geogr.*, **15**, 32–8

Chandler, T. J. (1961b) 'The changing form of London's heat

island', *Geography*, **46**, 295–307
Chandler, T. J. (1965) *The Climate of London*, London: Hutchinson
Chandler, T. J. (1967a) 'Absolute and relative humidities in towns', *Bull. Am. met. Soc.*, **48**, 394–9
Chandler, T. J. (1967b) 'Night-time temperatures in relation to Leicester's urban form', *Met. Mag.*, **96**, 244–50
Chandler, T. J. and Elsom, D. (1974a) *Meteorological controls upon ground level concentrations of smoke and sulphur dioxide in Great Britain: preliminary report*, unpublished report, Stevenage: Warren Spring Laboratory
Chandler, T. J. and Elsom, D. (1974b) *Meteorological controls upon ground level concentrations of smoke and sulphur dioxide in the Manchester area*, unpublished report, Stevenage: Warren Spring Laboratory
Clark, J. E. *et al* (1935) 'Report on the phenological observations in the British Isles from December 1933 to November 1934', *Q Jl R. met. Soc.*, **61**, 231–84
Clarke, P. C. (1964) 'The start of a prolonged cold spell', *Weather*, **19**, 366–9
Clarke, P. C. (1969) 'Snowfalls over south-east England, 1954–69', *Weather*, **24**, 438–47
Coleman, J. D. and Farrar, D. M. (1966) 'Soil moisture deficit and water table depth in London clay beneath natural grassland', *Q. Jl R. met. Soc.*, **92**, 162–8
Conrad, V. (1946) 'Usual formulas of continentality and their limits of validity', *Trans. Am. geophys. Un.*, **27**, 663–4
Conte, M. *et al.* (1973) 'A quasi-biennial rhythm of long lasting blocking systems over the Euro-Atlantic sector', *Ann. Met.*, **6**, 207–10
Cooper, J. P. (1964) 'Climatic variation in forage grasses, 1; leaf development in climatic races of *Lolium* and *Dactylis*', *J. appl. Ecol.*, **1**, 45–61
Corby, G. A. (1954) 'The airflow over mountains: a review of the state of current knowledge', *Q Jl R. met. Soc.*, **80**, 491–521
Corby, G. A. (1957) *Airflow Over Mountains*, London: H.M.S.O. (Met. Rep., no. 18)
Coulter, J. D. (1967) 'Mountain climate', *Proc. N. Z. ecol. Soc.*, **14**, 40–57
Court, A. (1948) 'Windchill', *Bull. Am. met. Soc.*, **29**, 487–93
Craddock, J. M. (1965a) 'Domestic fuel consumption and winter temperatures in London', *Weather*, **20**, 257–8
Craddock, J. M. (1965b) 'More rainfall statistics', *Weather*, **20**, 44–50
Craddock, J. M. (1970) 'Work in synoptic climatology with a digitized data bank' *Met. Mag.*, **99**, 221–31
Craddock, J. M. (1972) 'Urban development and its effects on the local temperature regime', *Statistician*, **21**, 63–75
Craddock, J. M. and Ward, R. (1962) *Some Statistical Relationships Between the Temperature Anomalies in Neighbouring Months in Europe and Western Siberia*, London: H.M.S.O.

(Meteorological Office Scientific Papers, no. 12)

Crisp, D. T. and Le Cren, E. D. (1970) 'The temperature of three small streams in northwest England', *Hydrobiologia,* **35**, 305–23

Crowe, P. R. (1940) 'A new approach to the study of the seasonal incidence of British rainfall', *Q. Jl R. met. Soc.,* **66**, 285–316

Crowe, P. R. (1962) 'Climate', *Manchester and its Region,* ed. C. F. Carter, Manchester: Manchester University Press, 17–46 (British Association Scientific Survey)

Crowe, P. R. (1971) *Concepts in Climatology*, London : Longman

Crumb (1973) : *Climatic Research Unit Monthly Bulletin,* University of East Anglia, 2(3)

Cutler, A. J. (1973) 'Seasonal indices – a review', *Weather,* **28** 59–64

Davenport, A. G. (1965) 'The relationship of wind structure to wind loading', *Wind Effects on Buildings and Structures*; proceedings of the conference held at the National Physical Laboratory, Teddington, 26–28 June, 1963, London: H.M.S.O., vol. 1, pp. 53–102 (N.P.L. Symposium no. 16)

Davis, N. E. (1967) 'The summers of north-west Europe', *Met. Mag.,* **96**, 178–87

Davis, N. E. (1968) 'An optimum summer weather index', *Weather,* **23**, 305–17

Davis, N. E. (1972a) 'Classified central England temperatures and England and Wales rainfall', *Met. Mag.,* **101**, 205–17

Davis, N. E. (1972b) 'The variability of the onset of spring in Britain', *Q. Jl R. met. Soc.,* **98**, 763–77

Day, G. J. (1961) 'Distribution of total solar radiation on a horizontal surface over the British Isles and adjacent areas', *Met. Mag.,* **90**, 269–84

De la Mothe, P. D. (1968) 'Middle latitude wavelength variation at 500 mb', *Met. Mag.,* **97**, 333–9

Deacon, E. L. (1955) 'Gust variation with height up to 150 m', *Q. Jl R. met. Soc.,* **81**, 562–73

Defant, F. (1949) 'Zur Theorie der Hangwinde', *Arch. Met. Geophys. Bioklim.,* **1**, 421–50

Derwent, R. G. and Stewart, H. N. M. (1973) 'Elevated ozone levels in the air of central London', *Nature,* **241**, 342

Dimbleby, G. W. (1953) 'Natural regeneration of pine and birch on the heather moors of north-east Yorkshire', *Forestry,* **26**, 41–52

Dinsdale, F. E. (1968) 'Fog frequencies at inland stations', *Met. Mag.,* **97**, 314–17

Dixon, F. E. (1939) 'Fog on the mainland and coasts of Scotland', *Prof. Notes met. Off., London,* **88**

Douglas, C. K. M. (1960) 'Some features of local weather in S. E. Devon', *Weather,* **15**, 14–17

Douglas, C. K. M. and Glasspoole, J. (1947) 'Meteorological conditions in heavy orographic rainfall in the British Isles', *Q. Jl R. met. Soc.,* **73**, 11–38

Dunsire, A. (1971) *Frequencies of Snow Depths and Days with Snow Lying at Stations in Scotland for Periods Ending Winter 1970/71,* Bracknell: Meteorological Office (Climatol. Memor., no. 70)

Dybeck, M. W. and Green, F. H. W. (1955) 'The Cairngorms weather survey', *Weather*, **10**, 41–8

E. S. S. A. (1966) *World Weather Records, 1951–60*, vol. 2, *Europe*, Washington: US Department of Commerce

Edington, J. M. (1966) 'Some observations on stream temperature', *Oikos*, **15**, 265–73

Edlin, H. L. (1957) 'Saltburn following a summer gale in south-east England', *Q. Jl For.*, **51**, 46–50

Edwards, R. S. (1968) 'Studies of airborne salt deposition in some North Wales forests', *Forestry*, **41**, 155–74

Eggleton, A. E. J. (1969) 'The chemical composition of atmospheric aerosols on Teesside and its relation to visibility', *Atmos. Environ.*, **3**, 355–72

Eide, O. (1942) 'On the temperature difference between mountain peak and free atmosphere at the same level', *Bergens Mus. Arb., Naturvitenskapelig rekke*, Nr. 2

Eire Meteorological Service (1971) *Monthly Seasonal and Annual Mean and Extreme Values of Duration of Bright Sunshine in Ireland, 1931–60*, Dublin: Department of Transport and Power, Meteorological Service

Eire Meteorological Service (1973) *Rainfall in Ireland, 1931–60*, Dublin: Department of Transport and Power, Meteorological Service

Ekhart, E. (1948) 'De la structure de l'atmosphere dans la montagne', *Météorologie*, 3–26

Elliott, A. (1964) 'Sea breezes at Porton Down', *Weather*, **19**, 147–50

Else, C. V. (1974) 'The Meteorological Office Mk. 5 wind system', *Met. Mag.*, **103**, 130–40

Erikson, W. (1971) 'Die Häufigkeit meteorologischer Fronten über Europa und ihre Bedeutung für die klimatische Gliederung des Kontinents', *Erdkunde*, **25**, 163–78

Evans, R. (1974) 'Infra-red linescan imagery and ground temperature', typescript of paper presented to Applied Environmental Science Symposium of the Institute of British Geographers, Norwich, University of East Anglia, Jan. 1974

Falconer, R. (1968) 'Windchill; a useful wintertime weather variable', *Weatherwise*, **21**, 227–9

Farmer, S. A. (1973) 'A note on the long term effects on the atmosphere of sea surface temperature anomalies in the North Pacific Ocean', *Weather*, **28**, 102–5

Faulkner, R. and Perry, A. H. (1974) 'A synoptic precipitation climatology of South Wales', *Cambria*, **1**, 127–38

Finch, C. R. (1972) 'Some heavy rainfalls in Great Britain, 1956–1971', *Weather*, **27**, 364–77

Findlater, J. (1963) 'Some aerial explorations of coastal airflow', *Met. Mag.*, **92**, 231–43

Findlater, J. (1964) 'The sea breeze and inland convection: an example of their interrelation', *Met. Mag.*, **93**, 82–9

Foord, H. V. (1968) 'An index of comfort for London', *Met. Mag.*,

97, 282–6
Francis, P. E. (1970) 'The effect of changes of atmospheric stability and surface roughness on off-shore winds over the east coast of Britain', *Met. Mag.*, **99**, 130–8
Freeman, M. H. (1962) 'North sea stratus over the Fens', *Met. Mag.*, **91**, 357–60
Freeman, M. H. (1968) 'Visibility statistics for London/Heathrow Airport', *Met. Mag.*, **97**, 214–18
Gameson, A. L. H. *et al.* (1957) 'Effects of heated discharges on the temperature of the Thames estuary', *Engineer, Lond.*, **204**, 3–12
Gameson, A. L. H. *et al.* (1959) 'A preliminary temperature survey of a heated river', *Wat. & Wat. Engng*, **63**, 13–17
Garland, J. A. (1969) 'Condensation on ammonium sulphate particles and its effect on visibility', *Atmos. Environ.*, **3**, 347–54
Garnett, A. (1939) 'Diffused light and sunlight in relation to relief and settlement in high latitudes', *Scott. geogr. Mag.*, **55**, 271–84
Garnett, A. (1956) 'Climate', *Sheffield and Its Region: A Scientific and Historical Survey*, ed. D. L. Linton, Sheffield: British Association, 44–69
Garnett, A. (1963) 'Survey of air pollution under characteristic anticyclonic conditions', *Int. J. Air Wat. Pollut.*, **7**, 963–6
Garnett, A. (1967) 'Some climatological problems in urban geography with reference to air pollution', *Trans. Inst. Br. Geogr.*, **42**, 21–43
Garnett, A. (1971) 'Weather inversions and air pollution', *Clean Air*, **1**, 16–21
Garnett, A. and Bach, W. (1965) 'An estimation of the ratio of artificial heat generation to natural radiation heat in Sheffield', *Mon. Weath. Rev.*, **93**, 283–5
Garnett, A. and Read, P. (1972) *Natural gas as a factor in air pollution*, London: Institution of Gas Engineers (Communication no. 868)
Garratt, J. R. (1972) 'Studies of turbulence in the surface layer over water (Lough Neagh). Part 2. Production and dissipation of velocity and temperature fluctuations', *Q. Jl R. met. Soc.*, **98**, 642–57
Geiger, R. (1965) *The Climate Near the Ground*, translated from the 4th German edn of 1961 by Scripta Technica, Cambridge, Mass.: Havard University Press
George, D. J. (1963) 'Temperature variations in a Welsh valley', *Weather*, **18**, 270–4
Gilchrist Shirlaw, D. W. (1966) *An Agricultural Geography of Great Britain*, Oxford: Pergamon
Gill, D. S. (1968) 'The diurnal variation of the sea-breeze at three stations in north-east Scotland', *Met. Mag.*, **97**, 19–24
Glasspoole, J. (1922) 'A comparison of the fluctuations of annual rainfall over the British Isles', *Br. Rainf.*, **62**, 260–6
Glasspoole, J. (1925) 'The relation between annual rainfall over Europe and that at Oxford and at Glenquoich', *Br. Rainf.*, **65**, 254–69

Glasspoole, J. (1929–30) 'The areas covered by intense and widespread falls of rain', *Minut. Proc. Instn Civ. Engrs,* **229**, 137–94
Glasspoole, J. (1938) 'The rainfall of the British Isles: a review of the development of our knowledge', *Water,* **40**, 77–87
Glasspoole, J. (1944) *Variation of Temperature Over the British Isles,* London: Meteorological Office (Memoir no. 452)
Glasspoole, J. and Hancock, D. S. (1951) 'Sunshine in the British Isles', *Q. Jl R. met. Soc.,* **77**, 454
Gloyne, R. W. (1957) *Problems of Surface Airflow and Related Phenomena in Agriculture, Horticulture and Forestry,* London: Air Ministry, Meteorological Research Committee (M.R.P. no. 1045)
Gloyne, R. W. (1958) *On the Growing Season,* Bracknell: Meteorological Office (Agric. Memor., no. 18)
Gloyne, R. W. (1964) 'Some characteristics of the natural wind and their modification by natural and artificial obstructions', *Scient. Hort.,* **17**, 7–19
Gloyne, R. W. (1967) 'Wind as a factor in hill climates', *Weather and Agriculture,* ed. J. A. Taylor, Oxford: Pergamon, pp. 59–67
Gloyne, R. W. (1968) 'Some climatic influences affecting hill land productivity', *Hill Land Productivity*; British Grassland Society, Symposium no. 4, Aberdeen, 1968, ed. I. V. Hunt, pp. 9–15
Gloyne, R. W. (1971) 'A note on the average annual mean of daily earth temperature in the United Kingdom', *Met. Mag.,* **100**, 1–6
Godske, C. L. *et al.* (1957) *Dynamic Meteorology and Weather Forecasting,* Boston & Washington: American Meteorological Society & Carnegie Institute
Gold, E. (1920) *Aids to Forecasting Types of Pressure Distribution for years 1905–18,* London: H.M.S.O. (Meteorological Office, Geophys. Mem., no. 16)
Gold, E. (1936) 'Wind in Britain; the Dines anemometer and some notable records during the last 40 years', *Q. Jl R. met. Soc.,* **62**, 167–206. (Additional records, **65** (1939) p. 66)
Gorham, E. (1958) 'The physical limnology of northern Britain: an epitome of the Bathymetrical Survey of the Scottish Freshwater Lochs, 1897–1909', *Limnol. Oceanogr.,* **3**, 40–50
Graham, I. R. (1968) 'An analysis of turbulence statistics at Fort Wayne, Indiana', *J. appl, Met.,* **7**, 90–3
Green, F. H. W. (1964) 'A map of annual average potential water deficit in the British Isles', *J. appl. Ecol.,* **1**, 151–8
Green. F. H. W. (1973) 'Changing incidence of snow in the Scottish Highlands', *Weather,* **28**, 386–94
Gregory, S. (1953) 'Weather and climate', *Merseyside: A Scientific Survey,* ed. W. Smith, Liverpool: Liverpool University Press for British Association, pp. 53–68
Gregory, S. (1954) 'Accumulated temperature maps of the British Isles', *Trans. Inst. Br. Geogr.,* **20**, 59–73
Gregory, S. (1955) 'Some aspects of the variability of annual rainfall over the British Isles for the standard period 1901–1930', *Q. Jl R. met. Soc.,* **81**, 257–62
Gregory, S. (1956) 'Regional variations in the annual rainfall over the British Isles', *Geogrl J.,* **122**, 346–53

Gregory, S. (1957) 'Annual rainfall probability maps of the British Isles', *Q. Jl. R. met. Soc.*, **83**, 543–9

Gregory, S. (1959) 'Climate and water supply in Great Britain', *Weather*, **14**, 227–32

Gregory, S. (1964) 'Climate', *The British Isles: A Systematic Geography*, ed. J. W. Watson and J. B. Sissons, London: Nelson, 53–73

Gregory, S. and Smith, K. (1967) 'Local temperature and humidity contrasts around small lakes and reservoirs', *Weather*, **22**, 447–505

Grindley, J. (1969) 'Some highlights of 1968 rainfall over Britain', *Weather*, **24**, 362–9

Grindley, J. (1972) 'Estimation and mapping of evaporation', *World Water Balance: Proceedings of the Reading Symposium*, Gentbrugge/Paris/Geneva: I.A.S.H./UNESCO/W.M.O., pp. 200–13

Gumbel, E. J. (1954) 'Statistical theory of extreme values and some practical applications', *Appl. Math. Ser.*, no. 33, Nat. Bur. Stats., Washington, D.C.

Hall, A. D. (1945) *The Soil* (5th edn) revised by G. W. Robinson, London: Murray

Hannell, F. G. (1955) 'Climate', *Bristol and its Adjoining Counties*, ed. C. M. MacInnes and W. F. Whittard, Bristol: British Association, pp. 47–65

Hanwell, D. J. and Newson, M. D. (1973) *Techniques in Physical Geography*, London: Macmillan

Hardman, C. E. *et al.* (1973) *Extreme Wind Speeds Over the United Kingdom for Periods Ending 1971*, Bracknell: Meteorological Office (Climatol. Memor., no. 50A)

Hare, F. K. (1970) 'The tundra climate', *Trans R. Soc. Can.*, **8**, 393–9

Harris, R. O. (1970) 'Notable British gales of the past fifty years', *Weather*, **25**, 57–68

Harrison, S. J. (1973) *An ecoclimatic gradient in north Cardiganshire, west central Wales*, unpublished Ph.D. dissertation, University College of Wales, Aberystwyth, Department of Geography

Harrison, S. J. (1974) 'Problems in the measurement and evaluation of the climatic resources of upland Britain', *Climatic Resources and Economic Activity: A Symposium*, ed. J. A. Taylor, Newton Abbot: David and Charles, pp. 47–63 (Aberystwyth Memoranda in Agricultural Meteorology, no. 15)

Harrower, T. N. S. (1963), 'Runway visual range, slant visual range and meteorological visibility', *Met. Mag.*, **92**, 26–34

Hartley, G. E. W. (1955) 'Remote-recording electrical anemograph', *Met. Mag.*, **84**, 111–15

Hastenrath, S. (1963) 'Uber den Einfluss der Massenerhebung auf den Verlauf der Klima-und Vegetationsstuffen in Mittelamerika und im Südlichen Mexiko', *Geogr. Annlr*, **45**, 76–83

Hawke, E. L. (1942) 'Notable falls of rain during intervals of a few days in Great Britain', *Q. Jl R. met. Soc.*, **68**, 279–86

Hawke, E. L. (1944) 'Thermal characteristics of a Hertfordshire frost-hollow', *Q. Jl R. met. Soc.*, **70**, 23–48

Hay, R. F. M. (1967) 'The association between autumn and winter circulations near Britain', *Met. Mag.*, **96**, 167–78

Hay, R. F. M. (1968) 'Relations between summer and September temperature anomalies in central England', *Met. Mag.*, **97**, 76–90
Hay, R. F. M. (1969) 'An analysis of monthly mean pressure patterns near the British Isles with possible applications to seasonal forecasting', *Met. Mag.*, **98**, 357–64
Hay, R. F. M. (1970a) 'Further analysis of monthly mean pressure patterns near the British Isles', *Met. Mag.*, **99**, 189–97
Hay, R. F. M. (1970b) 'October daily pressures and pressure patterns near Iceland related to temperature quintiles of the following winter in central England', *Met. Mag.*, **99**, 49–55
Helliwell, N. (1970) 'Some open scale measurements of wind over central London', *Tech. Notes Wld met. Org.*, **108**, 46–8
Hess, M. (1968) 'A new method of determining climatic conditions in mountain regions', *Geogr. Polonica*, **13**, 57–77
Heywood, G. S. P. (1933) 'Katabatic winds in a valley', *Q. Jl R. met. Soc.*, **59**, 47–63
Hindley, D. R. (1972) 'The importance of low sea surface temperatures in inhibiting convection along the North Sea coast in summer', *Met. Mag.*, **101**, 155–6
Höhn, R. (1973) 'On the climatology of the North Sea', *North Sea Science*; NATO Science Conference, Nov. 1971, ed. E. D. Goldberg, London: M.I.T. Press, pp. 183–236
Hogg, W. H. (1965) 'Climatic factors and choice of site, with special reference to horticulture', *Climatic Changes in Britain*, ed. C. G. Johnson and L. P. Smith, London: Academic Press for Institute of Biology, pp. 141–55
Hogg, W. H. (1967a) *Atlas of Long-term Irrigation Needs for England and Wales*, London: Ministry of Agriculture, Fisheries and Food
Hogg, W. H. (1967b) 'Meteorological factors in early crop production', *Weather*, **22**, 84–94
Hogg, W. H. (1971) 'Regional and local environments', *Potential Crop Production*, ed. P. F. Wareing and J. P. Cooper, London: Heinemann, pp. 6–22
Holgate, M. W. and Reed, L. (1974) 'The fate of pollutants', *Fuel and the Environment: Conference Proceedings, vol. 1, Papers*, London: Institute of Fuel, pp. 113–22 (Eastbourne conference, 1973)
Holland, D. J. (1967) 'Evaporation', *Br. Rainf.*, **101**, part 3, 5–34
Hosking, K. J. (1968) 'Winters at Ryde, Isle of Wight, 1918–67: a temperature assessment', *Weather*, **23**, 80–1
Howe, G. M. (1953) 'Observations on local climatic conditions in the Aberystwyth area', *Met. Mag.*, **82**, 270–4
Howe, G. M. (1956) 'The moisture balance in England and Wales', *Weather*, **11**, 74–82
Howe, G. M. (1962) 'Windchill, absolute humidity and the cold spell of Christmas, 1961', *Weather*, **17**, 349–58
Howe, G. M. (1972), *Man, Environment and Disease in Britain*, Newton Abbot: David and Charles
Hughes, G. H. (1972) 'Periodicity of 34 years in even summers', *Weather*, **27**, 241–6

Hunter, N. F. and Grant, S. A. (1971) 'The effect of altitude on grass growth in east Scotland', *J. appl. Ecol.*, **8**, 1–19

Hurst, G. W. (1967) 'Frost: an Aberystwyth symposium', *Weather*, **22**, 445–9

Hurst, G. W. and Smith, L. P. (1967) 'Grass growing days', *Weather and Agriculture*', ed. J. A. Taylor, Oxford: Pergamon, pp. 147–55

Ingham, B. (1967) 'Some comparisons between temperatures from *Met. Mag.*, **96**, 363–6
moorland stations and corresponding radio-sonde temperatures',

Institute of Fuel (1974) *Fuel and the Environment: Conference Proceedings, vol. 1, Papers*, London: Institute of Fuel (Eastbourne conference, 1973)

Irvine, S. G. (1968) 'An outline of the climate of Shetland', *Weather*, **23**, 392–403

Jackson, I. J. (1969) *Pressure Types and Precipitation Over N. E. England*, Newcastle: University of Newcastle Department of Geography (Research Ser. no. 5)

Jacobs, L. (1964) 'The validity of solar radiation data for the British Isles', *Q. Jl R. met. Soc.*, **90**, 105–6

Jenkin, P. M. (1942) 'Seasonal changes in the temperature of Windermere (English Lake District)', *J. Anim. Ecol.*, **11**, 461–504

Jenkins, I. (1969) 'Increase in averages of sunshine in Greater London', *Weather*, **24**, 52–4

Jenkins, I. (1971) 'Decrease in frequency of fog in Central London', *Met. Mag.*, **100**, 317–22

Job, D. A. (1974) Personal communication

Johnson, A. I. (1960) 'The summer of 1959', *Weather*, **15**, 185–96

Johnson, N. K. and Davies, E. L. (1927) 'Some measures of temperature near the surface in various kinds of soil', *Q. Jl R. met. Soc.*, **53**, 45–59

Jones, G. E. (1972) *The effects of selected environmental hazards on the growth of* ***Picea sitchensis*** *in three forests in Wales*, unpublished Ph.D. dissertation, University College of Wales, Aberystwyth

Jones, P. M. *et al* (1971) 'The urban wind velocity profile', *Atmos. Environ.*, **5**, 89–102

Kalma, J. D. (1968) 'A comparison of methods for computing daily mean air temperature and humidity', *Weather*, **23**, 248–52

Kelly, T. (1971) 'Thick and dense fog at London/Heathrow Airport and Kingsway/Holborn during the two decades 1950–59 and 1960–69', *Met. Mag.*, **100**, 257–67

Köppen, W. (1936) *Das geographische System der Klimate*, Berlin: Bornträger (*Handbuch der Klimatologie* ed. W. Köppen and R. Geiger, vol. 1, part C)

Kondratyev, K. Y. (1969) *Radiation in the Atmosphere*, New York: Academic Press

Lake, J. V. (1956) 'The temperature profile above bare soil on clear nights', *Q. Jl R. met. Soc.*, **82**, 187–97

Lamb, H. H. (1943) *Haars or North Sea Fogs on the Coast of Great Britain*, (2nd edn) Meteorological Office Report, M.O. 504

Lamb, H. H. (1950) 'Types and spells of weather around the year in the British Isles: annual trends, seasonal structure of the year, singularities', *Q. Jl R. met. Soc.*, **76**, 393–429
Lamb, H. H. (1958) 'The occurrence of very high surface temperatures', *Met Mag*, **87**, 39–43
Lamb, H. H. (1964) *The English Climate,* (2nd edn) London: English Universities Press
Lamb, H. H. (1965) 'Frequency of weather types', *Weather,* **20**, 9–12
Lamb, H. H. (1966) *The Changing Climate,* London: Metheun
Lamb, H. H. (1967) 'Britain's changing climate', *Geogrl J,* **133**, 445–68
Lamb, H. H. (1971) 'Climates and circulation regimes developed over the northern hemisphere during and since the last Ice Age', *Palaeogeogr., Palaeoclim., Palaeoecol.*, **10**, 125–62
Lamb, H. H. (1972) *British Isles Weather Types and a Register of the Daily Sequence of Circulation Patterns 1861–1971*, London: H.M.S.O. (Meteorological Office, Geophys. Mem., no. 116)
Lamb, H. H. (1973a) *The Seasonal Progression of the General Atmospheric Circulation Affecting the North Atlantic and Europe,* Norwich: University of East Anglia (Climatic Research Unit Report no. 1)
Lamb, H. H. (1973b) 'Whither climate now?', *Nature,* **244**, 395–7
Lamb, H. H. (1974) 'Contributions to historical climatology: The Middle Ages and after: Christmas weather and other aspects', *Bonn. met. Abh.*, **17**, 549–68
Lamb, H. H. and Johnson, A. I. (1960) 'The use of monthly mean "Climat" charts for the study of large scale weather patterns and their seasonal development', *Weather,* **15**, 83–91
Lamb, H. H. *et al.* (1973) *Northern Hemisphere Monthly and Annual Mean-Sea-Level Pressure Distribution for 1951–66, and Changes of Pressure and Temperature Compared With Those of 1900–39,* London: H.M.S.O. (Meteorological Office, Geophys. Mem. no. 118)
Landsberg, H. E. (1956) 'The climate in towns', *Man's Role in Changing the Face of the Earth,* ed. W. L. Thomas, Chicago: University of Chicago Press, pp. 584–606
Langford, T. E. and Aston, R. J. (1972) 'The ecology of some British rivers in relation to warm water discharges from power stations', *Proc. R. Soc.*, B, **180**, 407–19
Lavis, M. E. and Smith, K. (1972) 'Reservoir storage and the thermal regime of rivers, with special reference to the river Lune, Yorkshire', *Sci. Total Environ.*, **1**, 81–90
Lawrence, E. N. (1953) 'Föhn temperature in Scotland', *Met. Mag.*, **82**, 74–9
Lawrence, E. N. (1954) 'Nocturnal winds', *Prof. Notes met. Off.. Lond.*, **111**
Lawrence, E. N. (1957) 'Estimation of the frequency of runs of dry days', *Met. Mag.*, **86**, 257–69 and 301–4
Lawrence, E. N. (1966) 'Sunspots: a clue to bad smog?', *Weather,* **21**, 367–70
Lawrence, E. N. (1967) 'Atmospheric pollution during spells

of low-level air temperature inversion', *Atmos. Environ.*, **1**, 561–76

Lawrence, E. N. (1969) 'High values of atmospheric pollution in summer at Kew and the associated weather', *Atmos. Environ.*, **3**, 123–33

Lawrence, E. N. (1971a) 'Clean Air Act', *Nature*, **229**, 334–5

Lawrence, E. N. (1971b) 'Synoptic type rainfall averages over England and Wales', *Met. Mag.*, **100**, 333–9

Lawrence, E. N. (1971c) 'Urban climate and day-of-the-week', *Atmos. Environ.*, **5**, 935–48

Lawrence, E. N. (1972) 'Westerly type rainfall and atmospheric mean-sea-level pressure over England and Wales', *Met. Mag.*, **101**, 129–37

Lawrence, E. N. (1973a) 'Cyclonic type rainfall and atmospheric mean-sea-level pressure over England and Wales', *Met. Mag.*, **102**, 51–7

Lawrence, E. N. (1973b) 'High values of daily areal rainfall over England and Wales and synoptic patterns', *Met. Mag.*, **102**, 361–6

Lee, D. O. (1975) 'Rural atmospheric stability and the intensity of London's heat island', *Weather*, **30**, 102–8

Levick, R. B. M. (1949) 'Fifty years of English weather', *Weather*, **4**, 206–11

Lewis, L. F. (1939) 'The seasonal and geographical distribution of absolute drought in England', *Q. Jl R. met. Soc.*, **65**, 367–80

Linzon, S. N. (1972) 'Effects of sulphur oxides on vegetation', *For. Chron.*, **48**, 182–6

Logue, J. J. (1971) *Extreme Wind Speeds in Ireland*, Dublin: Eire Meteorological Service (Tech. Notes, no. 35)

Lowndes, C. A. S. (1963) 'Cold spells at London', *Met. Mag.*, **92**, 165–76

Lowndes, C. A. S. (1968) 'Forecasting large 24-hour rainfall totals in the Dee and Clwyd river authority area from September to February', *Met. Mag.*, **97**, 226–35

Lowndes, C. A. S. (1969) 'Forecasting large 24-hour rainfall totals in the Dee and Clwyd river authority area from March to August', *Met. Mag.*, **98**, 325–40

Lowndes, C. A. S. (1971) 'Substantial snowfalls over the United Kingdom, 1954–69', *Met. Mag.*, **100**, 193–207

Lumb, F. E. (1964) 'The influence of cloud on hourly amounts of total solar radiation at the sea surface', *Q Jl R. met. Soc.*, **90**, 43–56

Lyall, I. T. (1971a) 'An exceptionally warm day in January', *Weather*. **26**, 541–5

Lyall, I. T. (1971b) 'Early warm spells since 1957', *Weather*, **26**, 46–54

Lyall, I. T. (1971c) 'English winters since 1950', *Weather*, **26**, 445

Lyall, I. T. (1973) 'Low temperatures in southern Britain, 1953–72', *Weather*, **28**, 134–40

McBoyle, G. R. (1972) 'Factor analytic approach to a climatic classification of Europe', *Climatol. Bull.*, **12**, 1–11

Macan, T. T. (1958) 'The temperature of a small stony stream',

Hydrobiologia, **12**, 89–106

Mackinder, H. J. (1902) *Britain and the British Seas,* London: Heinemann

Manley, G. (1936) 'The climate of the northern Pennines', *Q. Jl R. met. Soc.,* **62**, 103–15

Manley, G. (1938) 'High level records from the northern Pennines', *Met. Mag.,* **73**, 69–72

Manley, G. (1942) 'Meteorological observations on Dun Fell, a mountain station in northern England', *Q. Jl R. met. Soc.,* **68**, 151–65

Manley, G. (1943) 'Further climatological averages for the northern Pennines, with a note on topographical effects', *Q. Jl R. met. Soc.,* **69**, 251–61

Manley, G. (1944) 'Topographical features and the climate of Britain', *Geogrl J.,* **103**, 241–63

Manley, G. (1945a) 'The effective rate of altitudinal change in temperate Atlantic climates', *Geogrl Rev.,* **35**, 408–17

Manley, G. (1945b) 'The helm wind at Crossfell, 1937–39', *Q. Jl R. met. Soc.,* **71**, 197–219

Manley, G. (1951) 'The range of variation of the British climate', *Geogrl J.,* **117**, 43–68

Manley, G. (1952) *Climate and the British Scene,* London: Collins

Manley, G. (1953) 'The mean temperature of central England, 1698–1952', *Q. Jl R. met. Soc.,* **79**, 242–61

Manley, G. (1955) 'The climate of Malham Tarn', Annual Rept., Council for the Promotion of Field Studies, 1955–6, pp. 43–53

Manley, G. (1969) 'Snowfall in Britain over the past 300 years', *Weather,* **24**, 428–37

Manley, G. (1970a) 'Climate in Britain over 10,000 years', *Geogrl Mag.,* **43**, 100–7

Manley, G. (1970b) 'The climate of the British Isles', *Climates of Northern and Western Europe,* ed. C. C. Wallen, Amsterdam: Elsevier, 81–133 (*World Survey of Climatology,* vol. 5)

Manley, G. (1971) 'The mountain snows of Britain', *Weather,* **26**, 192–200

Manley, G. (1974) 'Central England temperatures: monthly means 1659 to 1973' *Q. Jl R. met. Soc.,* **100**, 389–405

Marsh, K. J. and Foster, M. D. (1967) 'An experimental study of the emissions from chimneys in Reading', *Atmos. Environ.,* **1**, 527–50

Marshall, N. (1968) 'The icefields around Iceland in Spring 1968', *Weather,* **23**, 368–76

Mason, B. J. (1969) 'Some outstanding problems in cloud physics: the interaction of microphysical and dynamical processes', *Q. Jl R. met. Soc.,* **95**, 449–85

Mason, B. J. (1970) 'Future developments in meteorology: an outlook to the year 2000', *Q. Jl R. met. Soc.,* **96**, 349–68

Matthews, R. P. (1971) 'Synoptic categories of precipitation over the English Midlands – a pluvial transect', *N. Staffs Jnl Field Stud.,* **11**, 1–13

Meetham, A. R. (1945) *Atmospheric Pollution in Leicester,* London:

H.M.S.O. (Tech. Pap. atmos. Pollut. Res. Comm., no. 1)

Meteorological Office (1952) *Climatological Atlas of the British Isles,* London: H.M.S.O.

Meteorological Office (1956) *Handbook of Meteorological Instruments, Part 1: Instruments For Surface Observations,* London: H.M.S.O.

Meteorological Office (1963) *Averages of Temperature for Great Britain and Northern Ireland, 1931–60,* London: H.M.S.O.

Meteorological Office, (1964, 1965) *Weather in Home Fleet Waters and the North East Atlantic,* London: H.M.S.O. (in 2 vols)

Meteorological Office (1968a) *Monthly and Annual Averages of Rainfall for the United Kingdom of Great Britain and Northern Ireland for the W.M.O. period 1931–60,* Bracknell: Meteorological Office (Hydrol. Memor., no. 37)

Meteorological Office (1968b) *Tables of Surface Wind Speed and Direction Over the United Kingdom,* London: H.M.S.O.

Meteorological Office (1969) *Observer's Handbook* (3rd edn) London: H.M.S.O.

Meteorological Office (1971) *Handbook of Aviation Meteorology* (2nd edn) London: H.M.S.O.

Meteorological Office, (1972) *Meteorological Glossary,* London: H.M.S.O.

Meterological Office (1974) *Averages of Bright Sunshine Over the United Kingdom, 1941–70,* Bracknell: Meteorological Office (Climatol. Memor., no. 72)

Meteorological Office, Agricultural Branch (1955) *Weather and the Land,* London: H.M.S.O. (Bull. Minist. Agric. Fish. Fd., no. 165)

Mill, H. R. (1908) 'Monthly rainfall of 1908', *Br. Rainf.,* **48**, 137–69

Mill, H. R. and Salter, M. de C. S. (1915) 'Isometric rainfall maps of the British Isles', *Q. Jl R. met. Soc.,* **41**, 1–44

Miller, A. A. (1931) *Climatology,* London: Methuen

Milner, J. S. (1968) 'Fluctuations in annual and seasonal rainfall over the leeside of the south Pennines, 1761–1965', *Weather,* **23**, 435–41

Ministry of Agriculture, Fisheries and Food (1954) *The Calculation of Irrigation Need,* London: H.M.S.O. (Tech. Bull., no. 4)

Ministry of Agriculture, Fisheries and Food (1964) *The Farmer's Weather* (2nd edn) London: H.M.S.O. (Bull. no. 165)

Ministry of Agriculture, Fisheries and Food (1967) *Potential Transpiration,* London: H.M.S.O. (Tech. Bull., no. 16)

Mochlinski, K. (1970) 'Soil temperatures in the United Kingdom', *Weather,* **25**, 192–200

Moffitt, B. J. (1956) 'Nocturnal wind at Thorney Island', *Met. Mag.,* **85**, 268–71

Monteith, J. L. (1962) 'Attenuation of solar radiation: a climatological study', *Q. Jl R. met. Soc.,* **88**, 508–21

Monteith, J. L. (1966) 'Local differences in attenuation of solar radiation over Britain', *Q. Jl R. met. Soc.,* **92**, 254–62

Monteith, J. L. and Sceicz, G. (1962) 'Radiative temperatures in the heat balance of natural surfaces', *Q. Jl R. met. Soc.,* **88**, 496–507

Morris, R. E. and Barry, R. G. (1963) 'Soil and air temperatures in a New Forest valley', *Weather,* **18**, 325–30

Munn, R. E. (1966) *Descriptive Micrometeorology,* New York: Academic Press

Munn, R. E. (1970) 'Airflow in urban areas', *Tech. Notes Wld met. Org.,* **108**, 15–39

Munro, J. M. M. (1974) 'Ecology and agriculture in the hills and uplands', *Marginal Land Use: Integration or Competition?*; Potassium Institute colloquium proceedings, no. 4, Bangor, University College of North Wales, ed. D. A. Jenkins, *et al.*, pp. 39–53

Murray, R. (1966) 'A note on the large-scale features of the 1962–63 winter', *Met. Mag.,* **95**, 339–48

Murray, R. (1967a) 'Cyclonic Junes over the British Isles and the synoptic character of the following Septembers', *Met. Mag.,* **96**, 65–9

Murray, R. (1967b) 'Sequences in monthly rainfall over England and Wales', *Met. Mag.,* **96**, 129–35

Murray, R. (1972) 'On predicting seasonal weather for England and Wales from anomalous atmospheric circulation over the northern hemisphere', *Weather,* **27**, 396–402

Murray, R. and Benwell, P. R. (1970) 'PCSM indices in synoptic climatology and long-range forecasting', *Met. Mag.,* **99**, 232–44

Murray, R. and Lankester, J. D. (1974) 'Central England temperature quintiles and associated pressure anomalies on a monthly time scale', *Met. Mag.,* **103**, 3–14

Murray, R. and Lewis, R. P. W. (1966) 'Some aspects of the synoptic climatology of the British Isles as measured by simple indices', *Met. Mag.,* **95**, 192–203

Murray, R. and Moffitt, B. J. (1969) 'Monthly patterns of the quasi-biennial pressure oscillation', *Weather,* **24**, 382–9

Murray, R. and Ratcliffe, R. A. S. (1969) 'The summer weather of 1968: related atmospheric circulation and sea temperature patterns', *Met. Mag.,* **98**, 201–19

Nace, R. L. (1960) 'Water management, agriculture and groundwater supplies', *Circ. US geol. Surv.,* 415

Namias, J. (1964) 'Seasonal persistence and recurrence of European blocking during 1958–60', *Tellus,* **16**, 394–407

Newham, E. V. (1918) 'Notes on examples of katabatic winds in the valley of the Upper Thames', *Prof. Notes Met. Off., Lond.,* 2

Newson, M. D. (1974) Personal communication

Nicholls, J. M. (1973) 'Aircraft measurements of disturbed airflow over mountains', *Weather,* **28**, 141–52

Norman, J. T. *et al.* (1957) 'Winter temperatures in long and short grass', *Met. Mag.,* **86**, 148–52

Oke, T. R. (1973) 'City size and the urban heat island', *Atmos. Environ.,* **7**, 767–79

Oke, T. R. and Hannell, F. G. (1970) 'The form of the heat island in Hamilton, Canada', *Tech. Notes Wld met. Org.,* **108**, 113–26

Oliver, H. R. (1971) 'Wind profiles in and above a forest canopy', *Q. Jl R. met. Soc.,* **97**, 548–53

Oliver, H. R. *et al.* (1974) 'Hydrometeorological research in Thetford forest', typescript of paper presented to Applied Environmental Science Symposium of the Institute of British Geographers, Norwich, University of East Anglia, Jan. 1974

Oliver, J. (1960a) 'Upland climates in South Wales', *Hill Climates and Land Usage,* ed. J. A. Taylor, Aberystwyth: University College of Wales, Department of Geography, pp. 6–14 (Memor. no. 3)

Oliver, J. (1960b) 'Wind and vegetation in the Dale Peninsula', *Fld. Stud.,* 1 (2), 37–48

Oliver, J. (1961) 'Soil temperatures in upland peats', *Aspects of Soil Climate,* ed. J. A. Taylor, Aberystwyth: University College of Wales, Department of Geography, pp. 23–7 (Memor. no. 4)

Oliver, J. (1964) 'A study of upland temperatures and humidities in South Wales', *Trans. Inst. Br. Geogr.,* **35**, 37–54

Oliver, J. (1966) 'Low minimum temperatures at Santon Downham, Norfolk', *Met. Mag.,* **95**, 13–7

Ovington, J. D. (1954) 'A comparison of rainfall in different woodlands', *Forestry,* **27**, 41–53

Palutikof, J. (1973) *Precipitation trends over Greater Merseyside* unpublished Ph.D. dissertation, University of Liverpool, Department of Geography

Parry, M. (1956) 'Local temperature variations in the Reading area', *Q. Jl R. met. Soc.,* **82**, 45–57

Parry, M. L. (1975) 'Secular climatic change and marginal agriculture, *Trans. Inst. Br. Geogr.,* **64**, 1–13

Paton, J. (1954) 'Ben Nevis Observatory, 1883–1904', *Weather,* **9**, 291–308

Peacock, J. M. and Sheehy, J. E. (1974) 'The measurement and utilization of some climatic resources in agriculture', *Climatic Resources and Economic Activity: A Symposium,* ed. J. A. Taylor, Newton Abbot: David and Charles, pp. 87–108 (Aberystwyth Memoranda in Agricultural Meteorology, no. 15)

Pears, N. V. (1967) 'Wind as a factor in mountain ecology: some data for the Cairngorm mountains', *Scott. geogr. Mag.,* **83**, 118–24

Pearsall, W. H. (1950) *Mountains and Moorlands*, London: Collins

Pedgley, D. E. (1967) 'Why so much rain?', *Weather,* **27**, 478–82

Pedgley, D. E. (1970) 'Heavy rainfalls over Snowdonia', *Weather,* **25**, 340–50

Pedgley, D. E. (1974) 'Field studies of mountain weather in Snowdonia', *Weather,* **29**, 284–97

Penman, H. L. (1943) 'Daily and seasonal changes in the surface temperature of fallow soil at Rothamsted', *Q. Jl R. met. Soc.,* **69**, 1–16

Penman, H. L. (1948) 'Natural evaporation from open water, bare soil and grass' *Proc. R. Soc., A.,* **193**, 120–45

Penman, H. L. (1949) 'The dependence of transpiration on weather and soil conditions', *J. Soil Sci.,* **1**, 74–89

Penman, H. L. (1950) 'Evaporation over the British Isles', *Q. Jl R. met. Soc.,* **76**, 372–83

Penman, H. L. (1962) 'Woburn irrigation, 1951–1959; I. Purpose, design and weather; II. Results for grass; III. Results for rotation crops', *J. agric. Sci.,* **58**, 343–79

Penman, H. L. and Long, I. F. (1960) 'Weather in wheat: an essay in micrometeorology', *Q. Jl R. met. Soc.*, **86**, 16–50
Penman, H. L. and Schofield, R. K. (1941) 'Drainage and evaporation from fallow soil at Rothampsted', *J. agric. Sci.*, **31**, 74–109
Perry, A. H. (1967) 'Blocking patterns during hot weather in Britain', *Weather*, **22**, 420–2
Perry, A. H. (1968) ' "Summer days" in the British Isles', *Weather*, **23**, 212–4
Perry, A. H. (1970) 'Changes in duration and frequency of synoptic types over the British Isles', *Weather*, **25**, 123–6
Perry, A. H. (1972a) 'June 1972 – the coldest June of the century', *Weather*, **27**, 418–22
Perry, A. H. (1972b) 'Spatial and temporal characteristics of Irish precipitation', *Ir. Geogr.*, **6**, 428–42
Perry, A. H. (1973) 'The first warm spell of early summer at Kew 1881–1972', *Weather*, **28**, 516–19
Perry, A. H. and Barry, R. G. (1973) 'Recent temperature changes due to changes in the frequency and average temperature of weather types over the British Isles', *Met. Mag.*, **102**, 73–82
Petherbridge, P. (1969) *Sunpath Diagrams and Overlays, For Solar Heat Gain Calculations*, London: H.M.S.O.
Petterssen, S. (1950) 'Some aspects of the general circulation of the atmosphere', *Centenary Proceedings of the Royal Meteorological Society*, London: Royal Meteorological Society, pp. 120–55
Plant, J. A. (1971) *A Study of Intensities of Rainfall Recorded at Places In Scotland*, Bracknell: Meteorological Office (Hydrol. Memor., no. 40)
Poulter, R. M. (1962) 'The next few summers in London', *Weather*, **17**, 253–5
Pratt, K. A. (1968) *The 'Haar' of north-east England*, unpublished B.A. dissertation, University of Durham, Department of Geography
Prichard, R. J. (1973) 'Diurnal variation of thunder at Manchester Airport', *Weather*, **28**, 327–31
Pye, N. (1972) 'Weather and climate'. *Leicester and Its Region*, ed. N. Pye, Leicester: Leicester University Press for British Association, pp. 84–113
Ratcliffe, R. A. S. (1953) 'Differences in visibility between a weekday and Sunday near to an industrial area', *Met. Mag.*, **82**, 372
Ratcliffe, R. A. S. (1973) 'Recent work on sea surface temperature anomalies related to long range forecasting', *Weather*, **28**, 106–17
Ratcliffe, R. A. S. (1974) 'The use of 500 mb anomalies in long range forecasting', *Q. Jl R. met. Soc.*, **100**, 234–44
Ratcliffe, R. A. S. and Collison, P. (1969) 'Forecasting rainfall for the summer season in England and Wales', *Met. Mag.*, **98**, 33–9
Ratcliffe, R. A. S. and Murray, R. (1970) 'New lag associations between North Atlantic sea temperature and European pressure applied to long-range weather forecasting', *Q. Jl R. met. Soc.*, **96**, 41–9
Ratsey, S. (1973) 'The climate at and around Nettlecombe Court, Somerset', *Fld. Stud.*, **3**(5), 741–62
Reay, J. S. S. (1974) 'Monitoring of the environment', *Fuel and the*

Environment: Conference Proceedings, vol. 1, Papers, London: Institute of Fuel, pp. 105–10 (Eastbourne conference, 1973)

Reed, L. E. and Trott, P. E. (1971) 'Continuous measurement of carbon monoxide in streets, 1967–9', *Atmos. Environ.,* **5**, 27–39

Rees, A. R. (1968) 'Solar radiation on the south coast of England', *Q. Jl R. met. Soc.,* **94**, 397–401

Reid, I. (1973) 'The influence of slope aspect on precipitation receipt', *Weather,* **28**, 490–3

Reynolds, G. (1956a) 'Abrupt changes in rainfall regimes', *Weather,* **11**, 249–54

Reynolds, G. (1956b) 'Local temperature variations in Wirral', *Weather,* **11**, 15–16

Rider, G. C. and Simpson, J. E. (1968) 'Two crossing fronts on radar', *Met. Mag.,* **97**, 24–30

Rider, N. E. (1954) 'Eddy diffusion of momentum, water vapour and heat near the ground', *Phil. Trans. R. Soc.,* A, **246**, 481–501

Rider, N. E. and Robinson, C. D. (1951) 'A study of the transfer of heat and water vapour above a surface of short grass', *Q. Jl R. met. Soc.,* **77**, 375–401

Roach, W. T. *et al.* (1973) 'A field study of radiation fog', *Fogs and Smokes; Faraday Symposia of the Chemical Society,* no. 7, London: Chemical Society, 209–21

Roberts, D. G. (1972) 'The modification of geomorphic shelter belts', *Research Papers in Forest Meteorology: An Aberystwyth Symposium,* ed. J. A. Taylor, Aberystwyth: Cambrian News, pp. 134–46 (Aberystwyth Memoranda in Agricultural Meteorology, no. 13)

Robinson, T. R. (1850) 'Description of an improved anemometer for registering the direction of the wind and the space which it traverses in given intervals of time', *Trans. R. Ir. Acad.,* **22**, 155–78

Rodda, J. C. (1962) 'An objective method for the assessment of areal rainfall amounts', *Weather,* **17**, 54–9

Rodda, J. C. (1967) 'A country-wide study of intense rainfall for the United Kingdom', *J. Hydrol.,* **5**, 58–69

Rodda, J. C. (1970) 'Rainfall excesses in the United Kingdom', *Trans. Inst. Br. Geogr.,* **49**, 49–60

Rodda, J. C. (1971) 'Progress at Plynlimon: problems of investigating the effect of land use on the hydrological cycle', paper read to the annual meeting of the British Association for the Advancement of Science, Section K; appendix and diagrams mimeographed

Rodewald, M. (1973) *Der Trend der Meerestemperatur in Nordatlantik,* Beilagen zur Berliner Wetterkarte, no. 108/73 and 119/73

Rowles, K. (1972) 'Sea breeze front near the south coast of England', *Met. Mag.,* **101**, 153–4

Ruck, F. W. M. (1949) 'Mist over St. David's', *Weather,* **4**, 360–1

Rudloff, H. von (1967) *Die Schwankungen und Pendelungen des Klimas in Europa seit dem Beginn der regelmässigen Instrumenten-Beobachtungen (1670)* Braunschweig: Vieweg (*Die Wissenschaft,* vol. 122)

Russell, E. W. (1961) *Soil Conditions and Plant Growth,* London: Longman

Rutter, A. J. (1972) 'Evaporation from forests', *Research Papers in Forest Meteorology: An Aberystwyth Symposium*, ed. J. A. Taylor, Aberystwyth: Cambrian News, pp. 75–90 (Aberystwyth Memoranda in Agricultural Meteorology, no. 13)

Rutter, N. and Edwards, R. S. (1968) 'Deposition of air-borne marine salt at different sites over the College Farm, Aberystwyth (Wales), in relation to wind and weather', *Agric. Met.*, **5**, 235–54

Salter, M. de C. S. (1918) 'The relation of rainfall to configuration', *Br. Rainf.*, **58**, 40–56

Salter, M. de C. S. (1921) *The Rainfall of the British Isles*, London: University of London Press

Salter, M. de C. S. and Glasspoole, J. (1923) 'The fluctuation of annual rainfall in the British Isles considered cartographically', *Q. Jl R. met. Soc.*, **49**, 207–29

Samson, C. A. (1965) 'A comparison of mountain slope and radiosonde observations', *Mon. Weath. Rev.*, **93**, 327–30

Saunders, W. E. (1961) 'Diurnal variation of visibility with light winds at Plymouth in winter', *Met. Mag.*, **90**, 19–22

Sawyer, J. S. (1952) 'A study of the rainfall of two synoptic situations', *Q. Jl R. met. Soc.*, **78**, 231–46

Sawyer, J. S. (1956a) 'Rainfall of depressions which pass eastward over or near the British Isles', *Prof. Notes Met. Off., Lond.*, 118

Sawyer, J. S. (1956b) 'The physical and dynamical problems of orographic rain', *Weather*, **11**, 375–81

Schove, D. J. (1971) 'Biennial oscillation and solar cycles AD 1490–1970', *Weather*, **26**, 201–9

Scorer, R. S. (1949) 'Theory of waves in the lee of mountains', *Q. Jl R. met. Soc.*, **75**, 41–56

Scorer, R. S. (1951) 'Forecasting the occurrence of lee waves', *Weather*, **6**, 99–103

Scorer, R. S. (1953) 'Theory of airflow over mountains, II: the flow over a ridge', *Q. Jl R. met. Soc.*, **79**, 70–83

Scorer, R. S. (1954) 'Theory of airflow over mountains, III: airstream characteristics', *Q. Jl R. met. Soc.*, **80**, 417–28

Scorer, R. S. (1955) 'Theory of airflow over mountains, IV: a separation of flow from the surface', *Q. Jl R. met. Soc.*, **81**, 340–50

Sellers, W. D. (1965) *Physical Climatology*, Chicago: University of Chicago Press

Senior, M. R. (1969) 'Changes in the variability of annual rainfall over Britain', *Weather*, **24**, 354–9

Sharaf, A. E. A. T. (1954) 'The climate of the British Isles: a new classification', *Bull. Soc. Géogr. Égypte*, **27**, 209–45

Shaw, E. M. (1962) 'An analysis of the origins of precipitation in northern England, 1956–1960', *Q. Jl R. met. Soc.*, **88**, 539–47

Shaw, J. B. (1955) 'Vertical temperature gradient in the first 2,000 ft', *Met. Mag.*, **84**, 233–41

Shellard, H. C. (1958) 'Extreme wind speeds over the United Kingdom', *Met. Mag.*, **87**, 257–65

Shellard, H. C. (1959) 'Averages of accumulated temperature and standard deviation of monthly mean temperature over Britain, 1921–50', *Prof. Notes. met. Off., Lond.*, **125**

Shellard, H. C. (1962) 'Extreme wind speeds over the United Kingdom for periods ending 1959', *Met. Mag.*, **91**, 39–47

Shellard, H. C. (1965a) *Extreme Wind Speeds Over the United Kingdom For Periods Ending 1963*, Bracknell: Meteorological Office (Climatol. Memor. no. 50)

Shellard, H. C. (1965b) 'The estimation of design wind speeds', *Wind Effects on Buildings and Structures*; Proceedings of the Conference held at the National Physical Laboratory, Teddington, 26–28 June, 1963, London: H.M.S.O., vol. 1, pp. 29–51 (N.P.L. Symposium no. 16)

Shellard, H. C. (1968a) 'Results of some recent special measurements in the United Kingdom relevant to wind loading problems', *Wind Effects on Buildings and Structures*; Proceedings of the International Research Seminar, Ottawa, 1967, Toronto: University of Toronto Press, pp. 515–33

Shellard, H. C. (1968b) 'The winter of 1962–63 in the United Kingdom: a climatological survey', *Met. Mag.*, **97**, 129–41

Sheppard, P. A. *et al.* (1972) 'Studies of turbulence in the surface layer over water (Lough Neagh). Part 1. Instrumentation, programme, profiles', *Q. Jl R. met. Soc.*, **98**, 627–41

Simpson, J. E. (1964) 'Sea-breeze fronts in Hampshire', *Weather*, **19**, 208–20

Simpson, J. E. (1967) 'Aerial and radar observations of some sea-breeze fronts', *Weather*, **22**, 306–16

Siple, P. A. and Passel, C. F. (1945) 'Measurement of dry atmospheric cooling in sub-freezing temperatures', *Proc. Am. phil. Soc.*, **89**, 177

Smart, J. (1974) Personal communication

Smed, J. (1949) 'Monthly anomalies of the surface temperature in the sea round Iceland during the years 1876–1939 and 1945–1947', *Annls biol., Copenh.*, **6**, 11–26

Smith, C. G. (1974) 'Monthly, seasonal and annual fluctuations of rainfall at Oxford since 1815', *Weather*, **29**, 2–16

Smith, F. J. (1967) 'A comparison of the incidence of fog at a coastal station with that at an inland station', *Met. Mag.*, **96**, 77–81

Smith, I. R. (1973) 'The assessment of winds at Loch Leven, Kinross', *Weather*, **28**, 202–10

Smith, K. (1968) 'Some thermal characteristics of two rivers in the Pennine area of northern England', *J. Hydrol.*, **6**, 405–16

Smith, K. (1972) 'River water temperatures: an environmental review', *Scott. geogr. Mag.*, **88**, 211–20

Smith, K. and Lavis, M. E. (1975) 'Environmental influences on the temperature of a small upland stream', *Oikos*, **26**, 228–36

Smith, L. P. (1950) 'Variations of mean air temperature and hours of sunshine on the weather slope of a hill', *Met. Mag.*, **79**, 231–3

Smith, L. P. (1952) 'Variations in air temperature and humidity on the weather slope of a coastal hill', *Met. Mag.*, **81**, 102–4

Smith, L. P. (1961) 'Frequencies of poor afternoon visibilities in

England and Wales', *Met. Mag.,* **90**, 355–9

Smith, L. P. (1962) 'Meadow-hay yields', *Outl. Agric.,* **3**, 219–24

Smith, L. P. (1967) 'Meteorology and the pattern of British grassland farming', *Agric. Met.,* **4**, 321–38

Smith, M. F. (1974) 'A short note on a sea-breeze crossing East Anglia', *Met. Mag.,* **103**, 115–18

Smithson, P. A. (1969a) 'Effects of altitude on rainfall in Scotland', *Weather,* **24**, 370–6

Smithson, P. A. (1969b) 'Regional variations in the synoptic origin of rainfall across Scotland', *Scott. geogr. Mag.,* **85**, 182–95

Sowerby Wallis, H. (1902) 'On the rainfall at Camden Square for the 45 years 1858 to 1902', *Br. Rainf.,* **42**, 16–35

Sparks, W. R. (1962) 'The spread of low stratus from the North Sea across East Anglia', *Met. Mag.,* **91**, 361–5

Spence, M. T. (1936) 'Temperature changes over short distances in the Edinburgh district', *Q. Jl R. met. Soc.,* **62**, 25–31

Stagg, J. M. (1950) *Solar Radiation At Kew Observatory,* London: H.M.S.O. (Meteorological Office, Geophys. Mem., no. 86)

Stamp, L. D. ed. (1937–46) *The Land of Britain: The Report of the Land Utilisation Survey of Britain,* London: Geographical Publications, for Land Utilisation Survey (in 92 parts)

Stamp, L. D. and Beaver, S. H. (1971) *The British Isles: A Geographic and Economic Survey* (6th edn) London: Longman

Stephenson, P. M. (1967) 'Seasonal rainfall sequences over England and Wales', *Met. Mag.,* **96**, 335–42

Stephenson, P. M. (1971) 'Accumulated percentage departures of average rainfall', *British Rainfall, Supplement 1961–65,* London: H.M.S.O., vol. 1, 260–61

Stevenson, C. M. (1968) 'An analysis of the chemical composition of rain-water and air over the British Isles and Eire for the years 1959–1964', *Q. Jl R. met. Soc.,* **94**, 56–70

Stevenson, C. M. (1969) 'The dust fall and severe storms of 1 July 1968', *Weather,* **24**, 126–32

Stevenson, R. E. (1961) 'Sea-breezes along the Yorkshire coast in the summer of 1959', *Met. Mag.,* **90**, 153–62

Stewart, K. H. (1955) *Radiation Fog Investigations At Cardington,* London: Air Ministry, Meteorological Research Committee (M.R.P. no. 912)

Stone, R. G. (1934) 'The history of mountain meteorology in the United States and the Mount Washington Observatory', *Trans. Am. geophys. Un.,* **15**, 124–33

Sumner, E. J. (1959) 'Blocking anticyclones in the Atlantic–European sector of the northern hemisphere', *Met. Mag.,* **88**, 300–11

Tagg, J. R. (1957) *Wind Data Related to the Generation of Electricity by Wind Power,* Leatherhead: British Electrical and Allied Industries Research Association (Tech. Rep., C/T115)

Talman, C. F. (1934) 'Europe's mountain weather stations', *Bull. Am. met. Soc.,* **15**, 269

Tansley, A. G. (1939), *The British Isles and Their Vegetation,* Cambridge: Cambridge University Press

Taylor, J. A. (1961) 'The maritime uplands of Britain: inventory and prospect', *Nature,* **189**, 872–6

Taylor, J. A. (1965) 'Climatic change as related to altitudinal thresholds and soil variables', *The Biological Significance of Climatic Changes in Britain,* ed. C. G. Johnson and L. P. Smith, London: Academic Press, pp. 37–50 (Symposia of the Institute of Biology, no. 14)

Taylor, J. A. (1967) 'Soil climate: its definition and measurement', *Weather and Agriculture,* ed. J. A. Taylor, Oxford: Pergamon, pp. 37–47

Taylor, J. A. (1972) 'The revaluation of weather forecasts', *Weather Forecasting For Agriculture and Industry*: a symposium, ed. J. A. Taylor, Newton Abbot: David and Charles, pp. 1–81

Taylor, J. A. (1973) 'Chronometers and chronicles: a study of palaeo-environments in West Central Wales', *Prog. Geog.,* **5**, 250–334

Taylor, J. A. (1974a) 'The role of climatic factors in environmental and cultural changes in prehistoric times', *The Effect of Man on the Landscape: The Highland Zone,* ed. Evans, J. G. *et al.,* London: Council for British Archaeology, pp. 6–19 (Research report no. 11)

Taylor, J. A. (1974b) 'Marginal physical environments', *Marginal Land Use: Integration or Competition?*; Potassium Institute Colloquium proceedings, no. 4, Bangor, University College of North Wales, ed. D. A. Jenkins, *et al.,* pp. 10–29

Taylor, S. M. and Smith, L. P. (1961) 'Estimation of averages of radiation and illumination', *Met. Mag.,* **90**, 289–94

Thomas, T. M. (1960a) 'Some observations on the tracks of depressions over the eastern half of the North Atlantic', *Weather,* **15**, 325–36

Thomas, T. M. (1960b) 'Precipitation within the British Isles in relation to depression tracks', *Weather,* **15**, 361–73

Thompson, R. W. S. (1954) 'Stratification and overturn in lakes and reservoirs', *J. Instn. Wat. Engrs,* **8**, 19–52

Thomson, A. B. (1964) *Mean Winter Temperatures in Edinburgh, 1764–65 to 1962–63,* Bracknell: Meteorological Office (Climatol. Memor., no. 41)

Thornthwaite, C. W. (1933) 'The climates of the earth', *Geogrl Rev.,* **23**, 433–40

Thornthwaite, C. W. (1948) 'An approach toward a rational classification of climate', *Geogrl Rev.,* **38**, 55–94

Thran, P. and Broekhuizen, S. (1965) *Agro-Climatic Atlas of Europe,* Wageningen: Pudoc, Centre for Agricultural Publications and Documentation (*Agro-Ecological Atlas of Cereal Growing in Europe,* vol. 1)

Tout, D. G. (1973) 'Manchester sunshine', *Weather,* **28**, 164–6 (and correspondence by P. B. Wright and P. Woodward, *Weather,* **28**, 352)

Townsend, W. N. (1973) *An Introduction to the Scientific Study of the Soil,* London: Arnold

Trewartha, G. T. (1954) *An Introduction to Climate* (3rd edn), New

York : McGraw-Hill
Trewartha, G. T. (1966) *The Earth's Problem Climates,* London : Methuen
Troll, C. and Paffen, K. H. (1965) 'Seasonal climates of the earth', *World Maps of Climatology* (2nd edn) ed. E. Rodenwaldt and H. J. Jusatz, Berlin : Springer, 19–25
Tuller, S. E. (1973) 'Effects of vertical vegetation surfaces on the adjacent microclimate : the role of aspect', *Agric. Met.,* **12**, 407–24
Unsworth, M. H. (1974) 'Radiation measurements as indication of material and man-made atmospheric aerosol', *Observation and Measurement of Atmospheric Pollution* ; proceedings of the WMO/WHO technical conference, Helsinki, 30 July–4 Aug. 1973, Geneva : W.M.O., pp. 423–33
Unsworth, M. H. and Monteith, J. L. (1972) 'Aerosol and solar radiation in Britain', *Q. Jl R. met. Soc.,* **98**, 778–97
Unsworth, M. H. and Monteith, J. L. (1975) 'Long wave radiation at the ground', *Q. Jl R. met. Soc.,* **101**, 13–34
Unwin, D. J. (1969) 'The areal extension of rainfall records : an alternative model', *J. Hydrol.,* **7**, 404–14
Vries, D. V. de (1963) 'Thermal properties of soils', *Physics of the Plant Environment,* ed. W. R. van Wijk, Amsterdam : North-Holland Publishing Co., pp. 110–35
Wales-Smith, B. G. (1973) 'An analysis of monthly rainfall totals representative of Kew, Surrey, from 1697 to 1970, *Met. Mag.,* **102**, 157–71
Wallington, C. E. (1959) 'The structure of the sea-breeze front as revealed by gliding flights', *Weather,* **14**, 263–70
Wallington, C. E. (1961a) 'Airflow over broad mountain ranges : a study of five flights across the Welsh mountains', *Met. Mag.,* **90**, 213–22
Wallington, C. E. (1961b) 'An introduction to the sea-breeze front', *Schweizer Aerorevue,* **36**, 393–7
Walter, H. and Lieth, H. (1967) *Klimadiagramm Weltatlas,* Jena : Fischer
Ward, R. C. (1967) *Principles of hydrology,* London : McGraw-Hill
Ward, R. C. (1968) 'Some runoff characteristics of British rivers', *J. Hydrol.,* **6**, 358–72
Ward, R. C. (1971) 'Measuring evapotranspiration : a review', *J. Hydrol.,* **13**, 1–21
Warren Spring Laboratory (1972a) *National Survey of Air Pollution, 1961–71, Vol. 1: Introduction, United Kingdom, South East, Greater London,* London : H.M.S.O.
Warren Spring Laboratory (1972b) *National Survey of Air Pollution, 1961–71, Vol. 2: South West Region, Wales Region, North West Region,* London : H.M.S.O.
Warren Spring Laboratory (1973) *National Survey of Air Pollution, 1961–71, Vol 3: East Anglia, East Midlands, West Midlands,* London : H.M.S.O.
Water Resources Board (1971) *Surface Water Yearbook of Great Britain, 1965–66,* London : H.M.S.O.
Waterhouse, S. L. (1955) 'Microclimatological profiles in grass

cover in relation to biological problems', *Q. Jl R. met. Soc.*, **81**, 63–71

Watts, A. J. (1955) 'Sea-breeze at Thorney Island', *Met. Mag.*, **84**, 42–8

Weiss, I. and Lamb, H. H. (1970) 'On the problem of high waves in the North Sea and neighbouring waters and the possible future trend of the atmospheric circulation', *Die Zunahme der Wellenhöhen in jüngster Zeit in den Operationsgebieten der Bundesmarine, ihre vermutlichen Ursachen und ihre voraussichtliche weitere Entwicklung*, Porz-Wahn: Geophysikalischer Beteratungsdienst der Bundeswehr, Section 2, pp. 6–14 (Fachliche Mitteilungen, no. 160)

Wiggett, P. J. (1964) 'The year-to-year variation of the frequency of fog at London (Heathrow) Airport', *Met. Mag.*, **93**, 305–8

Williams, F. P. (1960) 'Pollution levels in cities', *Proceedings of the 27th Annual Conference of the National Society for Clean Air*: Halifax, 1960, London: N.S.C.A., pp. 83–6

Wilson, O. (1967) 'Objective evaluation of windchill index by records of frostbite in the Antarctic', *Int. J. Biomet.*, **11**, 29–32

Wood, C. M. (1973) 'Visibility and sunshine in Greater Manchester', *Clean Air*, **3**, 15–24

Wood, C. M. *et al.* (1974) *The Geography of Pollution: A Study in Greater Manchester*, Manchester: Manchester University Press

World Meteorological Organization (1966) *Climatic Change*, Geneva: W.M.O. (Tech. Notes, no. 79)

World Meteorological Organization (1969) *Air Pollutants, Meteorology and Plant Injury*, Geneva: W.M.O. (Tech. Notes, no. 96)

World Meteorological Organization (1970) *Urban Climates*, Geneva: W.M.O. (Tech. Notes, no. 108)

Wright, P. B. (1970) 'Warm and cold spells in May', *Weather*, **25**, 229–31

Zverev, A. A. (1972) 'Water temperature variations in the North Atlantic in 1948 through 1968', *Okeonologija*, **12**, 211–16 (in Russian)

STATION INDEX

Location of all stations referred to in the text

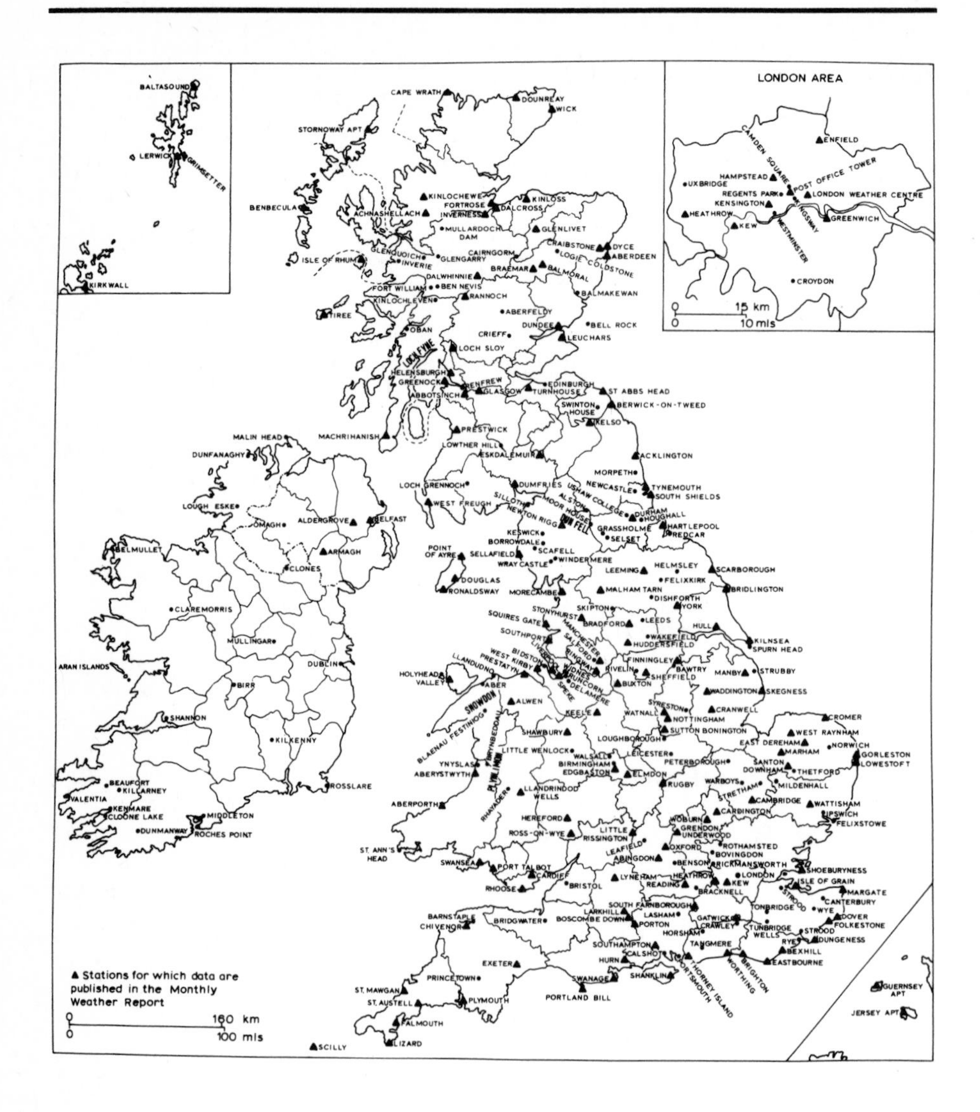

Station Index (Counties as of 1 January 1975)

Station	County	Reference
Abbotsinch	Renfrewshire	Ch. 6, 8; Fig. 8.2; Table 6.21
Aber	Gwynedd	Ch. 5; Table 5.2
Aberdeen (and Dyce)	Aberdeenshire	Ch. 2, 3, 5, 6, 7, 8, 15; Figs. 2.5, 3.1, 3.2, 3.3, 3.5, 5.7, 6 7, 15.1; Tables 3.3, 3.4, 3.6, 3.7, 3.8, 3.10, 3.11, 5.10, 6.1, 6.17, 6.19, 6.20, 6.21, 8.1
Aberfeldy	Perthshire	Ch. 6; Table 6.18
Aberporth	Dyfed	Ch. 2, 3, 4, 6, 11, 12; Figs. 2.5, 3.1, 3.2, 3.3, 3.5, 4.8; Tables 3.3, 3.4, 3.6, 3.7, 3.8, 3.10, 3.11, 6.1, 6.21, 12.1
Aberystwyth	Dyfed	Ch. 11, 12; Tables 12.5, 12.6
Abingdon	Oxfordshire	Ch. 3; Tables 3.8, 3.10
Achnashellach	Ross and Cromarty	Ch. 5; Table 5.2
Acklington	Northumberland	Ch. 6; Table 6.1
Aldergrove	Co. Antrim	Ch. 2, 3, 5, 6, 8, 12; Figs. 2.5, 3.1, 3.2, 3.3, 3.5, 5.2, 5.10, 8.2; Tables 3.3, 3.4, 3.5, 3.6, 3.7, 3.8, 3.10, 3.11, 5.1, 5.5, 5.6, 5.7, 6.1, 6.17, 6.21, 12.6
Alston	Cumbria	Ch. 5; Table 5.4
Alwen	Clwyd	Ch. 5; Table 5.4
Aran Islands	Co. Galway	Ch. 6
Armagh	Co. Armagh	Ch. 5, 6, 10; Fig. 10.9; Tables 6.19, 6.20
Ayre, Point of	Isle of Man	Ch. 3; Tables 3.8, 3.10
Balmakewan	Kincardineshire	Ch. 3; Tables 3.8, 3.10
Balmoral	Aberdeenshire	Ch. 6; Tables 6.19, 6.20
Baltasound	Zetland	Ch. 5, 6
Barnstaple	Devon	Ch. 5; Table 5.2
Bawtry	South Yorkshire	Ch. 5; Table 5.2
Beaufort	Co. Kerry	Ch. 6; Table 6.9
Belfast	Co. Antrim/Down	Ch. 15; Fig. 15.1
Bell Rock	off Angus	Ch. 3; Tables 3.8, 3.10
Belmullet	Co. Mayo	Ch. 3, 6; Figs. 3.1, 3.5; Tables 3.3, 3.4, 3.6, 3.7, 3.8, 3.10, 6.2, 6.7, 6.11, 6.13, 6.17, 6.19, 6.20
Benbecula	Inverness-shire	Ch. 3, 6, 8; Tables 3.9, 6.1, 6.21
Ben Nevis	Inverness-shire	Ch. 5, 6, 12; Fig. 12.5; Table 5.4
Benson	Oxfordshire	Ch. 13
Berwick on Tweed	Northumberland	Ch. 11
Bexhill	East Sussex	Ch. 5; Fig. 5.10; Table 5.7
Bidston (and Birkenhead)	Merseyside	Ch. 8, 10, 11; Figs. 10.9, 10.10, 10.11, 10.12; Table 8.1
Birmingham (and Edgbaston)	West Midlands	Ch. 3, 5, 6, 14, 15; Fig. 15.1; Tables 3.8, 3.10, 5.5, 5.6, 6.19, 6.20
Birr	Co. Offaly	Ch. 2, 6; Fig. 2.5; Tables 6.7, 6.17, 6.19, 6.20
Blackpool (and Squires Gate)	Lancashire	Ch. 5, 6, 11; Fig. 5.5; Tables 6.1, 6.17, 6.21
Blaenau Festiniog	Gwynedd	Ch. 6
Borrowdale	Cumbria	Ch. 6
Boscombe Down	Wiltshire	Ch. 6, 8, 9; Figs. 8.2, 9.1, 9.2; Tables 6.1, 6.17, 6.21
Bovingdon	Hertfordshire	Ch. 9; Fig. 9.3
Bracknell	Berkshire	Ch. 4; Fig. 4.9
Bradford	West Yorkshire	Ch. 10; Fig. 10.7, 10.8
Braemar	Aberdeenshire	Ch. 5, 10, 12; Fig. 5.10, 10.9, 10.10; Tables 5.4, 5.7, 12.6
Bridgwater	Somerset	Ch. 6; Table 6.1
Bridlington	Humberside	Ch. 11
Brighton	East Sussex	Ch. 2
Bristol	Avon	Ch. 15; Fig. 15.1
Brynbeddau	Dyfed	Ch. 12; Table 12.4
Buxton	Derbyshire	Ch. 6; Tables 6.19, 6.20
Cairngorm	Inverness-shire/Banffshire	Ch. 3, 12; Table 3.9
Calshot	Hampshire	Ch. 3; Tables 3.8, 3.10
Cambridge	Cambridgeshire	Ch. 4, 5, 6, 10, 14; Fig. 4.8; Tables 5.5, 5.6,6.18

Station Index (Counties as of 1 January 1975)—*contd.*

Station	County	Reference
Camden Square	(see London)	
Canterbury	Kent	Ch. 15; Fig. 15.1
Cape Wrath	Sutherlandshire	Ch. 2, 5, 6
Cardiff	South Glamorgan	Ch. 8, 12, 14, 15; Fig. 15.1; Tables 8.1, 12.6
Cardington	Bedfordshire	Ch. 3, 9, 12, 13; Tables 3.8, 3.10, 13.6
Chivenor	Devon	Ch. 6; Table 6.1
Claremorris	Co. Mayo	Ch. 5, 6; Tables 6.2, 6.7, 6.11, 6.13, 6.17, 6.19, 6.20
Clones	Co. Monaghan	Ch. 6; Tables 6.2, 6.7, 6.11, 6.13, 6.17, 6.19, 6.20
Cloone Lake	Co. Galway	Ch. 6; Table 6.9
Cockle Park	(see Morpeth)	
Craibstone	Aberdeenshire	Ch. 10
Cranwell	Lincolnshire	Ch. 2, 3, 6; Fig. 6.7; Tables 2.6, 3.8, 3.10, 6.1, 6.17
Crawley	West Sussex	Ch. 9
Crieff	Perthshire	Ch. 6
Cromer	Norfolk	Ch. 5; Fig. 5.10; Tables 5.2, 5.7
Croydon	(see London)	
Cwm Dyli	(see Snowdon)	
Dalcross	Inverness-shire	Ch. 6; Table 6.1
Dalwhinnie	Inverness-shire	Ch. 5, 10; Table 5.4
Delamere	Cheshire	Ch. 14
Dishforth	North Yorkshire	Ch. 8; Fig. 8.2
Douglas	Isle of Man	Ch. 5, 6, 11; Fig. 5.10; Tables 5.5, 5.6, 5.7, 6.18
Dounreay	Caithness	Ch. 3; Tables 3.6, 3.9
Dover	Kent	Ch. 3; Tables 3.6, 3.8, 3.10
Dublin	Co. Dublin	Ch. 3, 5, 6, 8, 12, 15; Figs. 3.1, 3.5, 5.10, 15.1; Tables 3.3, 3.4, 3.6, 3.7, 3.8, 3.10, 5.7, 6.2, 6.7, 6.11, 6.13, 6.17, 6.19, 6.20, 8.1, 12.6, 15.3
Dumfries	Dumfriesshire	Ch. 5, 11; Table 11.4
Dundee	Angus	Ch. 15; Fig. 15.1
Dunfanaghy Rd.	Co. Donegal	Ch. 3; Table 3.9
Dun Fell	Cumbria	Ch. 3, 9, 12; Fig. 9.3; Tables 3.9, 12.3
Dungeness	Kent	Ch. 3; Figs. 3.1, 3.2, 3.3, 3.5; Tables 3.3, 3.4, 3.7, 3.11
Dunmanway	Co. Cork	Ch. 6; Table 6.9
Durham	Durham	Ch. 3, 6, 11, 13, 15; Fig. 15.1; Tables 3.8, 3.10, 6.19, 6.20, 11.4, 15.1
Dyce	(see Aberdeen)	
Eastbourne	East Sussex	Ch. 5
East Dereham	Norfolk	Ch. 10; Figs. 10.11, 10.12
Edgbaston	(see Birmingham)	
Edinburgh	Midlothian	Ch. 3, 6, 10, 12, 15; Figs. 10.7, 10.8, 10.9, 10.10, 10.11, 10.12, 15.1; Tables 3.8, 3.10, 6.19, 6.20, 12.6
Eisteddfa Gurig	(see Plynlimon)	
Elmdon	West Midlands	Ch. 2, 3, 6, 8; Figs. 2.5, 3.1, 3.2, 3.3, 3.5, 8.2; Tables 2.6, 3.3, 3.4, 3.6, 3.7, 3.11, 6.1, 6.21
Enfield	(see London)	
Eskdalemuir	Dumfriesshire	Ch. 3, 4, 5, 6, 10, 12; Figs. 4.4, 4.7, 4.11, 5.2, 5.10; Tables 3.8, 3.10, 5.1, 5.5, 5.6, 5.7, 6.1, 6.5, 6.17, 6.19, 6.20, 6.21, 12.1, 12.3, 12.6
Exeter	Devon	Ch. 3, 6, 14, 15; Figs. 3.1, 3.2, 3.3, 3.5, 15.1; Tables 3.3, 3.4, 3.7, 6.1
Falmouth	Cornwall	Ch. 5, 6; Tables 5.5, 5.6, 6.18
Farnborough, South	Hampshire	Ch. 3, 6; Tables 3.8, 3.10, 6.1
Felixkirk	North Yorkshire	Ch. 6; Tables 6.1, 6.17
Felixstowe	Suffolk	Ch. 3; Tables 3.8, 3.10
Finningley	South Yorkshire	Ch. 6, 9; Table 6.21
Folkestone	Kent	Ch. 6; Table 6.1

Station Index (Counties as of 1 January 1975)—*contd.*

Station	County	Reference
Fortrose	Ross and Cromarty	Ch. 6; Table 6.18
Fort William	Inverness-shire	Ch. 6; Fig. 6.7
Gatwick	West Sussex	Ch. 8, 9, 11; Fig. 9.3; Table 9.4
Glasgow	Lanarkshire	Ch. 6, 8, 11, 14, 15; Fig. 15.1
Glengarry	Inverness-shire	Ch. 6
Glenlivet	Banffshire	Ch. 5; Table 5.4
Glenquoich	Inverness-shire	Ch. 6; Fig. 6.16
Gorleston	Norfolk	Ch. 2, 3, 5, 6, 8, 11; Fig. 2.5; Tables 3.8, 3.10, 5.1, 5.5, 5.6, 6.21, 8.1
Grain, Isle of	Kent	Ch. 6
Grassholme	Durham	Ch. 11; Fig. 11.5
Greenock	Renfrewshire	Ch. 6; Tables 6.1, 6.17
Greenwich	(see London)	
Grendon Underwood	Buckinghamshire	Ch. 6; Fig. 6.4
Grimsetter	Zetland	Ch. 6; Table 6.1
Guernsey	Channel Islands	Ch. 2; Fig. 2.5
Hampstead	(see London)	
Hartlepool	Cleveland	Ch. 11; Table 11.4
Heathrow (London Airport)	Greater London	Ch. 3, 5, 6, 8, 9, 14; Figs. 3.1, 3.2, 3.3, 3.5, 8.2, 8.3, 8.4, 9.1, 9.2; Tables 3.3, 3.4, 3.5, 3.6, 3.7, 6.21, 8.2, 9.2, 9.3, 9.4, 14.1
Helensburgh	Dunbartonshire	Ch. 6; Table 6.18
Helmsley	North Yorkshire	Ch. 13
Hereford	Hereford and Worcester	Ch. 14
Holyhead	Gwynedd	Ch. 6, 11; Tables 6.1, 6.17
Horsham	West Sussex	Ch. 5; Table 5.2
Houghall	Durham	Ch. 13
Huddersfield	West Yorkshire	Ch. 10
Hull	Humberside	Ch. 13
Hurn	Dorset	Ch. 3, 6, 8; Figs. 3.1, 3.2, 3.3, 3.5; Tables 3.3, 3.4, 3.6, 3.7, 6.21
Inverie	Inverness-shire	Ch. 6; Table 6.18
Inverness	Inverness-shire	Ch. 12; Table 12.6
Ipswich	Suffolk	Ch. 6; Table 6.1
Jersey	Channel Islands	Ch. 3, 6, 11; Tables 3.9, 6.1, 11.2
Keele	Staffordshire	Ch. 2; Table 2.6
Kelso	Roxburghshire	Ch. 5; Table 5.4
Kenmare	Co. Kerry	Ch. 6; Table 6.9
Kensington	(see London)	
Keswick	Cumbria	Ch. 6; Table 6.6
Kew	Greater London	Ch. 2, 3, 4, 5, 6, 7, 8, 9, 10, 13; Figs. 2.5, 3.4, 3.7, 4.3, 4.4, 4.7, 4.8, 4.11, 4.14, 4.15, 5.5, 5.9, 5.10, 7.3, 7.4, 10.9, 10.11, 10.12; Tables 3.6, 3.8, 3.10, 3.11, 5.1, 5.5, 5.6, 5.7, 6.1, 6.3, 6.19, 6.20, 8.1, 9.2
Kilkenny	Co. Kilkenny	Ch. 6; Tables 6.7, 6.17, 6.19, 6.20
Killarney	Co. Kerry	Ch. 6; Table 6.9
Kilnsea	Humberside	Ch. 6; Table 6.21
Kingsway	(see London)	
Kinlochewe	Ross and Cromarty	Ch. 10
Kinlochleven	Argyllshire	Ch. 11, 13
Kinlochquoich	(see Glenquoich)	
Kinloss	Morayshire	Ch. 8, 11; Figs. 8.2, 11.2
Kirkwall	Orkney	Ch. 3.6; Tables 3.9, 6.5
Larkhill	Wiltshire	Ch. 3, 8, 11; Tables, 3.8, 3.10
Lasham	Hampshire	Ch. 11; Fig. 11.4
Leafield	Oxfordshire	Ch. 13
Leeds	West Yorkshire	Ch. 6, 15; Fig. 15.1; Table 6.1
Leeming	North Yorkshire	Ch. 8; Fig. 8.2
Leicester	Leicestershire	Ch. 14, 15; Figs. 14.3, 14.5, 15.1

Station Index (Counties as of 1 January 1975)—*contd.*

Station	County	Reference
Lerwick	Zetland	Ch. 3, 4, 5, 6, 7, 8, 10, 11, 12; Figs. 3.1, 3.2, 3.3, 3.4, 3.5, 4.4, 4.8, 4.11, 4.14, 4.15, 5.5, 5.9, 5.10; Tables 3.1, 3.3, 3.4, 3.5, 3.6, 3.7, 3.8, 3.10, 3.11, 5.1, 5.5, 5.6, 5.7, 6.1, 6.17, 6.19, 6.20, 8.1, 11.2, 12.1, 12.6
Leuchars	Fife	Ch. 3, 6, 11; Tables 3.8, 3.10, 6.1, 6.5, 6.21, 11.3
Little Rissington	Gloucestershire	Ch. 6, 9; Fig. 9.3; Table 6.1
Little Wenlock	Salop	Ch. 14
Liverpool (and Speke)	Merseyside	Ch. 3, 6, 14, 15; Fig. 15.1; Tables 3.8, 3.10, 6.1
Lizard	Cornwall	Ch. 3; Tables 3.8, 3.10
Llandrindod Wells	Powys	Ch. 5; Fig. 5.10; Table 5.7
Llandudno	Gwynedd	Ch. 5; Table 5.2
Lliw, Upper	Gwynedd	Ch. 6; Table 6.6
Llyn Llydaw	(see Snowdon)	
Loch Fyne	Argyll	Ch. 6; Table 6.1
Loch Grennoch	Kirkcudbrightshire	Ch. 6; Fig. 6.4
Loch Leven	(see Kinlochleven)	
Loch Quoich	(see Glenquoich)	
Loch Sloy	Dunbartonshire	Ch. 6; Tables 6.1, 6.5
Logie Coldstone	Aberdeenshire	Ch. 5; Table 5.4
London	Greater London	Ch. 3, 5, 6, 8, 9, 10, 13, 14; Figs. 14.4; Tables 3.8, 3.10, 5.2, 6.1, 6.17, 9.2, 9.3, 14.1, 14.2, 14.3
Loughborough	Leicestershire	Ch. 6; Table 6.1
Lough Eske	Co. Donegal	Ch. 6; Table 6.9
Lowestoft	Suffolk	Ch. 6; Tables 6.19, 6.20
Lowther Hill	Dumfries-shire	Ch. 3; Table 3.9
Lyneham	Wiltshire	Ch. 9; Fig. 9.3
Machrihanish	Argyllshire	Ch. 6; Table 6.5
Malham Tarn	North Yorkshire	Ch. 5; Fig. 5.10; Table 5.7
Malin Head	Co. Donegal	Ch. 3, 5, 6; Tables 3.9, 6.2, 6.7, 6.11, 6.13, 6.17, 6.19, 6.20
Manby	Lincolnshire	Ch. 3; Figs. 3.1, 3.2, 3.3, 3.5; Tables 3.3, 3.4, 3.7, 3.11
Manchester (and Ringway)	Greater Manchester	Ch. 2, 3, 5, 6, 8, 10, 11, 14, 15; Figs. 2.5, 3.1, 3.2, 3.3, 3.5, 5.10, 15.1; Tables 3.3, 3.4, 3.5, 3.6, 3.7, 3.8, 3.10, 3.11, 5.7, 6.1, 6.21, 8.2
Margate	Kent	Ch. 5, 6
Marham	Norfolk	Ch. 6; Table 6.1
Midleton	Co. Cork	Ch. 6; Tables 6.2, 6.11, 6.13
Mildenhall	Suffolk	Ch. 3, 5, 6, 8; Figs. 3.1, 3.2, 3.3, 3.5, 8.2; Tables 3.3, 3.4, 3.6, 3.7, 3.8, 3.10, 3.11, 6.21
Moorhouse	Cumbria	Ch. 5, 12
Morecambe	Lancashire	Ch. 11; Tables 11.1, 11.4
Morpeth (Cockle Park)	Northumberland	Ch. 11; Table 11.4
Mullardoch Dam	Inverness-shire	Ch. 6; Table 6.5
Mullingar	Co. Westmeath	Ch. 5, 6; Fig. 5.10; Tables 5.7, 6.2, 6.7, 6.11, 6.13, 6.17, 6.19, 6.20
Newcastle-upon-Tyne	Tyne and Wear	Ch. 5, 15; Fig. 15.1
Newton Rigg	Cumbria	Ch. 5; Table 5.4
Norwich	Norfolk	Ch. 15; Fig. 15.1
Nottingham	Nottinghamshire	Ch. 15; Fig. 15.1
Oban	Argyllshire	Ch. 6; Tables 6.19, 6.20
Omagh	Co. Tyrone	Ch. 6; Table 6.18
Oxford	Oxfordshire	Ch. 5, 6, 10, 15; Figs. 5.10, 6.16, 10.9, 10.10, 15.1; Tables 5.7, 6.14, 6.16, 6.18
Peterborough	Cambridgeshire	Ch. 6; Table 6.1
Plymouth (Hoe and Mount Batten)	Devon	Ch. 2, 3, 5, 6, 8, 10, 11, 12; Figs. 2.5, 5.2, 8.2, 10.9, 10.10, 10.11, 10.12; Tables 3.8, 3.10, 3.11, 5.1, 6.1, 6.19, 6.20, 6.21, 8.1, 12.6

Station Index (Counties as of 1 January 1975)—*contd.*

Station	County	Reference
Plynlimon	Dyfed	Ch. 12; Table 12.5
Portland Bill	Dorset	Ch. 5, 11
Porton Down	Wiltshire	Ch. 11, 13; Fig. 11.3
Portsmouth	Hampshire	Ch. 14
Port Talbot	West Glamorgan	Ch. 3; Tables 3.8, 3.10
Prestatyn	Clwyd	Ch. 5; Table 5.2
Prestwick	Ayrshire	Ch. 3, 6, 8; Fig. 3.5, 8.2; Tables 3.4, 3.6, 3.7, 3.8, 3.10, 3.11, 6.1, 6.21
Princetown	Devon	Ch. 6; Table 6.6
Rannoch	Perthshire	Ch. 12; Table 12.3
Reading	Berkshire	Ch. 9, 11, 14
Redcar	Cleveland	Ch. 5
Regents Park	(see London)	
Renfrew	Renfrewshire	Ch. 2, 3, 6, 8; Fig. 2.5; Tables 3.8, 3.10, 6.19, 6.20, 8.1
Rhayader	Powys	Ch. 6, 13; Tables 6.19, 6.20
Rhoose	South Glamorgan	Ch. 8; Fig. 8.2
Rhum, Isle of	Inverness-shire	Ch. 6; Table 6.5
Rickmansworth	Hertfordshire	Ch. 9, 13
Ringway	(see Manchester)	
Rivelin	South Yorkshire	Ch. 6; Table 6.18
Roches Point	Co. Cork	Ch. 6; Tables 6.7, 6.17, 6.19, 6.20
Ronaldsway	Isle of Man	Ch. 3, 6; Figs. 3.1, 3.2, 3.3, 3.5; Tables 3.3, 3.4, 3.6, 3.7, 3.11, 6.1, 6.21
Ross-on-Wye	Hereford and Worcester	Ch. 6, 10; Tables 6.1, 6.17, 6.21
Rosslare	Co. Wexford	Ch. 3, 6; Figs. 3.1, 3.5; Tables 3.3, 3.4, 3.6, 3.7, 3.8, 3.10, 6.7, 6.17, 6.19, 6.20
Rothamsted	Hertfordshire	Ch. 6, 9, 10, 13; Table 6.1
Rugby	Warwickshire	Ch. 5; Fig. 5.10; Table 5.7
Runcorn	Cheshire	Ch. 14
Rye	East Sussex	Ch. 13
St Abbs Head	Berwickshire	Ch. 5
St Ann's Head	Dyfed	Ch. 3; Table 3.9
St Austell	Cornwall	Ch. 6; Fig. 6.7
St Mawgan	Cornwall	Ch. 6; Tables 6.1, 6.17, 6.21
St Peter	(see Jersey)	
Salford	Greater Manchester	Ch. 14
Santon Downham	Norfolk	Ch. 5, Tables 5.2, 5.4
Scafell	Cumbria	Ch. 6, 12
Scarborough	North Yorkshire	Ch. 5
Scilly	Cornwall	Ch. 2, 3, 5, 6, 7; Figs. 3.1, 3.2, 3.3, 3.5, 5.9, 5.10; Tables 2.2, 3.2, 3.3, 3.4, 3.5, 3.6, 3.7, 3.8, 3.10, 3.11, 5.7, 6.21
Seathwaite	(see Borrowdale)	
Sellafield	Cumbria	Ch. 3; Tables 3.8, 3.10
Selset	Durham	Ch. 11; Figs. 11.5, 11.6
Shanklin	Hampshire	Ch. 5; Fig. 5.10; Table 5.7
Shannon	Co. Clare	Ch. 3, 6; Fig. 3.1, 3.5; Tables 3.3, 3.4, 3.6, 3.7, 3.8, 3.10, 6.2, 6.7, 6.11, 6.13, 6.17, 6.19, 6.20
Shawbury	Salop	Ch. 2, 6; Tables 2.6, 6.1, 6.21
Sheffield	South Yorkshire	Ch. 2, 10, 14, 15; Fig. 15.1
Shoeburyness	Essex	Ch. 3; Tables 3.8, 3.10
Silloth	Cumbria	Ch. 6; Table 6.1
Skegness	Lincolnshire	Ch. 5
Skipton	North Yorkshire	Ch. 6; Table 6.1
Snowdon	Gwynedd	Ch. 2, 6, 12; Table 2.6
Southampton	Hampshire	Ch. 2, 5, 10, 14, 15; Figs. 2.6, 10.8, 15.1; Tables 5.5, 5.6
Southport	Merseyside	Ch. 3; Tables 3.8, 3.10
South Shields	Tyne and Wear	Ch. 3; Fig. 3.4; Tables 3.6, 3.8, 3.10
Speke	(see Liverpool)	
Sprinkling Tarn	(see Scafell)	
Spurn Head	Humberside	Ch. 3; Tables 3.8, 3.10
Squires Gate	(see Blackpool)	

Station Index (Counties as of 1 January 1975)—*contd.*

Station	County	Reference
Stonyhurst	Lancashire	Ch. 12; Table 12.6
Stornoway	Ross and Cromarty	Ch. 2, 3, 5, 6, 8, 12; Figs. 2.5, 3.1, 3.2, 3.3, 3.5, 8.2; Tables 2.2, 3.3, 3.4, 3.6, 3.7, 3.8, 3.10, 3.11, 5.1, 5.5, 5.6, 6.1, 6.5, 6.17, 6.18 6.19, 6.20, 6.21, 8.1, 12.6
Stretham	Cambridgeshire	Ch. 6
Strood	Kent	Ch. 6; Table 6.1
Strubby	Lincolnshire	Ch. 11; Fig. 11.1
Sutton Bonington	Nottinghamshire	Ch. 7; Fig. 7.1
Swanage	Dorset	Ch. 5
Swansea	West Glamorgan	Ch. 5, 6, 15; Figs. 5.10, 15.1; Tables 5.7, 6.1, 6.17, 6.18
Swinton House	Berwickshire	Ch. 6; Table 6.1
Syerston	Nottinghamshire	Ch. 11; Fig. 11.1
Tangmere	West Sussex	Ch. 2; Fig. 2.5
Tanllwyth	(see Plynlimon)	
Thetford	Norfolk	Ch. 13; Table 13.8
Thorney Island	West Sussex	Ch. 3, 8, 11; Fig. 8.2
Tiree	Argyllshire	Ch. 3, 5, 6, 8; Figs. 3.1, 3.2, 3.3, 3.4, 3.5, 5.10, 8.2, 8.4; Tables 3.3, 3.4, 3.6, 3.7, 3.8, 3.10, 3.11, 5.7, 6.1, 6.21
Tonbridge	Kent	Ch. 5; Table 5.2
Tunbridge Wells	Kent	Ch. 2, 5; Table 5.2
Turnhouse	Midlothian	Ch. 3, 6, 8; Figs. 3.1, 3.2, 3.3, 3.5, 8.2; Tables 3.3, 3.4, 3.6, 3.7, 3.11, 6.1
Tynemouth	Tyne and Wear	Ch. 2, 3, 5, 6, 11; Fig. 2.5, 3.1, 3.2, 3.3, 3.5, 5.5, 5.10; Tables 3.3, 3.4, 3.5, 3.7, 3.11, 5.7, 6.18, 6.21, 11.4
Ushaw College	Durham	Ch. 13
Uxbridge	(see London)	
Valentia	Co. Kerry	Ch. 2, 3, 4, 5, 6, 7, 8, 10, 11; Figs. 2.5, 3.1, 3.4, 3.5, 4.4, 5.10, 7.1, 7.3, 7.4, 10.9, 10.10, 10.11, 10.12; Tables 3.3, 3.4, 3.6, 3.7, 3.8, 3.10, 5.7, 6.2, 6.7, 6.11, 6.13, 6.17, 6.19, 6.20, 8.1
Valley	Gwynedd	Ch. 2, 3, 5, 6, 9; Figs. 2.5, 9.1, 9.2; Tables 2.6, 3.8, 3.10, 5.5, 5.6, 5.7, 6.21
Waddington	Lincolnshire	Ch. 5, 8, 9; Figs. 5.10, 8.2, 9.3; Table 5.7
Wakefield	West Yorkshire	Ch. 5; Table 5.2
Walsall	West Midlands	Ch. 14
Warboys	Cambridgeshire	Ch. 13
Watnall	Nottinghamshire	Ch. 2, 6; Tables 2.6, 6.1, 6.21
Wattisham	Suffolk	Ch. 9; Fig. 9.3
West Freugh	Wigtownshire	Ch. 6; Table 6.1
West Kirby	Merseyside	Ch. 11
Westminster	(see London)	
West Raynham	Norfolk	Ch. 6; Table 6.21
Wick	Caithness	Ch. 3, 5, 6, 8; Figs. 3.1, 3.2, 3.3, 3.5, 5.10, 8.2, 8.3; Tables 3.3, 3.4, 3.7, 3.11, 5.7, 6.1, 6.21
Widnes	Cheshire	Ch. 14
Windermere	Cumbria	Ch. 11
Woburn	Bedfordshire	Ch. 5
Worthing	West Sussex	Ch. 6; Table 6.1
Wray Castle	Cumbria	Ch. 11
Wye	Kent	Ch. 6, 13; Table 6.18
Ynyslas	Dyfed	Ch. 12; Table 12.4
York	North Yorkshire	Ch. 10, 15; Fig. 15.1
Ystwyth	(see Aberystwyth)	

General Index

(Note : page references in *italics* refer to figures.)